METHODS
OF REAL
ANALYSIS

RICHARD R.
GOLDBERG
The University
of Iowa

METHODS OF REAL ANALYSIS

SECOND EDITION

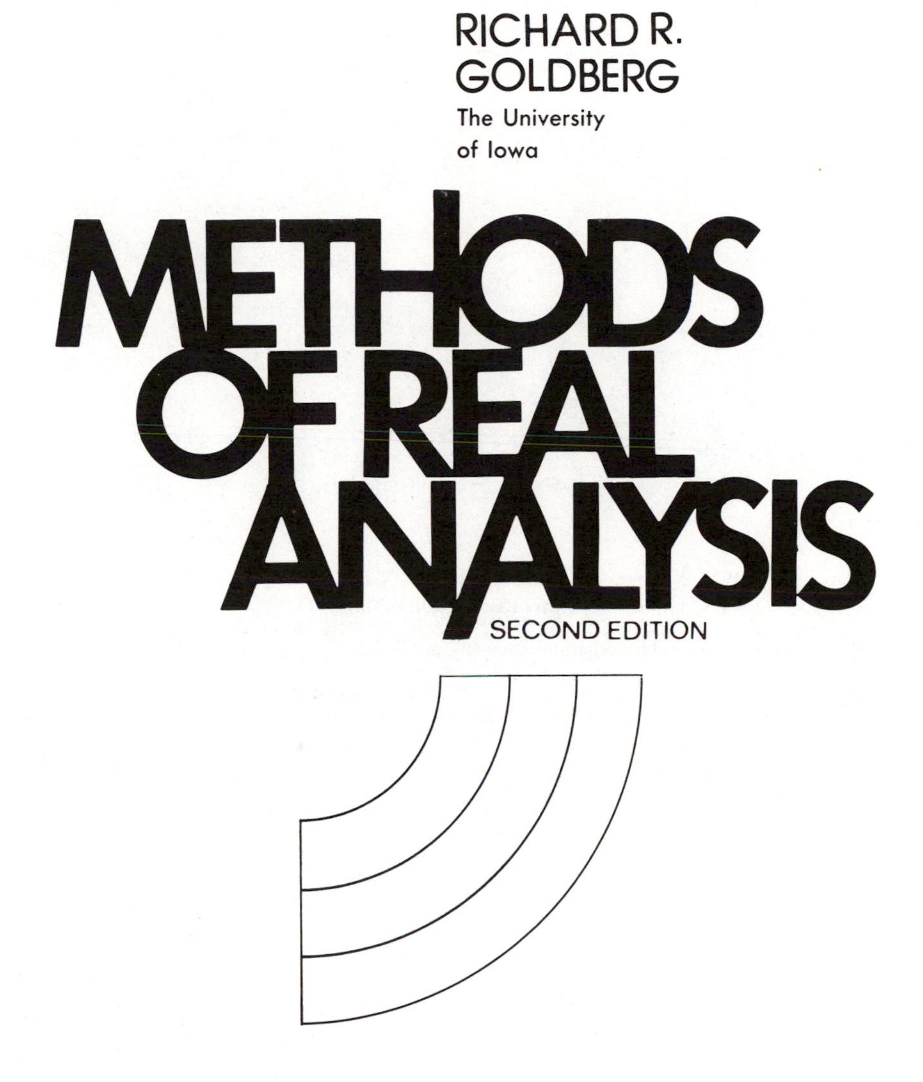

JOHN WILEY & SONS
New York • Chichester • Brisbane • Toronto

Library of Congress Cataloging in Publication Data:

Goldberg, Richard R.
Methods of real analysis.

Includes index.
1. Functions of real variables. 2. Mathematical analysis. I. Title.

QA331.5.G58 1976 515'.8 75-30615
ISBN 0-471-31065-4

Printed in the United States of America

10 9 8 7 6 5

PREFACE

This book is intended as a one-year course for students who have completed an ordinary sequence of courses in elementary calculus. It presents in rigorous fashion basic material on the fundamental concepts and tools of analysis—functions, limits, continuity, derivatives and integrals, sequences, and series. Most of the difficult points usually glossed over in elementary courses are dealt with in detail, as well as many more advanced topics designed to give a good background for (and, hopefully, a taste of) modern analysis and topology. In particular, there are treatments of metric spaces and Lebesgue integration, topics that are often reserved for more advanced courses. Also included are many smaller but interesting topics not usually presented in courses at this level; these topics include Baire category and discontinuous functions, summability of series, the Weierstrass theorem on approximation of continuous functions by polynomials, and a proof of the standard existence theorem for differential equations from the point of view of fixed-point theory.

The book is written at the same level as texts for traditional "advanced calculus" courses, but does not consider topics in "several variables." Material on differentials and vector calculus, in our opinion, can be understood best from the point of view of modern differential geometry and belong in a separate course.

REMARKS ON THE SECOND EDITION

Many changes in, additions to, and some deletions from the first edition have been made in accordance with thoughtful criticism from many colleagues at large and small institutions.

A major feature of this new edition is the addition of sections called "Notes and Additional Exercises," which include a variety of material. There are famous theorems related to the material in the body of the text—for example, the Schröder-Bernstein theorem from set theory, the Tietze extension theorem from topology, and Stone's generalization of the Weierstrass approximation theorem. The proofs are given in outline only, with a great deal left to the student as exercises. In these new sections there are also miscellaneous exercises (many of which are quite challenging) and an occasional historical note. An instructor's solutions manual for the problems in the new material can be obtained from the author.

I also added an appendix that contains an axiomatic treatment of the real number system. This was a compromise between no treatment at all in the first edition and a lengthy development from basic principles that I think would retard the reader's progress into the core of the book. All the assumptions about the real numbers and the necessary results that can be derived from these assumptions are carefully presented.

There are a number of pictorial illustrations—also a departure from the first edition—and new exercises in many of the chapters, and new proofs.

Richard R. Goldberg

CONTENTS

INTRODUCTION

ASSUMPTIONS AND NOTATIONS

A. The book does not begin with an extended development of the real numbers. However, the reader who wishes to proceed in strictly logical order should first digest the basic definitions and theorems about sets and functions in Sections 1.1 through 1.3, and then turn to the Appendix for the algebra and order axioms of the reals and the theorems on arithmetic and inequalities that are derived from those axioms. After the Appendix the reader should go to Section 1.7 where the least upper bound axiom is presented. At this point the reader will have seen a careful treatment of all the basic assumptions about the reals. Anyone choosing this approach may skip now to paragraph C.

B. There are some who feel, however, that it is preferable at first to be less formal about the real numbers so that the reader can get to the meat of the book more quickly. From this point of view it is better to delay reading the Appendix and simply proceed directly through the main body of the text. For those who wish to take this approach we mention briefly some facts about the reals.

An integer is a "whole number." Thus $6, 0, -3$ are integers. A rational number is a real number that can be expressed as a quotient of integers. Thus $3/2$ and $-9/276$ are rational numbers. Any integer k is thus a rational number since we may write $k = k/1$. An irrational number is a real number that is not a rational number. For example, a solution to the equation $x^2 = 2$ must be an irrational number.

The reader should have some facility in handling inequalities. He should know that if x and y are real numbers and $x < y$, then $-x > -y$. Also, if $0 < x < y$, then $0 < 1/y < 1/x$.

For $x > 0$ we define $|x|$ to be x. For $x < 0$ we define $|x|$ to be $-x$. Finally we define $|0| = 0$. Thus for any real numbers x, $|x|$ is the "numerical value" of x. We call $|x|$ the absolute value of x. By considering various cases according to the sign of x and of y, the reader should have no difficulty in proving the immensely important results

$$|x+y| \leqslant |x| + |y| \tag{1}$$

and

$$|xy| = |x| \cdot |y|.$$

If a and b are real numbers, then the geometric interpretation of $|a-b|$ is the distance from a to b (or from b to a). This interpretation is especially important for the understanding of the essential ideas in many of the proofs. If a, b, c are real numbers, the geometric meaning of the inequality

$$|a-b| \leqslant |a-c| + |c-b| \tag{2}$$

is that the distance from a to b is no greater than the distance from a to c plus the distance from c to b. This should seem quite reasonable. See if you can prove (2). [Let $x=a-c$, $y=c-b$ and use (1).]

C. We assume the truth of laws of exponents such as $a^{x+y}=a^x a^y$ for $a>0$ and for *rational* values of x and y. In Chapter 8 we define a^x for *any* real number x and then prove the familiar laws of exponents for arbitrary exponents. The notations $a^{1/2}$ and $\sqrt{a}$ both mean the positive square root of a. (The existence of a positive square root for any positive number is presented in exercise 8 of Section 6.2.)

D. Here are some notations. If a and b are real numbers with $a<b$, we denote by (a,b) the set of all real numbers x such that $a<x<b$. By (a,∞) we mean the set of all real x such that $x>a$. By $(-\infty,a)$ we mean the set of all real x with $x<a$. The set (a,b) is called a bounded open interval, while $(-\infty,a)$ and (a,∞) are called unbounded open intervals. The set of all real numbers is sometimes denoted by $(-\infty,\infty)$. Note that we are *not* defining the symbol ∞.

If $a \leqslant b$, then $[a,b]$ denotes the set of real numbers x such that $a \leqslant x \leqslant b$. This set is called a bounded closed interval. A closed interval may thus contain only one point (if $a=b$). We occasionally need to use "half-open" intervals. For example, $[0,1)$ denotes the interval of numbers x such that $0 \leqslant x < 1$.

We do not use the notation (a,b) to denote a point in the plane. As we will see, the point whose "x-coordinate" is a and whose "y-coordinate" is b will be denoted by $\langle a,b \rangle$.

It is often convenient to write in parentheses, to the right of a displayed statement, the values of the "variable" or "variables" for which the statement is true. For example,

$$f(x)<7 \qquad (0 \leqslant x \leqslant 3)$$

means that the number $f(x)$ is less than 7 for all x in $[0,3]$.

E. The material in this book is logically independent of courses in elementary geometry, trigonometry, and calculus. That is, we use no result from these elementary courses in any definition or in the statement or proof of a theorem unless we have previously established the result ourselves. Nevertheless, we use freely results and concepts from elementary calculus to *illustrate* our definitions and theorems. Thus, for example, we do not define the sine function until Chapter 8, but we do use familiar results about the sine function in examples and exercises in earlier chapters.

There are some pictorial illustrations in the text, but not a great many. We believe that the reader should learn to draw his own pictures as early as possible. Presumably, the instructor will help with the rough spots in this task.

1

SETS AND FUNCTIONS

1.1 SETS AND ELEMENTS

By a *set* we mean a collection of objects of any type whatsoever. The objects in a set are called its *elements* or *points*. Note that we have not really defined the terms set and element (since we did not define "collection" or "object"); rather, we have taken them as intuitive notions on which all our other notions will be based. Instead of "set" we sometimes use one of the following: class, family, aggregate. All these words (in this book) have the same meaning. We indicate in Section 3.12 something of what is involved in a more sophisticated approach to sets.

It is often useful to denote a set by putting braces around its elements. For example, $\{a,b,c\}$ denotes the set consisting of the three elements a, b, and c. With judicious use of dots we can even illustrate in this way sets with infinitely many elements (whatever that means—see Section 1.5D). For example, the set of all positive integers may be denoted by $\{1,2,3,\dots\}$. Another kind of set notation consists of braces around a description of the set. The first quadrant of the Cartesian plane may thus be denoted $\{\langle x,y\rangle | x \geqslant 0, y \geqslant 0\}$—the *set* of all points $\langle x,y\rangle$ such that x is nonnegative and y is nonnegative. Similarly, $[0,1]=\{x|0 \leqslant x \leqslant 1\}$.

DEFINITION. If b is an element of the set A, we write $b \in A$. If b is not an element of A, we write $b \not\in A$.

Thus $a \in \{a,b,c\}$ but $d \not\in \{a,b,c\}$. As another illustration, suppose we define a baseball team as the set of its players and define the American League to be the set of its twelve member teams. Then, in the notations we have just introduced,

$$\text{American League} = \{\text{A's, Tigers}, \dots, \text{Rangers}\},$$
$$\text{A's} = \{\text{Jackson, Bando}, \dots, \text{Campaneris}\},$$
$$\text{A's} \in \text{American League},$$
$$\text{Jackson} \in \text{A's}.$$

Note that the elements of the American League are themselves sets, which illustrates the fact that a set can be an element of another set. Note also that although Jackson plays in the American League, he is not an element of the American League as we have defined it. Hence

$$\text{Jackson} \notin \text{American League.}$$

Exercises 1.1

1. Describe the following sets of real numbers geometrically:

$$A = \{x \mid x < 7\},$$
$$B = \{x \mid |x| \geqslant 2\},$$
$$C = \{x \mid |x| = 1\}.$$

2. Describe the following sets of points in the plane geometrically:

$$A = \{\langle x, y \rangle \mid x^2 + y^2 = 1\},$$
$$B = \{\langle x, y \rangle \mid x \leqslant y\},$$
$$C = \{\langle x, y \rangle \mid x + y = 2\}.$$

3. Let P be the set of prime integers. Which of the following are true?
   (a) $7 \in P$.
   (b) $9 \in P$.
   (c) $11 \notin P$.
   (d) $7{,}547{,}193 \cdot 65{,}317 \in P$.
4. Let $A = \{1, 2, \{3\}, \{4, 5\}\}$. Are the following true or false?
   (a) $1 \in A$.
   (b) $3 \in A$.
   How many elements does A have?

1.2 OPERATIONS ON SETS

In grammar-school arithmetic the "elementary operations" of addition, subtraction, multiplication, and division are used to make new numbers out of old numbers—that is, to combine two numbers to create a third. In set theory there are also elementary operations—union, intersection, complementation—which correspond, more or less, to the arithmetic operations of addition, multiplication, and subtraction.

1.2A. DEFINITION. If A and B are sets, then $A \cup B$ (read "A union B" or "the union of A and B") is the set of all elements in either A or B (or both). Symbolically,

$$A \cup B = \{x \mid x \in A \quad \text{or} \quad x \in B\}.$$

Thus if

$$A = \{1, 2, 3\}, \qquad B = \{3, 4, 5\}, \tag{1}$$

then $A \cup B = \{1, 2, 3, 4, 5\}$.

1.2B. DEFINITION. If A and B are sets, then $A \cap B$ (read "A intersection B" or "the intersection of A and B") is the set of all elements in both A and B. Symbolically,

$$A \cap B = \{x \mid x \in A \quad \text{and} \quad x \in B\}.$$

Thus if A, B are as in (1) of Section 1.2A, then $A \cap B = \{3\}$. (Note the distinction between $\{3\}$ and 3. Since $A \cap B$ is the *set* whose only element is 3, to be consistent we must write $A \cap B = \{3\}$. This distinction is rarely relevant, and we often ignore it.) See Figure 1.

When A and B are sets with no elements in common, $A \cap B$ has nothing in it at all. We would still like, however, to call $A \cap B$ a set. We therefore make the following definition.

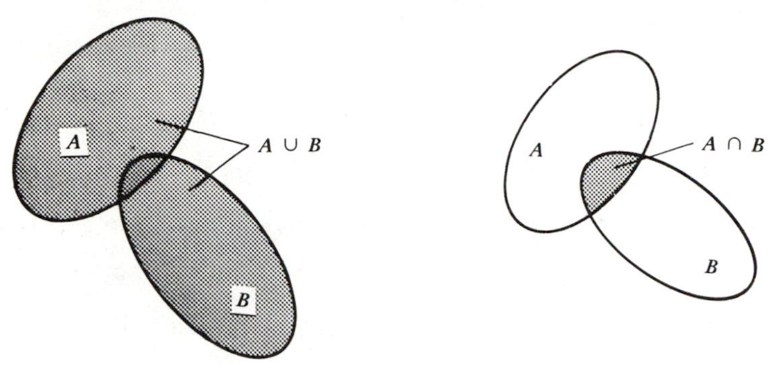

FIGURE 1.

1.2C. DEFINITION. We define the *empty set* (denoted by $\varnothing$) as the set which has no elements.

Thus $\{1,2\} \cap \{3,4\} = \varnothing$. Moreover, for any set A we have $A \cup \varnothing = A$ and $A \cap \varnothing = \varnothing$ (verify!).

1.2D. DEFINITION. If A and B are sets, then $B - A$ (read "B minus A") is the set of all elements of B which are not elements of A. Symbolically,

$$B - A = \{x \mid x \in B, x \notin A\}.$$

Thus if A, B are as in (1) of Section 1.2A, $B - A = \{4, 5\}$. See Figure 2.

There are relations for sets that correspond to the $\leqslant$ and $\geqslant$ signs in arithmetic. We now define them.

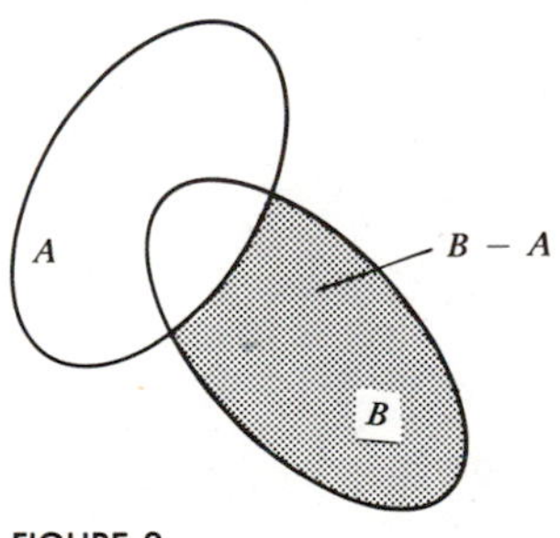

FIGURE 2.

1.2E. DEFINITION. If every element of the set A is an element of the set B, we write $A \subset B$ (read "A is contained in B" or "A is included in B") or $B \supset A$ (read "B contains A"). If $A \subset B$, we say that A is a subset of B. A proper subset of B is a subset $A \subset B$ such that $A \neq B$. (See Figure 3.)

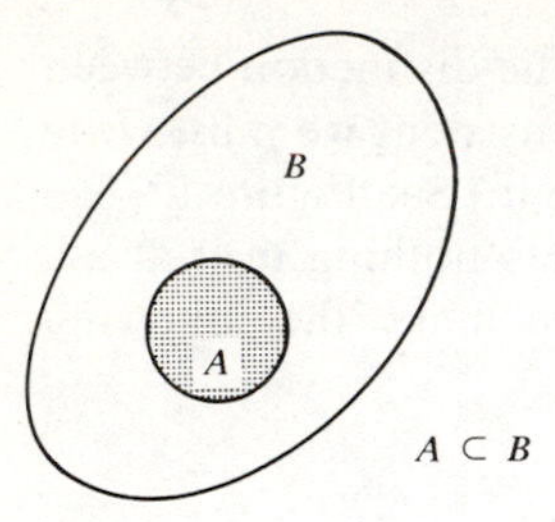

FIGURE 3.

Thus if

$$A=\{1,6,7\}, \qquad B=\{1,3,6,7,8\}, \qquad C=\{2,3,4,5,\ldots,100\}, \tag{1}$$

then $A \subset B$ but $B \not\subset C$ (even though C has 99 elements and B has only 5). Also $\varnothing \subset D$ and $D \subset D$ for any set D.

1.2F. DEFINITION. We say that *two sets are equal* if they contain precisely the same elements.

Thus $A=B$ if and only if $A \subset B$ and $B \subset A$ (verify!).

Note that for B and C in (1) of 1.2E, none of the relations $B \subset C, C \subset B, C=B$ hold.

1.2G. It is often the case that all sets $A, B, C, \ldots$ in a given discussion are subsets of a "big" set S. Then $S-A$ is called the *complement* of A (relative to S), the phrase in parentheses sometimes being omitted. For example, the set of rational numbers is the complement of the set of irrational numbers (relative to the reals). When there is no ambiguity as to what S is we write $S-A=A'$. Thus A'' [meaning $(A')'$] is equal to A. Moreover, $S=A \cup A'$. See Figure 4.

We now prove our first theorem.

1.2H. THEOREM. If A, B are subsets of S, then

$$(A \cup B)' = A' \cap B' \tag{1}$$

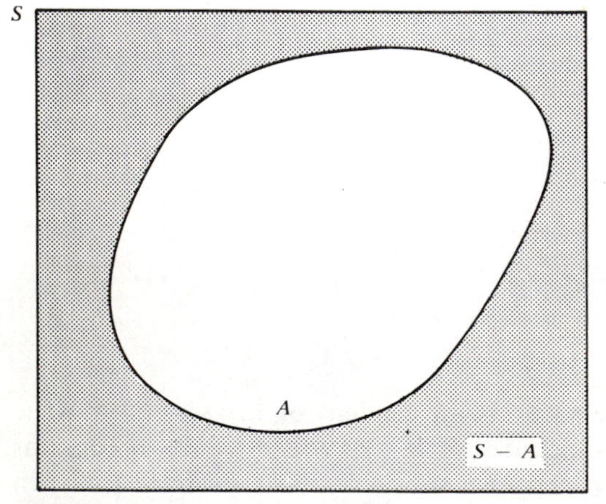

FIGURE 4.

and

$$(A \cap B)' = A' \cup B'. \tag{2}$$

These equations are sometimes called De Morgan's laws.

PROOF: If $x \in (A \cup B)'$, then $x \notin A \cup B$. Thus x is an element of neither A nor B so that $x \in A'$ and $x \in B'$. Thus $x \in A' \cap B'$. Hence $(A \cup B)' \subset A' \cap B'$. Conversely, if $y \in A' \cap B'$, then $y \in A'$ and $y \in B'$, so that $y \notin A$ and $y \notin B$. Thus $y \notin A \cup B$, and so $y \in (A \cup B)'$. Hence $A' \cap B' \subset (A \cup B)'$. This establishes (1).

Equation (2) may be proved in the same manner or it can be deduced from (1) as follows: In (1) replace A, B by A', B' respectively, so that A', B' are replaced by $A'' = A$ and $B'' = B$. We obtain $(A' \cup B')' = A \cap B$. Now take the complement of both sides.

Exercises 1.2

1. Let A be the set of letters in the word "trivial," $A = \{a, i, l, r, t, v\}$. Let B be the set of letters in the word "difficult." Find $A \cup B, A \cap B, A - B, B - A$. If S is the set of all 26 letters in the alphabet and $A' = S - A, B' = S - B$, find $A', B', A' \cap B'$. Then verify that $A' \cap B' = (A \cup B)'$.
2. For the sets A, B, C in Exercise 1 of Section 1.1, describe geometrically $A \cap B$, $B \cap C, A \cap C$.
3. Do the same for the sets A, B, C of Exercise 2 of Section 1.1.
4. For *any* sets A, B, C prove that
$$(A \cup B) \cup C = A \cup (B \cup C).$$
This is an associative law for union of sets and shows that $A \cup B \cup C$ may be written without parentheses.
5. Prove, for any sets A, B, C, that
$$(A \cap B) \cap C = A \cap (B \cap C).$$
This is an associative law for intersection of sets.
6. Prove the distributive law
$$A \cap (B \cup C) = (A \cap B) \cup (A \cap C).$$
See Figure 5.

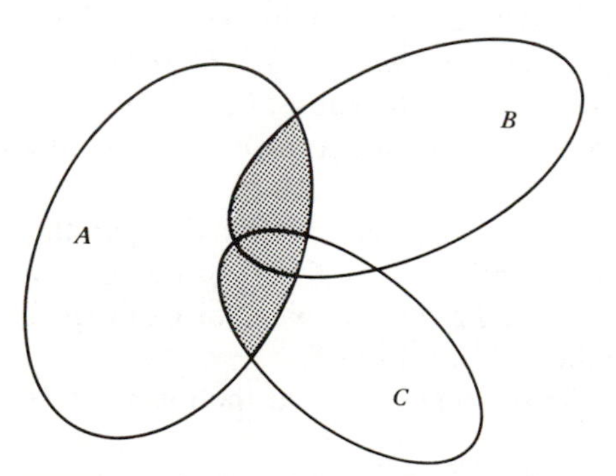

FIGURE 5. $A \cap (B \cup C) = (A \cap B) \cup (A \cap C)$

7. Prove
$$(A \cup B) - (A \cap B) = (A - B) \cup (B - A).$$

8. True or false (that is, prove true for all sets A, B, C, or give an example to show false):
   (a) $(A \cup B) - C = A \cup (B - C)$.
   (b) $(A \cup B) - A = B$.
   (c) $(A \cap B) \cup (B \cap C) \cup (A \cap C) = A \cap B \cap C$.
   (d) $(A \cup B) \cap C = A \cup (B \cap C)$.
9. True or false:
   (a) If $A \subset B$ and $B \subset C$, then $A \subset C$.
   (b) If $A \subset C$ and $B \subset C$, then $A \cup B \subset C$.
   (c) $[0,1] \supset (0,1)$.
   (d) $\{x \mid |x| \geqslant 4\} \cap \{y \mid |y| \geqslant 4\} = \{z \mid |z| \geqslant 4\}$.

1.3 FUNCTIONS

1.3A. In the cruder calculus texts we see the following definition: "If to each x (in a set S) there corresponds one and only one value of y, then we say that y is a function of x." This "definition," although it embodies the essential idea of the function concept, does not conform to our purpose of keeping undefined terms to a minimum. (What does "correspond" mean?)

In other places we see a function defined as a graph. Again, this is not suitable for us since "graph" is as yet undefined. However, since a plane graph (intuitively) is a certain kind of set of points, and each point is (given by) a pair of numbers, this will lead us to an acceptable definition of function in Section 1.3C.

1.3B. DEFINITION. If A, B are sets, then the *Cartesian product of A and B* (denoted $A \times B$) is the set of all ordered pairs* $\langle a, b\rangle$ where $a \in A$ and $b \in B$.

Thus the Cartesian product of the set of real numbers with itself gives the set of all ordered pairs of real numbers. We usually call this last set the plane (after we define the distance between pairs). See Figure 6.

The lateral surface of a right circular cylinder can be regarded as the Cartesian product of a line segment and a circle. (Why?)

We are now in a position to define function.

1.3C. DEFINITION. Let A and B be any two sets. A *function f from* (or on) *A into B* is a subset of $A \times B$ (and hence is a set of ordered pairs $\langle a, b\rangle$) with the property that each $a \in A$ belongs to precisely one pair $\langle a, b\rangle$. Instead of $\langle x, y\rangle \in f$ we usually write $y = f(x)$. Then y is called the *image of x under f*. The set A is called the *domain* of f. The *range* of f is the set $\{b \in B \mid b = f(a)$ for some $a\}$. That is, the range of f is the subset of B consisting of all images of elements of A. Such a function is sometimes called a *mapping* of A into B.

If $C \subset B$, then $f^{-1}(C)$ is defined as $\{a \in A \mid f(a) \in C\}$, the set of all points in the domain of f whose images are in C. If C has only one point in it, say $C = \{y\}$, we usually write $f^{-1}(y)$ instead of $f^{-1}(\{y\})$. The set $f^{-1}(C)$ is called the *inverse image* of C under f. (Note that no definition has been given for the symbol f^{-1} by itself.)

If $D \subset A$, then $f(D)$ is defined as $\{f(x) \mid x \in D\}$. The set $f(D)$ is called the *image* of D under f.

Consider Figure 7. The dots at the beginning of the arrows denote points in the domain of f. Thus the statement $f(a) = b$ is pictured by an arrow starting at a and ending

* To keep the record clear we had better define "ordered pair." What is needed is a set with a and b mentioned in an asymmetrical fashion. How about defining $\langle a, b\rangle$ to be $\{\{a\}, \{a, b\}\}$?

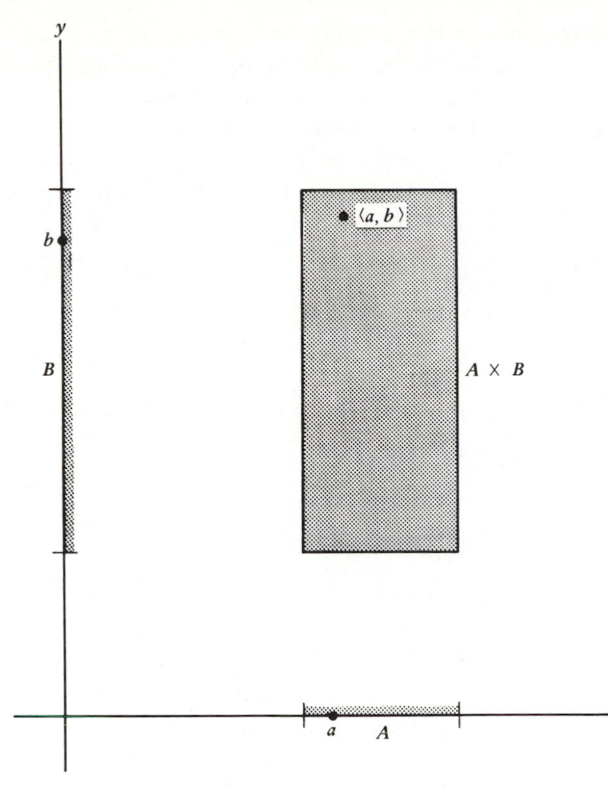

FIGURE 6. The Cartesian product of two intervals

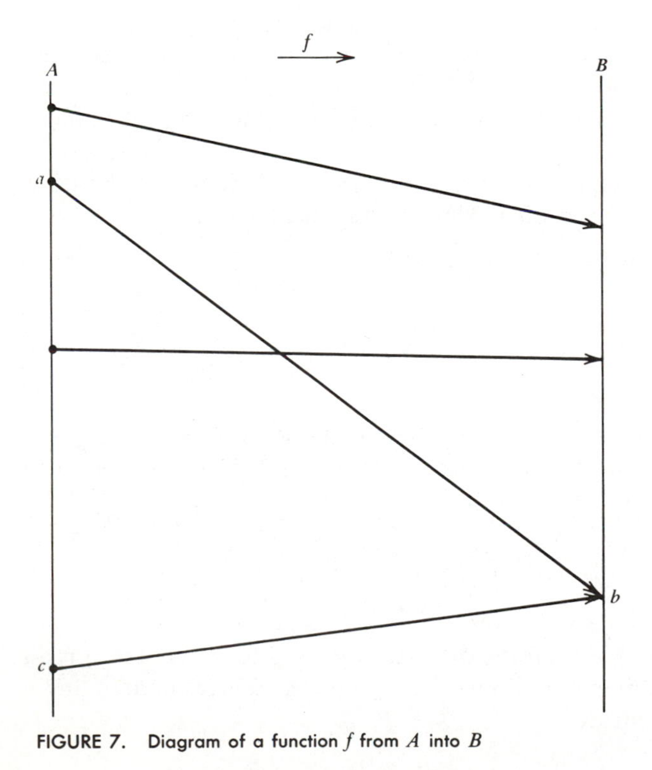

FIGURE 7. Diagram of a function f from A into B

at b. According to the definition of function, no two distinct arrows may begin at any $a \in A$, but two (or more) may end at some $b \in B$. We should think of f as sending points of A to points of B. Note that $f(a)=f(c)=b$ so that $f^{-1}(b)=\{a,c\}$.

If a function has domain and range both consisting of real numbers, then we can use the familiar method of graphing the function in the x-y plane. It is often possible to infer a good deal of information from such a graph. However, for understanding basic concepts about functions such as limit and continuity, a diagram like that in Figure 7 often is more helpful than an x-y graph.

For example, the set $f=\{\langle x,x^2\rangle \mid -\infty<x<\infty\}$ is the function usually described by the equation

$$f(x)=x^2 \qquad (-\infty<x<\infty).$$

The domain of this f is the entire real line. The range of f is $[0,\infty)$. In addition,

$$f(2)=4,$$
$$f^{-1}(4)=\{-2,2\},$$
$$f^{-1}(-7)=\varnothing,$$
$$f(\{x \mid x^2=9\})=\{9\},$$
$$f([0,3))=[0,9).$$

Draw, for this f, an x-y graph and a diagram like the one in Figure 7.

In the definition of function neither A nor B need be a set of numbers. For example, if A is the American League (see Section 1.1) and B is the set of the fifty states together with the District of Columbia, then the equation

$$f(x)=\text{state (or district) containing home ball park of } x\,(x\in A)$$

defines a function from A into B which consists of twelve ordered pairs.

Although an acceptable definition of function must be based on the set concept, the set notation is clearly more cumbersome than the classical notation. Note, however, that we make a notational distinction between f (the function) and $f(x)$ (the image of x under f).

It must be emphasized that an equation such as $f(x)=1+x^3$ does not define a function until the domain is explicitly specified. Thus the statements

$$f(x)=1+x^3 \qquad (1\leqslant x\leqslant 3)$$

and

$$g(x)=1+x^3 \qquad (1\leqslant x\leqslant 4)$$

define different functions according to our definition.

It is useful, however, to introduce terminology to describe pairs of functions that are related in the same way as f and g. In general, suppose f and g are two functions with respective domains X and Y. If

$$X\subset Y$$

and if

$$f(x)=g(x) \qquad (x\in X),$$

we say that g is an *extension* of f to Y or that f is the *restriction* of g to X. That is, g is an extension of f if the domain of g contains the domain of f *and* if the images under f and g of all points in the domain of f coincide.

1.3D. DEFINITION. If f is a function from A into B, we write $f:A\rightarrow B$. If the range of f is all of B, we say that f is a function from A *onto* B. In this case we sometimes write $f:A\Rightarrow B$.
Thus if $f(x)=x^2(-\infty<x<\infty)$ and $g(x)=x^3(-\infty<x<\infty)$, then

$$f:(-\infty,\infty)\rightarrow(-\infty,\infty),$$
$$g:(-\infty,\infty)\Rightarrow(-\infty,\infty).$$

We now give three theorems on images and inverse images of sets.

1.3E. THEOREM. If $f:A\rightarrow B$ and if $X\subset B, Y\subset B$, then

$$f^{-1}(X\cup Y)=f^{-1}(X)\cup f^{-1}(Y). \qquad (1)$$

In words, the inverse image of the union of two sets is the union of the inverse images.

PROOF: Suppose $a\in f^{-1}(X\cup Y)$. Then $f(a)\in X\cup Y$. Hence either $f(a)\in X$ or $f(a)\in Y$ so that either $a\in f^{-1}(X)$ or $a\in f^{-1}(Y)$. But this says $a\in f^{-1}(X)\cup f^{-1}(Y)$. Thus $f^{-1}(X\cup Y)\subset f^{-1}(X)\cup f^{-1}(Y)$. Conversely, if $b\in f^{-1}(X)\cup f^{-1}(Y)$, then either $b\in f^{-1}(X)$ or $b\in f^{-1}(Y)$. Hence either $f(b)\in X$ or $f(b)\in Y$ so that $f(b)\in X\cup Y$. Thus $b\in f^{-1}(X\cup Y)$, and so $f^{-1}(X)\cup f^{-1}(Y)\subset f^{-1}(X\cup Y)$. This proves (1).
The next theorem can be proved in exactly the same way.

1.3F. THEOREM. If $f:A\rightarrow B$ and if $X\subset B, Y\subset B$, then

$$f^{-1}(X\cap Y)=f^{-1}(X)\cap f^{-1}(Y).$$

In words, the inverse image of the intersection of two sets in the intersection of the inverse images.

PROOF: The proof is left as an exercise.

The last two results concerned inverse images. Here is one about images.

1.3G. THEOREM. If $f:A\rightarrow B$ and $X\subset A, Y\subset A$, then

$$f(X\cup Y)=f(x)\cup f(Y).$$

In words, the image of the union of two sets is the union of the images.

PROOF: If $b\in f(X\cup Y)$, then $b=f(a)$ for some $a\in X\cup Y$. Either $a\in X$ or $a\in Y$. Thus either $b\in f(X)$ or $b\in f(Y)$. Hence $b\in f(X)\cup f(Y)$ which shows $f(X\cup Y)\subset f(X)\cup f(Y)$. Conversely, if $c\in f(X)\cup f(Y)$ then either $c\in f(X)$ or $c\in f(Y)$. Then c is the image of some point in X or c is the image of some point in Y. Hence c is the image of some point in $X\cup Y$, that is $c\in f(X\cup Y)$. So $f(X)\cup f(Y)\subset f(X\cup Y)$.

1.3H. Conspicuously absent from this list of results is the relation

$$f(X\cap Y)=f(X)\cap f(Y) \qquad \text{for } X\subset A, Y\subset A.$$

Prove that this relation need *not* hold.

1.3I. DEFINITION (THE COMPOSITION OF FUNCTIONS). If $f:A\rightarrow B$ and $g:B\rightarrow C$, then we define the function $g\circ f$ by

$$g\circ f(x)=g[f(x)] \qquad (x\in A).$$

That is, the image of x under $g \circ f$ is defined to be the image of $f(x)$ under g. The function $g \circ f$ is called the composition of f with g. [Some people write $g(f)$ instead of $g \circ f$.]

Thus $g \circ f: A \to C$. For example, if

$$f(x) = 1 + \sin x \qquad (-\infty < x < \infty),$$

$$g(x) = x^2 \qquad (0 \leqslant x < \infty),$$

then

$$g \circ f(x) = 1 + 2\sin x + \sin^2 x \qquad (-\infty < x < \infty).$$

See Figure 8. Note that the range of f must be contained in the domain of g, but does not have to be equal to the domain of g.

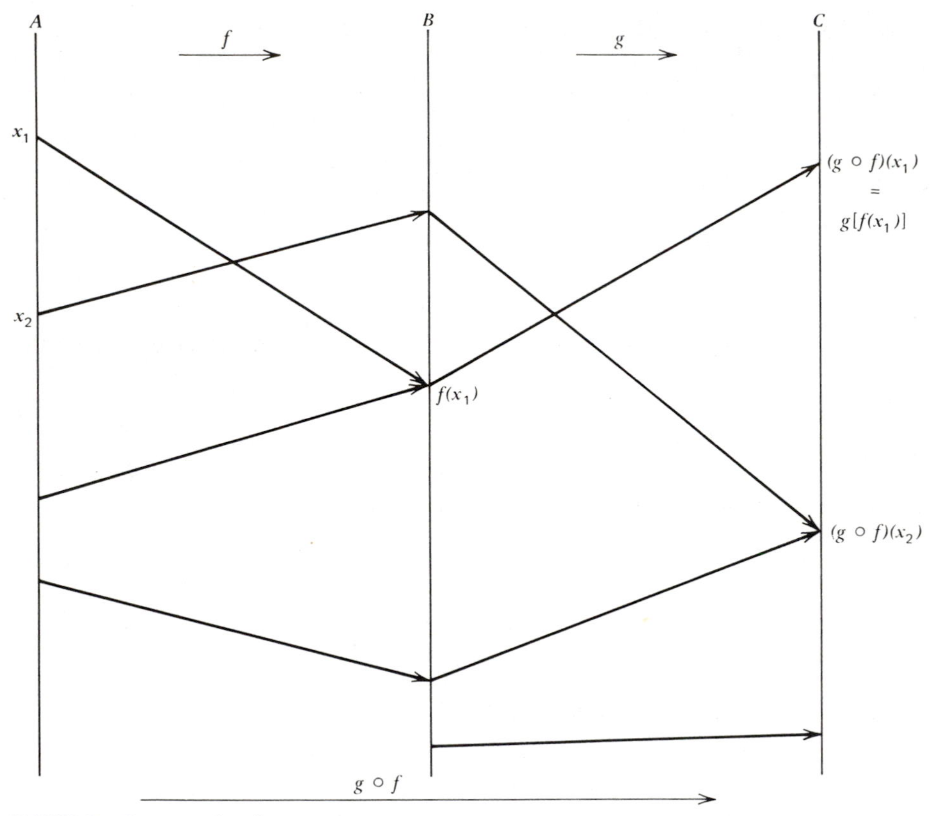

FIGURE 8. Diagram of $g \circ f$

Exercises 1.3

1. We have defined a function as a certain kind of set. Show that two functions f and g are equal (as sets) if and only if f and g have the same domain A and

$$f(x) = g(x) \qquad (x \in A).$$

   In other words, $f = g$ if and only if f is "identically equal to g" in the sense of functions.
2. Let $f: X \to Y$. If $A \subset X, B \subset X$, show that $f(A) - f(B) \subset f(A - B)$.

3. Let

$$f(x)=\log x \qquad (0<x<\infty).$$

(a) What is the range of f?
(b) If $A=[0,1]$ and $B=[1,2]$, find $f^{-1}(A), f^{-1}(B), f^{-1}(A\cup B), f^{-1}(A\cap B), f^{-1}(A)\cup f^{-1}(B)$, and $f^{-1}(A)\cap f^{-1}(B)$. Do your results agree with Sections 1.3E and 1.3F?

4. Consider the sine function defined by

$$f(x)=\sin x \qquad (-\infty<x<\infty).$$

(a) What is the image of $\pi/2$ under f?
(b) Find $f^{-1}(1)$.
(c) Find $f([0,\pi/6]), f([\pi/6,\pi/2]), f([0,\pi/2])$.
(d) Interpret the result of (c) using Section 1.3G.
(e) Let $A=[0,\pi/6], B=[5\pi/6,\pi]$. Does $f(A\cap B)=f(A)\cap f(B)$?

5. Consider the function f defined by

$$f(x)=\tan x \qquad \left(-\frac{\pi}{2}<x<\frac{\pi}{2}\right).$$

(a) What is the domain of f?
(b) What is the range of f?
(c) Let $A=(-\pi/2,-\pi/4), B=(\pi/4,\pi/2)$. Does $f(A\cap B)=f(A)\cap f(B)$?

6. Can you give a geometric interpretation for the Cartesian product of
(a) A line segment and a triangle?
(b) A large circle and a small circle?

7. Let $A=(-\infty,\infty)$ and let B be the plane. Let $f:A\to B$ be defined by

$$f(x)=\langle\cos x,\sin x\rangle \qquad (-\infty<x<\infty).$$

(a) What is the range of f?
(b) Find $f^{-1}[\langle 0,1\rangle]$.

8. Let $A=B=(-\infty,\infty)$. Which of the following functions map A *onto* B?
(a) $f(x)=3(-\infty<x<\infty)$,
(b) $f(x)=[x]=$ greatest integer not exceeding $x(-\infty<x<\infty)$,
(c) $f(x)=x^6+7x+1(-\infty<x<\infty)$,
(d) $f(x)=e^x(-\infty<x<\infty)$,
(e) $f(x)=\sinh x(-\infty<x<\infty)$.

9. Let $A=\{1,2,\ldots,n\}$ and let $B=\{0,1\}$. How many functions are there which map A into B? How many of these functions map A onto B?

10. If

$$f(x)=\arcsin x \qquad (-1<x<1),$$

$$g(x)=\tan x \qquad \left(-\frac{\pi}{2}<x<\frac{\pi}{2}\right),$$

and $h=g\circ f$, write a simple formula for h. What are the domain and range of h?

11. Let I denote the set of positive integers, $I=\{1,2,3,\ldots\}$. If

$$f(n)=n+7 \qquad (n\in I),$$

$$g(n)=2n \qquad (n\in I),$$

what is the range of $f\circ g$? What is the range of $g\circ f$?

12. If $f:A\to B, g:B\to C, h:C\to D$, prove that

$$h\circ(g\circ f)=(h\circ g)\circ f.$$

13. For which of the following pairs of functions f and g is g an extension of f?
    (a) $f(x)=x(0\leqslant x<\infty)$,
    $g(x)=|x|(-\infty<x<\infty)$,
    (b) $f(x)=1(-1\leqslant x\leqslant 1)$,
    $g(x)=1(0\leqslant x<\infty)$,
    (c) $f(x)=\sin x(0\leqslant x\leqslant 2\pi)$,
    $g(x)=\sqrt{1-cos^2 x}\ (-\infty<x<\infty)$.

1.4 REAL-VALUED FUNCTIONS

1.4A. In later chapters it is most often the case that the range of a given function f is contained in the set of all real numbers. (We henceforth denote the set of all real numbers by R.) If $f:A\rightarrow R$, we call f a *real-valued function*. If $x\in A$, then $f(x)$ (heretofore called the image of x under f) is also called the *value* of f at x.

We now define the sum, difference, product, and quotient of real-valued functions.

1.4B. DEFINITION. If $f:A\rightarrow R$ and $g:A\rightarrow R$, we define $f+g$ as the function whose value at $x\in A$ is equal to $f(x)+g(x)$. That is,

$$(f+g)(x)=f(x)+g(x) \qquad (x\in A).$$

In set notation

$$f+g=\{\langle x,f(x)+g(x)\rangle | x\in A\}.$$

It is clear that $f+g:A\rightarrow R$.

Similarly, we define $f-g$ and fg by

$$(f-g)(x)=f(x)-g(x) \qquad (x\in A),$$
$$(fg)(x)=f(x)g(x) \qquad (x\in A).$$

Finally, if $g(x)\neq 0$ for all $x\in A$, we can define f/g by

$$\left(\frac{f}{g}\right)(x)=\frac{f(x)}{g(x)} \qquad (x\in A).$$

The sum, difference, product, and quotient of two real-valued functions with the same domain are again real-valued functions. What permits us to define the sum of two real-valued functions is the fact that addition of real numbers is defined. In general, if $f:A\rightarrow B, g:A\rightarrow B$, there is no way to define $f+g$ unless there is a "plus" operation in B.

1.4C. DEFINITION. If $f:A\rightarrow R$ and c is a real number ($c\in R$), the function cf is defined by

$$(cf)(x)=c[f(x)] \qquad (x\in A).$$

Thus the value of $3f$ at x is 3 times the value of f at x.

1.4D. For a,b real numbers let $\max(a,b)$ denote the larger and $\min(a,b)$ denote the smaller of a and b. [If $a=b$, then $\max(a,b)=\min(a,b)=a=b$.] Then we can define $\max(f,g)$ and $\min(f,g)$ for real-valued functions f,g.

DEFINITION. If $f:A\rightarrow R, g:A\rightarrow R$, then $\max(f,g)$ is the function defined by

$$\max(f,g)(x)=\max[f(x),g(x)] \qquad (x\in A),$$

and $\min(f,g)$ is the function defined by

$$\min(f,g)(x)=\min[f(x),g(x)] \qquad (x\in A),$$

Thus if $f(x)=\sin x\,(0\leqslant x\leqslant \pi/2), g(x)=\cos x\,(0\leqslant x\leqslant \pi/2)$ and $h=\max(f,g)$, then

$$h(x)=\cos x \qquad \left(0\leqslant x\leqslant \frac{\pi}{4}\right),$$

$$h(x)=\sin x \qquad \left(\frac{\pi}{4}<x\leqslant \frac{\pi}{2}\right).$$

DEFINITION. If $f:A\rightarrow R$, then $|f|$ is the function defined by

$$|f|(x)=|f(x)| \qquad (x\in A).$$

If a,b are real numbers, the formulae

$$\max(a,b)=\frac{|a-b|+a+b}{2},$$

$$\min(a,b)=\frac{-|a-b|+a+b}{2},$$

are easy to verify. (Do so.) From them follow immediately the formulae

$$\max(f,g)=\frac{|f-g|+f+g}{2},$$

$$\min(f,g)=\frac{-|f-g|+f+g}{2},$$

for real-valued functions f,g.

1.4E. In this section we consider sets which are all subsets of a "big" set S. If $A\subset S$, then $A'=S-A$ (Section 1.2G). For each $A\subset S$ we define a function χ_A as follows.

DEFINITION. If $A\subset S$, then χ_A (called the *characteristic function of* A) is defined as

$$\chi_A(x)=1 \qquad (x\in A),$$

$$\chi_A(x)=0 \qquad (x\in A').$$

The reason for the name "characteristic function" is obvious—the set A is characterized (completely described) by χ_A. That is, $A=B$ if and only if $\chi_A=\chi_B$. The reader should verify the following useful equations for characteristic functions where A,B are subsets of S.

$$\chi_{A\cup B}=\max(\chi_A,\chi_B), \tag{1}$$

$$\chi_{A\cap B}=\min(\chi_A,\chi_B)=\chi_A\chi_B,$$

$$\chi_{A-B}=\chi_A-\chi_B \qquad (\text{provided } B\subset A),$$

$$\chi_{A'}=1-\chi_A,^{*}$$

$$\chi_S=1,$$

$$\chi_{\varnothing}=0.^{\dagger}$$

For example, to establish (1), suppose $x\in A\cup B$. Then $\chi_{A\cup B}(x)=1$. But either $x\in A$

* We are using 1 here to denote the real-valued function whose value at each $x\in S$ is equal to the number 1 (that is, here 1 is the "function identically 1"). Thus the symbol 1 has two different meanings—one a number, the other a function. The reader will be able to tell from the context which meaning to assign.

† The 0 denotes the function identically 0.

or $x \in B$ (or both), and so either $\chi_A(x)=1$ or $\chi_B(x)=1$. Thus $\max(\chi_A,\chi_B)(x)=1$. Hence

$$1=\chi_{A\cup B}(x)=\max(\chi_A,\chi_B)(x) \qquad (x\in A\cup B) \tag{2}$$

If $x \notin A\cup B$, then $\chi_{A\cup B}(x)=0$. But $x\in A'\cap B'$ by (1) of Section 1.2H and hence $x\in A'$ and $x\in B'$ so that $\chi_A(x)=0=\chi_B(x)$. Thus $\max(\chi_A,\chi_B)(x)=0$. Hence

$$0=\chi_{A\cup B}(x)=\max(\chi_A,\chi_B)(x) \qquad (x\notin A\cup B). \tag{3}$$

Equation (1) now follows from (2) and (3).

Exercises 1.4

1. Let $f(x)=2x(-\infty<x<\infty)$. Can you think of functions g and h which satisfy the two equations

$$g\circ f=2gh,$$
$$h\circ f=h^2-g^2?$$

2. If $f(x)=x^2(-\infty<x<\infty)$ and χ is the characteristic function of $[0,9]$, of what subset of R is $\chi\circ f$ the characteristic function?
3. If $f:A\to B$ and χ_E is the characteristic function of $E\subset B$, of what subset of A is $\chi_E\circ f$ the characteristic function?
4. Use whatever concept of continuity you possess to answer this question and the next one.
   Is there a characteristic function on R that is continuous?
   Do there exist three such functions?
5. Draw the graphs of two continuous functions f and g with the same domain.
   Would you guess the $\max(f,g)$ and $\min(f,g)$ are continuous?

1.5 EQUIVALENCE; COUNTABILITY

According to the definition of function, if $f:A\to B$, then each element $a\in A$ has precisely one image $f(a)\in B$. It often happens, however, that some element b in the range of f is the image of more than one element of A. For example, if $f(x)=x^2(-\infty<x<\infty)$, then 4 is the image of both -2 and $+2$. In this section we deal with functions f with the property that each b in the range of f is the image of precisely one a in the domain of f.

1.5A. DEFINITION. If $f:A\to B$, then f is called one-to-one (denoted 1–1) if

$$f(a_1)=f(a_2) \text{ implies } a_1=a_2 \qquad (a_1,a_2\in A).$$

Thus if f is 1–1 and $b=f(a_1)$, then $b\neq f(a_2)$ for any $a_2\in A$ distinct from a_1. Thus the function f defined by $f(x)=x^2(-\infty<x<\infty)$ is *not* 1–1 but the function g defined by $g(x)=x^2(0\leqslant x<\infty)$ is 1–1.

Stated otherwise, a function f is 1–1 if $f^{-1}(b)$ contains precisely one element for each b in the range of f. In this case, f^{-1} itself is a function. More precisely,

1.5B. DEFINITION. If $f:A\to B$ and f is 1–1, then the function f^{-1} (called the inverse function for f) is defined as follows:

$$\text{If } f(a)=b, \quad \text{then } f^{-1}(b)=a \qquad (b \text{ in range of } f). \tag{1}$$

Thus the domain of f^{-1} is the range of f and the range of f^{-1} is A (the domain of f). The definition of the function f^{-1} is consistent with the definition of inverse image in Section 1.3C. For if f is 1–1 and $f(a)=b$, then the inverse image of $\{b\}$ is $\{a\}$. That is, $f^{-1}(\{b\})=\{a\}$. If we omit the braces, we obtain (1).

For example, if $g(x)=x^2 (0 \leqslant x<\infty)$, then $g^{-1}(x)=\sqrt{x}\ (0 \leqslant x<\infty)$. For, if $b=g(a)=a^2$, then $a=\sqrt{b}=g^{-1}(b)$. Also, if $h(x)=e^x(-\infty<x<\infty)$, then $h^{-1}(x)=\log x (0<x<\infty)$. For, if $b=h(a)=e^a$, then $a=\log b=h^{-1}(b)$.

From the definition of inverse function it follows that

$$f^{-1}[f(a)]=a \qquad (a \in A),$$

$$f[f^{-1}(b)]=b \qquad (b \text{ in range of } f).$$

1.5C. A function that is both 1–1 and onto (Section 1.3D) has a special name.

DEFINITION. If $f: A \Rightarrow B$ and f is 1–1, then f is called a 1–1 correspondence (between A and B). If there exists a 1–1 correspondence between the sets A and B, then A and B are called *equivalent*.

Thus any two sets containing exactly seven elements are equivalent. The reader should not find it difficult to verify the following.

1. Every set A is equivalent to itself.
2. If A and B are equivalent, then B and A are equivalent.
3. If A and B are equivalent and B and C are equivalent, then A and C are equivalent.

We shall see presently that the set of all integers and the set of all rational numbers are equivalent, but that the set of all integers and the set of all real numbers are not equivalent. First let us talk a little bit about "infinite sets."

1.5D. The set A is said to be infinite if, for each positive integer n, A contains a subset with precisely n elements.*

Let us denote by I the set of all positive integers—

$$I=\{1,2,\ldots\}.$$

Then I is clearly an infinite set. The set R of all real numbers is also an infinite set. The reader should convince himself that if a set is not infinite, it contains precisely n elements for some nonnegative integer n. A set that is not infinite is called finite.

It will be seen that there are many "sizes" of infinite sets. The smallest size is called countable.

1.5E DEFINITION. The set A is said to be countable (or denumerable) if A is equivalent to the set I of positive integers. An uncountable set is an infinite set which is not countable.

Thus A is countable if there exists a 1–1 function f from I onto A. The elements of A are then the images $f(1), f(2), \ldots,$ of the positive integers—

$$A=\{f(1), f(2), \ldots\},$$

[where the $f(i)$ are all distinct from one another].

* If n is a positive integer, then the statement "B has n elements" means "B is equivalent to the set $\{1,2,\ldots,n\}$."

Hence, saying that A is countable means that its elements can be "counted" (arranged with "labels" $1,2,\ldots$). Instead of $f(1), f(2),\ldots$, we usually write $a_1, a_2, \ldots$.

For example, the set of *all* integers is countable. For by arranging the integers as $0, -1, +1, -2, +2, \ldots$, we give a scheme by which they can be counted. [The last sentence is an imprecise but highly intuitive way of saying that the function f defined by

$$f(n) = \frac{n-1}{2} \qquad (n = 1,3,5,\ldots),$$

$$f(n) = \frac{-n}{2} \qquad (n = 2,4,6,\ldots),$$

is a 1–1 correspondence between I and the set of all integers. For $f(1), f(2),\ldots$ is the same as $0, -1, 1, -2, 2, \ldots$.] See Figure 9.

This example shows that a set can be equivalent to a proper subset of itself.

The same reasoning shows that if A and B are countable, then so is $A \cup B$. For A can be expressed as $A = \{a_1, a_2, \ldots\}$ and similarly $B = \{b_1, b_2, \ldots\}$. Thus $a_1, b_1, a_2, b_2, a_3, b_3, \ldots$ is a scheme for "counting" the elements of $A \cup B$. (Of course, we must remove any b_i which occurs among the a_i's so that the same element in $A \cup B$ is not counted twice.)

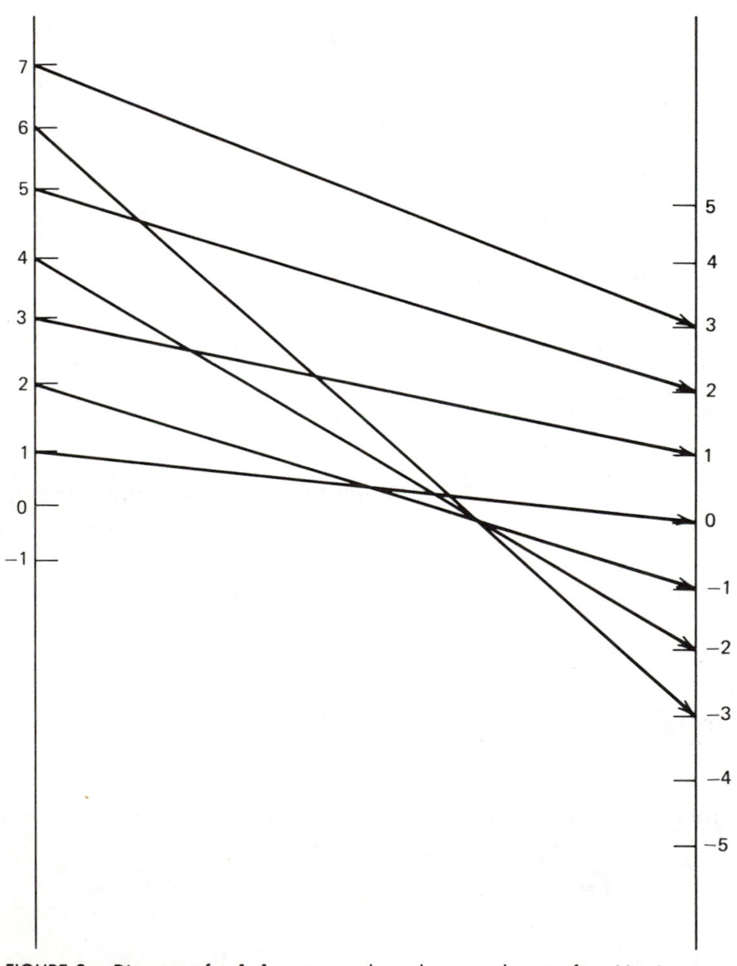

FIGURE 9. Diagram of a 1–1 correspondence between the set of positive integers and the set of all integers

The following theorem gives a much stronger result.

1.5F. THEOREM. If $A_1, A_2, \ldots$ are countable sets, then* $\bigcup_{n=1}^{\infty} A_n$ is countable. In words, the countable union of countable sets is countable.

PROOF: We may write $A_1 = \{a_1^1, a_2^1, a_3^1, \ldots\}, A_2 = \{a_1^2, a_2^2, a_3^2, \ldots\}, \ldots, A_n = \{a_1^n, a_2^n, a_3^n, \ldots\}$, so that a_k^j is the kth element of the set A_j. Define the height of a_k^j to be $j+k$. Then a_1^1 is the only element of height 2; likewise a_2^1 and a_1^2 are the only elements of height 3; and so on. Since for any positive integer $m \geqslant 2$ there are only $m-1$ elements of height m, we may arrange (count) the elements of $\bigcup_{n=1}^{\infty} A_n$ according to their height as

$$a_1^1, a_1^2, a_2^1, a_3^1, a_2^2, a_1^3, a_1^4, \ldots,$$

being careful to remove any a_k^j that has already been counted.

Pictorially, we are listing the elements of $\bigcup_{n=1}^{\infty} A_n$ in the following array and counting them in the order indicated by the arrows:

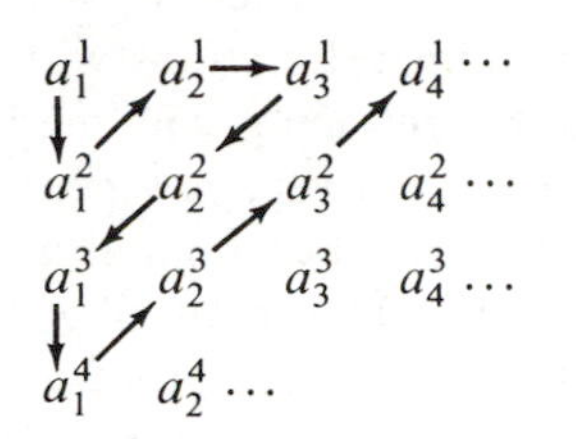

The fact that this counting scheme eventually counts every a_k^j proves that $\bigcup_{n=1}^{\infty} A_n$ is countable.

We obtain the following important corollary.

1.5G. COROLLARY. The set of all rational numbers is countable.

PROOF: The set of all rational numbers is the union $\bigcup_{n=1}^{\infty} E_n$ where E_n is the set of rationals which can be written with denominator n. That is, $E_n = \{0/n, -1/n, 1/n, -2/n, 2/n, \ldots\}$. Now each E_n is clearly equivalent to the set of all integers and is thus countable. (Why?) Hence the set of all rationals is the countable union of countable sets. Apply 1.5F.

It seems clear that if we can count the elements of a set we can count the elements of any subset. We make this precise in the next theorem.

1.5H. THEOREM. If B is an infinite subset of the countable set A, then B is countable.

PROOF: Let $A = \{a_1, a_2, \ldots\}$. Then each element of B is an a_i. Let n_1 be the smallest subscript for which $a_{n_1} \in B$, let n_2 be the next smallest, and so on. Then $B = \{a_{n_1}, a_{n_2}, \ldots\}$. The elements of B are thus labeled with $1, 2, \ldots$, and so B is countable.

1.5I. COROLLARY. The set of all rational numbers in $[0,1]$ is countable.

PROOF: The proof follows directly from 1.5G and 1.5H.

* We have not used the symbol $\bigcup_{n=1}^{\infty} A_n$ before. It means, of course, the set of all elements in at least one of the A_n.

Exercises 1.5

1. Which of the following define a 1–1 function?
   (a) $f(x)=e^x(-\infty<x<\infty)$,
   (b) $f(x)=e^{x^2}(-\infty<x<\infty)$,
   (c) $f(x)=\cos x(0\leqslant x\leqslant\pi)$,
   (d) $f(x)=ax+b(-\infty<x<\infty), a,b\in R$.
2. (a) If $f:A\to B$ and $g:B\to C$ and both f and g are 1–1, is $g\circ f$ also 1–1?
   (b) If f is not 1–1, is it still possible that $g\circ f$ is 1–1?
   (c) Give an example in which f is 1–1, g is not 1–1, but $g\circ f$ is 1–1.
3. Let P_n be the set of polynomial functions f of degree n,
$$f(x)=a_0x^n+a_1x^{n-1}+\cdots+a_{n-1}x+a_n,$$
   where n is a *fixed* nonnegative integer and the coefficients $a_0,a_1,\ldots,a_n$ are all integers. Prove that P_n is countable. (*Hint*: Use induction.)
4. Prove that the set of all polynomial functions with integer coefficients is countable.
5. Prove that the set of all polynomial functions with rational coefficients is countable. (*Hint*: This can be done by retracing the methods used in the preceding two problems.) However, also try this: Every polynomial g with rational coefficients can be written $g=(1/N)f$ where f is a polynomial with integer coefficients and N is a suitable positive integer. (Verify.) The set of all g that go with a given N is countable (by Exercise 4 of Section 1.5). Finish the proof.
6. We are assuming that every (nonempty) open interval (a,b) contains a rational (Introduction). Using this assumption, prove that every open interval contains infinitely many (and hence countably many) rationals.
7. Show that the intervals $(0,1)$ and $[0,1]$ are equivalent. (*Hint*: Consider separately the rationals and irrationals in the intervals.)
8. Prove that any infinite set contains a countable subset.
9. Prove that if A is an infinite set and $x\in A$, then A and $A-\{x\}$ are equivalent. (This shows that any infinite set is equivalent to a proper subset. This property is often taken as the definition of infinite sets.)
10. Show that the set of all ordered pairs of integers is countable.
11. Show that if A and B are countable sets, then the Cartesian product $A\times B$ is countable.
12. Prove that the family of all finite subsets of a countable set is itself countable.
13. (a) If f is a 1–1 function from A onto B, show that
$$f^{-1}\circ f(x)=x \qquad (x\in A), \quad \text{and} \quad f\circ f^{-1}(y)=y \qquad (y\in B).$$
    (b) If $g:C\to A$ and $h=f\circ g$, show that $g=f^{-1}\circ h$.

1.6 REAL NUMBERS

This section is out of logical order. We shall not at this time define the terms "decimal expansion," "binary expansion," and so on; rather, we rely here on the reader's experience and intuition. These terms, and the assumptions concerning them, are discussed carefully in Chapter 3. Insofar as the logical development of this book is concerned, this section could be ignored. Insofar as examples and understanding are concerned, however, this section should definitely not be ignored.

We have not as yet given an example of an infinite set that is not countable. We shall soon see that the set R of all real numbers provides such an example.

We shall *assume* that every real number x can be written in decimal expansion.

$$x = b.a_2a_2a_3\cdots = b + \frac{a_1}{10} + \frac{a_2}{10^2} + \frac{a_3}{10^3} + \cdots,$$

where the a_i are integers, $0 \leqslant a_i \leqslant 9$. This expansion is unique except for cases such as $x = \frac{1}{2}$ which can be expanded

$$\tfrac{1}{2} = 0.500000\cdots \quad \text{and} \quad \tfrac{1}{2} = 0.49999\cdots.$$

Every number $x \in [0,1]$ can thus be expanded $x = 0.a_1a_2a_3\cdots$. Conversely, we *assume* that every decimal of the form

$$b.a_1a_2a_3\cdots$$

is the decimal expansion for some real number. (We have not defined the real numbers. Hence we now take these relations between decimal expansion and real numbers as assumptions. As we presently show, however, they are consequences of the more basic axiom 1.7D.)

1.6A. THEOREM. The set $[0,1] = \{x \mid 0 \leqslant x \leqslant 1\}$ is uncountable.

PROOF: Suppose $[0,1]$ were countable. Then $[0,1] = \{x_1, x_2, \ldots\}$ where every number in $[0,1]$ occurs among the x_i. Expanding each x_i in decimals we have

$$\begin{aligned} x_1 &= 0.a_1^1a_2^1a_3^1\cdots \\ x_2 &= 0.a_1^2a_2^2a_3^2\cdots \\ &\vdots \\ x_n &= 0.a_1^na_2^na_3^n\cdots a_n^n\cdots. \\ &\vdots \end{aligned}$$

Let b_1 be any integer from 1 to 8 such that $b_1 \neq a_1^1$. Then let b_2 be any integer from 1 to 8 such that $b_2 \neq a_2^2$. In general, for each $n = 1, 2, \ldots$, let b_n be any integer from 1 to 8 such that $b_n \neq a_n^n$. Let $y = 0.b_1b_2\cdots b_n\cdots$. Then, for any n, the decimal expansion for y differs from the decimal expansion for x_n since $b_n \neq a_n^n$. Moreover, the decimal expansion for y is unique since no b_n is equal to 0 or 9. Hence $y \neq x_n$ for every n. Since $0 \leqslant y \leqslant 1$, this contradicts the assumption that every number in $[0,1]$ occurs among the x_i. This contradiction proves the theorem.

1.6.B. COROLLARY. The set R of all real numbers is uncountable.

PROOF: By 1.5H, if R were countable, then $[0,1]$ would be countable, contradicting 1.6A. Hence R is uncountable.

Here is another proof of 1.6B. Suppose R were countable—$R = \{x_1, x_2, \ldots\}$. Let I_1 be the interval $(x_1 - \frac{1}{4}, x_1 + \frac{1}{4})$, let I_2 be the interval $(x_2 - \frac{1}{8}, x_2 + \frac{1}{8})$, and in general, for each positive integer n, let I_n denote the interval $(x_n - 2^{-n-1}, x_n + 2^{-n-1})$. Then the length of I_n is 2^{-n} so the sum of the lengths of all the I_n is $2^{-1} + 2^{-2} + 2^{-3} + \cdots = 1$. But $x_n \in I_n$ so that $R = \cup_{n=1}^{\infty}\{x_n\} \subset \cup_{n=1}^{\infty} I_n$. But then the whole real line (whose length is infinite) would be covered by (contained in) a union of intervals whose lengths add up to 1. This seems to be a contradiction. Is it?

1.6C. In addition to decimal expansions it is useful to consider binary and ternary expansions for real numbers.

The binary expansion for a real number x uses only the digits 0 and 1. For example, $0.a_1a_2a_3\cdots$ means $a_1/2+a_2/2^2+a_3/2^3+\cdots$ so that

$$\tfrac{1}{2}=0.10000\cdots \quad (2),$$
$$\tfrac{1}{4}=0.01000\cdots \quad (2),$$
$$\tfrac{1}{16}=0.00010\cdots \quad (2),$$
$$\tfrac{13}{16}=\tfrac{1}{2}+\tfrac{1}{4}+\tfrac{1}{16}=0.1101000\cdots \quad (2),$$

where the (2) denotes binary expansion.

Similarly, the ternary expansion of a real x uses the digits $0,1,2$. Thus

$$x=0.b_1b_2b_3\cdots \quad (3)$$

means

$$x=\frac{b_1}{3}+\frac{b_2}{3^2}+\frac{b_3}{3^3}+\cdots.$$

For example,

$$\tfrac{1}{3}=0.1000\cdots \quad (3),$$
$$\tfrac{1}{3}=0.0222\cdots \quad (3),$$
$$\tfrac{1}{2}=0.111111\cdots \quad (3),$$
$$\tfrac{5}{6}=\tfrac{1}{2}+\tfrac{1}{3}=0.21111\cdots \quad (3).$$

The ternary expansion for a real number x is unique except for numbers such as $\tfrac{1}{3}$ with two expansions, one ending in a string of 2's, the other in a string of 0's.

1.6D. The following set serves as a useful example later on.

DEFINITION. The Cantor set K is the set of all numbers x in $[0,1]$ which have a ternary expansion without the digit 1.

Thus the numbers $\tfrac{1}{3}=0.0222\cdots$ (3) and $\tfrac{2}{3}=0.20000\cdots$ (3) are in K, but any x such that $\tfrac{1}{3}<x<\tfrac{2}{3}$ is not in K. [For such an x can only be expanded $x=0.1b_2b_3\cdots$ (3).]

For $x=0.b_1b_2b_3\cdots$ (3) in K (where each b_i is 0 or 2), let $f(x)=y=0.a_1a_2a_3\cdots$ (2) where $a_i=b_i/2$. For example, if $x=\tfrac{1}{3}=0.0222\cdots$ (3), then $f(x)=y=0.0111\cdots$ (2) $=\tfrac{1}{2}$. Then $0\leqslant y\leqslant 1$, and f is a function from K into $[0,1]$. It is not difficult to see that f is actually onto $[0,1]$, and it follows immediately that K *is not countable*. (See Exercise 1 of Section 1.6.)

On the other hand, we have already observed that $(\tfrac{1}{3},\tfrac{2}{3})\subset K'$ where $K'=[0,1]-K$. Similarly, the interval $I_1=(\tfrac{1}{9},\tfrac{2}{9})$ (which is the open middle third of $[0,\tfrac{1}{3}]$) and the interval $I_2=(\tfrac{7}{9},\tfrac{8}{9})$ (which is the open middle third of $[\tfrac{2}{3},1]$) are subsets of K' since any number in I_1 or I_2 must have a 1 as the second digit in its ternary expansion. Thus the Cantor set K can be obtained in the following way.

1. From $[0,1]$ remove the open middle third leaving $[0,\tfrac{1}{3}]$ and $[\tfrac{2}{3},1]$.
2. From each of $[0,\tfrac{1}{3}]$ and $[\tfrac{2}{3},1]$ remove the open middle third leaving $[0,\tfrac{1}{9}]$, $[\tfrac{2}{9},\tfrac{3}{9}]$, $[\tfrac{6}{9},\tfrac{7}{9}]$, $[\tfrac{8}{9},\tfrac{9}{9}]$.

n. Continue in this manner so that, at the nth step the open middle third is removed from each of 2^{n-1} intervals of length 3^{-n+1}. The total of the lengths removed at the nth step is thus $2^{n-1}\cdot\tfrac{1}{3}\cdot 3^{-n+1}=2^{n-1}/3^n$. There then remain 2^n intervals each of length 3^{-n}. During this nth step the numbers removed are precisely those with a 1 as the nth digit in their ternary expansion.

It is clear that what remains of $[0,1]$ after this process is continued indefinitely is precisely the set K. Note that the sum of the lengths of the intervals in K' is $\frac{1}{3}+2\cdot\frac{1}{9}+\cdots+2^{n-1}/3^n+\cdots=1$. Thus $K\subset[0,1]$ and is *the complement of the union of open intervals whose lengths add up to* 1. (This seems to say that K is "small" in contrast to the uncountability of K which seems to say that K is "big." That is why K is interesting.)

1.6E. We have seen that the set R is "bigger" than the set I in the sense that I is (equivalent to) a subset of R but I is not equivalent to R itself. It is natural to ask whether there exists a set that is "bigger" than R. We shall now show that the class S of all subsets of R is "bigger" than R.

The elements of S are thus the subsets of R—that is, $A\in S$ if and only if $A\subset R$. In particular, if $r\in R$, then $\{r\}\in S$, and so S contains as a subclass the class $\{\{r\}|r\in R\}$ of subsets of R containing one element. Clearly, R is equivalent to this subclass.

On the other hand, R is not equivalent to S. For suppose the contrary. Then there would be a 1–1 function f from R onto S. For each $x\in R$, then, $f(x)$ is a subset of R and every subset of R is equal to $f(x)$ for some $x\in R$. A given $x\in R$ may or may not be an element of the image subset $f(x)$. Let

$$A=\{x\in R|x\not\in f(x)\}.$$

Then $A\subset R$ and so $A\in S$. Hence $A=f(x_0)$ for some $x_0\in R$. Now we arrive at a contradiction. For either $x_0\in A$ or $x_0\not\in A$. But

1. If $x_0\in A$, then $x_0\not\in f(x_0)$ (by definition of A), and so $x_0\not\in A$ [since $A=f(x_0)$].
2. If $x_0\not\in A$, then $x_0\not\in f(x_0)$ [since $A=f(x_0)$], and so $x_0\in A$ (by definition of A).

Thus both $x_0\in A$ and $x_0\not\in A$ are impossible. The contradiction proves that R is not equivalent to S.

It is clear that no properties special to R were used. The argument therefore applies to any set B. We have thus shown that B is not equivalent to the class of subsets of B. In particular, there is no "biggest possible" set.

Exercises 1.6

1. If $f:A\to B$ and the range of f is uncountable, prove that the domain of f is uncountable.
2. Prove that if B is a countable subset of the uncountable set A, then $A-B$ is uncountable.
3. Prove that the set of all irrational numbers is uncountable.
4. If $a,b\in R$ and $a<b$, show that $[a,b]$ is equivalent to $[0,1]$.
5. Prove that between any two distinct real numbers there is an irrational number.
6. Prove that the set of all characteristic functions on I is uncountable.
7. A real number x is said to be an *algebraic* number if x is a root of some polynomial function f with rational coefficients [that is, $f(x)=0$]. A transcendental number is a real number that is not an algebraic number.

   Assume that a polynomial of degree n has at most n roots. Prove that the set of all transcendental numbers is uncountable. (See Exercise 5 of Section 1.5.)
8. For the function f in 1.6D show that $f(\frac{1}{3})=f(\frac{2}{3})$. More generally, show that if (a,b) is any one of the open intervals removed in the construction K, show that $f(a)=f(b)$. (*Hint:* Show that a and b can be written $a=0.a_1a_2\cdots a_n1$, $b=0.a_1a_2\cdots a_n2$, where each a_i is 0 or 2. Then rewrite the expansion for a using only 0 and 2.)

9. Show that if $x,y \in K$, $x<y$, and $f(x)=f(y)$ (where f is as in 1.6D), then (x,y) is one of the intervals (a,b) of the preceding exercise. (This shows that if we removed all such b's from the Cantor set, then f would be a 1–1 function from what remains of K *onto* $[0,1]$.)
10. Prove that the Cantor set is equivalent to $[0,1]$.
11. For each $t \in R$, let E_t be a subset of R. Suppose that if $s<t$, then E_s is a proper subset of E_t. (That is, $E_s \subset E_t$, $E_s \neq E_t$.) Must $\cup_{t \in R} E_t$ be uncountable? (*Answer:* No.)

1.7 LEAST UPPER BOUNDS

The proofs of many of the basic theorems of elementary calculus—existence of maxima and minima, the intermediate value theorem, Rolle's theorem, the meanvalue theorem, and so on—depend strongly on the so-called completeness property of the real numbers R. There are many ways to formulate this property. We do so in 1.7D with the "least upper bound axiom." First we have to define bounded sets and upper bounds.

1.7A. DEFINITION. The subset $A \subset R$ is said to be bounded above if there is a number $N \in R$ such that $x \leqslant N$ for every $x \in A$. The subset $A \subset R$ is said to be bounded below if there is a number $M \in R$ such that $M \leqslant x$ for every $x \in A$. If A is both bounded below and bounded above, we say that A is bounded.

Thus A is bounded if and only if $A \subset [M,N]$ for some interval $[M,N]$ of finite length. The set I of positive integers is bounded below but not above. Hence I is not bounded. The interval $[0,1]$ is bounded. This shows that the boundedness of a set has nothing to do with countability.

1.7B. DEFINITION. If $A \subset R$ is bounded above, then N is called an upper bound for A if $x \leqslant N$ for all $x \in A$. If $A \subset R$ is bounded below, then M is called a lower bound for A if $M \leqslant x$ for every $x \in A$.

We often abbreviate upper bound and lower bound by u.b. and l.b. respectively. Thus -7 is an l.b. for I. The number 1 is an u.b. for the set $B=\{\frac{1}{2},\frac{3}{4},\frac{7}{8},\ldots,(2^n-1)/2^n,\ldots\}$. Note that infinitely many numbers greater than -7 are lower bounds for I, but that there is no number less than 1 which is an upper bound for B. This leads us to the concept of least upper bound and greatest lower bound.

1.7C. DEFINITION. Let the subset A of R be bounded above. The number L is called the least upper bound for A if (1) L is an upper bound for A, and (2) no number smaller than L is an upper bound for A. See Figure 10.

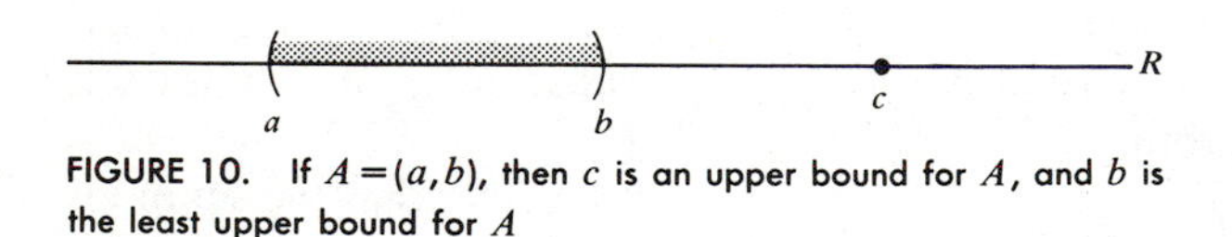

FIGURE 10. If $A=(a,b)$, then c is an upper bound for A, and b is the least upper bound for A

Similarly, l is called the greatest lower bound for the set A bounded below, if l is a lower bound for A and no number greater than l is a lower bound for A.

We abbreviate "least upper bound" as l.u.b. (or l.u.b.$_{x \in A}\, x$), and "greatest lower bound" as g.l.b. It is immediate that a set A can have no more than one l.u.b. For if $L =$ l.u.b. for A and $M < L$, then, by (2), M is not an upper bound for A. Moreover, if $M > L$, then M cannot be a l.u.b. for A since L is an u.b. and $L < M$. Similarly, no set can have more than one g.l.b.

It is not at all obvious that a nonempty set A which is bounded above necessarily has a l.u.b. This is the subject of the least upper bound axiom to be given shortly. First we give some examples.

If $B = \{\frac{1}{2}, \frac{3}{4}, \ldots, (2^n - 1)/2^n, \ldots\}$, then g.l.b.$_{x \in B}\, x = \frac{1}{2}$ and l.u.b.$_{x \in B}\, x = 1$. (Verify!) Note that the g.l.b. for B is an element of B but that the l.u.b. for B is *not* an element of B. The set $(3,4)$ (open interval) does not contain either its g.l.b. or its l.u.b., which are 3 and 4 respectively.

The g.l.b. for I is 1. There is no l.u.b. since I is not bounded above.

The g.l.b. and the l.u.b. for $\{0\}$ are both equal to 0.

According to our definitions, the empty set $\varnothing$ is bounded since $\varnothing \subset [M,N]$ for any interval $[M,N]$. Thus every number $N \in R$ is an u.b. for $\varnothing$ and so $\varnothing$ does not have a l.u.b.

The following axiom would be a theorem if we were to develop set theory carefully and then construct the real numbers from the definition. Since we are not doing this we call it an axiom.

1.7D. LEAST UPPER BOUND AXIOM. If A is any nonempty subset of R that is bounded above, then A has a least upper bound in R.

This axiom says roughly that R (visualized as a set of points on a line) has no holes in it. The set of all rational numbers does have holes in it. (That is, the l.u.b. axiom does not hold if R is replaced by the set of all rationals.) For example, if $A = \{1, 1.4, 1.41, 1.414, \ldots\}$, then (in R) the l.u.b. for A is $\sqrt{2}$ which is not in the set of rationals. Thus if we had never heard of irrational numbers, we would say that A had no l.u.b.

Our assumptions about the relation between real numbers and decimal expansions are consequences of the l.u.b. axiom 1.7D. We show how to deduce them in the next chapter.

The statement for g.l.b. corresponding to 1.7D need not be taken as an axiom. It can be deduced from 1.7D.

1.7E. THEOREM. If A is any nonempty subset of R that is bounded below, then A has a greatest lower bound in R.

PROOF: Let $B \subset R$ be the set of all $x \in R$ such that $-x \in A$. (That is, the elements of B are the negatives of the elements of A.) If M is a lower bound for A, then $-M$ is an upper bound for B. For if $x \in B$, then $-x \in A$ and so $M \leqslant -x, x \leqslant -M$. Hence B is bounded above so that, by 1.7D, B has a l.u.b. If Q is the l.u.b. for B, then $-Q$ is the g.l.b. for A. (Verify.)

One interesting consequence of the least upper bound axiom is the following result called the Archimedean property of the real numbers.

1.7F. THEOREM. If a and b are positive numbers, then there exists $n \in I$ such that $na > b$. (Thus no matter how small a is or how large b is, there is an integral multiple of a that is greater than b.)

PROOF: Let

$$A = \{na \mid n \in I\}.$$

If the theorem were false, then b would be an upper bound for A. Hence by 1.7D, the set A would have a least upper bound. Let $B = \text{l.u.b.}A$. Then, since B is the *least* upper bound, the number $B - a$ is not an upper bound for A so that $B - a < na$ for some $n \in I$. But then $B < (n+1)a$ and, since $(n+1)a \in A$, this shows that B is not an upper bound for A. This is a contradiction, and the theorem is proved.

Exercises 1.7

1. Find the g.l.b. for the following sets.
   (a) $(7,8)$.
   (b) $\{\pi + 1, \pi + 2, \pi + 3, \dots\}$.
   (c) $\{\pi + 1, \pi + \frac{1}{2}, \pi + \frac{1}{3}, \pi + \frac{1}{4}, \dots\}$.
2. Find the l.u.b. for the following sets.
   (a) $(7,8)$.
   (b) $\{\pi + 1, \pi + \frac{1}{2}, \pi + \frac{1}{3}, \dots\}$.
   (c) The complement in $[0,1]$ of the Cantor set.
3. Give an example of a countable bounded subset A of R whose g.l.b. and l.u.b. are both in $R - A$.
4. If A is a nonempty bounded subset of R, and B is the set of all upper bounds for A, prove

$$\underset{y \in B}{\text{g.l.b.}}\, y = \underset{x \in A}{\text{l.u.b.}}\, x.$$

5. If A is a nonempty bounded subset of R, and the g.l.b. for A is equal to the l.u.b. for A, what can you say about A?

2

SEQUENCES OF REAL NUMBERS

2.1 DEFINITION OF SEQUENCE AND SUBSEQUENCE

Our intuitive concept of the term "sequence of numbers" involves not only a set of numbers but also an order—there is a first number, a second and so on. That is, for each positive integer 1, 2, 3,..., there is "associated" a number in the sequence. We make this precise in the following definition.

2.1A. DEFINITION. A sequence $S=\{s_i\}_{i=1}^{\infty}$ of real numbers is a function from I (the set of positive integers) into R (the set of real numbers).

The notation* $\{s_i\}_{i=1}^{\infty}$ is classical. The real number s_i is (by definition) $S(i)$. Instead of S or $\{s_i\}_{i=1}^{\infty}$, we sometimes write $s_1, s_2, \ldots$. The number $s_i (i=1,2,\ldots)$ is called the ith term of the sequence.

Sequences such as $\{s_i\}_{i=-\infty}^{\infty}$ or $\{s_i\}_{i=1}^{M}$ (M a positive integer) may also be defined. We leave these definitions to the reader. Unless otherwise stated, we use sequence to mean $\{s_i\}_{i=1}^{\infty}$—a function with domain I.

There is, of course, no reason to restrict the definition of sequence to sequences of real numbers. In later sections we use sequences of sets and sequences of functions. For example, if X is any set, then a sequence $\{E_i\}_{i=1}^{\infty}$ of subsets of X is defined as a function from I into the class of all subsets of X.

2.1B. DEFINITION. Having defined sequence we would now like to formulate the concept of subsequence. Even though the idea of subsequence is easy to visualize, it is not quite obvious how to define subsequence making sure that a subsequence is still a

* It does not matter what subscript is used. We sometimes use j, k, l, m, or n. That is, $\{s_i\}_{i=1}^{\infty}=\{s_j\}_{j=1}^{\infty}=\{s_n\}_{n=1}^{\infty}=\cdots$.
In addition, do not confuse the sequence $\{s_n\}_{n=1}^{\infty}$ with the set $\{s_1,s_2,\ldots\}$. For example, if $s_n=2 (n\in I)$, then the set $A=\{s_1,s_2,\ldots\}$ contains only one element. That is, $A=\{2\}$. The sequence $\{s_n\}_{n=1}^{\infty}$, considered as a set (since it is a function), has infinitely many elements. In set notation

$$\{s_n\}_{n=1}^{\infty}=\{\langle n,2\rangle | n\in I\}.$$

sequence. (Try it.) In reading the next paragraph, use whatever intuitive idea of "subsequence" you have.

If $s_1, s_2, \ldots$ is a sequence, a subsequence is usually written $s_{n_1}, s_{n_2}, \ldots$. That is, from the original sequence we "keep" only those terms whose subscripts are $n_1, n_2, \ldots$. But these numbers $n_1, n_2, \ldots$ themselves form a subsequence of the sequence of positive integers 1, 2, 3,.... Thus we first define "subsequence of the sequence of positive integers" and then define subsequences of an arbitrary sequence of real numbers.

2.1C. DEFINITION. A subsequence N of $\{n\}_{n=1}^{\infty}$ (the sequence of positive integers) is a function from I (the set of positive integers) into I such that

$$N(i) < N(j) \quad \text{if} \quad i < j \quad (i,j \in I).$$

Since $N: I \to I$ it follows that $N: I \to R$. Therefore N is a sequence. Roughly, then, a subsequence of $\{n\}_{n=1}^{\infty}$ is a sequence of integers whose terms get larger and larger. For example, the sequence of primes 2, 3, 5, 7, 11,... is a subsequence of $\{n\}$. Other examples are 2, 4, 6, 8,... and 1, 3, 5, 7,....

2.1D. DEFINITION. If $S = \{s_n\}_{n=1}^{\infty}$ is a sequence of real numbers and $N = \{n_i\}_{i=1}^{\infty}$ is a subsequence of the sequence of positive integers, then the composite function $S \circ N$ is called a subsequence of S.

Note that for $i \in I$ we have

$$N(i) = n_i,$$
$$S \circ N(i) = S[N(i)] = S(n_i) = S_{n_i},$$

and hence

$$S \circ N = \{s_{n_i}\}_{i=1}^{\infty}.$$

Thus our definition 2.1D conforms to the accepted notation $s_{n_1}, s_{n_2}, \ldots$ for subsequences. In effect, N tells us which terms of S to keep.

For example, let us denote the sequence 1, 0, 1, 0,... by B, and define $N = \{n_i\}_{i=1}^{\infty}$ by

$$n_i = 2i - 1 \quad (i \in I)$$

so that $n_1 = 1$, $n_2 = 3$, $n_3 = 5, \ldots$. Then $B \circ N$ is the subsequence 1, 1, 1,... of B. For another example, if $C = \{c_n\}_{n=1}^{\infty} = \{\sqrt{n}\}_{n=1}^{\infty}$, and $N = \{n_i\}_{i=1}^{\infty} = \{i^4\}_{i=1}^{\infty}$, then

$$C \circ N = \{c_{n_i}\}_{i=1}^{\infty} = \{\sqrt{i^4}\}_{i=1}^{\infty} = \{i^2\}_{i=1}^{\infty}.$$

Exercises 2.1

1. Let $\{s_n\}_{n=1}^{\infty}$ be the sequence defined by

$$s_1 = 1,$$
$$s_2 = 1,$$
$$s_{n+1} = s_n + s_{n-1} \quad (n = 2, 3, 4, \ldots).$$

Find s_8. (The numbers s_n are called the Fibonacci numbers.)

2. Write a formula or formulae for s_n for each of the following sequences. [For example,

the sequence $2,1,4,3,6,5,8,7,\ldots$ can be described by $s_n = n+1$ $(n=1,3,5,\ldots)$, $s_n = n-1$ $(n=2,4,6,\ldots)$.]

(a) $1,0,1,0,1,0,\ldots$.

(b) $1,3,6,10,15,\ldots$.

(c) $1,-4,9,-16,25,-36,\ldots$.

(d) $1,1,1,2,1,3,1,4,1,5,1,6,\ldots$.

3. Which of the sequences (a), (b), (c), (d) in the previous exercise are subsequences of $\{n\}_{n=1}^{\infty}$?
4. If $S=\{s_n\}_{n=1}^{\infty}=\{2n-1\}_{n=1}^{\infty}$ and $N=\{n_i\}_{i=1}^{\infty}=\{i^2\}_{i=1}^{\infty}$, find s_5, s_9, n_2, s_{n_3}. Is N a subsequence of $\{k\}_{k=1}^{\infty}$?
5. Let S be a sequence. Prove that every subsequence of a subsequence of S is itself a subsequence of S.
6. If $\{s_{n_k}\}_{k=1}^{\infty}$ is a subsequence of $\{s_n\}_{n=1}^{\infty}$, prove that

$$n_k \geqslant k \qquad (k \in I).$$

2.2 LIMIT OF A SEQUENCE

The concept of limit is one of the most important (and conceivably *the* most difficult) in analysis. In this section we define the limit of a sequence (function on I). Limits for other functions are discussed in the fourth chapter.

Roughly speaking, the sequence $\{s_n\}_{n=1}^{\infty}$ has the limit L if $s_n - L$ is "small" for all sufficiently large values of n. From this crude description, we would expect that the sequence $1,1,1,\ldots,$has the limit 1, that the sequence $1,\frac{1}{2},\frac{1}{3},\ldots,$ has the limit 0, and the sequence $1,-2,3,-4,\ldots,$ does not have a limit. We shall see that our intuition in these cases is correct.

In other cases, for example $\{n\sin(\pi/n)\}_{n=1}^{\infty}$, our intuition is not sharp enough to tell if a given sequence has a "limit" or to compute the "limit" if there is one. We need a precise definition of "limit of a sequence" and enough theorems about the definition to make computations easy.

2.2A. DEFINITION. Let $\{s_n\}_{n=1}^{\infty}$ be a sequence of real numbers. We say that s_n approaches the limit L (as n approaches infinity),* if for every $\epsilon > 0$ there is a positive integer N such that

$$|s_n - L| < \epsilon \quad (n \geqslant N). \tag{1}$$

If s_n approaches the limit L we write

$$\lim_{n\to\infty} s_n = L$$

or

$$s_n \to L \quad (n\to\infty).$$

Instead of "s_n approaches the limit L" we often say that the sequence $\{s_n\}_{n=1}^{\infty}$ has the limit L.

We emphasize the fact that our definition requires that L be a real number.

Thus $\lim_{n\to\infty} s_n = L$ means that for any $\epsilon > 0$, the inequality $|s_n - L| < \epsilon$ must hold for

* The phrase "as n approaches infinity" is part of the definition. We are not defining "infinity."

all values of n except at most a finite number—namely, $n=1,2,\ldots,N-1$. The value of N will, in general, depend on the value of ϵ. Thus for a given $\{s_n\}_{n=1}^{\infty}$, the proof that $\lim_{n\to\infty} s_n = L$ consists, upon being given an $\epsilon>0$, of finding a value of N such that

$$|s_n - L| < \epsilon \quad (n \geqslant N).$$

There is no need to find the smallest value of N for which (1) holds. If, for each $\epsilon>0$, *any* N for which (1) is true has been found, this proves $\lim_{n\to\infty} s_n = L$.

Consider Figure 11. All of the s_n, except for at most a finite number of n, must be inside the parentheses.

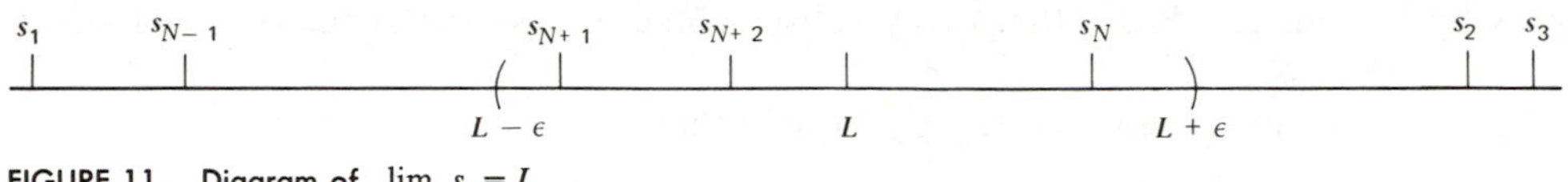

FIGURE 11. Diagram of $\lim_{n\to\infty} s_n = L$

For example, consider the sequence $1, \frac{1}{2}, \frac{1}{3}, \ldots$. That is, consider $\{s_n\}_{n=1}^{\infty}$ where $s_n = 1/n$ $(n=1,2,\ldots)$. We would naturally guess that this sequence has the limit $L=0$. Let us prove this. Given $\epsilon>0$ we must find N so that (1) holds. In this case (1) reads

$$\left|\frac{1}{n} - 0\right| < \epsilon \qquad (n \geqslant N),$$

or

$$\frac{1}{n} < \epsilon \qquad (n \geqslant N). \tag{2}$$

Thus if we choose N so that $1/N < \epsilon$, then certainly (2), and thus (1), will hold since $1/n \leqslant 1/N$ if $n \geqslant N$. Now $1/N<\epsilon$ if and only if $N > 1/\epsilon$. Hence if we take any $N \in I$ such that $N > 1/\epsilon$, then (1) will hold for this sequence $\{s_n\}_{n=1}^{\infty}$ with $L=0$. This proves $\lim_{n\to\infty} 1/n = 0$. Note that the limit 0 is not equal to any term of the sequence.

Let us now examine the sequence $\{s_n\}_{n=1}^{\infty}$ where $s_n = 1$ $(n=1,2,\ldots)$. We have previously guessed that this sequence has the limit $L=1$. To prove this we note that $s_n - L = 1 - 1 = 0$ so that for any $\epsilon>0$,

$$|s_n - L| < \epsilon \qquad (n \geqslant 1).$$

Thus in this case, for any $\epsilon>0$ we can make (1) hold by taking $N=1$. (This is one of the rare cases where N does not depend on ϵ.) This proves $\lim_{n\to\infty} 1 = 1$.

For a third example, consider $\{s_n\}_{n=1}^{\infty}$ where $s_n = n$ $(n=1,2,\ldots)$—that is, consider the sequence $1,2,3,\ldots$. We shall prove this sequence does not have a limit. Assume the contrary. Then $\lim_{n\to\infty} s_n = L$ for some $L \in R$. Then for any ϵ there is an N for which (1) holds. In particular, for $\epsilon=1$ there is an N for which (1) holds:

$$|s_n - L| < 1 \qquad (n \geqslant N).$$

This is equivalent to

$$-1 < s_n - L < 1 \qquad (n \geqslant N)$$

or

$$-1 < n - L < 1 \qquad (n \geqslant N)$$

or

$$L - 1 < n < L + 1 \qquad (n \geqslant N).$$

The last statement says that all values of n greater than N lie between $L-1$ and $L+1$. This is clearly false and the contradiction shows that $\{s_n\}_{n=1}^{\infty}=\{n\}_{n=1}^{\infty}$ does not have a limit.

The last example shows that a sequence whose terms get "too big" cannot have a limit. This is not the only kind of sequence that does not have a limit. Consider the sequence $\{s_n\}_{n=1}^{\infty}$ where $s_n=(-1)^n$ $(n=1,2,\ldots)$. The terms of this sequence are $-1,1,-1,1,\ldots$. Suppose there were an $L\in R$ for which $\lim_{n\to\infty}s_n=L$. Then for $\epsilon=\frac{1}{2}$ there would be an $N\in I$ such that (1) holds. That is,

$$|(-1)^n-L|<\tfrac{1}{2} \qquad (n\geqslant N). \tag{3}$$

For n even (3) says

$$|1-L|<\tfrac{1}{2}, \tag{4}$$

while for n odd (3) says

$$|-1-L|<\tfrac{1}{2}. \tag{5}$$

Since $|a-b|$ is the distance from a to b, (4) implies that L is less than $\frac{1}{2}$ unit away from 1, while (5) implies that L is less than $\frac{1}{2}$ unit away from -1. This is a contradiction. [To deduce a contradiction from (4) and (5) without geometry, reason as follows: The inequality (5) is equivalent to $|1+L|<\frac{1}{2}$. But then

$$2=|2|=|1+1|=|(1+L)+(1-L)|\leqslant|1+L|+|1-L|<\tfrac{1}{2}+\tfrac{1}{2}=1,$$

which is a contradiction.]

Hence no limit L exists for the sequence $\{(-1)^n\}_{n=1}^{\infty}$ (even though the terms of the sequence all have absolute value 1 and hence are not "too big").

We emphasize that at this early stage in our development of limit, if we wish to show that a given sequence has a limit, then we must first guess what the limit is! We have as yet developed no general criteria that will tell us if a limit exists for a given sequence. Here is an example indicating how to go about guessing under a typical set of circumstances.

Let

$$\{s_n\}_{n=1}^{\infty}=\left\{\frac{2n}{n+4n^{1/2}}\right\}_{n=1}^{\infty}.$$

When n is "large," then n is "much bigger than" $n^{1/2}$. We thus guess that, for purposes of establishing what the limit is (if it exists), the $4n^{1/2}$ term can be ignored. That is, we have $s_n=2n/(n+\theta)$ where, as n becomes larger, the quantity θ becomes more and more negligible compared to the other quantities (even though $\theta=4n^{1/2}$ itself is "large" when n is "large"). Thus, for "large" n, s_n must be near 2. We therefore guess that $\{s_n\}_{n=1}^{\infty}$ has the limit 2.

From a slightly different, and a more algebraic point of view, we note that $s_n=2/(1+4/n^{1/2})$. As n gets "large" the 2 and the 1 are not affected but $4/n^{1/2}$ gets "small." In the "long run" s_n is *roughly* $2/(1+0)$. So again we guess $\lim_{n\to\infty}s_n=2$.

Now let us *prove* that $\lim_{n\to\infty}s_n=2$. Given $\epsilon>0$ we must find (calculate) $N\in I$ such that

$$\left|\frac{2n}{n+4n^{1/2}}-2\right|<\epsilon \qquad (n\geqslant N). \tag{6}$$

The inequality (6) is equivalent to

$$\left|\frac{2n-2n-8n^{1/2}}{n+4n^{1/2}}\right|<\epsilon \qquad (n\geqslant N),$$

or

$$\frac{8n^{1/2}}{n+4n^{1/2}}<\epsilon \qquad (n \geqslant N). \tag{7}$$

Now the left side of (7) is less than $8n^{1/2}/n=8/n^{1/2}$. (Why?) Hence (7) will be true if

$$\frac{8}{n^{1/2}}<\epsilon \qquad (n \geqslant N). \tag{8}$$

If we choose N so that $8/N^{1/2}<\epsilon$, that is, choose $N>64/\epsilon^2$, then (8) will certainly be true. (For $8/n^{1/2} \leqslant 8/N^{1/2}$ if $n \geqslant N$.) We have thus shown that if N is any positive integer greater than $64/\epsilon^2$, then (8) and hence (7) and finally (6) will be true. This *proves* $\lim_{n\to\infty} s_n=2$.

Our intuition tells us that a sequence of nonnegative numbers cannot have a negative limit. This we now prove.

2.2B. THEOREM. If $\{s_n\}_{n=1}^{\infty}$ is a sequence of nonnegative numbers and if $\lim_{n\to\infty} s_n = L$, then $L \geqslant 0$.

PROOF: Suppose the contrary, namely that $L<0$. Then for $\epsilon=-L/2$ there exists $N \in I$ such that

$$|s_n - L|<\frac{-L}{2} \qquad (n \geqslant N).$$

In particular

$$|s_N - L|<\frac{-L}{2},$$

which implies

$$s_N - L<\frac{-L}{2}$$

or

$$s_N<\frac{L}{2}.$$

But, by hypothesis, $s_N \geqslant 0$. This implies $L>0$, contradicting our supposition that $L<0$. Hence $L \geqslant 0$.

This proof is a precise way of saying the following: If s_n gets "arbitrarily close" to L when n is "large," and $L<0$, then $s_n<0$ for sufficiently large n.

Exercises 2.2

1. If $\{s_n\}_{n=1}^{\infty}$ is a sequence of real numbers, if $s_n \leqslant M$ $(n \in I)$, and if $\lim_{n\to\infty} s_n = L$, prove $L \leqslant M$.
2. If $L \in R$, $M \in R$, and $L \leqslant M+\epsilon$ for every $\epsilon>0$, prove $L \leqslant M$.
3. If $\{s_n\}_{n=1}^{\infty}$ is a sequence of real numbers and if, for every $\epsilon>0$,

$$|s_n - L|<\epsilon \qquad (n \geqslant N)$$

where N does *not* depend on ϵ, prove that all but a finite number of terms of $\{s_n\}_{n=1}^{\infty}$ are equal to L.

4. (a) Find $N \in I$ such that

$$\left|\frac{2n}{n+3}-2\right|<\frac{1}{5} \qquad (n \geqslant N).$$

(b) Prove $\lim_{n\to\infty} 2n/(n+3)=2$.

5. (a) Find $N \in I$ so that $1/\sqrt{n+1} < 0.03$ when $n \geqslant N$.

(b) Prove that $\lim_{n\to\infty} 1/\sqrt{n+1} = 0$.

6. If θ is a rational number, prove that the sequence $\{\sin n!\theta\pi\}_{n=1}^{\infty}$ has a limit.

7. For each of the following sequences, prove either that the sequence has a limit or that the sequence does not have a limit.

(a) $\left\{\dfrac{n^2}{n+5}\right\}_{n=1}^{\infty}$.

(b) $\left\{\dfrac{3n}{n+7n^{1/2}}\right\}_{n=1}^{\infty}$.

(c) $\left\{\dfrac{3n}{n+7n^2}\right\}_{n=1}^{\infty}$.

8. (a) Prove that the sequence $\{10^7/n\}_{n=1}^{\infty}$ has limit 0.

(b) Prove that $\{n/10^7\}_{n=1}^{\infty}$ does not have a limit.

(c) Note that the first 10^7 terms of the sequence in (a) are greater than the corresponding terms of the sequence in (b). This emphasizes that the existence of a limit for a sequence does not depend on the first "few" ("few" = "any finite number") terms.

9. Prove that $\{n-1/n\}_{n=1}^{\infty}$ does not have a limit.

10. If $s_n = 5^n/n!$, show that $\lim_{n\to\infty} s_n = 0$. (*Hint*: Prove that $s_n \leqslant (5^5/5!)(5/n)$ if $n>5$.)

11. If P is a polynomial function of third degree,

$$P(x)=ax^3+bx^2+cx+d \qquad (a,b,c,d,x \in R),$$

prove that

$$\lim_{n\to\infty} \frac{P(n+1)}{P(n)}=1.$$

2.3 CONVERGENT SEQUENCES

2.3A. DEFINITION. If the sequence of real numbers $\{s_n\}_{n=1}^{\infty}$ has the limit L, we say that $\{s_n\}_{n=1}^{\infty}$ is convergent to L. If $\{s_n\}_{n=1}^{\infty}$ does not have a limit, we say that $\{s_n\}_{n=1}^{\infty}$ is divergent.

From the examples of the last section we see that the sequences $1,1,1,\ldots$ and $1,\frac{1}{2},\frac{1}{3},\ldots$ are convergent (to the limits 1 and 0, respectively) and that the sequences $1,2,3,\ldots$ and $-1,+1,-1,+1,\ldots$ are divergent.

We now prove that a sequence cannot converge to more than one limit.

2.3B. THEOREM. If the sequence of real numbers $\{s_n\}_{n=1}^{\infty}$ is convergent to L, then $\{s_n\}_{n=1}^{\infty}$ cannot also converge to a limit distinct from L. That is, if $\lim_{n\to\infty} s_n = L$ and $\lim_{n\to\infty} s_n = M$, then $L=M$.

PROOF: Assume the contrary. Then $L \neq M$ so that $|M-L|>0$. Let $\epsilon = \frac{1}{2}|M-L|$. By the hypothesis $\lim_{n\to\infty} s_n = L$ there exists $N_1 \in I$ such that

$$|s_n - L| < \epsilon \qquad (n \geqslant N_1).$$

Similarly, since $\lim_{n\to\infty} s_n = M$ there exists $N_2 \in I$ such that

$$|s_n - M| < \epsilon \qquad (n \geqslant N_2).$$

Let $N = \max(N_1, N_2)$. Then $N \geqslant N_1$ and $N \geqslant N_2$ so that both $|s_N - L|$ and $|s_N - M|$ are less than ϵ. Thus,

$$|M-L| = |(s_N - L) - (s_N - M)| \leqslant |s_N - L| + |s_N - M| < 2\epsilon = |M - L|,$$

which implies $|M-L|<|M-L|$. This contradiction shows $M=L$, which is what we wished to show.

The next result is almost obvious. In fact it *is* obvious, so the proof is left to the reader.

2.3C. THEOREM. If the sequence of real numbers $\{s_n\}_{n=1}^{\infty}$ is convergent to L, then any subsequence of $\{s_n\}_{n=1}^{\infty}$ is also convergent to L.

There is a useful corollary.

2.3D. COROLLARY. All subsequences of a convergent sequence of real numbers converge to the same limit.

PROOF: If the sequence S converges to L then, by 2.3B, S converges to no other limit. By 2.3C, then, all subsequences of S converge to L (and to no other limit).

This corollary yields an easy proof that $S = \{(-1)^n\}_{n=1}^{\infty}$ is divergent. For both $1, 1, 1, \ldots$ and $-1, -1, -1, \ldots$ are subsequences of S and converge to different limits.

The example $\{(-1)^n\}_{n=1}^{\infty}$ shows that a divergent sequence may have a convergent subsequence. The example $\{n\}_{n=1}^{\infty}$ shows that a divergent sequence need have no convergent subsequence.

Here are some more examples. If θ is a rational number, $0<\theta<1$, and $S = \{\sin n\theta\pi\}_{n=1}^{\infty}$, then S is divergent. For we can write $\theta = a/b$ where a and b are integers and $b \geqslant 2$. The terms of S for $n = b, 2b, 3b, \ldots$ are $\sin a\pi, \sin 2a\pi, \sin 3a\pi, \ldots$. Thus S contains the subsequence $0, 0, 0, \ldots$. But the terms of S for which $n = b+1, 2b+1, 3b+1, \ldots$ are $\sin(a\pi + a\pi/b), \sin(2a\pi + a\pi/b), \sin(3a\pi + a\pi/b), \ldots$ or $(-1)^a \sin(a\pi/b), (-1)^{2a}\sin(a\pi/b), (-1)^{3a}\sin(a\pi/b), \ldots$. These terms all have absolute value $\sin(a\pi/b)$ and hence do not approach 0. Thus S contains a subsequence which has the limit 0 *and* a subsequence which (may or may not converge but certainly) does not have the limit 0. By 2.3D, S is divergent.

For $\theta = 0$ or $\theta = 1$ the sequence $\{\sin n\theta\pi\}_{n=1}^{\infty}$ is clearly convergent to 0.

It may be shown that if θ is irrational, then $\{\sin n\theta\pi\}_{n=1}^{\infty}$ is divergent. This, however, is somewhat more difficult.

Exercises 2.3

1. For any $a, b \in R$ show that

$$\big||a| - |b|\big| \leqslant |a - b|.$$

Then prove that $\{|s_n|\}_{n=1}^{\infty}$ converges to $|L|$ if $\{s_n\}_{n=1}^{\infty}$ converges to L.

2. Give an example of a sequence $\{s_n\}_{n=1}^{\infty}$ of real numbers for which $\{|s_n|\}_{n=1}^{\infty}$ converges but $\{s_n\}_{n=1}^{\infty}$ does not.
3. Prove that if $\{|s_n|\}_{n=1}^{\infty}$ converges to 0, then $\{s_n\}_{n=1}^{\infty}$ converges to 0.
4. Can you find a sequence of real numbers $\{s_n\}_{n=1}^{\infty}$ which has no convergent subsequence and yet for which $\{|s_n|\}_{n=1}^{\infty}$ converges?
5. If $\{s_n\}_{n=1}^{\infty}$ is a sequence of real numbers and if

$$\lim_{m\to\infty} s_{2m} = L, \qquad \lim_{m\to\infty} s_{2m-1} = L,$$

prove that $s_n \to L$ as $n \to \infty$. (That is, if the subsequence of $\{s_n\}_{n=1}^{\infty}$ of terms with even subscripts converges to L, as well as the subsequence with odd subscripts, then $\{s_n\}_{n=1}^{\infty}$ converges to L also.)

2.4 DIVERGENT SEQUENCES

From the examples in Section 2.3 we see that the sequences $\{n\}_{n=1}^{\infty}$ and $\{(-1)^n\}_{n=1}^{\infty}$ are both divergent. As we have observed before, however, these sequences behave very differently. For $\{n\}_{n=1}^{\infty}$ diverges because its terms get "too big," whereas $\{(-1)^n\}_{n=1}^{\infty}$ diverges because its terms "oscillate too much." In this section we make a classification of divergent sequences.

2.4A. DEFINITION. Let $\{s_n\}_{n=1}^{\infty}$ be a sequence of real numbers. We say that s_n approaches infinity as n approaches infinity if for any real number $M>0$ there is a positive integer N such that

$$s_n \geqslant M \qquad (n \geqslant N).$$

In this case we write $s_n \to \infty$ as $n \to \infty$. Instead of "s_n approaches infinity" we sometimes say $\{s_n\}_{n=1}^{\infty}$ diverges to infinity.

Just as we think of ϵ in definition 2.2A as being "small," we think of the M in this definition as being "large." Thus if $s_n \to \infty$ as $n \to \infty$, then all but the first "few" of the s_n are "large."

It is obvious that $\{n\}_{n=1}^{\infty}$ diverges to infinity. For given $M>0$, just choose $N \in I$ such that $N \geqslant M$. Then certainly

$$n \geqslant M \qquad (n \geqslant N).$$

The reader should be sure to verify that if $s_n \to \infty$ as $n \to \infty$, then s_n definitely does not approach a limit. (This justifies our use of the phrase "*diverges* to infinity.") We never refer to "infinity" as a limit of a sequence. A limit of a sequence must be a real number.

2.4B. DEFINITION. Let $\{s_n\}_{n=1}^{\infty}$ be a sequence of real numbers. We say that s_n approaches minus infinity as n approaches infinity if, for any real number $M>0$, there is a positive integer N such that

$$s_n < -M \qquad (n \geqslant N).$$

We then write $s_n \to -\infty$ as $n \to \infty$ and say $\{s_n\}_{n=1}^{\infty}$ diverges to minus infinity.

Again we think of M as "large" so that $-M$ is "large negative."

For example, the sequence $\{\log(1/n)\}_{n=1}^{\infty}$ diverges to minus infinity. To prove this, given $M>0$ we must find $N \in I$ such that

$$\log \frac{1}{n} < -M \qquad (n \geqslant N). \tag{1}$$

But this is equivalent to

$$\log n > M \qquad (n \geqslant N),$$

$$n > e^M \qquad (n \geqslant N). \tag{2}$$

Thus if we choose $N \geqslant e^M$, then (2) and hence (1) will hold.

The sequence $1, -2, 3, -4, \ldots$ does not approach either infinity or minus infinity. However, this sequence has the subsequence $1, 3, 5, \ldots$ which approaches infinity and also has the subsequence $-2, -4, -6, \ldots$ which approaches minus infinity.

It is easy to show that if the sequence $\{s_n\}_{n=1}^{\infty}$ diverges to infinity, then so does any subsequence of $\{s_n\}_{n=1}^{\infty}$. (This is analogous to the result of 2.3C.)

Some divergent sequences neither diverge to infinity nor diverge to minus infinity—they "oscillate."

2.4C. DEFINITION. If the sequence $\{s_n\}_{n=1}^{\infty}$ of real numbers diverges but does not diverge to infinity and does not diverge to minus infinity, we say that $\{s_n\}_{n=1}^{\infty}$ oscillates.

An example of a sequence which oscillates is $\{(-1)^n\}_{n=1}^{\infty}$. Another example is the sequence $1, 2, 1, 3, 1, 4, 1, 5, \ldots$. For, by 2.3D, this sequence diverges since it has the divergent subsequence $1, 2, 3, 4, \ldots$. Moreover, the sequence does not diverge to infinity since there is no $N \in I$ for which the statement

$$s_n > 2 \qquad (n \geqslant N)$$

is true. The sequence obviously does not diverge to minus infinity. Hence it oscillates.

We emphasize that "oscillate" does not mean "the terms go up and down." The sequence $1, -\frac{1}{2}, +\frac{1}{3}, -\frac{1}{4}, \ldots$ converges to zero. Hence, by definition, *it does not oscillate* even though its terms "go up and down." Oscillate is a term applied only to certain *divergent* sequences. Roughly speaking, a sequence oscillates if its terms "go up and down too much."

Exercises 2.4

1. Label each of the following sequences either (A) convergent, (B) divergent to infinity, (C) divergent to minus infinity, or (D) oscillating. (Use your intuition or information from your calculus course. Do not try to prove anything.)
   (a) $\{\sin(n\pi/2)\}_{n=1}^{\infty}$.
   (b) $\{\sin n\pi\}_{n=1}^{\infty}$.
   (c) $\{e^n\}_{n=1}^{\infty}$.
   (d) $\{e^{1/n}\}_{n=1}^{\infty}$.
   (e) $\{n \sin(\pi/n)\}_{n=1}^{\infty}$.
   (f) $\{(-1)^n \tan(\pi/2 - 1/n)\}_{n=1}^{\infty}$.
   (g) $\{1 + \frac{1}{2} + \frac{1}{3} + \cdots + 1/n\}_{n=1}^{\infty}$.
   (h) $\{-n^2\}_{n=1}^{\infty}$.
2. Prove that $\{\sqrt{n}\}_{n=1}^{\infty}$ diverges to infinity.
3. Prove that $\{\sqrt{n+1} - \sqrt{n}\}_{n=1}^{\infty}$ is convergent. (*Hint:* Recall how to find dy/dx by the Δx process when $y = \sqrt{x}$.)
4. Prove that if the sequence of real numbers $\{s_n\}_{n=1}^{\infty}$ diverges to infinity, then $\{-s_n\}_{n=1}^{\infty}$ diverges to minus infinity.
5. Suppose $\{s_n\}_{n=1}^{\infty}$ converges to 0. Prove that $\{(-1)^n s_n\}_{n=1}^{\infty}$ converges to 0.
6. Suppose $\{s_n\}_{n=1}^{\infty}$ converges to $L \neq 0$. Prove that $\{(-1)^n s_n\}_{n=1}^{\infty}$ oscillates.
7. Suppose $\{s_n\}_{n=1}^{\infty}$ diverges to infinity. Prove that $\{(-1)^n s_n\}_{n=1}^{\infty}$ oscillates.

2.5 BOUNDED SEQUENCES

Recalling that a sequence of real numbers $\{s_n\}_{n=1}^{\infty}$ is a function from I into R, we see that the range of $\{s_n\}_{n=1}^{\infty}$ (namely $\{s_1, s_2, \ldots\}$) is a subset of R.

2.5A. DEFINITION. We say that the sequence $\{s_n\}_{n=1}^{\infty}$ is bounded above if the range of $\{s_n\}_{n=1}^{\infty}$ is bounded above (see 1.7A). Similarly, we say that $\{s_n\}_{n=1}^{\infty}$ is bounded below or bounded if the range of $\{s_n\}_{n=1}^{\infty}$ is respectively bounded below or bounded.

Thus $\{s_n\}_{n=1}^{\infty}$ is bounded if and only if there exists $M \in R$ such that

$$|s_n| \leqslant M \qquad (n \in I).$$

If a sequence diverges to infinity (or to minus infinity) the sequence is not bounded. (Verify.) A sequence that diverges to infinity must, however, be bounded below. (For such a sequence can have only a finite number of negative terms.)

An oscillating sequence may or may not be bounded. The sequence $1, -2, 3, -4, \ldots$ oscillates and is neither bounded above nor bounded below. The sequence $-1, 1, -1, 1, \ldots$ oscillates and *is* bounded. The sequence $1, 2, 1, 3, 1, 4, \ldots$ oscillates and is bounded below but is not bounded above.

2.5B. THEOREM. If the sequence of real numbers $\{s_n\}_{n=1}^{\infty}$ is convergent, then $\{s_n\}_{n=1}^{\infty}$ is bounded.

PROOF: Suppose $L = \lim_{n\to\infty} s_n$. Then, given $\epsilon = 1$, there exists $N \in I$ such that

$$|s_n - L| < 1 \qquad (n \geqslant N).$$

This implies

$$|s_n| < |L| + 1 \qquad (n \geqslant N). \tag{1}$$

(For $|s_n| = |L + (s_n - L)| \leqslant |L| + |s_n - L|$.)
If we let $M = \max\{|s_1|, |s_2|, \ldots, |s_{N-1}|\}$, then we have

$$|s_n| < M + |L| + 1 \qquad (n \in I),$$

which shows that $\{s_n\}_{n=1}^{\infty}$ is bounded.

Thus in summary, all convergent sequences are bounded; all sequences diverging to infinity (or minus infinity) are *not* bounded; some oscillating sequences are bounded and some are not.

Exercises 2.5

1. True or false? If a sequence of positive numbers is not bounded, then the sequence diverges to infinity.
2. Give an example of a sequence $\{s_n\}_{n=1}^{\infty}$ which is not bounded but for which $\lim_{n\to\infty} s_n/n = 0$.
3. Prove that if $\lim_{n\to\infty} s_n/n = L \neq 0$, then $\{s_n\}_{n=1}^{\infty}$ is not bounded.
4. If $\{s_n\}_{n=1}^{\infty}$ is a bounded sequence of real numbers, and $\{t_n\}_{n=1}^{\infty}$ converges to 0, prove that $\{s_n t_n\}_{n=1}^{\infty}$ converges to 0.
5. If the sequence $\{s_n\}_{n=1}^{\infty}$ is bounded, prove that for any $\epsilon > 0$ there is a closed interval $J \subset R$ of length ϵ such that $s_n \in J$ for infinitely many values of n.

2.6 MONOTONE SEQUENCES

In the preceding section we saw that a sequence may be bounded and still not be convergent. In this section we consider a condition which, together with boundedness, will ensure that a sequence is convergent.

2.6A. DEFINITION. Let $\{s_n\}_{n=1}^{\infty}$ be a sequence of real numbers. If $s_1 \leqslant s_2 \leqslant \cdots \leqslant s_n \leqslant s_{n+1} \leqslant \cdots$, then $\{s_n\}_{n=1}^{\infty}$ is called nondecreasing. Similarly, if $s_1 \geqslant s_2 \geqslant \cdots \geqslant s_n \geqslant s_{n+1} \geqslant \cdots$, then $\{s_n\}_{n=1}^{\infty}$ is called nonincreasing. A monotone sequence is a sequence which is either nonincreasing or nondecreasing (or both).

The sequence $1, 1\frac{1}{2}, 1\frac{3}{4}, 1\frac{7}{8}, \ldots$ (that is, $\{2-1/2^{n-1}\}_{n=1}^{\infty}$) is nondecreasing (and bounded). The sequence $\{n\}_{n=1}^{\infty}$ is nondecreasing (and not bounded). These sequences exemplify the results of the next two theorems, the first of which (2.6B) is of tremendous importance.

2.6B. THEOREM. A nondecreasing sequence which is bounded above* is convergent.

PROOF: Suppose $\{s_n\}_{n=1}^{\infty}$ is nondecreasing and bounded above. Then the set

$$A = \{s_1, s_2, \ldots\}$$

is a nonempty subset of R which is bounded above. By 1.7D this set has a l.u.b. Let

$$M = \text{l.u.b.}\{s_1, s_2, \ldots\} = \text{l.u.b. for } A.$$

We will prove that $s_n \to M$ as $n \to \infty$. Given $\epsilon > 0$ the number $M - \epsilon$ is not an u.b. for A. Hence, for some $N \in I, s_N > M - \epsilon$. But, since $\{s_n\}_{n=1}^{\infty}$ is nondecreasing, this implies

$$s_n > M - \epsilon \qquad (n \geqslant N). \tag{1}$$

On the other hand, since M is an u.b. for A,

$$M \geqslant s_n \qquad (n \in I).$$

From (1) and (2) we conclude

$$|s_n - M| < \epsilon \qquad (n \geqslant N).$$

This proves $\lim_{n\to\infty} s_n = M$ which is what we wished to show.

For example, the sequence $\{2-1/2^{n-1}\}_{n=1}^{\infty}$ converges to 2.

Theorem 2.6B is our first important application of the least upper bound axiom 1.7D.

Theorem 2.6B gives us our first set of criteria that will enable us to prove that a sequence converges without first guessing its limit. Here is an interesting application.

2.6C. COROLLARY. The sequence $\{(1+1/n)^n\}_{n=1}^{\infty}$ is convergent.

PROOF: Let $s_n = (1+1/n)^n$. By the binomial theorem

$$s_n = 1 + n \cdot \frac{1}{n} + \frac{n(n-1)}{1 \cdot 2} \frac{1}{n^2} + \cdots + \frac{n(n-1)\cdots 1}{1 \cdot 2 \cdots n} \cdot \frac{1}{n^n}.$$

For $k = 1, \ldots, n$, the $(k+1)$st term on the right is

$$\frac{n(n-1)\cdots(n-k+1)}{1 \cdot 2 \cdots k} \cdot \frac{1}{n^k}$$

* A nondecreasing sequence $\{s_n\}_{n=1}^{\infty}$ is always bounded below (by s_1).

which equals

$$\frac{1}{1\cdot 2\cdots k}\left(1-\frac{1}{n}\right)\left(1-\frac{2}{n}\right)\cdots\left(1-\frac{k-1}{n}\right). \tag{1}$$

If we expand s_{s+1}, we obtain $n+2$ terms (one more than for s_n) and, for $k=1,\dots,n$, the $(k+1)$st term is

$$\frac{1}{1\cdot 2\cdots k}\left(1-\frac{1}{n+1}\right)\left(1-\frac{2}{n+1}\right)\cdots\left(1-\frac{k-1}{n+1}\right),$$

which is greater than or equal to the quantity (1). This shows that $s_n \leqslant s_{n+1}$ (that is, $\{s_n\}_{n=1}^{\infty}$ is nondecreasing). But also,

$$\begin{aligned} s_n &\leqslant 1+1+\frac{1}{1\cdot 2}+\frac{1}{1\cdot 2\cdot 3}+\cdots+\frac{1}{1\cdot 2\cdots n} \\ &\leqslant 1+1+\frac{1}{2}+\frac{1}{2^2}+\cdots+\frac{1}{2^{n-1}} \\ &=1+\frac{1-(\frac{1}{2})^n}{1-\frac{1}{2}}<1+\frac{1}{1-\frac{1}{2}}=3. \end{aligned}$$

Thus $\{s_n\}_{n=1}^{\infty}$ is bounded above (by 3). Hence, by theorem 2.6B, $\{s_n\}_{n=1}^{\infty}$ is convergent.

It is customary to denote $\lim_{n\to\infty} s_n$ by e. That is,

$$e=\lim_{n\to\infty}\left(1+\frac{1}{n}\right)^n. \tag{2}$$

The proof of 2.6C shows that $2<e\leqslant 3$. (Actually this number e, familiar from calculus, is a transcendental number whose decimal expansion is $2.718\cdots$.)

We know that a convergent sequence is bounded (2.5B). Therefore we know that an unbounded sequence is divergent. It seems intuitively obvious that a nondecreasing sequence does not oscillate. This would imply that an unbounded nondecreasing sequence must diverge to infinity. This we now prove.

2.6D. THEOREM. A nondecreasing sequence which is not bounded above diverges to infinity.

PROOF: Suppose $\{s_n\}_{n=1}^{\infty}$ is nondecreasing but not bounded above. Given $M>0$ we must find $N\in I$ such that

$$s_n>M \qquad (n\geqslant N). \tag{1}$$

Now, since M is not an upper bound for $\{s_1,s_2,\dots\}$ there must exist $N\in I$ such that $s_N>M$. Then, for this N, (1) follows from the hypothesis that $\{s_n\}_{n=1}^{\infty}$ is nondecreasing. This proves the theorem.

The proof of the following theorem follows the proofs of 2.6B and 2.6D exactly, with all upper bounds and least upper bounds replaced by lower bounds and greatest lower bounds. We leave the details to the reader.

2.6E. THEOREM. A nonincreasing sequence which is bounded below is convergent. A nonincreasing sequence which is not bounded below diverges to minus infinity.

2.6F We close this section by showing that monotone subsequences always occur.

THEOREM. Let $S=\{s_n\}_{n=1}^{\infty}$ be a sequence of real numbers. Then S has a monotone subsequence.

PROOF: Let T_1 be the sequence s. Let T_2 denote the sequence $s_2, s_3, s_4, \ldots$. Indeed, for each $n \in I$, let T_n denote the sequence $s_n, s_{n+1}, s_{n+2}, \ldots$. We divide the proof into two cases.

CASE I. Suppose every sequence T_n has a greatest term. Let s_{n_1} be the greatest term in the sequence T_1. (If there is more than one greatest term, pick any one for s_{n_1}.) Let s_{n_2} be the greatest term in the sequence T_{n_1+1}. Then $n_2 > n_1$ and $s_{n_2} \leqslant s_{n_1}$. Let s_{n_3} be the greatest term in the sequence T_{n_2+1}. Then $n_3 > n_2$ and $s_{n_3} \leqslant s_{n_2}$. Continuing in this fashion we can construct $\{s_{n_j}\}_{j=1}^{\infty}$—a nonincreasing subsequence of S.

CASE II. If case I does not hold, then for some $n_1 \in I$, the sequence T_{n_1} has no greatest term. Since s_{n_1} is a term of T_{n_1}, there is a term s_{n_2} of T_{n_1} that is greater than s_{n_1}. Then there is a term s_{n_3} of T_{n_1} that is greater than s_{n_2}. Moreover, we may pick s_{n_3} with $n_3 > n_2$ (why?). Continuing in this fashion, we can construct a nondecreasing subsequence $\{s_{n_j}\}_{j=1}^{\infty}$ of S.

Hence in either case, S has a monotonic subsequence, and the proof is complete.

Exercises 2.6

1. Which of the following sequences are monotone?
   (a) $\{\sin n\}_{n=1}^{\infty}$. (b) $\{\tan n\}_{n=1}^{\infty}$.
   (c) $\left\{\dfrac{1}{1+n^2}\right\}_{n=1}^{\infty}$. (d) $\{2n+(-1)^n\}_{n=1}^{\infty}$.
2. If $\{s_n\}_{n=1}^{\infty}$ is nondecreasing and bounded above, and $L=\lim_{n\to\infty} s_n$, prove that $s_n \leqslant L (n \in I)$.
   Formulate the corresponding statement for nonincreasing sequences.
3. If $\{s_n\}_{n=1}^{\infty}$ and $\{t_n\}_{n=1}^{\infty}$ are nondecreasing bounded sequences, and if $s_n \leqslant t_n$ $(n \in I)$, prove that $\lim_{n\to\infty} s_n \leqslant \lim_{n\to\infty} t_n$.
4. Find the limit of $\{n^{-n}(n+1)^n\}_{n=1}^{\infty}$.
5. If $s_n = 10^n/n!$, find $N \in I$ such that
$$s_{n+1} < s_n \qquad (n \geqslant N).$$
6. For $n \in I$, let
$$s_n = \frac{1\cdot 3\cdot 5\cdots(2n-1)}{2\cdot 4\cdot 6\cdots 2n}.$$
   Prove that $\{s_n\}_{n=1}^{\infty}$ is convergent and $\lim_{n\to\infty} s_n \leqslant \frac{1}{2}$.
7. For $n \in I$, let
$$s_n = \frac{2\cdot 4\cdot 6\cdots 2n}{1\cdot 3\cdot 5\cdots(2n-1)}\cdot\frac{1}{n^2}.$$
   Verify that $s_1 > s_2 > s_3$. Prove that $\{s_n\}_{n=1}^{\infty}$ is nonincreasing.
8. Let
$$s_n = \frac{1+2+\cdots+n}{n^2} \qquad (n \in I).$$
   Show that $\{s_n\}_{n=1}^{\infty}$ is monotone and bounded, and that $\lim_{n\to\infty} s_n = \frac{1}{2}$.

9. Let $\{s_n\}_{n=1}^{\infty}$ be a sequence of real numbers, and let

$$t_n = \frac{s_1 + s_2 + \cdots + s_n}{n} \qquad (n \in I).$$

If $\{s_n\}_{n=1}^{\infty}$ is monotone and bounded, show that $\{t_n\}_{n=1}^{\infty}$ is monotone and bounded.

10. For $n \in I$, let

$$t_n = 1 + \frac{1}{1!} + \frac{1}{2!} + \cdots + \frac{1}{n!}.$$

(a) Prove that $\{t_n\}_{n=1}^{\infty}$ is nondecreasing.

(b) Using only facts established in the proof of 2.6C, prove that $\{t_n\}_{n=1}^{\infty}$ is bounded above and then prove that $\lim_{n\to\infty} t_n \geqslant \lim_{n\to\infty}(1+1/n)^n$.

11. Let $\mathcal{L}$ denote the class of all sequences of real numbers. Let $\mathcal{C}$ denote the class of all convergent sequences and $\mathcal{D}$ the class of all divergent sequences. Further let $\mathcal{D}_P$ and $\mathcal{D}_M$ denote the classes of sequences that diverge to (plus) infinity and minus infinity, respectively. Let $\mathcal{O}$ denote the class of oscillating sequences. Finally, let $\mathcal{B}$ denote the class of all bounded sequences and let $\mathcal{M}$ denote the class of all monotone sequences.

By citing the proper difinitions or theorems, verify the following statements.

(a) $\mathcal{L} = \mathcal{C} \cup \mathcal{D}$.

(b) $\mathcal{D} = \mathcal{D}_P \cup \mathcal{D}_M \cup \mathcal{O}$.

(c) $\mathcal{C} \subset \mathcal{B}$.

(d) $\mathcal{M} \cap \mathcal{B} \subset \mathcal{C}$.

(e) $\mathcal{M} \cap \mathcal{B}' \subset \mathcal{D}_P \cup \mathcal{D}_{\mathcal{M}}$.

(f) $\mathcal{B} \cap \mathcal{D}_P = \varnothing$.

2.7 OPERATIONS ON CONVERGENT SEQUENCES

Since sequences of real numbers are real-valued functions, the definition of the sum, difference, product, and quotient of sequences follows from definition 1.4B. Thus if $\{s_n\}_{n=1}^{\infty}$ and $\{t_n\}_{n=1}^{\infty}$ are sequences of real numbers, then $\{s_n\}_{n=1}^{\infty} + \{t_n\}_{n=1}^{\infty}$ is the sequence $\{s_n + t_n\}_{n=1}^{\infty}$, and $\{s_n\}_{n=1}^{\infty} \cdot \{t_n\}_{n=1}^{\infty}$ is the sequence $\{s_n \cdot t_n\}_{n=1}^{\infty}$, and so on. Also, if $c \in R$, then $c\{s_n\}_{n=1}^{\infty}$ is the sequence $\{cs_n\}_{n=1}^{\infty}$.

From the next theorem it follows that the sum of two convergent sequences is convergent.

2.7A. THEOREM. If $\{s_n\}_{n=1}^{\infty}$ and $\{t_n\}_{n=1}^{\infty}$ are sequences of real numbers, if $\lim_{n\to\infty} s_n = L$, and if $\lim_{n\to\infty} t_n = M$, then $\lim_{n\to\infty}(s_n + t_n) = L + M$. In words, the limit of the sum (of two convergent sequences) is the sum of the limits.

PROOF: Given $\epsilon > 0$ we must find $N \in I$ such that

$$|(s_n + t_n) - (L + M)| < \epsilon \qquad (n \geqslant N). \tag{1}$$

Now $|(s_n + t_n) - (L + M)| = |(s_n - L) + (t_n - M)| \leqslant |s_n - L| + |t_n - M|$. Hence (1) will certainly hold if

$$|s_n - L| + |t_n - M| < \epsilon \qquad (n \geqslant N). \tag{2}$$

We thus try to make both $|s_n - L|$ and $|t_n - M|$ less than $\epsilon/2$ by taking n sufficiently large.

Since $\lim_{n\to\infty}s_n=L$, there exists $N_1\in I$ such that

$$|s_n-L|<\frac{\epsilon}{2} \qquad (n\geqslant N_1).$$

Also, since $\lim_{n\to\infty}t_n=M$, there exists $N_2\in I$ such that

$$|t_n-M|<\frac{\epsilon}{2} \qquad (N_2\in I).$$

Hence if we let $N=\max(N_1,N_2)$, then the terms on the left of (2) are each less than $\epsilon/2$ when $n\geqslant N$. Thus for this N, (2) and hence (1) hold and the proof is complete.

The next theorem is easier to prove.

2.7B. THEOREM. If $\{s_n\}_{n=1}^{\infty}$ is a sequence of real numbers, if $c\in R$, and if $\lim_{n\to\infty}s_n=L$, then $\lim_{n\to\infty}cs_n=cL$.

PROOF: If $c=0$, the theorem is obvious. We therefore assume $c\neq 0$. Given $\epsilon>0$ we must find $N\in I$ such that

$$|cs_n-cL|<\epsilon \qquad (n\geqslant N). \tag{1}$$

Now, since $\lim_{n\to\infty}s_n=L$, there exists $N\in I$ such that

$$|s_n-L|<\frac{\epsilon}{|c|} \qquad (n\geqslant N).$$

But then

$$|c|\cdot|s_n-L|<\epsilon \qquad (n\geqslant N),$$

which is equivalent to (1).

Theorem 2.7B is used in the proof of the following useful result.

2.7C. THEOREM. (a) If $0<x<1$, then $\{x^n\}_{n=1}^{\infty}$ converges to 0. (b) If $1<x<\infty$, then $\{x^n\}_{n=1}^{\infty}$ diverges to infinity.

PROOF: (a) If $0<x<1$, then $x^{n+1}=x\cdot x^n<x^n$. Hence $\{x^n\}_{n=1}^{\infty}$ is nonincreasing. Since $x^n>0$ for $n\in I$, $\{x^n\}_{n=1}^{\infty}$ is bounded below. By 2.6E, $\{x^n\}_{n=1}^{\infty}$ is convergent. Let $L=\lim_{n\to\infty}x^n$. From 2.7B (with $c=x$) it follows that $\lim_{n\to\infty}x\cdot x^n=xL$. That is, $\{x^{n+1}\}_{n=1}^{\infty}$ converges to xL. But $\{x^{n+1}\}_{n=1}^{\infty}$ is a subsequence of $\{x^n\}_{n=1}^{\infty}$. By 2.3D, $L=xL$ and so $L(1-x)=0$. Since $x\neq 1$, this shows $L=0$, and part (a) if proved.

(b) If $x>1$, then $x^{n+1}=x\cdot x^n>x^n$ so that $\{x^n\}_{n=1}^{\infty}$ is nondecreasing. We will show that $\{x^n\}_{n=1}^{\infty}$ is not bounded above. For if $\{x^n\}_{n=1}^{\infty}$ were bounded above, then by 2.6B $\{x^n\}_{n=1}^{\infty}$ would converge to some $L\in R$. But the same reasoning as in (a) would show that $L=Lx$, so that $L=0=\lim_{n\to\infty}x^n$. But $x^n\geqslant 1$ and so $\{x^n\}_{n=1}^{\infty}$ obviously cannot converge to 0. This contradiction proves that $\{x^n\}_{n=1}^{\infty}$ is not bounded above. Conclusion (b) follows from 2.6D.

We now treat the limit of the difference of two convergent sequences.

2.7D. THEOREM. If $\{s_n\}_{n=1}^{\infty}$ and $\{t_n\}_{n=1}^{\infty}$ are sequences of real numbers, if $\lim_{n\to\infty}s_n=L$, and if $\lim_{n\to\infty}t_n=M$, then $\lim_{n\to\infty}(s_n-t_n)=L-M$.

PROOF: Since $\lim_{n\to\infty}t_n=M$, it follows from 2.7B (with $c=-1$) that $\lim_{n\to\infty}(-t_n)$

$= -M$. But then, using 2.7A (at the second equals sign),

$$\lim_{n\to\infty}(s_n - t_n) = \lim_{n\to\infty}[s_n + (-t_n)] = \lim_{n\to\infty} s_n + \lim_{n\to\infty}(-t_n) = L + (-M) = L - M,$$

which is what we wished to prove.

A useful consequence of 2.7D and 2.2B is the following.

2.7E. COROLLARY. If $\{s_n\}_{n=1}^{\infty}$ and $\{t_n\}_{n=1}^{\infty}$ are convergent sequences of real numbers if

$$s_n \leqslant t_n \qquad (n \in I),$$

and if $\lim_{n\to\infty} s_n = L$, $\lim_{n\to\infty} t_n = M$, then $L \leqslant M$.

PROOF: By 2.7D, $M - L = \lim_{n\to\infty}(t_n - s_n)$. But $t_n - s_n \geqslant 0 (n \in I)$. Hence by 2.2B, $M - L \geqslant 0$, which establishes our result.

This corollary, of course, remains true even if $s_n > t_n$ for a finite number of values of n.

We now show that the limit of the product of two convergent sequences is the product of their respective limits. We give two proofs of this result; each uses a technique useful in many other contexts. The first proof requires a lemma.

2.7F. LEMMA. If $\{s_n\}_{n=1}^{\infty}$ is a sequence of real numbers which converges to L, then $\{s_n^2\}_{n=1}^{\infty}$ converges to L^2.

PROOF: We must prove $\lim_{n\to\infty} s_n^2 = L^2$. That is, given $\epsilon > 0$ we must find $N \in I$ such that

$$|s_n^2 - L^2| < \epsilon \qquad (n \geqslant N)$$

or, equivalently,

$$|s_n - L| \cdot |s_n + L| < \epsilon \qquad (n \geqslant N). \tag{1}$$

Now, by 2.5B $\{s_n\}_{n=1}^{\infty}$ is bounded. Thus for some $M > 0$

$$|s_n| \leqslant M \qquad (n \in I),$$

so that

$$|s_n + L| \leqslant |s_n| + |L| \leqslant M + |L| \qquad (n \in I). \tag{2}$$

Since $\lim_{n\to\infty} s_n = L$, there exists $N \in I$ such that

$$|s_n - L| < \frac{\epsilon}{M + |L|} \qquad (n \geqslant N). \tag{3}$$

But then, using (2) and (3),

$$|s_n - L| \cdot |s_n + L| < \frac{\epsilon}{M + |L|} \cdot (M + |L|) = \epsilon \qquad (n \geqslant N).$$

Thus for this N, (1) holds and the proof is complete.

2.7G. THEOREM. If $\{s_n\}_{n=1}^{\infty}$ and $\{t_n\}_{n=1}^{\infty}$ are sequences of real numbers, if $\lim_{n\to\infty} s_n = L$, and if $\lim_{n\to\infty} t_n = M$, then $\lim_{n\to\infty} s_n t_n = LM$.

FIRST PROOF: We use the identity

$$ab = \tfrac{1}{4}\left[(a+b)^2 - (a-b)^2\right] \qquad (a, b \in R). \tag{1}$$

Now, as $n \to \infty$,

$$s_n + t_n \to L + M \qquad \text{by 2.7A,}$$

$$(s_n + t_n)^2 \to (L+M)^2 \qquad \text{by 2.7F;} \tag{2}$$

also

$$s_n - t_n \to L - M \qquad \text{by 2.7D,}$$

$$(s_n - t_n)^2 \to (L-M)^2 \qquad \text{by 2.7F.} \tag{3}$$

From (2), (3), and 2.7D,

$$(s_n + t_n)^2 - (s_n - t_n)^2 \to (L+M)^2 - (L-M)^2 = 4LM. \tag{4}$$

Finally, using (1), (4), and 2.7B,

$$s_n t_n = \tfrac{1}{4}\left[(s_n + t_n)^2 - (s_n - t_n)^2\right] \to \tfrac{1}{4}(4LM) = LM.$$

Note that this proof uses no ϵ. The technique of using the identity (1) to deal with the product is called polarization.

SECOND PROOF: Given $\epsilon > 0$ we must find $N \in I$ such that

$$|s_n t_n - LM| < \epsilon \qquad (n \geqslant N). \tag{1}$$

The problem here is to do something algebraic [as we did in going from (1) to (2) of 2.7A] that will enable us to use our hypotheses $\lim_{n\to\infty} s_n = L$ and $\lim_{n\to\infty} t_n = M$. The trick of adding and subtracting the same quantity (in this case Lt_n) will be used many times in this book.

We have

$$s_n t_n - LM = s_n t_n - Lt_n + Lt_n - LM = t_n(s_n - L) + L(t_n - M),$$

$$|s_n t_n - LM| \leqslant |t_n| \cdot |s_n - L| + |L| \cdot |t_n - M|.$$

Hence (1) will certainly hold if

$$|t_n| \cdot |s_n - L| + |L| \cdot |t_n - M| < \epsilon \qquad (n \geqslant N). \tag{2}$$

By 2.5B, $\{t_n\}_{n=1}^{\infty}$ is bounded, so that $|t_n| \leqslant Q\,(n \in I)$ for some $Q > 0$. Then (2) will certainly hold if

$$Q|s_n - L| + |L| \cdot |t_n - M| < \epsilon \qquad (n \geqslant N). \tag{3}$$

Thus if we choose $N_1 \in I$ so that

$$Q|s_n - L| < \frac{\epsilon}{2} \qquad (n \geqslant N_1),$$

and choose $N_2 \in I$ such that

$$|L||t_n - M| < \frac{\epsilon}{2} \qquad (n \geqslant N_2),$$

then (3) will hold for $N = \max(N_1, N_2)$. Hence (2) and finally (1) will hold for this N, and we are done.

Now we turn our attention to the quotient of convergent sequences.

2.7H. LEMMA. If $\{t_n\}_{n=1}^{\infty}$ is a sequence of real numbers, if $\lim_{n\to\infty} t_n = M$ where $M \neq 0$, then* $\lim_{n\to\infty}(1/t_n) = 1/M$.

* The hypothesis implies that t_n can be equal to zero for at most a finite number of n. Thus $1/t_n$ is defined for all but at most a finite number of n.

PROOF: Either $M>0$ or $M<0$. We will prove the lemma in the case $M>0$. (The case $M<0$ can be proved by applying the first case to $\{-t_n\}_{n=1}^{\infty}$.)

So we assume $M>0$. Given $\epsilon>0$ we must find $N\in I$ such that

$$\left|\frac{1}{t_n}-\frac{1}{M}\right|<\epsilon \qquad (n\geqslant N),$$

or

$$\frac{|t_n-M|}{|t_nM|}<\epsilon \qquad (n\geqslant N). \tag{1}$$

Now there exists $N_1\in I$ such that $|t_n-M|<M/2$ $(n\geqslant N_1)$. This implies

$$t_n>\frac{M}{2} \qquad (n\geqslant N_1).$$

In addition, there exists $N_2\in I$ such that

$$|t_n-M|<\frac{M^2\epsilon}{2} \qquad (n\geqslant N_2).$$

Thus if $N=\max(N_1,N_2)$, we have, for $n\geqslant N$,

$$\frac{|t_n-M|}{|t_nM|}=\frac{1}{|t_nM|}\cdot|t_n-M|<\frac{1}{M^2/2}\cdot\frac{M^2\epsilon}{2}=\epsilon.$$

Hence (1) holds for this N. This completes the proof.

2.7I. THEOREM. If $\{s_n\}_{n=1}^{\infty}$ and $\{t_n\}_{n=1}^{\infty}$ are sequences of real numbers, if $\lim_{n\to\infty}s_n=L$, and if $\lim_{n\to\infty}t_n=M$ where $M\neq 0$, then $\lim_{n\to\infty}(s_n/t_n)=L/M$.

PROOF: Using 2.7H and 2.7G we have

$$\lim_{n\to\infty} s_n\cdot\frac{1}{t_n}=L\cdot\frac{1}{M},$$

which is what we wished to show.

2.7J. In Section 1.9 we proved directly from the definition of limit that

$$\lim_{n\to\infty}\frac{2n}{n+4n^{1/2}}=2.$$

We now do a similar problem illustrating the results of this section.

PROBLEM: Prove

$$\lim_{n\to\infty}\frac{3n^2-6n}{5n^2+4}=\frac{3}{5}.$$

First we write

$$\frac{3n^2-6n}{5n^2+4}=\frac{3-6/n}{5+4/n^2}.$$

We proved in Section 1.9 that $\lim_{n\to\infty}(1/n)=0$. Hence

$$\lim_{n\to\infty}\frac{6}{n}=6\cdot 0=0 \qquad \text{by 2.7B.}$$

We also proved in Section 1.9 that $\lim_{n\to\infty} 1=1$. Hence

$$\lim_{n\to\infty} 3=3 \quad \text{by 2.7B.}$$

Then

$$\lim_{n\to\infty}\left(3-\frac{6}{n}\right)=3-0=3 \quad \text{by 2.7D.} \tag{1}$$

Since we know $\lim_{n\to\infty}(1/n)=0$, we have

$$\lim_{n\to\infty}\frac{1}{n^2}=\lim_{n\to\infty}\frac{1}{n}\cdot\lim_{n\to\infty}\frac{1}{n}=0\cdot 0=0 \qquad \text{by 2.7G.}$$

Thus

$$\lim_{n\to\infty}\frac{4}{n^2}=0 \qquad \text{by 2.7B.}$$

Reasoning as before, we conclude

$$\lim_{n\to\infty}\left(5+\frac{4}{n^2}\right)=5+0=5 \qquad \text{by 2.7A.} \tag{2}$$

But then, from (1), (2), and 2.7I,

$$\lim_{n\to\infty}\frac{3-6/n}{5+4/n}=\frac{\lim_{n\to\infty}(3-6/n)}{\lim_{n\to\infty}(5+4/n^2)}=\frac{3}{5},$$

which is what we wished to show.

Exercises 2.7

1. Prove
   (a) $\lim_{n\to\infty}\dfrac{2n^3+5n}{4n^3+n^2}=\frac{1}{2}$.
   (b) $\lim_{n\to\infty}\dfrac{n^2}{(n-7)^2-6}=1$.
2. Prove that if $\{s_n\}_{n=1}^{\infty}$ converges to 1, then $\{s_n^{1/2}\}_{n=1}^{\infty}$ converges to 1.
3. Evaluate $\lim_{n\to\infty}\sqrt{n}\,(\sqrt{n+1}-\sqrt{n}\,)$.
4. Suppose $\{s_n\}_{n=1}^{\infty}$ is a sequence of positive numbers and $0<x<1$. If $s_{n+1}<xs_n$ $(n\in I)$, prove $\lim_{n\to\infty}s_n=0$.
5. Suppose

$$\lim_{n\to\infty}\frac{s_n-1}{s_n+1}=0.$$

   Prove $\lim_{n\to\infty}s_n=1$. [*Hint*: Let $\epsilon_n=(s_n-1)/(s_n+1)$ and solve for s_n.] Which theorems from this section did you use?
6. Prove that $\lim_{n\to\infty}(1+1/n)^{n+1}=e$. Also, prove that

$$\lim_{n\to\infty}\left[1+\frac{1}{n+1}\right]^n=e.$$

   Which theorems from this section did you use?

7. Using the identity $1+2/n=[1+1/(n+1)](1+1/n)$, prove that

$$\lim_{n\to\infty}\left(1+\frac{2}{n}\right)^n=e^2.$$

8. If $c>1$, prove that $\lim_{n\to\infty}c^{1/n}=1$. [*Hint*: Write $c^{1/n}=1+s_n$ and take the nth power of both sides to show that $\{ns_n\}_{n=1}^{\infty}$ is bounded. Then conclude that $s_n\to 0$ as $n\to\infty$.]
9. Let $s_1=\sqrt{2}$ and let $s_{n+1}=\sqrt{2}\cdot\sqrt{s_n}$ for $n\geqslant 1$.
   (a) Prove, by induction, that $s_n\leqslant 2$ for all n.
   (b) Prove that $s_{n+1}\geqslant s_n$ for all n.
   (c) Prove that $\{s_n\}_{n=1}^{\infty}$ is convergent.
   (d) Prove that $\lim_{n\to\infty}s_n=2$.
10. Suppose $s_1>s_2>0$, and let $s_{n+1}=\frac{1}{2}(s_n+s_{n-1})$ $(n\geqslant 2)$. Prove that
   (a) $s_1,s_3,s_5,\ldots$ is nonincreasing,
   (b) $s_2,s_4,s_6,\ldots$ is nondecreasing,
   (c) $\{s_n\}_{n=1}^{\infty}$ is convergent.
11. If $r_n\leqslant s_n\leqslant t_n$ for all $n\in I$, and if both $\{r_n\}_{n=1}^{\infty}$ and $\{t_n\}_{n=1}^{\infty}$ converge to s, prove that $\{s_n\}_{n=1}^{\infty}$ converges to s.

2.8 OPERATIONS ON DIVERGENT SEQUENCES

In the preceding section we saw that the sum, difference, product, and quotient (if defined) of convergent sequences are again convergent. No such statement can be made in general about divergent sequences. Indeed, if $\{s_n\}_{n=1}^{\infty}$ is a divergent sequence, then $\{-s_n\}_{n=1}^{\infty}$ is also divergent, and the sum of these two sequences is clearly not divergent. Moreover, the product of the divergent sequence $\{(-1)^n\}_{n=1}^{\infty}$ with itself is not divergent.

For sequences that diverge to infinity, however, some positive results can be proved.

2.8A. THEOREM. If $\{s_n\}_{n=1}^{\infty}$ and $\{t_n\}_{n=1}^{\infty}$ are sequences of real numbers that diverge to infinity, then so do their sum and product. That is, $\{s_n+t_n\}_{n=1}^{\infty}$ and $\{s_nt_n\}_{n=1}^{\infty}$ diverge to infinity.

PROOF: Given $M>0$, choose $N_1\in I$ such that

$$s_n>M \qquad (n\geqslant N_1),$$

and choose $N_2\in I$ such that

$$t_n>1 \qquad (n\geqslant N_2).$$

(The above is possible since both $s_n\to\infty$ and $t_n\to\infty$ as $n\to\infty$.) Then, for $N=\max(N_1,N_2)$ we have

$$s_n+t_n>M+1>M \quad (n\geqslant N),$$

and

$$s_nt_n>M\cdot 1=M \quad (n\geqslant N).$$

Since M was an arbitrary positive number, this proves the theorem.

2.8B. THEOREM. If $\{s_n\}_{n=1}^{\infty}$ and $\{t_n\}_{n=1}^{\infty}$ are sequences of real numbers, if $\{s_n\}_{n=1}^{\infty}$ diverges to infinity, and if $\{t_n\}_{n=1}^{\infty}$ is bounded, then $\{s_n+t_n\}_{n=1}^{\infty}$ diverges to infinity.

PROOF: By hypothesis there exists $Q>0$ such that

$$|t_n| \leqslant Q \qquad (n \in I).$$

Given $M>0$ choose $N \in I$ such that

$$s_n > M+Q \qquad (n \geqslant N).$$

Then, for $n \geqslant N$,

$$s_n + t_n > s_n - |t_n| > (M+Q) - Q = M.$$

That is,

$$s_n + t_n > M \qquad (n \geqslant N),$$

which shows that $s_n + t_n \to \infty$ as $n \to \infty$.

2.8C. COROLLARY. If $\{s_n\}_{n=1}^{\infty}$ diverges to infinity and if $\{t_n\}_{n=1}^{\infty}$ converges, then $\{s_n + t_n\}_{n=1}^{\infty}$ diverges to infinity.

PROOF: The proof follows directly from 2.5B and 2.8B.

It is easy to show that 2.8A, 2.8B, and 2.8C remain true if "infinity" is replaced by "minus infinity."

2.8D. Almost any kind of sequence can be formed from the sum of two properly chosen oscillating sequences. For example, the sum of the oscillating sequences $0,1,0,2,0,3,\ldots$ and $1,0,2,0,3,0,\ldots$ is the sequence $1,1,2,2,3,3,\ldots$ which diverges to infinity. The sum of the oscillating sequences $1,0,1,0,1,0,\ldots$ and $0,1,0,1,0,1,\ldots$ is a convergent sequence. The sum of an oscillating sequence and itself is oscillating.

Exercises 2.8

1. Give an example of sequences $\{s_n\}_{n=1}^{\infty}$ and $\{t_n\}_{n=1}^{\infty}$ for which, as $n \to \infty$,
   (a) $s_n \to \infty$, $t_n \to -\infty$, $s_n + t_n \to \infty$,
   (b) $s_n \to \infty$, $t_n \to \infty$, $s_n - t_n \to 7$.
2. Suppose that $\{s_n\}_{n=1}^{\infty}$ is a divergent sequence of real numbers and $c \in R$, $c \neq 0$. Prove that $\{cs_n\}_{n=1}^{\infty}$ diverges.
3. True or false? If $\{s_n\}_{n=1}^{\infty}$ is oscillating and not bounded, and $\{t_n\}_{n=1}^{\infty}$ is bounded, then $\{s_n + t_n\}_{n=1}^{\infty}$ is oscillating and not bounded.

2.9 LIMIT SUPERIOR AND LIMIT INFERIOR

If $\{s_n\}_{n=1}^{\infty}$ is a convergent sequence, then $\lim_{n\to\infty} s_n$ measures, roughly, "the size of s_n when n is large." Of course, $\lim_{n\to\infty} s_n$ is a concept used only in connection with convergent sequences. In this section we introduce the related concepts of limit superior and limit inferior which can be applied to all sequences. Roughly, the limit superior of a sequence $\{s_n\}_{n=1}^{\infty}$ is a measure of "how big s_n can be when n is large," and the limit inferior or $\{s_n\}_{n=1}^{\infty}$ is a measure of "how small s_n can be when n is large." If $\lim_{n\to\infty} s_n$ exists, it is then plausible that the limit, limit superior, and limit inferior of $\{s_n\}_{n=1}^{\infty}$ are all equal, and this turns out to be the case. The real application of limit superior and limit inferior, however, is to sequences which are not known to be convergent.

2.9A. First let us consider a sequence $\{s_n\}_{n=1}^{\infty}$ that is bounded above—say

$$s_n \leqslant M \quad (n \in I).$$

Then, for fixed $n \in I$, the set $\{s_n, s_{n+1}, s_{n+2}, \ldots\}$ is clearly bounded above and hence (1.7D) has a least upper bound

$$M_n = \text{l.u.b.}\{s_n, s_{n+1}, s_{n+2}, \ldots\}. \tag{1}$$

Moreover, it is easy to see that $M_n \geqslant M_{n+1}$ since $M_{n+1} = \text{l.u.b.}\ \{s_{n+1}, s_{n+2}, \ldots\}$ is the l.u.b. of a subset of $\{s_n, s_{n+1}, s_{n+2}, \ldots\}$. Thus the sequence $\{M_n\}_{n=1}^{\infty}$ is nonincreasing and thus either converges or diverges to minus infinity.

DEFINITION. Let $\{s_n\}_{n=1}^{\infty}$ be a sequence of real numbers that is bounded above, and let $M_n = \text{l.u.b.}\ \{s_n, s_{n+1}, s_{n+2}, \ldots\}$.

(a) If $\{M_n\}_{n=1}^{\infty}$ converges, we define $\limsup_{n\to\infty} s_n$ to be $\lim_{n\to\infty} M_n$.
(b) If $\{M_n\}_{n=1}^{\infty}$ diverges to minus infinity we write

$$\limsup_{n\to\infty} s_n = -\infty.$$

For example, let $s_n = (-1)^n$ $(n \in I)$. Then $\{s_n\}_{n=1}^{\infty}$ is bounded above. In this case $M_n = 1$ for every $n \in I$ and hence $\lim_{n\to\infty} M_n = 1$. Thus $\limsup_{n\to\infty}(-1)^n = 1$.

Consider next the sequence $1, -1, 1, -2, 1, -3, 1, -4, \ldots$. Again $M_n = 1$ for every n, and so the limit superior of this sequence is 1.

If $s_n = -n$ $(n \in I)$, then $M_n = \text{l.u.b.}\{-n, -n-1, -n-2, \ldots\} = -n$. Hence $M_n \to -\infty$ as $n \to \infty$, and so $\limsup_{n\to\infty}(-n) = -\infty$.

2.9B. DEFINITION. If $\{s_n\}_{n=1}^{\infty}$ is a sequence of real numbers that is not bounded above, we write $\limsup_{n\to\infty} s_n = \infty$.

Obviously, $\limsup_{n\to\infty} n = \infty$.

At this point the reader should verify the following statements. (1) If $\{s_n\}_{n=1}^{\infty}$ is bounded above and has a subsequence that is bounded below by A, then $\limsup_{n\to\infty} s_n \geqslant A$; (2) if $\{s_n\}_{n=1}^{\infty}$ has no subsequence that is bounded below, then $\limsup_{n\to\infty} s_n = -\infty$.

We note that changing a finite number of terms of the sequence $\{s_n\}_{n=1}^{\infty}$ does not change $\limsup_{n\to\infty} s_n$. Thus the limit superior of the sequence $10^{100}, 1, -1, 1, -1, 1, -1, 1, \ldots$ is 1.

2.9C. THEOREM. If $\{s_n\}_{n=1}^{\infty}$ is a convergent sequence of real numbers, then

$$\limsup_{n\to\infty} s_n = \lim_{n\to\infty} s_n.$$

PROOF: Let $L = \lim_{n\to\infty} s_n$. Then given $\epsilon > 0$ there exists $N \in I$ such that

$$|s_n - L| < \epsilon \qquad (n \geqslant N),$$

or

$$L - \epsilon < s_n < L + \epsilon \qquad (n \geqslant N).$$

Thus if $n \geqslant N$, then $L + \epsilon$ is an u.b. for $\{s_n, s_{n+1}, s_{n+2}, \ldots\}$ and $L - \epsilon$ is not an u.b. Hence

$$L - \epsilon < M_n = \text{l.u.b.}\{s_n, s_{n+1}, s_{n+2}, \ldots\} \leqslant L + \epsilon,$$

and so, by 2.7E,

$$L-\epsilon \leqslant \lim_{n\to\infty} M_n \leqslant L+\epsilon.$$

But $\lim_{n\to\infty} M_n = \limsup_{n\to\infty} s_n$. Thus

$$L-\epsilon \leqslant \limsup_{n\to\infty} s_n \leqslant L+\epsilon.$$

Since ϵ was arbitrary, this implies $\limsup_{n\to\infty} s_n = L$, which is what we wished to prove. Note that we have used Exercise 2 of Section 2.2.

We now define limit inferior.

2.9D. If the sequence of real numbers $\{s_n\}_{n=1}^{\infty}$ is bounded below, then the set $\{s_n, s_{n+1}, s_{n+2}, \ldots\}$ has a g.l.b. If we let

$$m_n = \text{g.l.b.}\{s_n, s_{n+1}, s_{n+2}, \ldots\},$$

then $\{m_n\}_{n=1}^{\infty}$ is a nondecreasing sequence (verify) and hence either converges or diverges to infinity.

DEFINITION. Let $\{s_n\}_{n=1}^{\infty}$ be a sequence of real numbers that is bounded below, and let $m_n = \text{g.l.b.}\{s_n, s_{n+1}, s_{n+2}, \ldots\}$.

(a) If $\{m_n\}_{n=1}^{\infty}$ converges, we define $\liminf_{n\to\infty} s_n$ to be $\lim_{n\to\infty} m_n$.
(b) If $\{m_n\}_{n=1}^{\infty}$ diverges to infinity, we write $\liminf_{n\to\infty} s_n = \infty$.

2.9E. DEFINITION. If $\{s_n\}_{n=1}^{\infty}$ is a sequence of real numbers that is not bounded below, we write $\liminf_{n\to\infty} s_n = -\infty$.

Thus $\liminf_{n\to\infty}(-1)^n = -1$, $\liminf_{n\to\infty} n = \infty$, $\liminf_{n\to\infty}(-n) = -\infty$. The sequence $1, -1, 1, -2, 1, -3, 1, -4, \ldots$ has $\liminf = -\infty$.

2.9F. THEOREM. If $\{s_n\}_{n=1}^{\infty}$ is a convergent sequence of real numbers, then

$$\liminf_{n\to\infty} s_n = \lim_{n\to\infty} s_n .$$

PROOF: The proof of this theorem is very similar to the proof of 2.9C and is omitted.

2.9G. If we make the notation convention for the symbols $-\infty$ and ∞ that

$$\begin{aligned} -\infty &< x \qquad (x\in R), \\ x &< \infty \qquad (x\in R), \\ -\infty &< \infty, \end{aligned} \tag{1}$$

the following theorem is easy to prove.

THEOREM. If $\{s_n\}_{n=1}^{\infty}$ is a sequence of real numbers, then

$$\liminf_{n\to\infty} s_n \leqslant \limsup_{n\to\infty} s_n . \tag{2}$$

PROOF: If $\{s_n\}_{n=1}^{\infty}$ is bounded, then

$$m_n = \text{g.l.b.}\{s_n, s_{n+1}, s_{n+2}, \ldots\} \leqslant \text{l.u.b.}\{s_n, s_{n+1}, s_{n+2}, \ldots\} = M_n.$$

Thus $m_n \leqslant M_n$ and so, by 2.7E, (2) holds. If $\{s_n\}^{\infty}d_{n=1}$ is not bounded, then either $\limsup_{n\to\infty} = \infty$ or $\liminf_{n\to\infty} s_n = -\infty$ and (2) follows from (1).

From 2.9C and 2.9F we see that if $\lim_{n\to\infty} s_n = L$ then

$$\limsup_{n\to\infty} s_n = \liminf_{n\to\infty} s_n = L.$$

We now prove the converse of this statement.

2.9H. THEOREM. If $\{s_n\}_{n=1}^{\infty}$ is a sequence of real numbers, and if $\limsup_{n\to\infty} s_n = \liminf_{n\to\infty} s_n = L$ where $L \in R$, then $\{s_n\}_{n=1}^{\infty}$ is convergent and $\lim_{n\to\infty} s_n = L$.

PROOF: By hypothesis we have

$$L = \limsup_{n\to\infty} s_n = \lim_{n\to\infty} \text{l.u.b.}\{s_n, s_{n+1}, s_{n+2}, \ldots\}.$$

Thus given $\epsilon > 0$ there exists $N_1 \in I$ such that

$$|\text{l.u.b.}\{s_n, s_{n+1}, s_{n+2}, \ldots\} - L| < \epsilon \qquad (n \geqslant N_1).$$

This implies

$$s_n < L + \epsilon \qquad (n \geqslant N_1). \tag{1}$$

Similarly, since $\liminf_{n\to\infty} s_n = L$, there exists $N_2 \in I$ such that

$$|\text{g.l.b.}\{s_n, s_{n+1}, s_{n+2}, \ldots\} - L| < \epsilon \qquad (n \geqslant N_2),$$

which implies

$$s_n > L - \epsilon \qquad (n \geqslant N_2). \tag{2}$$

If $N = \max(N_1, N_2)$, then from (1) and (2) we conclude

$$|s_n - L| < \epsilon \qquad (n \geqslant N).$$

This proves $\lim_{n\to\infty} s_n = L$.

There is a similar result on sequences diverging to infinity.

2.9I. THEOREM. If $\{s_n\}_{n=1}^{\infty}$ is a sequence of real numbers and if $\limsup_{n\to\infty} s_n = \infty = \liminf_{n\to\infty} s_n$, then s_n diverges to infinity.

PROOF: Since $\liminf_{n\to\infty} s_n = \infty$, given $M > 0$ there exists an $N \in I$ such that

$$\text{g.l.b.}\{s_n, s_{n+1}, s_{n+2}, \ldots\} > M \qquad (n \geqslant N).$$

This implies that M is a lower bound (but not the g.l.b.) for $\{s_n, s_{n+1}, \ldots\}$, so that

$$s_n > M \qquad (n \geqslant N),$$

which establishes the required conclusion.

There is an obvious analogue of 2.9I for sequences diverging to minus infinity which the reader should formulate and prove. The converse of 2.9I is exercise 4 of this section.

We now prove a result for limit superior corresponding to 2.7E.

2.9J. THEOREM. If $\{s_n\}_{n=1}^{\infty}$ are bounded sequences of real numbers, and if

$$s_n \leqslant t_n \qquad (n \in I),$$

then $\limsup_{n\to\infty} s_n \leqslant \limsup_{n\to\infty} t_n$ and $\liminf_{n\to\infty} s_n \leqslant \liminf_{n\to\infty} t_n$.

PROOF: From the hypothesis $s_n \leqslant t_n$ it is clear that

$$\text{l.u.b.}\{s_n, s_{n+1}, s_{n+2}, \dots\} \leqslant \text{l.u.b.}\{t_n, t_{n+1}, t_{n+2}, \dots\},$$

and

$$\text{g.l.b.}\{s_n, s_{n+1}, s_{n+2}, \dots\} \leqslant \text{g.l.b.}\{t_n, t_{n+1}, t_{n+2}, \dots\}.$$

(Can you prove this?) Taking the limit as $n \to \infty$ on both sides of these inequalities, and using 2.7E, we prove the theorem.

Theorem 2.9J, of course, remains true even if $s_n > t_n$ for a finite number of n.

It is not always true that

$$\limsup_{n\to\infty}(s_n + t_n) = \limsup_{n\to\infty} s_n + \limsup_{n\to\infty} t_n,$$

even for bounded sequences $\{s_n\}_{n=1}^{\infty}$ and $\{t_n\}_{n=1}^{\infty}$. For example, if $s_n = (-1)^n$ $(n \in I)$ and $t_n = (-1)^{n+1}$ $(n \in I)$, then $s_n + t_n = 0$ $(n \in I)$. Here

$$\limsup_{n\to\infty} s_n = 1 = \limsup_{n\to\infty} t_n$$

but $\limsup_{n\to\infty}(s_n + t_n) = 0$. There are, however, important inequalities that can be proved.

2.9K. THEOREM. If $\{s_n\}_{n=1}^{\infty}$ and $\{t_n\}_{n=1}^{\infty}$ are bounded sequences of real numbers, then

$$\limsup_{n\to\infty}(s_n + t_n) \leqslant \limsup_{n\to\infty} s_n + \limsup_{n\to\infty} t_n; \tag{a}$$

$$\liminf_{n\to\infty}(s_n + t_n) \geqslant \liminf_{n\to\infty} s_n + \liminf_{n\to\infty} t_n. \tag{b}$$

PROOF: (a) Let

$$M_n = \text{l.u.b.}\{s_n, s_{n+1}, s_{n+2}, \dots\},$$

$$P_n = \text{l.u.b.}\{t_n, t_{n+1}, t_{n+2}, \dots\}.$$

Then

$$s_k \leqslant M_n \quad (k \geqslant n), \qquad t_k \leqslant P_n \quad (k \geqslant n),$$

and so

$$s_k + t_k \leqslant M_n + P_n \qquad (k \geqslant n).$$

Thus $M_n + P_n$ is an u.b. for $\{s_n + t_n, s_{n+1} + t_{n+1}, s_{n+2} + t_{n+2}, \dots\}$, so that

$$\text{l.u.b.}\{s_n + t_n, s_{n+1} + t_{n+1}, s_{n+2} + t_{n+2}, \dots\} \leqslant M_n + P_n.$$

By 2.7E and 2.7A,

$$\lim_{n\to\infty} \text{l.u.b.}\{s_n + t_n, s_{n+1} + t_{n+1}, s_{n+2} + t_{n+2}, \dots\} \leqslant \lim_{n\to\infty}(M_n + P_n) = \lim_{n\to\infty} M_n + \lim_{n\to\infty} P_n$$

or

$$\limsup_{n\to\infty}(s_n + t_n) \leqslant \limsup_{n\to\infty} s_n + \limsup_{n\to\infty} t_n,$$

which is precisely conclusion (a). The proof of (b) is very much the same and is left to the reader. [Note that the inequality sign in (b) is the reverse of that in (a).]

There are other ways to define limit superior and limit inferior. The following theorem indicates one such approach.

2.9L. THEOREM. Let $\{s_n\}_{n=1}^{\infty}$ be a bounded sequence of real numbers.

1. If $\limsup_{n\to\infty} s_n = M$, then for any $\epsilon > 0$, (a) $s_n < M + \epsilon$ for all but a finite number of values of n; (b) $s_n > M - \epsilon$ for infinitely many values of n.
2. If $\liminf_{n\to\infty} s_n = m$, then for any $\epsilon > 0$, (c) $s_n > m - \epsilon$ for all but a finite number of values of n; (d) $s_n < m + \epsilon$ for infinitely many values of n.

PROOF: We prove part 2 only. If (c) were false, then, for some $\epsilon > 0$, we would have $s_n \leqslant m - \epsilon$ for infinitely many n. But then, for *any* $n \in I$, the set $\{s_n, s_{n+1}, s_{n+2}, \ldots\}$ would contain a number $\leqslant m - \epsilon$. This would imply

$$\text{g.l.b.}\{s_n, s_{n+1}, s_{n+2}, \ldots\} \leqslant m - \epsilon \qquad (n \in I)$$

and, on taking limits we would obtain, by 2.7E, $\liminf_{n\to\infty} s_n \leqslant m - \epsilon$ which contradicts the hypothesis. Thus (c) is true.

Now suppose (d) is false. Then, for some $\epsilon > 0$, $s_n < m + \epsilon$ for only a finite number of values of n. But then there exists $N \in I$ such that

$$s_n \geqslant m + \epsilon \qquad (n \geqslant N).$$

By 2.9J,

$$\liminf_{n\to\infty} s_n \geqslant m + \epsilon,$$

which again contradicts the hypothesis. Thus (d) is true.

Although we do not prove it, the converse of 2.9L is true. That is, if $\{s_n\}_{n=1}^{\infty}$ is a bounded sequence of real numbers and if $M \in R$ is such that (a) and (b) hold for every $\epsilon > 0$, then $\limsup_{n\to\infty} s_n = M$; similarly, if (c) and (d) hold for every $\epsilon > 0$, then $\liminf_{n\to\infty} s_n = m$.

Using 2.9L we can prove the following useful result.

2.9M. THEOREM. Any bounded sequence of real numbers has a convergent subsequence.

PROOF: Suppose $\{s_n\}_{n=1}^{\infty}$ is a bounded sequence of real numbers and let $M = \limsup_{n\to\infty} s_n$. We shall construct a subsequence $\{s_{n_k}\}_{k=1}^{\infty}$ which converges to M. By (b) of 2.9L there are infinitely many values of n such that $s_n > M - 1$. Let n_1 be one such value. That is, $n_1 \in I$ and $s_{n_1} > M - 1$. Similarly, since there are infinitely many values of n such that $s_n > M - \frac{1}{2}$, we can find $n_2 \in I$ such that $n_2 > n_1$ and $s_{n_2} > M - \frac{1}{2}$. Continuing then, for each integer $k > 1$ we can find $n_k \in I$ such that $n_k > n_{k-1}$ and

$$s_{n_k} > M - \frac{1}{k}. \tag{1}$$

Given $\epsilon > 0$, by (a) of 2.9L we can find $N \in I$ such that

$$s_n < M + \epsilon \qquad (n \geqslant N). \tag{2}$$

Now, choose $K \in I$ so that $1/K < \epsilon$ and $n_K > N$. Then, if $k \geqslant K$, we have $1/k < \epsilon$ and $n_k > N$. Hence using (1) and (2),

$$M - \epsilon < M - 1/k < s_{n_k} < M + \epsilon \qquad (k \geqslant K),$$

which implies

$$|s_{n_k} - M| < \epsilon \qquad (k \geqslant K).$$

This proves $s_{n_k} \to M$ as $k \to \infty$, which is what we wished to show.

Exercises 2.9

1. Find the limit superior and the limit inferior for the following sequences.
   (a) $1,2,3,1,2,3,1,2,3,\ldots$.
   (b) $\{\sin(n\pi/2)\}_{n=1}^{\infty}$.
   (c) $\{(1+1/n)\cos n\pi\}_{n=1}^{\infty}$.
   (d) $\{(1+1/n)^n\}_{n=1}^{\infty}$.
2. If the lim sup of the sequence $\{s_n\}_{n=1}^{\infty}$ is equal to M, prove that the lim sup of any subsequence of $\{s_n\}_{n=1}^{\infty}$ is $\leqslant M$.
3. If $\{s_n\}_{n=1}^{\infty}$ is a bounded sequence of real numbers and $\liminf_{n\to\infty} s_n = m$, prove there is a subsequence of $\{s_n\}_{n=1}^{\infty}$ which converges to m.
   Also, prove that no subsequence of $\{s_n\}_{n=1}^{\infty}$ can converge to a limit less than m.
4. If $\{s_n\}_{n=1}^{\infty}$ is a sequence of real numbers diverging to infinity, prove that
   $$\limsup_{n\to\infty} s_n = \infty = \liminf_{n\to\infty} s_n.$$
   (This is the converse of theorem 2.9I.) Formulate and prove the corresponding statement for sequences diverging to minus infinity.
5. Write the set of all rational numbers in (0, 1) as $\{r_1, r_2, r_3, \ldots\}$. Calculate $\limsup_{n\to\infty} r_n$ and $\liminf_{n\to\infty} r_n$.
6. Prove that if the sequence $\{s_n\}_{n=1}^{\infty}$ has no convergent subsequence, then $\{|s_n|\}_{n=1}^{\infty}$ diverges to infinity.
7. If $\{s_n\}_{n=1}^{\infty}$ is a sequence of real numbers and if
   $$\sigma_n = \frac{s_1 + s_2 + \cdots + s_n}{n} \qquad (n \in I),$$
   prove that
   $$\limsup_{n\to\infty} \sigma_n \leqslant \limsup_{n\to\infty} s_n,$$
   and
   $$\liminf_{n\to\infty} \sigma_n \geqslant \liminf_{n\to\infty} s_n.$$
   (*Hint:* Use 2.9L.)

2.10 CAUCHY SEQUENCES

The most important criterion for proving that a sequence converges without knowing its limit is called the Cauchy criterion.

2.10A. DEFINITION. Let $\{s_n\}_{n=1}^{\infty}$ be a sequence of real numbers. Then $\{s_n\}_{n=1}^{\infty}$ is called a Cauchy sequence if for any $\epsilon > 0$ there exists an $N \in I$ such that

$$|s_m - s_n| < \epsilon \qquad (m, n \geqslant N).$$

Roughly, a sequence $\{s_n\}_{n=1}^{\infty}$ is Cauchy if s_m and s_n are close together when m and n are large. First we show that a convergent sequence must be Cauchy.

2.10B. THEOREM. If the sequence of real numbers $\{s_n\}_{n=1}^{\infty}$ converges, then $\{s_n\}_{n=1}^{\infty}$ is a Cauchy sequence.

PROOF: Let $L = \lim_{n\to\infty} s_n$. Then, given $\epsilon > 0$, there exists an $N \in I$ such that

$$|s_k - L| < \frac{\epsilon}{2} \qquad (k \geqslant N).$$

Thus if $m, n \geqslant N$, we have

$$|s_m - s_n| = |(s_m - L) + (L - s_n)| \leqslant |s_m - L| + |L - s_n| < \frac{\epsilon}{2} + \frac{\epsilon}{2}$$

so that

$$|s_m - s_n| < \epsilon \qquad (m, n \geqslant N),$$

which proves that $\{s_n\}_{n=1}^{\infty}$ is Cauchy.

Theorem 2.10B says roughly that if the terms $\{s_n\}_{n=1}^{\infty}$ get close to "something," then they get close to each other. It is the converse of 2.10B that is really important. The converse tells that, once we establish that a given sequence is Cauchy, the sequence must be convergent. We first prove a lemma.

2.10C. THEOREM. If $\{s_n\}_{n=1}^{\infty}$ is a Cauchy sequence of real numbers, then $\{s_n\}_{n=1}^{\infty}$ is bounded.

PROOF: Given $\epsilon = 1$, choose $N \in I$ such that

$$|s_m - s_n| < 1 \qquad (m, n \geqslant N).$$

Then

$$|s_m - s_N| < 1 \qquad (m \geqslant N). \tag{1}$$

Hence, if $m \geqslant N$, we have

$$|s_m| = |(s_m - s_N) + s_N| \leqslant |s_m - s_N| + |s_N|$$

and so, using (1)

$$|s_m| \leqslant 1 + |s_N| \qquad (m \geqslant N).$$

If $M = \max(|s_1|, \ldots, |s_{N-1}|)$, then

$$|s_m| < M + 1 + |s_N| \qquad (m \in I),$$

so that $\{s_n\}_{n=1}^{\infty}$ is bounded.

2.10D. THEOREM. If $\{s_n\}_{n=1}^{\infty}$ is a Cauchy sequence of real numbers, then $\{s_n\}_{n=1}^{\infty}$ is convergent.

FIRST PROOF: By lemma 2.10C we know that $\limsup_{n\to\infty} s_n$ and $\liminf_{n\to\infty} s_n$ are (finite) real numbers. By 2.9H, then, to prove the theorem it is sufficient to show that

$$\limsup_{n\to\infty} s_n = \liminf_{n\to\infty} s_n.$$

But, by the theorem in 2.9G, we know that $\limsup_{n\to\infty} s_n \geqslant \liminf_{n\to\infty} s_n$. Thus all we need to prove is that

$$\limsup_{n\to\infty} s_n \leqslant \liminf_{n\to\infty} s_n. \tag{1}$$

Since $\{s_n\}_{n=1}^{\infty}$ is Cauchy, given $\epsilon > 0$ there exists $N \in I$ such that

$$|s_m - s_n| < \frac{\epsilon}{2} \qquad (m, n \geqslant N)$$

and so

$$|s_N - s_n| < \frac{\epsilon}{2} \qquad (n \geqslant N).$$

It follows that $s_N + \epsilon/2$ and $s_N - \epsilon/2$ are, respectively, upper and lower bounds for the set $\{s_N, s_{N+1}, s_{N+2}, \ldots\}$. Hence if $n \geqslant N$, $s_N + \epsilon/2$ and $s_N - \epsilon/2$ are upper and lower bounds for $\{s_n, s_{n+1}, s_{n+2}, \ldots\}$. This implies, for $n \geqslant N$

$$s_N - \frac{\epsilon}{2} \leqslant \text{g.l.b.}\{s_n, s_{n+1}, s_{n+2}, \ldots\}$$

$$\leqslant \text{l.u.b.}\{s_n, s_{n+1}, s_{n+2}, \ldots\} \leqslant s_N + \frac{\epsilon}{2}.$$

Since the left and right ends of this inequality differ by ϵ, we must have

$$\text{l.u.b.}\{s_n, s_{n+1}, s_{n+2}, \ldots\} - \text{g.l.b.}\{s_n, s_{n+1}, s_{n+2}, \ldots\} \leqslant \epsilon,$$

$$\text{l.u.b.}\{s_n, s_{n+1}, s_{n+2}, \ldots\} \leqslant \text{g.l.b.}\{s_n, s_{n+1}, s_{n+2}, \ldots\} + \epsilon.$$

Taking the limit on both sides and using 2.7E, we obtain

$$\limsup_{n\to\infty} s_n \leqslant \liminf_{n\to\infty} s_n + \epsilon.$$

Since ϵ was arbitrary, this establishes (1), which is what we wished to prove.

We emphasize how strongly this proof of theorem 2.10D depends on axiom 1.7D, both in the use of l.u.b. and g.l.b. for $\{s_n, s_{n+1}, s_{n+2}, \ldots\}$ and in the use of limits superior and inferior. The existence of lim sup and lim inf for a bounded sequence depends on the existence of limit for a bounded monotone sequence. This in turn depends on 1.7D. In fact, it may be shown that 2.10D and 1.7D are equivalent. In some developments, 2.10D is taken as the fundamental axiom and 1.7D is a theorem.

Here is a second proof of 2.10D.

SECOND PROOF: By 2.6F, $\{s_n\}_{n=1}^{\infty}$ has a monotonic subsequence $\{s_{n_j}\}_{j=1}^{\infty}$. By 2.10C, $\{s_n\}_{n=1}^{\infty}$ is bounded—hence $\{s_{n_j}\}_{j=1}^{\infty}$ is bounded. Thus $\{s_{n_j}\}_{j=1}^{\infty}$ converges to some $s \in R$. We will show that $\{s_n\}_{n=1}^{\infty}$ itself converges to s. Fix $\epsilon > 0$. Since $\{s_{n_j}\}_{j=1}^{\infty}$ converges to s, there exists $J \in I$ such that

$$|s_{n_j} - s| < \frac{\epsilon}{2} \qquad (j \geqslant J). \tag{1}$$

Since $\{s_n\}_{n=1}^{\infty}$ is Cauchy, there exists $K \in I$ such that

$$|s_m - s_n| < \frac{\epsilon}{2} \qquad (m, n \geqslant K). \tag{2}$$

We may choose K so that $K \geqslant J$.

Now suppose $k \in I$ and $k \geqslant K$. Then $k \geqslant J$, so (1) implies

$$|s_{n_k} - s| < \frac{\epsilon}{2}.$$

Also, $n_k \geqslant k \geqslant K$, so (2) implies

$$|s_k - s_{n_k}| < \frac{\epsilon}{2}.$$

Therefore

$$|s_k - s| < \epsilon \qquad (k \geqslant K)$$

and the theorem follows.

We now present a famous result about the set of real numbers. It is called the nested-interval theorem [because of hypothesis (a)].

2.10E. THEOREM. For each $n \in I$ let $I_n = [a_n, b_n]$ be a (nonempty) closed bounded interval of real numbers such that

$$I_1 \supset I_2 \supset \cdots \supset I_n \supset I_{n+1} \supset \cdots, \tag{a}$$

and

$$\lim_{n\to\infty} (b_n - a_n) = \lim_{n\to\infty} (\text{length of } I_n) = 0. \tag{b}$$

Then $\cap_{n=1}^{\infty} I_n$ contains precisely one point.

PROOF: By hypothesis (a) we have $I_n \supset I_{n+1}$ and so $a_n \leqslant a_{n+1} \leqslant b_{n+1} \leqslant b_n$. This shows that the sequences $\{a_n\}_{n=1}^{\infty}$ and $\{b_n\}_{n=1}^{\infty}$ are respectively nondecreasing and nonincreasing. Moreover, by (a) again, all terms of both these sequences lie in I_1 and so the sequences are both bounded. By 2.6B and 2.6E both sequences are convergent. Let $x = \lim_{n\to\infty} a_n$ and let $y = \lim_{n\to\infty} b_n$. Then for any n we have $a_n \leqslant x$ and $y \leqslant b_n$ (why?). But by 2.7D and hypothesis (b) we have

$$y - x = \lim_{n\to\infty} b_n - \lim_{n\to\infty} a_n = \lim_{n\to\infty} (b_n - a_n) = 0.$$

Thus $x = y$. But then $a_n \leqslant x \leqslant b_n$ for each n, which shows that $x \in \cap_{n=1}^{\infty} I_n$. Clearly, no $z \neq x$ can lie in $\cap_{n=1}^{\infty} I_n$ since, by hypothesis (b), $|z - x|$ is greater than the length of I_n for n sufficiently large. Hence $\cap_{n=1}^{\infty} I_n$ contains x and no other point, and the theorem is proved.

2.10F. The nested-interval theorem has an important generalization which we discuss in Chapter 6. This generalization is proved with Cauchy sequences. Therefore, it will be very instructive for the reader to give a different proof of 2.10E using the information from this section on Cauchy sequences. We give the outline of such a proof.

1. Show that, for any $N \in I$, the points $a_N, a_{N+1} a_{N+2}, \ldots$ all lie in I_N.
2. Use hypothesis (b) to infer that $\{a_n\}_{n=1}^{\infty}$ is Cauchy.
3. Then $\{a_n\}_{n=1}^{\infty}$ is convergent (why?).
4. Similarly $\{b_n\}_{n=1}^{\infty}$ is convergent.
5. The rest of the proof is the same as in 2.10E.

Exercises 2.10

1. If $\{s_n\}_{n=1}^{\infty}$ is a Cauchy sequence of real numbers which has a subsequence converging to L, prove that $\{s_n\}_{n=1}^{\infty}$ itself converges to L.
2. For each $n \in I$ let $s_n = 1 + \frac{1}{2} + \cdots + \frac{1}{n}$. By considering $s_{2n} - s_n$, prove that $\{s_n\}_{n=1}^{\infty}$ is not Cauchy.
3. Prove that every subsequence of a Cauchy sequence is a Cauchy sequence.
4. Let $\{s_n\}_{n=1}^{\infty}$ be a sequence of real numbers. If $c \in R, 0 < r < 1$, and
$$|s_{n+1} - s_n| \leqslant cr^n \qquad (n \in I),$$
show that $\{s_n\}_{n=1}^{\infty}$ converges.
5. Find a sequence of closed intervals $I_1 \supset I_2 \supset \cdots \supset I_n \supset \cdots$ whose end points are rational numbers and such that
$$\bigcap_{n=1}^{\infty} I_n = \{e\}.$$

6. Let $\{a_n\}_{n=1}^{\infty}$ be a sequence of real numbers, and, for each $n \in I$, let
$$s_n = a_1 + a_2 + \cdots + a_n,$$
$$t_n = |a_1| + |a_2| + \cdots + |a_n|.$$
Prove that if $\{t_n\}_{n=1}^{\infty}$ is a Cauchy sequence, then so is $\{s_n\}_{n=1}^{\infty}$.
7. Show by example that the conclusion of 2.10E need not hold if the intervals I_n are not assumed to be closed.
8. Use the nested interval theorem to give a new proof that $[0, 1]$ is not countable. (Begin this way: Let $J = [0, 1]$. Suppose that J were countable, say $J = \{x_1, x_2, \cdots\}$. At least one of the three closed intervals $[0, \frac{1}{3}]$, $[\frac{1}{3}, \frac{2}{3}]$, $[\frac{2}{3}, 1]$ does not contain x_1. Call one such interval J_1. Now divide J_1 into three closed intervals and let J_2 be one of the three that does not contain x_2.)

2.11 SUMMABILITY OF SEQUENCES

In the next chapter we take up infinite series. One important branch of the field of infinite series is the study of summability of divergent series. This study is an attempt to attach a value to series that may not converge—that is, an attempt to generalize the concept of the sum of a convergent series. Many (but not all) of the well-known methods of summability deal exclusively with the *sequence* of partial sums of an infinite series. These methods, then, are in reality concerned with sequences as opposed to series, and we examine some of them now.

2.11A. We have seen that the sequences $\{(-1)^n\}_{n=1}^{\infty}$ and $\{n\}_{n=1}^{\infty}$ are both divergent but of very different character; the former is oscillating and bounded, the latter diverges to infinity. Writing $\{(-1)^n\}_{n=1}^{\infty}$ as $-1, 1, -1, 1, \ldots$, we feel intuitively that the terms of this sequence have an "average size" of 0. The simplest kind of summability for sequences, called $(C, 1)$ summability (C for Cesaro), makes precise this concept of average size.

DEFINITION. Let $\{s_n\}_{n=1}^{\infty}$ be a sequence of real numbers and let
$$\sigma_n = \frac{s_1 + s_2 + \cdots + s_n}{n} \qquad (n \in I).$$
We shall say that $\{s_n\}_{n=1}^{\infty}$ is $(C, 1)$ summable to L if the sequence $\{\sigma_n\}_{n=1}^{\infty}$ converges to L. In this case we write
$$\lim_{n\to\infty} s_n = L \qquad (C, 1).$$
Note that σ_n is precisely the average of the first n terms of the sequence $\{s_k\}_{n=1}^{\infty}$. Thus $\sigma_1 = s_1, \sigma_2 = (s_1 + s_2)/2$, and so on.

For example, if $s_n = (-1)^n (n \in I)$, then
$$\sigma_n = \frac{(-1) + (-1)^2 + \cdots + (-1)^n}{n},$$
and so
$$\sigma_n = 0 \qquad (n = 2, 4, 6, \ldots),$$
$$\sigma_n = \frac{-1}{n} \qquad (n = 1, 3, 5, \ldots).$$

Obviously $\lim_{n\to\infty}\sigma_n=0$, which shows that $\{(-1)^n\}_{n=1}^{\infty}$ is $(C,1)$ summable to 0. That is,

$$\lim_{n\to\infty}(-1)^n=0 \qquad (C,1).$$

This shows that a divergent sequence may be $(C,1)$ summable.

Consider the convergent sequence $1,1,1,\ldots$. Here $s_n=1(n\in I)$ so that, also, $\sigma_n=n^{-1}(s_1+\cdots+s_n)=1$. Hence $\{\sigma_n\}_{n=1}^{\infty}$ converges to 1 and so

$$\lim_{n\to\infty}1=1 \qquad (C,1).$$

The last example is a very special case of the important result that if a sequence $\{s_n\}_{n=1}^{\infty}$ is convergent to L then $\{s_n\}_{n=1}^{\infty}$ is also $(C,1)$ summable to L.

2.11B. THEOREM. If

$$\lim_{n\to\infty}s_n=L,$$

then

$$\lim_{n\to\infty}s_n=L \qquad (C,1).$$

PROOF: Case I, $L=0$. In this case $\lim_{n\to\infty}s_n=0$. We wish to prove that $\lim_{n\to\infty}\sigma_n=0$. Given $\epsilon>0$ there exists $N_1\in I$ such that $|s_n|<\epsilon/2(n\geqslant N_1)$. If we let $M=\max(|s_1|,|s_2|,\ldots,|s_{N_1-1}|)$, then we have, for $n\geqslant N_1$,

$$|\sigma_n|\leqslant\frac{(|s_1|+\cdots+|s_{N_1-1}|)+(|s_{N_1}|+\cdots+|s_n|)}{n}$$

$$\leqslant\frac{(N_1-1)M+(n-N_1+1)\epsilon/2}{n},$$

and hence

$$|\sigma_n|\leqslant\frac{(N_1-1)M}{n}+\frac{\epsilon}{2} \qquad (n\geqslant N_1). \tag{1}$$

Now choose $N_2\in I$ so that $\epsilon N_2>2(N_1-1)M$. Then

$$\frac{(N_1-1)M}{n}<\frac{\epsilon}{2} \quad (n\geqslant N_2). \tag{2}$$

If $N=\max(N_1,N_2)$, then (1) and (2) imply

$$|\sigma_n|<\epsilon \quad (n\geqslant N),$$

and hence $\lim_{n\to\infty}|\sigma_n|=0$. Thus $\lim_{n\to\infty}\sigma_n=0$, which is what we wished to show. This proves case I.

Case II, $L\neq0$. We have $\lim_{n\to\infty}s_n=L$ so that $\lim_{n\to\infty}(s_n-L)=0$. Hence by case I, $\{s_n-L\}_{n=1}^{\infty}$ is $(C,1)$ summable to 0. That is,

$$\lim_{n\to\infty}(s_n-L)=0 \qquad (C,1),$$

which means

$$\lim_{n\to\infty}\frac{(s_1-L)+(s_2-L)+\cdots+(s_n-L)}{n}=0. \tag{3}$$

But

$$\frac{(s_1-L)+(s_2-L)+\cdots+(s_n-L)}{n}=\frac{s_1+s_2+\cdots+s_n}{n}-L=\sigma_n-L.$$

Thus from (3)

$$\lim_{n\to\infty}(\sigma_n - L)=0,$$

and so

$$\lim_{n\to\infty}\sigma_n = L.$$

This, by definition, implies $\lim_{n\to\infty}s_n = L(C,1)$, which proves case II.

We have now seen that all convergent sequences are $(C,1)$ summable (to their respective limits), and that the divergent sequence $\{(-1)^n\}_{n=1}^{\infty}$ is also $(C,1)$ summable.

Not all divergent sequences are $(C,1)$ summable. For example, if $s_n = n(n\in I)$, then $\{s_n\}_{n=1}^{\infty}$ is not $(C,1)$ summable. For, in this case,

$$\sigma_n = \frac{s_1+s_2+\cdots+s_n}{n} = \frac{1+2+\cdots+n}{n} = \frac{n(n+1)}{2n} = \frac{n+1}{2},$$

and so $\{\sigma_n\}_{n=1}^{\infty}$ does not converge. In an exercise the reader is asked to show that no sequence that diverges to infinity can be $(C,1)$ summable.

The sequence $1, -1, 2, -2, 3, -3,\ldots$ is an oscillating sequence. We shall show that it is not $(C,1)$ summable. [However, when we take up $(C,2)$ summability, we shall see that this sequence is $(C,2)$ summable.] For this sequence,

$$s_n = \frac{n+1}{2} \qquad (n=1,3,5,\ldots),$$

$$s_n = \frac{-n}{2} \qquad (n=2,4,6,\ldots).$$

Obviously, if n is even, then $(s_1+s_2)+(s_3+s_4)+\cdots+(s_{n-1}+s_n)=0$. Thus

$$\sigma_n = 0 \qquad (n=2,4,6,\ldots).$$

If n is odd, however, then $n-1$ is even and $s_1+s_2+\cdots+s_{n-1}+s_n = s_n$. Hence

$$\sigma_n = \frac{s_1+s_2+\cdots+s_n}{n} = \frac{s_n}{n} = \frac{n+1}{2n} \qquad (n=1,3,5,\ldots).$$

Since $(n+1)/2n \to \frac{1}{2}$ as $n\to\infty$, the sequence $\{\sigma_n\}_{n=1}^{\infty}$ has the subsequence $\sigma_1, \sigma_3, \sigma_5,\ldots$ converging to $\frac{1}{2}$ and the subsequence $\sigma_2, \sigma_4, \sigma_6,\ldots$ converging to 0. By 2.3D, $\{\sigma_n\}_{n=1}^{\infty}$ is not convergent, and hence $\{s_n\}_{n=1}^{\infty}$ is not $(C,1)$ summable.

To keep the record even we give one more example of a divergent sequence that *is* $(C,1)$ summable. In Section 2.3 we saw that if θ is a rational number in $(0,1)$, then $\{\sin n\theta\pi\}_{n=1}^{\infty}$ diverges. We shall show, however, that this sequence is $(C,1)$ summable to 0. For, from the identity

$$\sin x + \sin 2x + \cdots + \sin nx = \frac{\cos\frac{1}{2}x - \cos(n+\frac{1}{2})x}{2\sin\frac{1}{2}x} \qquad (0<x<\pi)$$

which will be proved in Section 8.4, we see that

$$\sigma_n = \frac{\sin\theta\pi + \sin 2\theta\pi + \cdots + \sin n\theta\pi}{n} = \frac{\cos\frac{1}{2}\theta\pi - \cos(n+\frac{1}{2})\theta\pi}{2n\sin\frac{1}{2}\theta\pi},$$

and hence

$$|\sigma_n| \leqslant \frac{1}{n\sin(\theta\pi/2)}.$$

It follows easily that $\sigma_n \to 0$ as $n\to\infty$, which proves that $\{\sin n\theta\pi\}_{n=1}^{\infty}$ is $(C,1)$ summable to 0. Note that the argument applies equally well when θ is irrational.

The reader should have no difficulty in proving the following result.

2.11C. THEOREM. If $\{s_n\}_{n=1}^{\infty}$ and $\{t_n\}_{n=1}^{\infty}$ are $(C,1)$ summable to L and M, respectively, then $\{s_n+t_n\}_{n=1}^{\infty}$ and $\{s_n-t_n\}_{n=1}^{\infty}$ are $(C,1)$ summable to $L+M$ and $L-M$, respectively.

We now turn to $(C,2)$ summability.

2.11D. DEFINITION. Let $\{s_n\}_{n=1}^{\infty}$ be a sequence of real numbers and for each $n\in I$ let

$$\tau_n=\frac{ns_1+(n-1)s_2+(n-2)s_3+\cdots+2s_{n-1}+s_n}{1+2+\cdots+n}=\frac{2(ns_1+\cdots+s_n)}{n(n+1)}.$$

We shall say that $\{s_n\}_{n=1}^{\infty}$ is $(C,2)$ summable to L if the sequence $\{\tau_n\}_{n=1}^{\infty}$ converges to L. In this case we write

$$\lim_{n\to\infty} s_n=L \qquad (C,2).$$

We have

$$\tau_1=s_1, \qquad \tau_2=\frac{2s_1+s_2}{1+2}=\frac{2s_1+s_2}{3}, \qquad \tau_3=\frac{3s_1+2s_2+s_3}{6},$$

etc.

It is clear that $\lim_{n\to\infty}1=1(C,2)$.

Now consider $s_n=(-1)^n(n\in I)$. [We have already seen that $\lim_{n\to\infty}(-1)^n=0(C,1)$.] Then

$$\tau_n=\frac{-n+(n-1)-(n-2)+\cdots+(-1)^n}{1+2+\cdots+n}.$$

Now suppose n is even. Then the numerator for τ_n is

$$[-n+(n-1)]+[-(n-2)+(n-3)]+\cdots+[-2+1]=-\frac{n}{2},$$

since there are $n/2$ brackets and each quantity in brackets is equal to -1. If n is odd, then

$$-n+[(n-1)-(n-2)]+\cdots+[2-1]=-n+\frac{n-1}{2}=-\left(\frac{n+1}{2}\right).$$

Thus

$$\tau_n=\frac{-n/2}{n(n+1)/2}=\frac{-1}{n+1} \qquad (n=2,4,6,\ldots),$$

and

$$\tau_n=\frac{-(n+1)/2}{n(n+1)/2}=\frac{-1}{n} \qquad (n=1,3,5,\ldots),$$

which shows that $\tau_n\to 0$ as $n\to\infty$. Hence

$$\lim_{n\to\infty}(-1)^n=0 \qquad (C,2).$$

In the last example, $(C,2)$ summability gave the same "value" for $\{(-1)^n\}_{n=1}^{\infty}$ as did $(C,1)$ summability but was harder to apply. We shall now show by example that a sequence may be $(C,2)$ summable even if it is not $(C,1)$ summable. Indeed, consider $1,-1,2,-2,3,-3,\ldots$, which we have already shown is not $(C,1)$ summable. [Here $s_n=(n+1)/2$ if n is odd and $s_n=-n/2$ if n is even.] Thus if n is even then $n-1$ is odd and

$$s_{n-1}=\frac{(n-1)+1}{2}=\frac{n}{2}, \qquad s_n=\frac{-n}{2}$$

and so

$$\tau_n = \frac{ns_1 + (n-1)s_2 + \cdots + 2s_{n-1} + s_n}{n(n+1)/2}$$

$$= \frac{[n-(n-1)] + [(n-2)\cdot 2 - (n-3)\cdot 2] + \cdots + [2\cdot(n/2) - (n/2)]}{n(n+1)/2}$$

$$= \frac{1 + 2 + \cdots + n/2}{n(n+1)/2}.$$

Since $n/2$ is an integer if n is even, we have

$$1 + 2 + \cdots + \frac{n}{2} = \frac{(n/2)(n/2+1)}{2} = \frac{n(n+2)}{8}$$

and hence

$$\tau_n = \frac{n+2}{4(n+1)} \qquad (n = 2, 4, 6, \ldots).$$

If n is odd, we have

$$s_{n-2} = \frac{n-2+1}{2} = \frac{n-1}{2}, \qquad s_{n-1} = \frac{-(n-1)}{2}, \qquad s_n = \frac{n+1}{2}$$

and so

$$\tau_n = \frac{ns_1 + (n-1)s_2 + \cdots + 3s_{n-2} + 2s_{n-1} + s_n}{n(n+1)/2}$$

$$= \frac{\left\{[n-(n-1)] + \cdots + \left[3\left(\frac{n-1}{2}\right) - 2\left(\frac{n-1}{2}\right)\right]\right\} + \frac{n+1}{2}}{n(n+1)/2}$$

$$= \frac{\{1 + 2 + \cdots + (n-1)/2\} + (n+1)/2}{n(n+1)/2} = \frac{(n^2-1)/8 + (n+1)/2}{n(n+1)/2}$$

$$= \frac{n^2 + 4n + 3}{4(n^2 + n)}.$$

(Verify this for $n = 5$. You should obtain

$$\tau_5 = \frac{5s_1 + 4s_2 + \cdots + s_5}{15} = \frac{5 - 4 + 6 - 4 + 3}{15} = \frac{2}{5}.)$$

Thus $\tau_n \to \frac{1}{4}$ as $n \to \infty$, and so $1, -1, 2, -2, 3, -3, \ldots$ is $(C,2)$ summable to $\frac{1}{4}$.

Thus $(C,2)$ can do some things that $(C,1)$ cannot. The next theorem shows that anything $(C,1)$ can do $(C,2)$ can do also.

2.11E. THEOREM. If

$$\lim_{n\to\infty} s_n = L \qquad (C,1),$$

then

$$\lim_{n\to\infty} s_n = L \qquad (C,2).$$

PROOF:

CASE I, $L=0$. We have $\lim_{n\to\infty}\sigma_n=0$ where

$$\sigma_n=\frac{s_1+s_2+\cdots+s_n}{n}.$$

We wish to prove that $\lim_{n\to\infty}\tau_n=0$. Now

$$\begin{aligned}\tau_n&=\frac{ns_1+(n-1)s_2+\cdots+s_n}{1+2+\cdots+n}\\&=\frac{\begin{array}{c}(s_1+s_2+\cdots+s_n)+(s_1+s_2+\cdots+s_{n-1})\\+(s_1+s_2+\cdots+s_{n-2})+\cdots+(s_1+s_2)+s_1\end{array}}{1+2+\cdots+n}\\&=\frac{n\sigma_n+(n-1)\sigma_{n-1}+(n-2)\sigma_{n-2}+\cdots+2\sigma_2+\sigma_1}{1+2+\cdots+n}\\&=\frac{\sigma_1+2\sigma_2+\cdots+n\sigma_n}{1+2+\cdots+n}.\end{aligned}$$

Since $\sigma_n\to 0$ as $n\to\infty$, given $\epsilon>0$ there exists $N_1\in I$ such that

$$|\sigma_n|<\frac{\epsilon}{2}\qquad (n\geqslant N_1).$$

Let $M=\max(|\sigma_1|,|\sigma_2|,\ldots,|\sigma_{N_1-1}|)$. Then, for $n\geqslant N_1$,

$$\begin{aligned}|\tau_n|&\leqslant\frac{[|\sigma_1|+2|\sigma_2|+\cdots+(N_1-1)|\sigma_{N_1-1}|]+(N_1|\sigma_{N_1}|+\cdots+n|\sigma_n|)}{1+2+\cdots+n}\\&<\frac{M[1+2+\cdots+(N_1-1)]+(\epsilon/2)(N_1+\cdots+n)}{1+2+\cdots+n},\end{aligned}$$

and so

$$|\tau_n|<\frac{MN_1(N_1-1)}{n(n+1)}+\frac{\epsilon}{2}\qquad (n\geqslant N_1).$$

The remainder of the proof is exactly as in case I of the proof of 2.11B.

CASE II, $L\neq 0$. We have $\lim_{n\to\infty}s_n=L(C,1)$. Hence, by 2.11C,

$$\lim_{n\to\infty}(s_n-L)=0\qquad (C,1).$$

But then, by case I of this proof, $\lim_{n\to\infty}(s_n-L)=0\ (C,2)$. That is,

$$\lim_{n\to\infty}\frac{n(s_1-L)+(n-1)(s_2-L)+\cdots+(s_n-L)}{1+2+\cdots+n}=0. \tag{1}$$

But, removing parentheses in the numerator of (1), we have

$$\lim_{n\to\infty}\left[\frac{ns_1+(n-1)s_2+\cdots+s_n}{1+2+\cdots+n}-L\right]=0$$

or

$$\lim_{n\to\infty}(\tau_n-L)=0.$$

Thus $\lim_{n\to\infty}\tau_n=L$, which shows $\lim_{n\to\infty}s_n=L(C,2)$, and the proof is complete.

2.11F. COROLLARY. If $\{s_n\}_{n=1}^{\infty}$ converges to L, then $\{s_n\}_{n=1}^{\infty}$ is $(C,2)$ summable to L.

PROOF: The proof follows directly from 2.11B and 2.11E.

2.11G. Although we will not give details, we mention that the sequence $1, -2, 3, -4, 5, -6, \ldots$ is *not* $(C,2)$ summable but is $(C,3)$ summable. For the record we give the definition of (C,k) summability for any $k \in I$.

DEFINITION. Let $\{s_n\}_{n=1}^{\infty}$ be a sequence of real numbers, let k denote any fixed positive integer, and for $n \in I$ let

$$\lambda_n = \left[\binom{n+k-2}{n-1}s_1 + \binom{n+k-3}{n-2}s_2 + \cdots + \binom{k}{1}s_{n-1} + s_n\right] \Big/ \binom{n+k-1}{n-1},$$

where

$$\binom{n}{k} = \frac{n!}{k!(n-k)!}.$$

[In summation notation,

$$\lambda_n = \frac{1}{\binom{n+k-1}{n-1}} \sum_{j=1}^{n} \binom{n+k-1-j}{n-j} s_j.]$$

Then $\{s_n\}_{n=1}^{\infty}$ is said to be (C,k) summable to L if $\{\lambda_n\}_{n=1}^{\infty}$ converges to L.

The reader should verify that the special cases $k=1$ and $k=2$ in this definition actually coincide with $(C,1)$ and $(C,2)$ summability as previously defined.

It may be shown that if $k>1$ and $\{s_n\}_{n=1}^{\infty}$ is $(C,k-1)$ summable to L, then $\{s_n\}_{n=1}^{\infty}$ is (C,k) summable to L. Moreover, there will be a sequence which is (C,k) summable but not $(C,k-1)$ summable.

2.11H. In general, the term "summability method" can be defined as a real-valued function T whose domain is a set of sequences. A point (sequence) is in the domain of T if T "sums" the sequence—that is, if the sequence is assigned a real number by T.

Thus the domain of $(C,1)$ is a proper subset of the domain of $(C,2)$. Since $(C,1)$ and $(C,2)$ agree at a sequence where they are both defined, we may say that, from a function point of view, $(C,2)$ is an extension of $(C,1)$.

The (C,k) summability methods are an important but very minute part of the class of summation methods. The various methods differ greatly in their ability to sum divergent sequences and in the ease with which they can be applied. However, we almost always insist on one minimum requirement for a summability method T, namely, that any convergent sequence be T summable to its limit.

DEFINITION. Let T denote a summability method for sequences [for example, $(C,1), (C,2), \ldots$]. Then T is said to be *regular* if, whenever $\{s_n\}_{n=1}^{\infty}$ converges to L, then $\{s_n\}_{n=1}^{\infty}$ is also T summable to L.

2.11I. THEOREM. $(C,1)$ summability and $(C,2)$ summability are regular.

PROOF: The proof follows directly from 2.11B and 2.11F.

It is interesting to note that if a method of summability sums too many sequences, it cannot be regular. Indeed, it may be shown that if every bounded sequence is T summable, then T is *not* regular.

A famous, very general summability theorem will be presented in Section 3.12, Notes and Additional Exercises for Chapters 1–3.

Exercises 2.11

1. Prove that the following sequences are $(C,1)$ summable.
   (a) $1,0,1,0,1,0,\ldots$
   (b) $1,0,0,1,0,0,1,0,0,\ldots$
   (c) $-1,2,2,-1,2,2,-1,2,2,\ldots$
2. If $s_1,s_2,s_3,\ldots$ is $(C,1)$ summable to s, and if $t\in R$, prove that $t,s_1,s_2,s_3,\ldots$ is $(C,1)$ summable to s.
3. Prove that a sequence that diverges to infinity cannot be $(C,1)$ summable.
4. Let $\{s_n\}_{n=1}^{\infty}$ be a sequence of real numbers, $\sigma_n=n^{-1}(s_1+s_2+\cdots+s_n)$. Prove that if $\{s_n\}_{n=1}^{\infty}$ is $(C,1)$ summable, then $\lim_{n\to\infty}(s_n/n)=0$. [*Hint:* Compute $n\sigma_n-(n-1)\sigma_{n-1}$.] Deduce that $1,-1,2,-2,3,-3,\ldots$ is *not* $(C,1)$ summable.
5. Let $\{s_n\}_{n=1}^{\infty}$ be a sequence of positive numbers with $\lim_{n\to\infty}s_n=s$, where $s>0$. Prove
$$\lim_{n\to\infty}\sqrt[n]{s_1s_2\cdots s_n}=s.$$
(*Hint:* Take logarithms.)
6. If $\{s_n\}_{n=1}^{\infty}$ is a sequence of positive numbers and if $\lim_{n\to\infty}(s_n/s_{n-1})=L$, prove $\lim_{n\to\infty}{}^{n}\sqrt{s_n}=L$. (*Hint:* Let $t_1=s_1,t_2=s_2/s_1,\ldots,t_n=s_n/s_{n-1}$. Apply the preceding exercise to $\{t_n\}_{n=1}^{\infty}$.)
7. Without using $(C,1)$ summability, prove that $1,0,1,0,1,0,\ldots$ is $(C,2)$ summable to $\frac{1}{2}$.
8. If $\{s_n\}_{n=1}^{\infty}$ is monotone, prove that $\{\sigma_n\}_{n=1}^{\infty}$ is monotone where
$$\sigma_n=\frac{s_1+\cdots+s_n}{n}.$$
9. Prove theorem 2.11B by using the result of exercise 7 of Section 2.9.

2.12 LIMIT SUPERIOR AND LIMIT INFERIOR FOR SEQUENCES OF SETS

2.12A Suppose $E_1,E_2,\ldots$ are subsets of a set S. For each $n\in I$ let χ_n denote the characteristic function of E_n. Then, if $x\in S$, the terms of the sequence $\{\chi_n(x)\}_{n=1}^{\infty}$ consist of 0's and 1's. It is then clear that either $\limsup_{n\to\infty}\chi_n(x)=0$ or $\limsup_{n\to\infty}\chi_n(x)=1$, and similarly for $\liminf_{n\to\infty}\chi_n(x)$. We have the following theorem.

THEOREM. Let $\{E_n\}_{n=1}^{\infty}$ be a sequence of subsets of a set S, and let χ_n be the characteristic function of E_n $(n\in I)$. Let x be any point in S. Then

(a) $\limsup_{n\to\infty}\chi_n(x)=1$ if $x\in E_n$ for infinitely many values of n, while $\lim\sup_{n\to\infty}\chi_n(x)=0$ if $x\in E_n$ for only a finite number of n.

Also

(b) $\liminf_{n\to\infty}\chi_n(x)=1$ if $x\in E_n$ for all but a finite number of values of n, while $\lim\inf_{n\to\infty}\chi_n(x)=0$ if there are infinitely many values of n such that $x\not\in E_n$.

PROOF: We shall prove (b). If x is in E_n for all but a finite number of values of n, then there exists $N \in I$ such that $x \in E_n$ $(n \geq N)$. Hence $\chi_n(x) = 1$ $(n \geq N)$ and so $\liminf_{n\to\infty} \chi_n(x) = \lim_{n\to\infty} \chi_n(x) = 1$.

However, if there are infinitely many n such that E_n does not contain x, then $\chi_n(x) = 0$ for infinitely many values of n. Hence

$$\text{g.l.b.}\{\chi_n(x), \chi_{n+1}(x), \ldots\} = 0$$

for all n, and so $\liminf_{n\to\infty} \chi_n(x) = 0$. This proves (b). The proof of (a) is left to the reader.

It is then natural to make the following definition.

2.12B. DEFINITION. Let $\{E_n\}_{n=1}^{\infty}$ be a sequence of subsets of a set S. Then we define $\limsup_{n\to\infty} E_n$ to be the set of all $x \in S$ such that x is in E_n for infinitely many values of n. We also define $\liminf_{n\to\infty} E_n$ to be the set of all $x \in S$ such that x is in E_n for all but a finite number of values of n.

From part (a) of the theorem it then follows that if $\chi^*(x) = \limsup_{n\to\infty} \chi_n(x)$ $(x \in S)$, and $E^* = \limsup_{n\to\infty} E_n$, then χ^* is the characteristic function of E^*. Similarly, part (b) of the theorem shows that if $\chi_*(x) = \liminf_{n\to\infty} \chi_n(x)$ $(x \in S)$, and $E_* = \liminf_{n\to\infty} E_n$, then χ_* is the characteristic function of E_*.

Briefly, then, the characteristic function of $\limsup_{n\to\infty} E_n$ is $\limsup_{n\to\infty} \chi_n$, and similarly for lim inf.

Exercises 2.12

1. Prove that, if $\{E_n\}_{n=1}^{\infty}$ is a sequence of subsets of S, then

$$\liminf_{n\to\infty} E_n \subset \limsup_{n\to\infty} E_n.$$

2. If $E_1 = E_3 = E_5 = \cdots = S$ and $E_2 = E_4 = E_6 = \cdots = \varnothing$, compute

$$\limsup_{n\to\infty} E_n \quad \text{and} \quad \liminf_{n\to\infty} E_n.$$

3. Let $\{E_n\}_{n=1}^{\infty}$ be a sequence of subsets of S.
   (a) If $x \in \limsup_{n\to\infty} E_n$, prove that $x \in \bigcup_{k=n}^{\infty} E_k$ for every $n \in I$.
   (b) Prove that

$$\limsup_{n\to\infty} E_n = \bigcap_{n=1}^{\infty} \left(\bigcup_{k=n}^{\infty} E_k \right).$$

4. (a) If $x \in \liminf_{n\to\infty} E_n$, prove there exists $n \in I$ such that $x \in \bigcap_{k=n}^{\infty} E_k$.
   (b) Prove that

$$\liminf_{n\to\infty} E_n = \bigcup_{n=1}^{\infty} \left(\bigcap_{k=n}^{\infty} E_k \right).$$

5. (a) If $E_1 \subset E_2 \subset E_3 \subset \cdots$, prove that

$$\limsup_{n\to\infty} E_n = \liminf_{n\to\infty} E_n = \bigcup_{n=1}^{\infty} E_n.$$

(b) If $E_1 \supset E_2 \supset E_3 \supset \cdots$, prove that

$$\limsup_{n\to\infty} E_n = \liminf_{n\to\infty} E_n = \bigcap_{n=1}^{\infty} E_n.$$

6. Give a definition of $\lim_{n\to\infty} E_n$.
7. If E_n denotes the closed interval $[n, 2n]$, find $\lim_{n\to\infty} E_n$. What can you say about the length of E_n as $n\to\infty$? Do the answers to the first two parts of this question agree with your intuition?

3

SERIES OF REAL NUMBERS

3.1 CONVERGENCE AND DIVERGENCE

We recall that the sum of the infinite series $a_1+a_2+\cdots+a_n+\cdots$ is defined as $\lim_{n\to\infty}(a_1+\cdots+a_n)$, provided that the limit exists. This, however, is the definition of the *sum* of an infinite series and is not the definition of "infinite series" itself.

Like the term "ordered pair," the term "infinite series" is a highly intuitive one whose proper definition is not very illuminating.

3.1A. DEFINITION. The infinite series $\sum_{n=1}^{\infty} a_n$ is an ordered pair $\langle \{a_n\}_{n=1}^{\infty}, \{s_n\}_{n=1}^{\infty} \rangle$ where $\{a_n\}_{n=1}^{\infty}$ is a sequence of real numbers and

$$s_n = a_1 + a_2 + \cdots + a_n \qquad (n \in I).$$

The number a_n is called the nth term of the series. The number s_n is called the nth partial sum of the series.

In addition to $\sum_{n=1}^{\infty} a_n$, we sometimes denote a series by $a_1+a_2+\cdots+a_n+\cdots$ or simply by $a_1+a_2+\cdots$. Thus the nth partial sum of the series $1-1+\cdots+(-1)^{n+1}+\cdots$ is 1 if n is odd and 0 if n is even.

It is often convenient to index the terms of a series beginning with $n=0$. That is, we write some series as $\sum_{n=0}^{\infty} a_n$. (In this case we let $s_n = a_0 + a_1 + \cdots + a_n$.) Thus the series $1+x+x^2+\cdots$ can be written $\sum_{n=0}^{\infty} x^n$. It is always trivial to verify that any definition or theorem about series written $\sum_{n=1}^{\infty} a_n$ has an exact analog for series written $\sum_{n=0}^{\infty} a_n$ or $\sum_{n=p}^{\infty} a_n$ for any integer $p \geqslant 0$. We shall not further belabor this point.

The definition of convergence or divergence of the series $\sum_{n=1}^{\infty} a_n$ depends on the convergence or divergence of the sequence $\{s_n\}_{n=1}^{\infty}$ of partial sums.

3.1B. DEFINITION. Let $\sum_{n=1}^{\infty} a_n$ be a series of real numbers with partial sums $s_n = a_1 + \cdots + a_n$ $(n \in I)$. If the *sequence* $\{s_n\}_{n=1}^{\infty}$ converges to $A \in R$, we say that the series $\sum_{n=1}^{\infty} a_n$ converges to A. If $\{s_n\}_{n=1}^{\infty}$ diverges, we say that $\sum_{n=1}^{\infty} a_n$ diverges.

If $\Sigma_{n=1}^{\infty} a_n$ converges to A, we often write $\Sigma_{n=1}^{\infty} a_n = A$. Thus we use $\Sigma_{n=1}^{\infty} a_n$ not only to denote a series, but also (in case the series converges) its sum. With this warning we leave it to the reader to convince himself that no ambiguities arise.

From theorems on convergent sequences follows the next result.

3.1C. THEOREM. If $\Sigma_{n=1}^{\infty} a_n$ converges to A and $\Sigma_{n=1}^{\infty} b_n$ converges to B, then the series $\Sigma_{n=1}^{\infty}(a_n + b_n)$ converges to $A + B$. Also, if $c \in R$, then $\Sigma_{n=1}^{\infty} ca_n$ converges to cA.

PROOF: If $s_n = a_1 + \cdots + a_n$ and $t_n = b_1 + \cdots + b_n$, then, by hypothesis, $\lim_{n\to\infty} s_n = A$ and $\lim_{n\to\infty} t_n = B$. But the nth partial sum of $\Sigma_{n=1}^{\infty}(a_n + b_n)$ is $(a_1 + b_1) + \cdots + (a_n + b_n) = s_n + t_n$ which, by 2.7A, approaches $A + B$ as $n \to \infty$. This proves $\Sigma_{n=1}^{\infty}(a_n + b_n) = A + B$. The second part of the theorem follows from 2.7B.

An obvious consequence of 3.1C is that $\Sigma_{n=1}^{\infty}(a_n - b_n) = A - B$.

The following theorem gives a necessary (but not sufficient!) condition that a series be convergent.

3.1D. THEOREM. If $\Sigma_{n=1}^{\infty} a_n$ is a convergent series, then $\lim_{n\to\infty} a_n = 0$.

PROOF: Suppose $\Sigma_{n=1}^{\infty} a_n = A$. Then $\lim_{n\to\infty} = A$ where $s_n = a_1 + \cdots + a_n$. But then $\lim_{n\to\infty} s_{n-1} = A$. Since $a_n = s_n - s_{n-1}$ we have, by 2.7D, $\lim_{n\to\infty} a_n = \lim_{n\to\infty} s_n - \lim_{n\to\infty} s_{n-1} = A - A = 0$, which is what we wished to show.

Thus we see immediately that the series $\Sigma_{n=1}^{\infty}(1-n)/(1+2n)$ must diverge. Here, $a_n = (1-n)/(1+2n)$, and so $\lim_{n\to\infty} a_n = -\frac{1}{2} \neq 0$. Thus by 3.1D, $\Sigma_{n=1}^{\infty} a_n$ cannot be convergent. Similarly, the series $\Sigma_{n=1}^{\infty}(-1)^n$ must diverge since $\lim_{n\to\infty}(-1)^n$ does not even exist.

We emphasize that the condition $\lim_{n\to\infty} a_n = 0$ is *not* sufficient to ensure that $\Sigma_{n=1}^{\infty} a_n$ be convergent. In the next section we will see that $\Sigma_{n=1}^{\infty}(1/n)$ is *not* convergent even though $\lim_{n\to\infty} a_n = \lim_{n\to\infty}(1/n) = 0$.

Exercises 3.1

1. Prove that if $a_1 + a_2 + \cdots$ converges to s, then $a_2 + a_3 + \cdots$ converges to $s - a_1$.
2. Prove that the series $\Sigma_{n=1}^{\infty}[1/n(n+1)]$ converges. [*Hint:* Write
$$\frac{1}{n(n+1)} = \frac{1}{n} - \frac{1}{n+1}$$
and compute the partial sums of the series.]
3. For what values of x does the series $(1-x) + (x - x^2) + (x^2 - x^3) + \cdots$ converge?
4. Prove that the series $(a_1 - a_2) + (a_2 - a_3) + (a_3 - a_4) + \cdots$ converges if and only if the sequence $\{a_n\}_{n=1}^{\infty}$ converges.
5. Does the series $\Sigma_{n=1}^{\infty} \log(1 + 1/n)$ converge or diverge?
6. Prove that for any $a, b \in R$ the series $a + (a+b) + (a+2b) + (a+3b) + \cdots$ diverges unless $a = b = 0$.
7. Show that $\Sigma_{k=1}^{\infty} a_k$ converges if and only if given $\epsilon > 0$ there exists $N \in I$ such that
$$\left| \sum_{k=m+1}^{n} a_k \right| < \epsilon \qquad (n > m \geqslant N).$$

8. Prove that if $a_1+a_2+a_3+\cdots$ converges to A, then $\frac{1}{2}(a_1+a_2)+\frac{1}{2}(a_2+a_3)+\frac{1}{2}(a_3+a_4)+\cdots$ converges. What is the sum of the second series?
9. Does $\sum_{n=1}^{\infty}[(n+1)/(n+2)]$ converge or diverge? Does

$$\sum_{n=1}^{\infty}\frac{n+1}{10^{10}(n+2)}$$

converge or diverge?
10. Show that if $a_1+a_2+a_3+\cdots$ converges to L, then so does $a_1+0+a_2+0+a_3+0+\cdots$. More generally, show that any number of 0 terms may be inserted anywhere (or removed anywhere) in a convergent series without affecting its convergence or its sum.
11. Prove that if $\sum_{n=1}^{\infty}a_n$ converges and $\sum_{n=1}^{\infty}b_n$ diverges, then $\sum_{n=1}^{\infty}(a_n+b_n)$ diverges.
12. Let $\sum_{n=1}^{\infty}a_n$ be a convergent series. Let $\{n_i\}_{i=1}^{\infty}$ be any subsequence of the sequence of positive integers. Finally, let

$$\begin{aligned} b_1 &= a_1+a_2+\cdots+a_{n_1}\\ b_2 &= a_{n_1+1}+\cdots+a_{n_2}\\ &\vdots\\ b_k &= a_{n_{k-1}+1}+\cdots+a_{n_k} \qquad (k\in I). \end{aligned}$$

Prove that $\sum_{k=1}^{\infty}b_k$ converges and has the same sum as $\sum_{n=1}^{\infty}a_n$.
13. Verify that the preceding exercise yields the following important result. If $\sum_{n=1}^{\infty}a_n$ converges, then any series formed from $\sum_{n=1}^{\infty}a_n$ by inserting parentheses [for example, $(a_1+a_2)+(a_3+\cdots+a_7)+(\cdots)\cdots$] converges to the same sum.
14. Give an example of a series $\sum_{n=1}^{\infty}a_n$ such that $(a_1+a_2)+(a_3+a_4)+\cdots$ converges but $a_1+a_2+a_3+a_4+\cdots$ diverges. (This shows that *removing* parentheses *may* cause difficulties.)

3.2 SERIES WITH NONNEGATIVE TERMS

The easiest series to deal with are those with nonnegative terms. For these series, all theory on convergence and divergence is embodied in the following theorem.

3.2A. THEOREM. If $\sum_{n=1}^{\infty}a_n$ is a series of nonnegative numbers with $s_n=a_1+\cdots+a_n$ $(n\in I)$, then (a) $\sum_{n=1}^{\infty}a_n$ converges if the sequence $\{s_n\}_{n=1}^{\infty}$ is bounded; (b) $\sum_{n=1}^{\infty}a_n$ diverges if $\{s_n\}_{n=1}^{\infty}$ is not bounded.

PROOF: (a) Since $a_{n+1}\geqslant 0$ we have $s_{n+1}=a_1+\cdots+a_n+a_{n+1}=s_n+a_{n+1}\geqslant s_n$. Thus $\{s_n\}_{n=1}^{\infty}$ is nondecreasing and (by hypothesis) bounded. By 2.6B, $\{s_n\}_{n=1}^{\infty}$ is convergent, and thus $\sum_{n=1}^{\infty}a_n$ converges.

(b) If $\{s_n\}_{n=1}^{\infty}$ is not bounded then, by 2.5B, $\{s_n\}_{n=1}^{\infty}$ diverges. Hence so does $\sum_{n=1}^{\infty}a_n$.

We now give two important examples of series with nonnegative terms. The first is the geometric series $1+x+x^2+\cdots$.

3.2B. THEOREM. (a) If $0<x<1$, then $\sum_{n=0}^{\infty}x^n$ converges to $1/(1-x)$.

(b) If $x\geqslant 1$, then $\sum_{n=0}^{\infty}x^n$ diverges.

PROOF: Conclusion (b) is an immediate consequence of theorem 3.1D since, if $x \geqslant 1$, then $\{x^n\}_{n=1}^{\infty}$ does not converge to 0. To prove (a) we have $s_n = 1 + x + \cdots + x^n$ and so

$$s_n = \frac{1 - x^{n+1}}{1 - x} = \frac{1}{1 - x} - \frac{x^{n+1}}{1 - x} \qquad (n \in I).$$

But if $0 < x < 1$, then $\lim_{n \to \infty} x^{n+1} = 0$ by 2.7C. Hence $\lim_{n \to \infty} s_n = 1/(1 - x)$. This proves (a).

The second example is the series $1 + \frac{1}{2} + \cdots + 1/n + \cdots$, known as the *harmonic series*.

3.2C. THEOREM. The series $\sum_{n=1}^{\infty}(1/n)$ is divergent.

PROOF: We examine the subsequence $s_1, s_2, s_4, s_8, \ldots, s_{2^{n-1}}, \cdots$ of $\{s_n\}_{n=1}^{\infty}$ where, in this case, $s_n = 1 + \frac{1}{2} + \cdots + 1/n$. We have

$$\begin{aligned}
s_1 &= 1,\\
s_2 &= 1 + \tfrac{1}{2} = \tfrac{3}{2},\\
s_4 &= s_2 + \tfrac{1}{3} + \tfrac{1}{4} > \tfrac{3}{2} + \tfrac{1}{4} + \tfrac{1}{4} = 2,\\
s_8 &= s_4 + \tfrac{1}{5} + \tfrac{1}{6} + \tfrac{1}{7} + \tfrac{1}{8} > 2 + \tfrac{1}{8} + \tfrac{1}{8} + \tfrac{1}{8} + \tfrac{1}{8} = \tfrac{5}{2};
\end{aligned}$$

in general, it may be shown by induction that $s_{2^n} \geqslant (n+2)/2$. Thus $\{s_n\}_{n=1}^{\infty}$ contains a divergent subsequence and hence, by 2.3D, diverges. This proves the theorem.

We repeat that the divergence of the harmonic series shows that $\sum_{n=1}^{\infty} a_n$ may diverge even if $\lim_{n \to \infty} a_n = 0$.

3.2D. For series with nonnegative terms only we introduce the following notation.

If $\sum_{n=1}^{\infty} a_n$ is a convergent series of nonnegative numbers, we sometimes write $\sum_{n=1}^{\infty} a_n < \infty$. If $\sum_{n=1}^{\infty} a_n$ is a divergent series of nonnegative numbers, we sometimes write $\sum_{n=1}^{\infty} a_n = \infty$. Thus

$$\sum_{n=0}^{\infty} \left(\tfrac{1}{2}\right)^n < \infty,$$

$$\sum_{n=1}^{\infty} \frac{1}{n} = \infty.$$

3.2E. It is very interesting to note that there is no series that diverges "as slowly as possible." More precisely,

THEOREM. If $\sum_{n=1}^{\infty} a_n$ is a divergent series of positive numbers, then there is a sequence $\{\epsilon_n\}_{n=1}^{\infty}$ of positive numbers which converges to zero but for which $\sum_{n=1}^{\infty} \epsilon_n a_n$ still diverges.

PROOF: Let $s_n = a_1 + a_2 + \cdots + a_n$. We first show that the series $\sum_{k=1}^{\infty}(s_{k+1} - s_k)/s_{k+1}$ diverges. For any $m \in I$ choose $n \in I$ such that $s_{n+1} > 2s_m$. (This is possible since by hypothesis $\{s_k\}_{k=1}^{\infty}$ diverges to infinity.) Now $\{s_k\}_{k=1}^{\infty}$ is nondecreasing. Hence

$$\sum_{k=m}^{n} \frac{s_{k+1}-s_k}{s_{k+1}} \geqslant \sum_{k=m}^{n} \frac{s_{k+1}-s_k}{s_{n+1}}$$

$$= \frac{1}{s_{n+1}}[(s_{m+1}-s_m)+(s_{m+2}-s_{m+1})+\cdots+(s_{n+1}-s_n)]$$

$$= \frac{s_{n+1}-s_m}{s_{n+1}} > \frac{s_{n+1}-\frac{1}{2}s_{n+1}}{s_{n+1}} = \frac{1}{2}.$$

Thus for any $m \in I$ there exists $n \in I$ such that

$$\sum_{k=m}^{n} \frac{s_{k+1}-s_k}{s_{k+1}} \geqslant \frac{1}{2}.$$

The partial sums of the series $\sum_{k=1}^{\infty}(s_{k+1}-s_k)/s_{k+1}$ thus do not form a Cauchy sequence and hence

$$\sum_{k=1}^{\infty} \frac{s_{k+1}-s_k}{s_{k+1}} = \infty.$$

(See Exercise 7 of Section 3.1.)

But $s_{k+1}-s_k = a_{k+1}$. Thus

$$\sum_{k=1}^{\infty} \frac{a_{k+1}}{s_{k+1}} = \sum_{k=2}^{\infty} \frac{a_k}{s_k} = \infty.$$

Let $\epsilon_k = 1/s_k$. Then $\epsilon_k \to 0$ as $k \to \infty$ and $\sum_{k=2}^{\infty} \epsilon_k a_k = \infty$. This completes the proof.

Exercises 3.2

1. If $\sum_{n=1}^{\infty} a_n$ is a convergent series of positive numbers, and if $\{a_{n_i}\}_{i=1}^{\infty}$ is a subsequence of $\{a_n\}_{n=1}^{\infty}$, prove that $\sum_{i=1}^{\infty} a_{n_i}$ converges.
2. Prove that

$$1 + \frac{1}{2!} + \frac{1}{4!} + \frac{1}{6!} + \cdots$$

converges.
3. If $0 \leqslant a_n \leqslant 1$ $(n \geqslant 0)$ and if $0 \leqslant x < 1$, then prove that $\sum_{n=0}^{\infty} a_n x^n$ converges, and that its sum is not greater than $1/(1-x)$.
4. If $\{s_n\}_{n=1}^{\infty}$ is nondecreasing, and $s_n \geqslant 0$ $(n \in I)$, prove that there exists a series $\sum_{k=1}^{\infty} a_k$ with $a_k \geqslant 0$ $(k \in I)$ and

$$s_n = a_1 + a_2 + \cdots + a_n \qquad (n \in I).$$

5. Prove that $1 + \frac{1}{3} + \frac{1}{5} + \frac{1}{7} + \cdots$ is divergent.
6. For what values of $x \in R$ does the series

$$1 + \frac{1-x}{1+x} + \left(\frac{1-x}{1+x}\right)^2 + \left(\frac{1-x}{1+x}\right)^3 + \cdots$$

converge, and what is its sum?

3.3 ALTERNATING SERIES

An alternating series is an infinite series whose terms alternate in sign. For example, the series $1-\frac{1}{2}+\frac{1}{4}-\frac{1}{8}+\cdots$, $1-2+3-4+\cdots$, $1-\frac{1}{2}+\frac{1}{3}-\frac{1}{4}+\cdots$ are all alternating series. An alternating series may thus be written as $\sum_{n=1}^{\infty}(-1)^{n+1}a_n$ where each a_n is positive [or as $\sum_{n=1}^{\infty}(-1)^n a_n$ if the first term in the series is negative]. We now demonstrate the fundamental result on alternating series.

3.3A. THEOREM. If $\{a_n\}_{n=1}^{\infty}$ is a sequence of positive numbers such that

(a) $a_1 \geqslant a_2 \geqslant \cdots \geqslant a_n \geqslant a_{n+1} \geqslant \cdots$ (that is, $\{a_n\}_{n=1}^{\infty}$ is nonincreasing), and
(b) $\lim_{n\to\infty} a_n = 0$,

then the alternating series $\sum_{n=1}^{\infty}(-1)^{n+1}a_n$ is convergent.

PROOF: Consider first the partial sums with odd index $s_1, s_3, s_5, \ldots$. We have $s_3 = s_1 - a_2 + a_3$. Since, by (a), $a_3 \leqslant a_2$, this implies $s_3 \leqslant s_1$. Indeed, for any $n \in I$ we have $s_{2n+1} = s_{2n-1} - a_{2n} + a_{2n+1} \leqslant s_{2n-1}$. Thus $s_1 \geqslant s_3 \geqslant \cdots \geqslant s_{2n-1} \geqslant s_{2n+1} \geqslant \cdots$ so that $\{s_{2n-1}\}_{n=1}^{\infty}$ is nonincreasing. But $s_{2n-1} = (a_1 - a_2) + (a_3 - a_4) + \cdots + (a_{2n-3} - a_{2n-2}) + a_{2n-1}$. Since each quantity in parentheses is nonnegative and $a_{2n-1} > 0$, we have $s_{2n-1} > 0$. Hence by 2.6E, $\{s_{2n-1}\}_{n=1}^{\infty}$ is convergent. Similarly, the sequence $s_2, s_4, \ldots, s_{2n}, \ldots$ is convergent. For $s_{2n+2} = s_{2n} + a_{2n+1} - a_{2n+2} \geqslant s_{2n}$, and so $\{s_{2n}\}_{n=1}^{\infty}$ is nondecreasing. But also $s_{2n} = a_1 - (a_2 - a_3) - \cdots - (a_{2n-2} - a_{2n-1}) - a_{2n}$. Thus $s_{2n} \leqslant a_1$ so that $\{s_{2n}\}_{n=1}^{\infty}$ is bounded above. Now, let $M = \lim_{n\to\infty} s_{2n-1}$ and let $L = \lim_{n\to\infty} s_{2n}$. Then since $a_{2n} = s_{2n} - s_{2n-1}$ we have, by hypothesis (b),

$$0 = \lim_{n\to\infty} a_{2n} = \lim_{n\to\infty} s_{2n} - \lim_{n\to\infty} s_{2n-1} = L - M.$$

Thus $L = M$, and so both $\{s_{2n}\}_{n=1}^{\infty}$ and $\{s_{2n-1}\}_{n=1}^{\infty}$ converge to L. From this it is easy to show that $\{s_n\}_{n=1}^{\infty}$ converges to L, and hence that $\sum_{n=1}^{\infty}(-1)^{n+1}a_n$ is convergent to L, which completes the proof.

Note that the proof shows that $s_{2n-1} \geqslant L$ and $s_{2n} \leqslant L$. Thus $0 \leqslant s_{2n-1} - L \leqslant s_{2n-1} - s_{2n} = a_{2n}$, and so $|s_{2n-1} - L| \leqslant a_{2n}$. Similarly, $0 \leqslant L - s_{2n} \leqslant s_{2n+1} - s_{2n} = a_{2n+1}$, so that $|s_{2n} - L| \leqslant a_{2n+1}$. That is, whether k is odd or even we have shown that $|s_k - L| \leqslant a_{k+1}$. We thus have the following corollary, which enables us to estimate the sum of this kind of convergent alternating series.

3.3B. COROLLARY. If the alternating series $\sum_{n=1}^{\infty}(-1)^{n+1}a_n$ satisfies the hypotheses of theorem 3.3A, and hence converges to some $L \in R$, then

$$|s_k - L| \leqslant a_{k+1} \qquad (k \in I).$$

Thus the difference between the sum of $\sum_{n=1}^{\infty}(-1)^{n+1}a_n$ and any partial sum will be no greater than the magnitude of the first term not included in the partial sum.

Let us now illustrate 3.3A and 3.3B. We saw in 3.2C that $\sum_{n=1}^{\infty} 1/n$ diverges. However, since $\{1/n\}_{n=1}^{\infty}$ is a nonincreasing sequence and $\lim_{n\to\infty} 1/n = 0$, it follows from 3.3A that $\sum_{n=1}^{\infty}(-1)^{n+1}/n$ converges. That is, for some $L \in R$,

$$1 - \frac{1}{2} + \frac{1}{3} - \frac{1}{4} + \cdots + \frac{(-1)^{n+1}}{n} + \cdots = L.$$

Of course, we do not know what L is, but we can estimate it using 3.3B. For, by 3.3B, for any $n \in I$ we have

$$\left|\left[1-\frac{1}{2}+\cdots+\frac{(-1)^{n+1}}{n}\right]-L\right| \leqslant \frac{1}{n+1}.$$

If we take $n=9$, this yields

$$|0.7456-L| \leqslant \tfrac{1}{10},$$

so that $0.6456 \leqslant L \leqslant 0.8456$. (In fact we know $s_9 \geqslant L$, and so we can conclude $0.6456 \leqslant L \leqslant 0.7456$.) Actually, it may be shown that $L=\log 2=0.6932\cdots$.

If $0<x<1$, then 3.3A implies $1-x+x^2-\cdots$ converges. The method of 3.2B may also be used to show that

$$1-x+x^2-\cdots=\frac{1}{1+x} \qquad (0<x<1).$$

As a final example consider the series

$$\sum_{n=0}^{\infty} \frac{(-1)^n}{n!}=1-\frac{1}{1!}+\frac{1}{2!}-\frac{1}{3!}+\cdots.$$

This series converges by 3.3A. If $L=\sum_{n=0}^{\infty}(-1)^n/n!$, then

$$\left|\left(1-\frac{1}{1!}+\frac{1}{2!}-\frac{1}{3!}+\frac{1}{4!}-\frac{1}{5!}\right)-L\right| \leqslant \frac{1}{6!}$$

From this we conclude $|L-0.3666| \leqslant 0.0014$. (From elementary calculus you should recall that $L=e^{-1}=0.3679\cdots$.)

Exercises 3.3

1. For what values of p does the series $1/1^p-1/2^p+1/3^p-1/4^p+\cdots$ converge?
2. If x is not an integer, prove that $1/(x+1)-1/(x+2)+1/(x+3)-\cdots$ converges.
3. Prove that
   (a) $2-2^{1/2}+2^{1/3}-2^{1/4}+\cdots$ diverges,
   (b) $(1-2)-(1-2^{1/2})+(1-2^{1/3})-(1-2^{1/4})+\cdots$ converges.
4. Show that if

$$a_n=\frac{1}{\sqrt{n}}+\frac{(-1)^{n-1}}{n},$$

   then $\sum_{n=1}^{\infty}(-1)^{n+1}a_n$ diverges. (Here $a_n>0$ and $\lim_{n\to\infty}a_n=0$. Why doesn't theorem 3.3A apply?)
5. Show that $\sum_{n=1}^{\infty}(-1)^{n+1}n/(2n-1)$ diverges.

3.4 CONDITIONAL CONVERGENCE AND ABSOLUTE CONVERGENCE

We saw in the preceding section that the series

$$1-\tfrac{1}{2}+\tfrac{1}{4}-\tfrac{1}{8}+\cdots \tag{1}$$

and the series

$$1-\tfrac{1}{2}+\tfrac{1}{3}-\tfrac{1}{4}+\cdots \tag{2}$$

both converge. However, these two series differ in the following respect. If we take the absolute value of each term in (1), we obtain

$$1+\tfrac{1}{2}+\tfrac{1}{4}+\tfrac{1}{8}+\cdots, \tag{3}$$

which *converges*, whereas if we take the absolute value of each term in (2) we obtain

$$1+\tfrac{1}{2}+\tfrac{1}{3}+\tfrac{1}{4}+\cdots \tag{4}$$

which *diverges*. This leads us to the following definition, which divides convergent series into two classes.

3.4A. DEFINITION. Let $\sum_{n=1}^{\infty} a_n$ be a series of real numbers.

(a) If $\sum_{n=1}^{\infty}|a_n|$ converges, we say that $\sum_{n=1}^{\infty} a_n$ converges absolutely.
(b) If $\sum_{n=1}^{\infty} a_n$ converges but $\sum_{n=1}^{\infty}|a_n|$ diverges, we say that $\sum_{n=1}^{\infty} a_n$ converges conditionally.

Thus the series (1) converges absolutely while the series (2) converges conditionally.

We must justify the use of the word "converges" in the phrase "converges absolutely." This is done in the following theorem.

3.4B. THEOREM. If $\sum_{n=1}^{\infty} a_n$ converges absolutely, then $\sum_{n=1}^{\infty} a_n$ converges.

PROOF: Let $s_n = a_1 + \cdots + a_n$. We wish to prove that $\{s_n\}_{n=1}^{\infty}$ converges. By 2.10D, it is enough to show that $\{s_n\}_{n=1}^{\infty}$ is Cauchy. By hypothesis $\sum_{n=1}^{\infty}|a_n| < \infty$ (see 3.2D) and thus $\{t_n\}_{n=1}^{\infty}$ converges where $t_n = |a_1| + \cdots + |a_n|$. By 2.10B, $\{t_n\}_{n=1}^{\infty}$ is Cauchy. Thus given $\epsilon > 0$ there exists $N \in I$ such that

$$|t_m - t_n| < \epsilon \qquad (m,n \geqslant N).$$

But (if $m > n$, say), $|s_m - s_n| = |a_{n+1} + \cdots + a_m| \leqslant |a_{n+1}| + \cdots + |a_m| = |t_m - t_n|$. Thus

$$|s_m - s_n| < \epsilon \qquad (m,n \geqslant N).$$

This proves that $\{s_n\}_{n=1}^{\infty}$ is Cauchy, which is what we wished to show.

3.4C. If we separate a series $\sum_{n=1}^{\infty} a_n$ into the series of positive a_n and the series of negative a_n, we can show up an important distinction between absolutely convergent and conditionally convergent series.

More precisely, if $\sum_{n=1}^{\infty} a_n$ is a series of real numbers, let

$$p_n = a_n \quad \text{if} \quad a_n > 0,$$
$$p_n = 0 \quad \text{if} \quad a_n \leqslant 0.$$

[Thus for the series $1-\tfrac{1}{2}+\tfrac{1}{3}-\cdots$, $p_1 = 1$, $p_3 = \tfrac{1}{3}$, $p_{2n-1} = 1/(2n-1)$, while $p_2 = p_4 = \cdots = 0$.] Similarly, let

$$q_n = a_n \quad \text{if} \quad a_n \leqslant 0,$$
$$q_n = 0 \quad \text{if} \quad a_n > 0.$$

The p_n are thus the positive terms of $\sum_{n=1}^{\infty} a_n$ (along with some 0's) while the q_n are the negative terms. It is easy to see that

$$p_n = \max(a_n, 0), \qquad q_n = \min(a_n, 0)$$

and hence by 1.4D,

$$2p_n = a_n + |a_n| \quad \text{and} \quad 2q_n = a_n - |a_n|. \tag{$*$}$$

Also,

$$a_n = p_n + q_n.$$

It is now not difficult to prove the following interesting result.

THEOREM. (a) If $\sum_{n=1}^{\infty} a_n$ converges absolutely, then both $\sum_{n=1}^{\infty} p_n$ and $\sum_{n=1}^{\infty} q_n$ converge. However,

(b) If $\sum_{n=1}^{\infty} a_n$ converges conditionally, then both $\sum_{n=1}^{\infty} p_n$ and $\sum_{n=1}^{\infty} q_n$ diverge. Finally,

(c) If $\sum_{n=1}^{\infty} p_n$ and $\sum_{n=1}^{\infty} q_n$ both converge, then $\sum_{n=1}^{\infty} a_n$ converges absolutely.

PROOF: (a) If $\sum_{n=1}^{\infty} a_n$ and $\sum_{n=1}^{\infty} |a_n|$ both converge, then, by 3.1C, so does $\sum_{n=1}^{\infty}(a_n + |a_n|)$. Thus from ($*$), $\sum_{n=1}^{\infty} 2p_n$ converges. By 3.1C again, this implies the convergence of $\sum_{n=1}^{\infty} p_n$. The series $\sum_{n=1}^{\infty} q_n$ may be proved convergent by similar reasoning.

(b) We now assume that $\sum_{n=1}^{\infty} a_n$ converges but that $\sum_{n=1}^{\infty} |a_n|$ diverges. From ($*$) we have $|a_n| = 2p_n - a_n$. If $\sum_{n=1}^{\infty} p_n$ converged, then, by 3.1C, so would

$$\sum_{n=1}^{\infty} (2p_n - a_n) = \sum_{n=1}^{\infty} |a_n|,$$

contradicting our assumption. Hence $\sum_{n=1}^{\infty} p_n$ diverges. Again, $\sum_{n=1}^{\infty} q_n$ may be handled in the same way.

(c) Since $p_n = (a_n + |a_n|)/2$ and $q_n = (a_n - |a_n|)/2$, we have $|a_n| = p_n - q_n$. Hence if $\sum_{n=1}^{\infty} p_n$ and $\sum_{n=1}^{\infty} q_n$ both converge, then so does $\sum_{n=1}^{\infty} |a_n|$, which shows that $\sum_{n=1}^{\infty} a_n$ converges absolutely.

Thus since $1 - \frac{1}{2} + \frac{1}{3} - \frac{1}{4} + \cdots$ is a conditionally convergent series, it follows that $1 + 0 + \frac{1}{3} + 0 + \frac{1}{5} + 0 + \cdots$ diverges, and hence that $1 + \frac{1}{3} + \frac{1}{5} + \cdots$ diverges.

The last theorem tells us, roughly, that an absolutely convergent series converges because its terms are "small" while a conditionally convergent series converges because of "cancellation" between its positive and negative terms.

Exercises 3.4

1. Classify as to divergent, conditionally convergent, or absolutely convergent:
   (a) $1 - \dfrac{1}{1!} + \dfrac{1}{2!} - \dfrac{1}{3!} + \cdots,$
   (b) $1 - \frac{1}{3} + \frac{1}{5} - \frac{1}{7} + \cdots,$
   (c) $\frac{1}{2} - \frac{2}{3} + \frac{3}{4} - \frac{4}{5} + \cdots,$
   (d) $1 - 1 + \frac{1}{2} - \frac{1}{2} + \frac{1}{3} - \frac{1}{3} + \cdots,$
   (e) $1 - \dfrac{1}{2} + \dfrac{1}{2} - \dfrac{1}{2^2} + \dfrac{1}{3} - \dfrac{1}{2^3} + \dfrac{1}{4} - \dfrac{1}{2^4} + \cdots.$
2. Can a series of nonnegative numbers converge conditionally?
3. Prove that if $\sum_{n=1}^{\infty} |a_n| < \infty$, then $|\sum_{n=1}^{\infty} a_n| \leqslant \sum_{n=1}^{\infty} |a_n|$.
4. If $\sum_{n=1}^{\infty} a_n$ converges absolutely, and if $\epsilon_n = \pm 1$ for every $n \in I$, prove that $\sum_{n=1}^{\infty} \epsilon_n a_n$ converges.
5. If $\sum_{n=1}^{\infty} \epsilon_n a_n$ converges for every sequence $\{\epsilon_n\}_{n=1}^{\infty}$ such that $\epsilon_n = \pm 1$ $(n \in I)$, prove that $\sum_{n=1}^{\infty} a_n$ converges absolutely.

3.5 REARRANGEMENTS OF SERIES

3.5A. Roughly speaking, a rearrangement of a series $\Sigma_{n=1}^{\infty}a_n$ is a series $\Sigma_{n=1}^{\infty}b_n$ whose terms are the same as those of $\Sigma_{n=1}^{\infty}a_n$ but occur in different order. (A precise definition of rearrangement is given later in this section.) We shall see that rearranging an absolutely convergent series has no effect on its sum but that rearranging a conditionally convergent series can have drastic effect.

We have seen that the series $\Sigma_{n=1}^{\infty}(-1)^{n+1}/n$ converges conditionally to some $L\in R$ (where we have stated but not proved that $L=\log 2$). In addition, we know $0.6\leqslant L\leqslant 0.8$ so that $L\neq 0$. We have

$$L=1-\tfrac{1}{2}+\tfrac{1}{3}-\tfrac{1}{4}+\tfrac{1}{5}-\tfrac{1}{6}+\tfrac{1}{7}-\tfrac{1}{8}+\cdots. \tag{1}$$

By 3.1C,

$$\tfrac{1}{2}L=\tfrac{1}{2}-\tfrac{1}{4}+\tfrac{1}{6}-\tfrac{1}{8}+\tfrac{1}{10}-\tfrac{1}{12}+\cdots,$$

and so, certainly,

$$\tfrac{1}{2}L=0+\tfrac{1}{2}-0-\tfrac{1}{4}+0+\tfrac{1}{6}-0-\tfrac{1}{8}+\cdots. \tag{2}$$

If we then add (2) to (1) we obtain, again by 3.1C,

$$\tfrac{3}{2}L=(1+0)+\left(\frac{-1}{2}+\frac{1}{2}\right)+\left(\frac{1}{3}-0\right)+\left(\frac{-1}{4}+\frac{-1}{4}\right)$$
$$+\left(\frac{1}{5}+0\right)+\left(\frac{-1}{6}+\frac{1}{6}\right)+\left(\frac{1}{7}+0\right)+\left(\frac{-1}{8}+\frac{-1}{8}\right)+\cdots$$

or

$$\tfrac{3}{2}L=1+\tfrac{1}{3}-\tfrac{1}{2}+\tfrac{1}{5}+\tfrac{1}{7}-\tfrac{1}{4}+\tfrac{1}{9}+\tfrac{1}{11}-\tfrac{1}{6}+\cdots. \tag{3}$$

The series on the right of (3) is a rearrangement of the series on the right of (1), but they converge to different sums!

3.5B. Indeed, we can find a rearrangement of $\Sigma_{n=1}^{\infty}(-1)^{n+1}/n$ that will converge to any preassigned real number—say, for example, 512. From 3.4C we know that $1+\frac{1}{3}+\frac{1}{5}+\cdots$ diverges. By 3.2A the partial sums of this series must be unbounded. Thus $1+\frac{1}{3}+\frac{1}{5}+\cdots+1/N$ will be greater than 512 for all sufficiently large odd integers N. Let N_1 be the *smallest* odd integer such that

$$1+\frac{1}{3}+\frac{1}{5}+\cdots+\frac{1}{N_1}>512.$$

Then

$$1+\frac{1}{3}+\frac{1}{5}+\cdots+\frac{1}{N_1}-\frac{1}{2}\leqslant 512$$

(why?). Now let N_2 be the smallest odd integer greater than N_1 such that

$$1+\frac{1}{3}+\frac{1}{5}+\cdots+\frac{1}{N_1}-\frac{1}{2}+\frac{1}{N_1+2}+\cdots+\frac{1}{N_2}>512.$$

Then

$$1+\frac{1}{3}+\frac{1}{5}+\cdots+\frac{1}{N_1}-\frac{1}{2}+\frac{1}{N_1+2}+\cdots+\frac{1}{N_2}-\frac{1}{4}\leqslant 512.$$

Continuing in this fashion we may construct a rearrangement of $\Sigma_{n=1}^{\infty}(-1)^{n+1}/n$ that converges to 512. You supply the details.

Now let us define "rearrangement" precisely.

3.5C. DEFINITION. Let $N=\{n_i\}_{i=1}^{\infty}$ be a sequence of positive integers where each positive integer occurs exactly once among the n_i. (That is, N is a 1-1 function from I onto I.) If $\Sigma_{n=1}^{\infty} a_n$ is a series of real numbers and if

$$b_i = a_{n_i} \qquad (i \in I),$$

then $\Sigma_{i=1}^{\infty} b_i$ is called a rearrangement of $\Sigma_{n=1}^{\infty} a_n$.

If $A=\{a_i\}_{i=1}^{\infty}$ and $B=\{b_i\}_{i=1}^{\infty}$, then, in the definition 3.5C, we have $A \circ N = B$. If $N^{-1}=\{m_i\}_{i=1}^{\infty}$ is the inverse function for N, then (by Exercise 12 of Section 1.5) $A = B \circ N^{-1}$ so that $a_i = b_{m_i}$. This shows that $\Sigma_{i=1}^{\infty} a_i$ is also a rearrangement of $\Sigma_{n=1}^{\infty} b_n$ if $\Sigma_{n=1}^{\infty} b_n$ is a rearrangement of $\Sigma_{n=1}^{\infty} a_n$.

In the exercises the reader is asked to supply the proof of the following theorem, which consists of an imitation of the method in 3.5B.

3.5D THEOREM. Let $\Sigma_{n=1}^{\infty} a_n$ be a conditionally convergent series of real numbers. Then for any $x \in R$ there is a rearrangement of $\Sigma_{n=1}^{\infty} a_n$ which converges to x.

For absolutely convergent series the story is entirely different. We first treat the case of a series of nonnegative terms.

3.5E LEMMA. If $\Sigma_{n=1}^{\infty} a_n$ is a series of nonnegative numbers which converges to $A \in R$, and $\Sigma_{n=1}^{\infty} b_n$ is a rearrangement of $\Sigma_{n=1}^{\infty} a_n$, then $\Sigma_{n=1}^{\infty} b_n$ converges and $\Sigma_{n=1}^{\infty} b_n = A$.

PROOF: For each $N \in I$, let $s_N = b_1 + \cdots + b_N$. Since $b_i = a_{n_i}$ for some sequence $\{n_i\}_{i=1}^{\infty}$, we have

$$b_1 = a_{n_1}, \ldots, b_N = a_{n_N}.$$

Let $M = \max(n_1, \ldots, n_N)$. Then, certainly, $s_N \leqslant a_1 + \cdots + a_M \leqslant A$. Thus by 3.2A, $\Sigma_{n=1}^{\infty} b_n$ converges to some $B \in R$. But $B = \lim_{n\to\infty} s_N$ and so by 2.7E, $B \leqslant A$. (That is, $\Sigma_{n=1}^{\infty} b_n \leqslant \Sigma_{n=1}^{\infty} a_n$.) But, since $\Sigma_{n=1}^{\infty} a_n$ is also a rearrangement of $\Sigma_{n=1}^{\infty} b_n$, the same reasoning with the roles of $\Sigma_{n=1}^{\infty} a_n$ and $\Sigma_{n=1}^{\infty} b_n$ reversed would show $A \leqslant B$. Hence $B = A$ and the proof is complete.

The result in 3.5E clearly holds also for a series of nonpositive numbers. The lemma is a special case of the following theorem.

3.5F THEOREM. If $\Sigma_{n=1}^{\infty} a_n$ converges absolutely to A, then any rearrangement $\Sigma_{n=1}^{\infty} b_n$ of $\Sigma_{n=1}^{\infty} a_n$ also converges absolutely to A.

PROOF: Define p_n and q_n as in 3.4C, so that $a_n = p_n + q_n$. Then, by the theorem in 3.4C, both $\Sigma_{n=1}^{\infty} p_n$ and $\Sigma_{n=1}^{\infty} q_n$ converge. Say $\Sigma_{n=1}^{\infty} p_n = P$ and $\Sigma_{n=1}^{\infty} q_n = Q$ (so that $Q \leqslant 0$). Then $A = P + Q$ by 3.1C. For some $\{n_i\}_{i=1}^{\infty}$, we have

$$b_i = a_{n_i} = p_{n_i} + q_{n_i}.$$

Moreover, $\Sigma_{i=1}^{\infty} p_{n_i}$ is a rearrangement of the series $\Sigma_{n=1}^{\infty} p_n$ of nonnegative terms. Hence by the lemma 3.5E, $\Sigma_{i=1}^{\infty} p_{n_i}$ converges and $\Sigma_{i=1}^{\infty} p_{n_i} = P$. Similarly, $\Sigma_{i=1}^{\infty} q_{n_i} = Q$. Since

$$b_i = p_{n_i} + q_{n_i},$$

theorem 3.1C implies that $\Sigma_{i=1}^{\infty} b_i$ converges and

$$\sum_{i=1}^{\infty} b_i = \sum_{i=1}^{\infty} p_{n_i} + \sum_{i=1}^{\infty} q_{n_i} = P + Q = A.$$

All that remains to be demonstrated is the *absolute* convergence of $\Sigma_{i=1}^{\infty} b_i$. But since $b_i = p_{n_i} + q_{n_i}$ we have

$$|b_i| \leqslant |p_{n_i}| + |q_{n_i}| = p_{n_i} - q_{n_i}.$$

Thus for any $N \in I$,

$$|b_1| + \cdots + |b_N| \leqslant \sum_{i=1}^{N} p_{n_i} - \sum_{i=1}^{N} q_{n_i} \leqslant \sum_{i=1}^{\infty} p_{n_i} - \sum_{i=1}^{\infty} q_{n_i} = P - Q.$$

The partial sums of $\Sigma_{i=1}^{\infty} |b_i|$ are thus all bounded above by $P-Q$ and hence $\Sigma_{i=1}^{\infty}|b_i| < \infty$.* This completes the proof.

3.5G. Our theorem on rearrangements gives us a theorem on the multiplication of series.

If we formally† take the product of two power series $\Sigma_{n=0}^{\infty} a_n x^n$ and $\Sigma_{n=0}^{\infty} b_n x^n$ and collect terms with the same power of x, we have

$$\left(a_0 + a_1 + a_2 x^2 + \cdots\right)\left(b_0 + b_1 x + b_2 x^2 + \cdots\right) = a_0 b_0 + (a_0 b_1 + a_1 b_0)x$$
$$+ (a_0 b_2 + a_1 b_1 + a_2 b_0)x^2 + \cdots .$$

That is,

$$\left(\sum_{n=0}^{\infty} a_n x^n\right)\left(\sum_{n=0}^{\infty} b_n x^n\right) = \sum_{n=0}^{\infty} c_n x^n \qquad (*)$$

where $c_n = \Sigma_{k=0}^{n} a_k b_{n-k} (n = 0, 1, 2, \ldots)$. For purposes of application, it is enough to examine $(*)$ in the case $x = 1$. We shall prove that

$$\left(\sum_{n=0}^{\infty} a_n\right)\left(\sum_{n=0}^{\infty} b_n\right) = \sum_{n=0}^{\infty} c_n$$

under the hypothesis that the two series on the left converge absolutely.

THEOREM. If the series $\Sigma_{n=0}^{\infty} a_n$ and $\Sigma_{n=0}^{\infty} b_n$ converge absolutely to A and B, respectively, then $AB = C$ where $C = \Sigma_{n=0}^{\infty} c_n$ (the series converging absolutely) and

$$c_n = \sum_{k=0}^{n} a_k b_{n-k} \qquad (n = 0, 1, 2, \ldots).$$

PROOF: For $k = 0, 1, 2, \ldots$ we have $|c_k| \leqslant |a_0 b_k| + |a_1 b_{k-1}| + \cdots + |a_k b_0|$. Thus for any n,

$$|c_0| + |c_1| + \cdots + |c_n|$$
$$\leqslant |a_0 b_0| + (|a_0 b_1| + |a_1 b_0|) + \cdots + (|a_0 b_n| + |a_1 b_{n-1}| + \cdots + |a_0 b_n|)$$
$$\leqslant (|a_0| + \cdots + |a_n|)(|b_0| + \cdots + |b_n|) \leqslant \left(\sum_{k=0}^{\infty} |a_k|\right)\left(\sum_{k=0}^{\infty} |b_k|\right).$$

The sequence of partial sums of $\Sigma_{k=0}^{\infty} |c_k|$ is thus bounded above, and hence

$$\sum_{k=0}^{\infty} |c_k| < \infty.$$

* See 3.2D.
† That is, without regard to rigor.

The foregoing inequalities also show the absolute convergence of the series

$$a_0b_0 + a_0b_1 + a_1b_0 + a_0b_2 + a_1b_1 + a_2b_0 + a_0b_3 + \cdots \tag{1}$$

(whose sum is $\sum_{k=0}^{\infty} c_k$).

By 3.5F we may rearrange the terms in (1) to obtain*

$$\sum_{k=0}^{\infty} c_k = [a_0b_0] + [a_0b_1 + a_1b_0 + a_1b_1] + [a_0b_2 + a_2b_0 + a_1b_2 + a_2b_1 + a_2b_2] + \cdots. \tag{2}$$

Inside the nth bracket ($n = 0, 1, 2, \ldots$) on the right of (2) are all products a_jb_k where either j or k is equal to n and neither j nor k is greater than n. Let us examine the sum of the terms in each bracket. If

$$A_n = a_0 + a_1 + \cdots + a_n, \quad \text{and} \quad B_n = b_0 + b_1 + \cdots + b_n,$$

we have

$$a_0b_0 = A_0B_0,$$

$$a_0b_1 + a_1b_0 + a_1b_1 = (a_0 + a_1)(b_0 + b_1) - a_0b_0 = A_1B_1 - A_0B_0,$$

$$a_0b_2 + a_2b_0 + a_1b_2 + a_2b_1 + a_2b_2 = (a_0 + a_1 + a_2)(b_0 + b_1 + b_2) - (a_0 + a_1)(b_0 + b_1) = A_2B_2 - A_1B_1,$$

and in general, for $n \geqslant 1$ the quantity in the nth bracket on the right of (2) is equal to $A_nB_n - A_{n-1}B_{n-1}$. The sum of the first n brackets on the right of (2) is therefore $[A_0B_0] + [A_1B_1 - A_0B_0] + \cdots + [A_nB_n - A_{n-1}B_{n-1}] = A_nB_n$, which approaches AB as $n \to \infty$. The right side of (2) is thus equal to AB and the proof is complete.

3.5H COROLLARY. If for some $x \in R$ the power series $\sum_{n=0}^{\infty} a_n x^n$ and $\sum_{n=0}^{\infty} b_n x^n$ are absolutely convergent, then

$$\left(\sum_{n=0}^{\infty} a_n x^n\right)\left(\sum_{n=0}^{\infty} b_n x^n\right) = \sum_{n=0}^{\infty} c_n x^n \tag{1}$$

where $c_n = \sum_{k=0}^{n} a_k b_{n-k}$.

PROOF: Let $A_n = a_n x^n$, $B_n = b_n x^n$. Then, by 3.5G

$$\left(\sum_{n=0}^{\infty} A_n\right)\left(\sum_{n=0}^{\infty} B_n\right) = \sum_{n=0}^{\infty} C_n, \tag{2}$$

where

$$C_n = \sum_{k=0}^{n} A_k B_{n-k} = \sum_{k=0}^{n} a_k x^k b_{n-k} x^{n-k} = x^n \sum_{k=0}^{n} a_k b_{n-k} = c_n x^n.$$

Equation (1) thus follows from (2).

Exercises 3.5

1. Prove that if $|x| < 1$, then

$$1 + x^2 + x + x^4 + x^6 + x^3 + x^8 + x^{10} + x^5 + \cdots = \frac{1}{1-x}.$$

* See Exercise 13 of Section 3.1.

2. Prove that if $a_1+a_2+a_3+\cdots$ is absolutely convergent, then $a_1+a_2+a_3+\cdots=(a_1+a_3+a_5+\cdots)+(a_2+a_4+a_6+\cdots)$. Is this true for all conditionally convergent series?
3. What, if anything, is wrong with the following?

$$\begin{aligned}1-\tfrac{1}{2}+\tfrac{1}{3}-\tfrac{1}{4}+\tfrac{1}{5}-\tfrac{1}{6}+\cdots\\ =1+(\tfrac{1}{2}-1)+\tfrac{1}{3}+(\tfrac{1}{4}-\tfrac{1}{2})+\tfrac{1}{5}+(\tfrac{1}{6}-\tfrac{1}{3})+\cdots\\ =(1+\tfrac{1}{2}+\tfrac{1}{3}+\tfrac{1}{4}+\tfrac{1}{5}+\tfrac{1}{6}+\cdots)-1-\tfrac{1}{2}-\tfrac{1}{3}-\cdots\\ =(1+\tfrac{1}{2}+\tfrac{1}{3}+\cdots)-(1+\tfrac{1}{2}+\tfrac{1}{3}+\cdots)=0.\end{aligned}$$

4. Show that there exists a rearrangement $\sum_{n=1}^{\infty}b_n$ of $1-\frac{1}{2}+\frac{1}{3}-\frac{1}{4}+\frac{1}{5}-\cdots$ such that, if $t_n=b_1+\cdots+b_n$, then

$$\limsup_{n\to\infty} t_n=100, \qquad \liminf_{n\to\infty} t_n=-100.$$

5. Show that any conditionally convergent series has a rearrangement that diverges.
6. Prove theorem 3.5D.
7. Write the following products in the form $\sum_{n=0}^{\infty}c_n x^n$:
   (a) $(\sum_{n=0}^{\infty}nx^n)(\sum_{n=0}^{\infty}x^n)$,
   (b) $(\sum_{n=0}^{\infty}x^n)(\sum_{n=0}^{\infty}(-1)^n x^n)$.
8. If $0\leqslant x<1$, prove that

$$\left(\frac{1}{1-x}\right)^2=\sum_{n=0}^{\infty}(n+1)x^n.$$

9. Let $L=\sum_{n=1}^{\infty}[(-1)^{n+1}]/n$. Show that each of the following series converges to the indicated sum:
   (a) $1-\frac{1}{2}-\frac{1}{4}+\frac{1}{3}-\frac{1}{6}-\frac{1}{8}+\frac{1}{5}-\frac{1}{10}-\frac{1}{12}+\ldots=\frac{L}{2}$.
   (b) $1+\frac{1}{3}+\frac{1}{5}-\frac{1}{2}-\frac{1}{4}-\frac{1}{6}+\frac{1}{7}+\frac{1}{9}+\frac{1}{11}-\ldots=L$.

3.6 TESTS FOR ABSOLUTE CONVERGENCE

In the last section we discussed the behavior of absolutely convergent and conditionally convergent series in general. In this section we take up methods (tests) used to decide whether or not a specific series converges absolutely.

3.6A. DEFINITION. Let $\sum_{n=1}^{\infty}a_n$ and $\sum_{n=1}^{\infty}b_n$ be two series of real numbers. We shall say that $\sum_{n=1}^{\infty}a_n$ is dominated by* $\sum_{n=1}^{\infty}b_n$ if there exists $N\in I$ such that

$$|a_n|\leqslant|b_n| \qquad (n\geqslant N).$$

(That is, $|a_n|\leqslant|b_n|$ except for a finite number of values of n.) In this case we write

$$\sum_{n=1}^{\infty}a_n\ll\sum_{n=1}^{\infty}b_n.$$

For example, $\sum_{n=1}^{\infty}(-1)^n/n^2\ll\sum_{n=1}^{\infty}1/(2n+1)$ since $|(-1)/n^2|\leqslant 1/(2n+1)$ for $n\geqslant 3$. Also $100+\frac{1}{2}+\frac{1}{4}+\cdots$ is dominated by $1+\frac{1}{2}+\frac{1}{4}+\cdots$. (This shows that $\sum_{n=1}^{\infty}a_n\ll\sum_{n=1}^{\infty}b_n$ does not necessarily imply $\sum_{n=1}^{\infty}a_n<\sum_{n=1}^{\infty}b_n$.)

At the end of Section 3.4 we mentioned that an absolutely convergent series converges because its terms are "small." If the series $\sum_{n=1}^{\infty}a_n$ is dominated by an absolutely

* Or that $\sum_{n=1}^{\infty}b_n$ dominates $\sum_{n=1}^{\infty}a_n$.

convergent series $\Sigma_{n=1}^{\infty} b_n$, then $\Sigma_{n=1}^{\infty} a_n$ certainly ought to converge absolutely since most of its terms are no larger than those of $\Sigma_{n=1}^{\infty} b_n$. This we now prove.

3.6B THEOREM. If $\Sigma_{n=1}^{\infty} a_n$ is dominated by $\Sigma_{n=1}^{\infty} b_n$ where $\Sigma_{n=1}^{\infty} b_n$ converges absolutely, then $\Sigma_{n=1}^{\infty} a_n$ also converges absolutely. Symbolically, if $\Sigma_{n=1}^{\infty} a_n \ll \Sigma_{n=1}^{\infty} b_n$ and $\Sigma_{n=1}^{\infty} |b_n| < \infty$, then $\Sigma_{n=1}^{\infty} |a_n| < \infty$.

PROOF: Let $M = \Sigma_{n=1}^{\infty} |b_n|$. We have $|a_n| \leqslant |b_n|$ for $n \geqslant N$. Hence if $s_n = |a_1| + \cdots + |a_n|$, we have, for $n \geqslant N$,

$$s_n \leqslant |a_1| + \cdots + |a_N| + |b_{N+1}| + \cdots + |b_n| \leqslant |a_1| + \cdots + |a_N| + M.$$

The sequence of partial sums of $\Sigma_{n=1}^{\infty} |a_n|$ is thus bounded above and the theorem follows from 3.2A.

Theorem 3.6B is called the *comparison test* for absolute convergence since it involves term-by-term comparison of $\Sigma_{n=1}^{\infty} |a_n|$ and $\Sigma_{n=1}^{\infty} |b_n|$. It is the basis for the other tests in this section.

3.6C. From 3.6B, it follows immediately that for any $x \in (-1, 1)$ the geometric series $\Sigma_{n=1}^{\infty} x^n$ converges absolutely. For

$$\sum_{n=0}^{\infty} x^n \ll \sum_{n=0}^{\infty} |x|^n$$

and the series on the right converges (absolutely) by 3.2B.

We emphasize that theorem 3.6B deals only with absolute convergence. Note that the series $\Sigma_{n=1}^{\infty} 1/n$ is dominated by the conditionally convergent series $\Sigma_{n=1}^{\infty} (-1)^n/n$ but the former series does not converge at all. The concept of "dominated" is usable only in connection with absolute values.

The following result may be deduced from part (b) of 3.2A. We omit the proof.

3.6D THEOREM. If $\Sigma_{n=1}^{\infty} a_n$ is dominated by $\Sigma_{n=1}^{\infty} b_n$ and $\Sigma_{n=1}^{\infty} |a_n| = \infty$, then $\Sigma_{n=1}^{\infty} |b_n| = \infty$. (That is, if $\Sigma_{n=1}^{\infty} a_n \ll \Sigma_{n=1}^{\infty} b_n$ and $\Sigma_{n=1}^{\infty} |a_n| = \infty$, then $\Sigma_{n=1}^{\infty} |b_n| = \infty$.)

For example, consider $\Sigma_{n=1}^{\infty} b_n = \Sigma_{n=1}^{\infty} 1/(2n+5)$. This series dominates $\Sigma_{n=1}^{\infty} 1/3n$, which diverges. Hence $\Sigma_{n=1}^{\infty} 1/(2n+5)$ diverges.

We now give the first important consequence of 3.6B.

3.6E. THEOREM. (a) If $\Sigma_{n=1}^{\infty} b_n$ converges absolutely and if $\lim_{n\to\infty} |a_n|/|b_n|$ exists, then $\Sigma_{n=1}^{\infty} a_n$ converges absolutely.

(b) If $\Sigma_{n=1}^{\infty} |a_n| = \infty$ and if $\lim_{n\to\infty} |a_n|/|b_n|$ exists, then $\Sigma_{n=1}^{\infty} |b_n| = \infty$.

PROOF: (a) By 2.5B, $\{|a_n/b_n|\}_{n=1}^{\infty}$ is bounded. Thus for some $M > 0$,

$$|a_n| \leqslant M|b_n| \qquad (n \in I).$$

This shows that $\Sigma_{n=1}^{\infty} a_n$ is dominated by the absolutely convergent series $\Sigma_{n=1}^{\infty} Mb_n$. By 3.6B, $\Sigma_{n=1}^{\infty} |a_n| < \infty$.

(b) As in the proof of (a) we have $|a_n| \leqslant M|b_n|$, so that $\Sigma_{n=1}^{\infty} |b_n|$ dominates $\Sigma_{n=1}^{\infty} (1/M) \cdot |a_n|$ which diverges. Apply 3.6D.

Thus the series $\Sigma_{n=1}^{\infty} 2n/(n^2 - 4n + 7)$ must diverge. For if we let $b_n = 2n/(n^2 - 4n + 7)$

and $a_n = 1/n$, then

$$\lim_{n\to\infty}\left|\frac{a_n}{b_n}\right| = \lim_{n\to\infty}\frac{n^2-4n+7}{2n^2} = \frac{1}{2}.$$

But

$$\sum_{n=1}^{\infty}|a_n| = \sum_{n=1}^{\infty}\frac{1}{n} = \infty.$$

Thus by (b) of 3.6E,

$$\sum_{n=1}^{\infty}\left|\frac{2n}{n^2-4n+7}\right| = \sum_{n=1}^{\infty}\frac{2n}{n^2-4n+7} = \infty.$$

The following result, called the ratio test, is very useful in treating specific power series.

3.6F. THEOREM. Let $\sum_{n=1}^{\infty}a_n$ be a series of nonzero real numbers and let

$$a = \liminf_{n\to\infty}\left|\frac{a_{n+1}}{a_n}\right|, \qquad A = \limsup_{n\to\infty}\left|\frac{a_{n+1}}{a_n}\right|,$$

(so that $a \leqslant A$). Then

(a) If $A<1$, then $\sum_{n=1}^{\infty}|a_n| < \infty$;
(b) If $a>1$ then $\sum_{n=1}^{\infty}a_n$ diverges;
(c) If $a \leqslant 1 \leqslant A$, then the test fails. (That is, no information about convergence may be deduced.)

PROOF: (a) If $A<1$, choose any B such that $A<B<1$. Then $B = A+\epsilon$ for some $\epsilon>0$ so, by 2.9L, there exists $N \in I$ such that

$$\left|\frac{a_{n+1}}{a_n}\right| \leqslant B \qquad (n \geqslant N).$$

Then $|a_{N+1}/a_N| \leqslant B$, $|a_{N+2}/a_{N+1} \leqslant B$, and so

$$\left|\frac{a_{N+2}}{a_N}\right| = \left|\frac{a_{N+2}}{a_{N+1}}\right|\cdot\left|\frac{a_{N+1}}{a_N}\right| \leqslant B^2.$$

For any $k \geqslant 0$ we have similarly

$$\left|\frac{a_{N+k}}{a_N}\right| = \left|\frac{a_{N+k}}{a_{N+k-1}}\right|\cdots\left|\frac{a_{N+1}}{a_N}\right| \leqslant B^k.$$

Thus

$$|a_{N+k}| \leqslant |a_N|B^k \qquad (k=0,1,2,\ldots).$$

But $\sum_{k=0}^{\infty}|a_N|.\ B^k$ converges, by 3.2B, since $0<B<1$. Thus by 3.6B, $\sum_{k=0}^{\infty}|a_{N+k}|$ converges. That is, $|a_N|+|a_{N+1}|+|a_{N+2}|+\cdots$ converges. It follows easily that $\sum_{n=1}^{\infty}|a_n| < \infty$. This proves (a).

(b) If $a>1$, then by 2.9L, $|a_{n+1}/a_n|>1$ for all $n \geqslant N$ (for some $N \in I$). But then $|a_N|<|a_{N+1}|<|a_{N+2}|<\cdots$, and so, certainly, $\{a_n\}_{n=1}^{\infty}$ does not converge to 0. Thus, by 3.1D, $\sum_{n=1}^{\infty}a_n$ diverges.

(c) To illustrate conclusion (c), consider first $\sum_{n=1}^{\infty} a_n = \sum_{n=1}^{\infty} 1/n$. Here

$$\lim_{n \to \infty} \frac{a_{n+1}}{a_n} = 1$$

so that $a = 1 = A$. The series diverges.

But we shall soon see that $\sum_{n=1}^{\infty} 1/n^2$ is a *convergent* series which also has $a = 1 = A$.

From 3.6F we see immediately that if $\lim_{n\to\infty} |a_{n+1}/a_n|$ exists (and is equal to L, say), then $\sum_{n=1}^{\infty} |a_n|$ converges if $L < 1$ and $\sum_{n=1}^{\infty} a_n$ diverges if $L > 1$, while if $L = 1$, we can conclude nothing.

Here are some examples to illustrate 3.6F. Consider first $\sum_{n=1}^{\infty} n^n/n!$. Here $a_n = n^n/n!$ so that

$$\frac{|a_{n+1}|}{|a_n|} = \frac{(n+1)^{n+1}}{(n+1)!} \cdot \frac{n!}{n^n} = \frac{(n+1)^n}{n^n} = \left(1 + \frac{1}{n}\right)^n.$$

But by 2.6C, $\lim_{n\to\infty}(1 + 1/n)^n = e > 2$ and so $a = e = A$. In particular, $a > 2$ so that $\sum_{n=1}^{\infty} n^n/n!$ diverges. These computations show that for the series $\sum_{n=1}^{\infty} n!/n^n$ we have $A = 1/e < \frac{1}{2} < 1$ so that $\sum_{n=1}^{\infty} n!/n^n$ converges.

Consider next the series $\sum_{n=0}^{\infty} x^n/n!$ for some $x \in R$, Here we have

$$\left| \frac{a_{n+1}}{a_n} \right| = \frac{|x|^{n+1}}{(n+1)!} \cdot \frac{n!}{|x|^n} = \frac{|x|}{n+1}.$$

Thus $\lim_{n\to\infty} |a_{n+1}/a_n| = 0$, which shows that the series in question converges absolutely for any real x. (Recall from calculus that the sum of the series is e^x.)

Finally, let us try to determine the values of x for which $\sum_{n=1}^{\infty} x^n/n$ converges absolutely. For this series we have $\lim_{n\to\infty} |a_{n+1}/a_n| = |x|$. Thus the series is absolutely convergent for $|x| < 1$ and is divergent for $|x| > 1$. The ratio test fails if $|x| = 1$—that is, if $x = 1$ or $x = -1$. But for these values of x the series becomes $\sum_{n=1}^{\infty} 1/n$ and $\sum_{n=1}^{\infty} (-1)^n/n$, neither of which is absolutely convergent. Thus $\sum_{n=1}^{\infty} x^n/n$ is absolutely convergent only for $-1 < x < 1$ (and is convergent for $-1 \leqslant x < 1$.)

The last test of this section is called the root test. It will yield an interesting general theorem about power series.

3.6G THEOREM. If $\limsup_{n\to\infty} \sqrt[n]{|a_n|} = A$, then the series of real numbers $\sum_{n=1}^{\infty} a_n$ (a) converges absolutely if $A < 1$, (b) diverges if $A > 1$. (This includes the case $\limsup_{n\to\infty} \sqrt[n]{|a_n|} = \infty$.)

If $A = 1$, the test fails.

PROOF: If $A < 1$, choose B so that $A < B < 1$. Then by 2.9L there exists $N \in I$ such that

$$\sqrt[n]{|a_n|} < B \qquad (n \geqslant N).$$

This implies $|a_n| < B^n (n \geqslant N)$. Thus $\sum_{n=1}^{\infty} |a_n|$ is dominated by $\sum_{n=1}^{\infty} B^n$, which is (absolutely) convergent. By 3.6B, $\sum_{n=1}^{\infty} |a_n| < \infty$. This proves (a). If

$$\limsup_{n\to\infty} \sqrt[n]{|a_n|} > 1$$

then, by 2.9L, $\sqrt[n]{|a_n|} > 1$ for infinitely many values of n. But this implies that $|a_n| > 1$ for infinitely many n, and so $\{a_n\}_{n=1}^{\infty}$ does not converge to 0. By 3.1D, $\sum_{n=1}^{\infty} a_n$ diverges. This proves (b).

Note that for the divergent series $\sum_{n=1}^{\infty} 1/n$ and the series $\sum_{n=1}^{\infty} 1/n^2$ (which will be shown to converge) we have $\lim_{n\to\infty} \sqrt[n]{|a_n|} = 1$. [Remembering from calculus that $\lim_{n\to\infty}(\log n/n) = 0$, we have

$$\lim_{n\to\infty} \sqrt[n]{1/n} = \lim_{n\to\infty} e^{(-\log n/n)} = e^0 = 1.$$

Thus also,

$$\lim_{n\to\infty} \sqrt[n]{1/n^2} = \lim_{n\to\infty} \left(\sqrt[n]{1/n}\right)^2 = 1^2 = 1.]$$

Here is a corollary on power series $\sum_{n=0}^{\infty} a_n x^n$.

3.6H. THEOREM. Let $\{a_n\}_{n=0}^{\infty}$ be a sequence of real numbers. Then

(a) If $\limsup_{n\to\infty} \sqrt[n]{|a_n|} = 0$, the series $\sum_{n=0}^{\infty} a_n x^n$ converges absolutely for all real x;

(b) If $\limsup_{n\to\infty} \sqrt[n]{|a_n|} = L > 0$, then $\sum_{n=0}^{\infty} a_n x^n$ converges absolutely for $|x| < 1/L$ and diverges for $|x| > 1/L$;

(c) If $\limsup_{n\to\infty} \sqrt[n]{|a_n|} = \infty$, then $\sum_{n=0}^{\infty} a_n x^n$ converges only for $x = 0$ and diverges for all other x.

PROOF: We have $\sqrt[n]{|a_n x^n|} = |x| \sqrt[n]{|a_n|}$. Thus if $\limsup_{n\to\infty} \sqrt[n]{|a_n|} = 0$, then, for any $x, \limsup_{n\to\infty} \sqrt[n]{|a_n x^n|} = |x| \cdot 0 = 0$ and by (a) of 3.6G, $\sum_{n=0}^{\infty} a_n x^n$ is absolutely convergent. This proves (a). To prove (b) we have $\limsup_{n\to\infty} \sqrt[n]{|a_n x^n|} = |Lx|$. Again by 3.6G, $\sum_{n=0}^{\infty} a_n x^n$ will converge absolutely if $|Lx| < 1$ and diverge if $|Lx| > 1$. This proves (b). The proof of (c) is left to the reader.

In any of the three cases in 3.6H the following is true.

3.6I. COROLLARY. If the power series $\sum_{n=0}^{\infty} a_n x^n$ converges for $x = x_0$, then it converges absolutely for all x such that $|x| < |x_0|$.

The following theorem shows that if the ratio test works, so will the root test.

3.6J. THEOREM. Let $\{a_n\}_{n=1}^{\infty}$ be a sequence of nonzero real numbers. Then

$$\limsup_{n\to\infty} \sqrt[n]{|a_n|} \leqslant \limsup_{n\to\infty} \left| \frac{a_{n+1}}{a_n} \right| \tag{1}$$

and

$$\liminf_{n\to\infty} \sqrt[n]{|a_n|} \geqslant \liminf_{n\to\infty} \left| \frac{a_{n+1}}{a_n} \right|. \tag{2}$$

Hence if the ratio test implies $\sum_{n=1}^{\infty} |a_n| < \infty$, so does the root test, and, if the ratio test implies $\sum_{n=1}^{\infty} |a_n| = \infty$, so does the root test.

PROOF: We will prove (1).

If $\limsup_{n\to\infty} |a_{n+1}/(a_n)| = \infty$, then (1) is obvious. Suppose $\limsup_{n\to\infty} |a_{n+1}/(a_n)| = A$ where $A \in R$. Then by 2.9L, if $\epsilon > 0$, there exists $N \in I$ such that

$$\left| \frac{a_{n+1}}{a_n} \right| \leqslant A + \epsilon \qquad (n \geqslant N).$$

Thus
$$|a_{N+1}| \leqslant (A+\epsilon)|a_N|,$$
$$|a_{N+2}| \leqslant (A+\epsilon)|a_{N+1}| \leqslant (A+\epsilon)^2 |a_N|.$$
Indeed, if $n \geqslant N$, we have
$$|a_n| = |a_{N+(n-N)}| \leqslant (A+\epsilon)^{n-N}|a_N|.$$
Thus
$$|a_n| \leqslant B(A+\epsilon)^n \qquad (n \geqslant N)$$
where $B = |a_N|/(A+\epsilon)^n$, and so
$$\sqrt[n]{a_n} \leqslant B^{1/n}(A+\epsilon) \qquad (n \geqslant N).$$
Since $B^{1/n} \to 0$ as $n \to \infty$, this implies
$$\limsup_{n\to\infty} \sqrt[n]{|a_n|} \leqslant A+\epsilon.$$
But ϵ was an arbitrary positive number, so
$$\limsup_{n\to\infty} \sqrt[n]{|a_n|} \leqslant A = \limsup_{n\to\infty} \left|\frac{a_{n+1}}{a_n}\right|.$$
This proves (1). The inequality (2) may be proved in similar fashion.

Now, suppose for a given sequence $\{a_n\}_{n=1}^{\infty}$ that the ratio test implies convergence for $\sum_{n=1}^{\infty}|a_n|$. Then the right side of (1) is <1. Hence so is the left side of (1), which shows that the root test also implies convergence. In like manner it follows from (2) that if the ratio test implies divergence, then so does the root test. This completes the proof.

The preceding theorem thus shows that if the ratio test gives definite information about $\sum_{n=1}^{\infty}|a_n|$, then the root test will also. Nevertheless, the ratio test is still valuable since often it is easier to apply than the root test.

Exercises 3.6

1. Prove true or false: If $\sum_{n=1}^{\infty} a_n$ is a convergent series of nonnegative numbers and $\sum_{n=1}^{\infty} b_n$ is a divergent series of nonnegative numbers, then $\sum_{n=1}^{\infty} a_n \ll \sum_{n=1}^{\infty} b_n$.
2. Do the following series converge?

   (a) $\displaystyle\sum_{n=0}^{\infty} \frac{n^4}{n!}$.

   (b) $\displaystyle\sum_{n=1}^{\infty} \frac{1+n}{1+n^2}$.

   (c) $\displaystyle\sum_{n=1}^{\infty} \frac{3}{4+2^n}$.
3. Show that if $|x|<1$, then $\sum_{n=1}^{\infty} n^{10{,}000} x^n$ converges absolutely.
4. For any $x>0$ prove that the series $1 - \dfrac{x^2}{2!} + \dfrac{x^4}{4!} - \ldots$ and $x - \dfrac{x^3}{3!} + \dfrac{x^5}{5!} - \ldots$ converge absolutely. Using theorem 3.5G find the first few terms in their product. Deduce $\sin 2x = 2\sin x \cos x$.
5. (a) Does the ratio test give any information about the series
$$\left(\tfrac{1}{2}\right)^0 + \left(\tfrac{1}{4}\right)^1 + \left(\tfrac{1}{2}\right)^2 + \left(\tfrac{1}{4}\right)^3 + \left(\tfrac{1}{2}\right)^4 + \ldots?$$

   (b) Does the series converge?

6. If $\{a_n\}_{n=1}^{\infty}$ is a sequence of real numbers, and if $\lim_{n\to\infty}|a_{n+1}/a_n| = L < 1$, prove that $\lim_{n\to\infty} a_n = 0$.
7. For what values of x does the series

$$x - \frac{x^3}{3} + \frac{x^5}{5} - \frac{x^3}{7} + \cdots$$

converge?
8. For what values of x does $1 + 2x + 3x^2 + 4x^3 + \cdots$ converge?
9. If $\sum_{n=1}^{\infty}|a_n| < \infty$ and if, for each $n \in I$, $|b_{n+1}/b_n| \leqslant |a_{n+1}/a_n|$, prove that $\sum_{n=1}^{\infty}|b_n| < \infty$. (*Hint*: First show $|b_n| \leqslant |b_1/a_1| \cdot |a_n|$.)
10. Test the convergence of $\sum_{n=1}^{\infty} a_n$ where $a_n = (3-e)(3-e^{1/2})(3-e^{1/3}) \cdots (3-e^{1/n})$.
11. Use the root test 3.6G on the following series and state what can be concluded.
    (a) $\displaystyle\sum_{n=1}^{\infty} \frac{x^n}{n}$.
    (b) $\displaystyle\sum_{n=1}^{\infty} \frac{1}{(\log n)^n}$.
    (c) $\displaystyle\sum_{n=1}^{\infty} \frac{(1+1/n)^{2n}}{e^n}$.
12. Show that $\sum_{n=1}^{\infty} x^n/n^n$ converges for all $x \in R$.

3.7 SERIES WHOSE TERMS FORM A NONINCREASING SEQUENCE

The tests of the previous section fail to give any information about the important series $\sum_{n=1}^{\infty} 1/n^2$. This series has the special property that its terms form a nonincreasing sequence. Such series are often treated by the integral test familiar from calculus. However, since we have not talked about integrals as yet, we use another very interesting test called the Cauchy condensation test.

3.7A. THEOREM. If $\{a_n\}_{n=1}^{\infty}$ is a nonincreasing sequence of positive numbers and if $\sum_{n=0}^{\infty} 2^n a_{2^n}$ converges, then $\sum_{n=1}^{\infty} a_n$ converges.

PROOF: We have

$$a_1 \leqslant a_1,$$
$$a_2 + a_3 \leqslant a_2 + a_2 = 2a_2,$$
$$a_4 + a_5 + a_6 + a_7 \leqslant 4a_4,$$

and, for any $n \in I$,

$$a_{2^n} + a_{2^n+1} + \cdots + a_{2^{n+1}-1} \leqslant 2^n a_{2^n}.$$

From these inequalities it follows that

$$\sum_{k=1}^{2^{n+1}-1} a_k \leqslant \sum_{k=0}^{n} 2^k a_{2^k} \leqslant \sum_{k=0}^{\infty} 2^k a_{2^k}.$$

Hence for any $m \in I$ we have

$$\sum_{k=1}^{m} a_k \leqslant \sum_{k=0}^{\infty} 2^k a_{2^k}$$

(why?). Since by hypothesis $\sum_{k=1}^{\infty} 2^k a_{2^k} < \infty$, the theorem follows from 3.2A.

The converse of 3.7A is also true.

3.7B. THEOREM. If $\{a_n\}_{n=1}^{\infty}$ is a nonincreasing sequence of positive numbers and if $\sum_{n=0}^{\infty} 2^n a_{2^n}$ diverges, then $\sum_{n=1}^{\infty} a_n$ diverges.

PROOF: We have

$$a_3 + a_4 \geqslant 2a_4,$$
$$a_5 + a_6 + a_7 + a_8 \geqslant 4a_8,$$

and in general

$$a_{2^n+1} + \cdots + a_{2^{n+1}} \geqslant 2^n a_{2^{n+1}} = \tfrac{1}{2}(2^{n+1} a_{2^{n+1}}),$$

so that

$$\sum_{k=3}^{2^{n+1}} a_k \geqslant \frac{1}{2} \sum_{k=1}^{n} 2^{k+1} a_{2^{k+1}} = \frac{1}{2} \sum_{k=2}^{n+1} 2^k a_{2^k}.$$

The remainder of the proof is left as an exercise. Note the similarity to the proof of 3.2C.

3.7C. COROLLARY. The series $\sum_{n=1}^{\infty} 1/n^2$ converges.

PROOF: For $a_n = 1/n^2$ we have

$$\sum_{n=1}^{\infty} 2^n a_{2^n} = \sum_{n=1}^{\infty} 2^n \cdot \frac{1}{(2^n)^2} = \sum_{n=1}^{\infty} \left(\tfrac{1}{2}\right)^n < \infty.$$

Hence by 3.7A,

$$\sum_{n=1}^{\infty} a_n = \sum_{n=1}^{\infty} \frac{1}{n^2} < \infty.$$

Note that for $\sum_{n=1}^{\infty} 1/n$ we have $a_n = 1/n$, and so $\sum_{n=1}^{\infty} 2^n a_{2^n} = \sum_{n=1}^{\infty} 2^n \cdot (1/2^n) = \infty$. Thus the divergence of $\sum_{n=1}^{\infty} 1/n$ follows from 3.7B.

The series $\sum_{n=4}^{\infty} 1/(n \log n)$ diverges. For here $a_n = 1/(n \log n)$ and so

$$\sum_{n=2}^{\infty} 2^n a_{2^n} = \sum_{n=2}^{\infty} 2^n \cdot \frac{1}{2^n \log 2^n} = \sum_{n=2}^{\infty} \left(\frac{1}{\log 2}\right) \cdot \frac{1}{n}$$

which diverges. By 3.7B, $\sum_{n=4}^{\infty} 1/(n \log n) = \infty$.

The series $\sum_{n=4}^{\infty} 1/[n(\log n)^2]$ converges. For

$$\sum_{n=2}^{\infty} 2^n a_2 n = \frac{1}{(\log 2)^2} \sum_{n=2}^{\infty} \frac{1}{n^2} < \infty.$$

If the terms of a convergent series form a nonincreasing sequence, they must approach zero "faster" than $1/n$. This result, called Pringsheim's theorem (although it was originally discovered by Abel), we now present.

3.7D. THEOREM. If $\{a_n\}_{n=1}^{\infty}$ is a nonincreasing sequence of positive numbers and if $\sum_{n=1}^{\infty} a_n$ converges, then $\lim_{n\to\infty} n a_n = 0$.

PROOF: Let $s_n = a_1 + \cdots + a_n$. If $\sum_{n=1}^{\infty} a_n = A$, then

$$\lim_{n\to\infty} s_n = A = \lim_{n\to\infty} s_{2n}.$$

Thus $\lim_{n\to\infty}(s_{2n} - s_n) = 0$. Now

$$s_{2n} - s_n = a_{n+1} + a_{n+2} + \cdots + a_{2n} \geqslant a_{2n} + a_{2n} + \cdots + a_{2n},$$

and so $0 \leqslant na_{2n} \leqslant s_{2n} - s_n$. Thus $\lim_{n\to\infty} na_{2n} = 0$ and so

$$\lim_{n\to\infty} 2na_{2n} = 0. \tag{1}$$

But $a_{2n+1} \leqslant a_{2n}$. Thus

$$(2n+1)a_{2n+1} \leqslant \left(\frac{2n+1}{2n}\right)(2na_{2n}).$$

By (1)

$$\lim_{n\to\infty} (2n+1)a_{2n+1} = 0. \tag{2}$$

The conclusion of the theorem follows from (1) and (2).

Theorem 3.7D is no longer true if we drop the hypothesis that $\{a_n\}_{n=1}^{\infty}$ is nonincreasing; consider the series $\sum_{n=1}^{\infty} a_n$ where

$$a_n = \frac{1}{n} \quad (n = 1, 4, 9, 16, \ldots),$$

$$a_n = \frac{1}{n^2} \quad \text{if } n \text{ is not a perfect square.}$$

Then

$$\sum_{n=1}^{\infty} a_n = \frac{1}{1} + \frac{1}{2^2} + \frac{1}{3^2} + \frac{1}{4} + \frac{1}{5^2} + \cdots + \frac{1}{8^2} + \frac{1}{9} + \cdots.$$

The partial sums of $\sum_{n=1}^{\infty} a_n$ are thus bounded above by

$$\left(\frac{1}{2^2} + \frac{1}{3^2} + \frac{1}{5^2} + \cdots\right) + \left(1 + \frac{1}{4} + \frac{1}{9} + \cdots\right)$$

which, in turn, is less than $2\sum_{n=1}^{\infty} 1/n^2$. Thus $\sum_{n=1}^{\infty} a_n$ converges. But na_n does not approach zero as $n \to \infty$, since $na_n = 1$ whenever n is a perfect square.

Note also that the converse of 3.7D is not true. That is, there exists a nonincreasing sequence of positive numbers $\{a_n\}_{n=1}^{\infty}$ such that $\lim_{n\to\infty} na_n = 0$ but such that $\sum_{n=1}^{\infty} a_n$ diverges. Indeed, let $a_1 = 2$ and let $a_n = 1/(n \log n)$ for $n \geqslant 2$.

Exercises 3.7

1. For what values of x does $\sum_{n=1}^{\infty} 1/n^x$ converge?
2. For what values of x does $\sum_{n=2}^{\infty} 1/[n(\log n)^x]$ converge?
3. Prove that for any real x the series $\sum_{n=3}^{\infty} 1/(\log n)^x$ diverges.
4. (a) If the terms of the convergent series $\sum_{n=1}^{\infty} a_n$ are positive and form a nonincreasing sequence, use 3.7B to prove that $\lim_{n\to\infty} 2^n a_{2^n} = 0$.
   (b) Deduce another proof of 3.7D.
5. Use 3.7D to give another proof of 3.2C.

3.8 SUMMATION BY PARTS

3.8A. THEOREM. Let $\{a_n\}_{n=1}^{\infty}$ and $\{b_n\}_{n=1}^{\infty}$ be two sequences of real numbers and let $s_n = a_1 + \cdots + a_n$. Then, for each $n \in I$,

$$\sum_{k=1}^{n} a_k b_k = s_n b_{n+1} - \sum_{k=1}^{n} s_k (b_{k+1} - b_k). \tag{1}$$

PROOF: Define $s_0=0$. Since $a_k=s_k-s_{k-1}$ we have

$$\begin{aligned}\sum_{k=1}^{n} a_k b_k &= \sum_{k=1}^{n} b_k(s_k-s_{k-1}) = b_1(s_1-s_0)+b_2(s_2-s_1)+\cdots\\ &\quad + b_{n-1}(s_{n-1}-s_{n-2})+b_n(s_n-s_{n-1})\\ &= s_1(b_1-b_2)+s_2(b_2-b_3)+\cdots+s_{n-1}(b_{n-1}-b_n)+s_nb_n\\ &= -\sum_{k=1}^{n-1} s_k(b_{k+1}-b_k)+s_nb_n-s_nb_{n+1}+s_nb_{n+1}\\ &= -\sum_{k=1}^{n} s_k(b_{k+1}-b_k)+s_nb_{n+1},\end{aligned}$$

which proves (1).

If we introduce the notation $\Delta a_k=a_{k+1}-a_k$ for any sequence $\{a_k\}_{k=1}^{\infty}$, then

$$\Delta a_k=\frac{a_{k+1}-a_k}{(k+1)-k}$$

resembles a "derivative of a_k with respect to k." The formula (1) becomes

$$\sum_{k=1}^{n} b_k\Delta s_{k-1}=s_nb_{n+1}-\sum_{k=1}^{n} s_k\Delta b_k \tag{2}$$

which resembles the formula

$$\int_c^d b\,ds = sb\Big|_c^d - \int_c^d s\,db$$

of integration by parts. The formula (1) is thus sometimes called summation by parts.

A consequence of 3.8A is the following result called Abel's lemma. It, in turn, yields a new test for convergence and, in a later section, a theorem on summation of series.

3.8B. ABEL'S LEMMA. If $\{a_n\}_{n=1}^{\infty}$ is a sequence of real numbers whose partial sums $s_n=\sum_{k=1}^{n}a_k$ satisfy

$$m \leqslant s_n \leqslant M \qquad (n \in I)$$

for some $m, M \in R$, and if $\{b_n\}_{n=1}^{\infty}$ is a nonincreasing sequence of nonnegative numbers, then

$$mb_1 \leqslant \sum_{k=1}^{n} a_kb_k \leqslant Mb_1 \qquad (n\in I). \tag{1}$$

PROOF: From (1) of 3.8A we have

$$\sum_{k=1}^{n} a_kb_k=\sum_{k=1}^{n} s_k(b_k-b_{k+1})+s_nb_{n+1}.$$

Since, by hypothesis, $b_k-b_{k+1}\geqslant 0$ and $s_k\leqslant M$, this implies

$$\begin{aligned}\sum_{k=1}^{n} a_kb_k &\leqslant M\sum_{k=1}^{n}(b_k-b_{k+1})+Mb_{n+1}\\ &= M[(b_1-b_2)+(b_2-b_3)+\cdots+(b_n-b_{n+1})+b_{n+1}]=Mb_1.\end{aligned}$$

This proves the right-hand inequality in (1). The left-hand inequality may be proved similarly.

Thus in 3.8B, if $|\Sigma_{k=1}^{n} a_k| \leqslant M$ for all n, then $|\Sigma_{k=1}^{n} a_k b_k| \leqslant M b_1$.

The following is called Dirichlet's test.

3.8C. THEOREM. Let $\{a_n\}_{n=1}^{\infty}$ be a sequence of real numbers whose partial sums $s_n = \Sigma_{k=1}^{n} a_k$ form a bounded sequence, and let $\{b_n\}_{n=1}^{\infty}$ be a nonincreasing sequence of nonnegative numbers which converges to 0. Then $\Sigma_{k=1}^{\infty} a_k b_k$ converges.

PROOF: It is sufficient to prove that the partial sums of $\Sigma_{k=1}^{\infty} a_k b_k$ form a Cauchy sequence. That is, given $\epsilon > 0$ we must find $N \in I$ such that

$$\left| \sum_{k=m}^{n} a_k b_k \right| < \epsilon \qquad (n \geqslant m \geqslant N).$$

Now, by hypothesis, there exists $M > 0$ such that $|s_n| \leqslant M$ $(n \in I)$. Hence for any $m, n \in I$,

$$\left| \sum_{k=m}^{n} a_k \right| = |s_n - s_{m-1}| \leqslant |s_n| + |s_{m-1}|,$$

and so

$$\left| \sum_{k=m}^{n} a_k \right| \leqslant 2M \qquad (m, n \in I; m \leqslant n).$$

By 3.8B (applied to $\{a_k\}_{k=m}^{\infty}$ and $\{b_k\}_{k=m}^{\infty}$) this implies

$$\left| \sum_{k=m}^{n} a_k b_k \right| \leqslant 2M b_m \qquad (m, n \in I; m \leqslant n).$$

But, by hypothesis, there exists $N \in I$ such that $b_n < \epsilon / 2M$ $(n \geqslant N)$,. Hence $2Mb_m < \epsilon$ $(m \geqslant N)$ so that

$$\left| \sum_{k=m}^{n} a_k b_k \right| < \epsilon \qquad (n \geqslant m \geqslant N),$$

which is what we wished to show.

3.8D. For example, from the identity

$$2 \sin \frac{x}{2} (\sin x + \sin 2x + \cdots + \sin nx) = \cos \frac{x}{2} - \cos \frac{2n+1}{2} x,$$

we see that if $\sin x/2 \neq 0$, then the partial sums $s_n = \Sigma_{k=1}^{n} \sin kx$ satsify $|s_n| \leqslant 1/|\sin x/2|$. Thus the series

$$\sum_{n=1}^{\infty} \frac{\sin nx}{n} \quad \text{and} \quad \sum_{n=3}^{\infty} \frac{\sin nx}{\log n}$$

both converge for all real x. (For $\{1/n\}_{n=1}^{\infty}$ and $\{1/\log n\}_{n=3}^{\infty}$ are nonincreasing and converge to zero.)

A somewhat different test, called Abel's test, can be obtained by assuming more about $\Sigma_{n=1}^{\infty} a_n$ but less about $\{b_n\}_{n=1}^{\infty}$.

3.8E. THEOREM. If $\Sigma_{n=1}^{\infty} a_n$ is a convergent series of real numbers and if the sequence $\{b_n\}_{n=1}^{\infty}$ is monotone and convergent, then $\Sigma_{n=1}^{\infty} a_n b_n$ also converges.

PROOF: Let us suppose that $\{b_n\}_{n=1}^{\infty}$ is nondecreasing and let $c_n = b - b_n$ where $b = \lim_{n \to \infty} b_n$. (If $\{b_n\}_{n=1}^{\infty}$ were nonincreasing, we would let $c_n = b_n - b$.) Then $c_n \geqslant 0$,

$\lim_{n\to\infty} c_n = 0$, and $\{c_n\}_{n=1}^{\infty}$ is nonincreasing. By 3.8C, then the series $\sum_{n=1}^{\infty} a_n c_n$ converges. By 3.1C, the series $\sum_{n=1}^{\infty} b a_n$ converges. But $a_n b_n = b a_n - a_n c_n$. Thus $\sum_{n=1}^{\infty} a_n b_n = \sum_{n=1}^{\infty} (b a_n - a_n c_n)$ converges (again by 3.1C), which is what we wished to show.

For example, the series

$$1 - 1 + \tfrac{1}{2} - \tfrac{1}{2} + \tfrac{1}{3} - \tfrac{1}{3} + \cdots$$

converges, whereas the sequence $0, \frac{1}{2}, \frac{1}{2}, \frac{2}{3}, \frac{2}{3}, \frac{3}{4}, \frac{3}{4}, \ldots$ is monotonic and convergent. Therefore, by 3.8E, the series

$$0 - \frac{1}{2} + \frac{1}{2^2} - \frac{1}{3} + \frac{2}{3^2} - \frac{1}{4} + \frac{3}{4^2} - \cdots$$

must converge. Note that theorem 3.3A does not apply to this last series. (Why?)

Exercises 3.8

1. Prove that if $\sum_{n=1}^{\infty} n a_n$ converges, then so does $\sum_{n=1}^{\infty} a_n$.
2. Given that

$$\cos x + \cos 3x + \cos 5x + \cdots + \cos(2n-1)x = \frac{\sin 2nx}{2 \sin x}$$

if $\sin x \neq 0$, prove that $\sum_{n=1}^{\infty} \cos(2n-1)x/(2n-1)$ converges if x is not a multiple of π.
3. Prove that $\sum_{n=1}^{\infty} (1/n)\log(1+1/n)$ converges. (*Hint:* Use (2) of 3.8A with $b_k = \log k$.)
4. Show that the series $1 - \frac{1}{2} - \frac{1}{3} + \frac{1}{4} + \frac{1}{5} - \frac{1}{6} - \frac{1}{7} + + - - \cdots$ is convergent.
5. Deduce theorem 3.3A as a special case of Dirichlet's test (3.8C).

3.9 $(C,1)$ SUMMABILITY OF SERIES

Just as the convergence of the series $\sum_{n=1}^{\infty} a_n$ is defined to mean the convergence of the sequence $\{s_n\}_{n=1}^{\infty}$ of partial sums, the $(C,1)$ summability of $\sum_{n=1}^{\infty} a_n$ will now be defined to mean the $(C,1)$ summability of $\{s_n\}_{n=1}^{\infty}$. (See 2.11A.)

3.9A. DEFINITION. Let $\sum_{n=1}^{\infty} a_n$ be a series of real numbers with partial sums $s_n = a_1 + \cdots + a_n$. We shall say that $\sum_{n=1}^{\infty} a_n$ is $(C,1)$ summable to A if

$$\lim_{n\to\infty} s_n = A \qquad (C,1).$$

In this case we write

$$\sum_{n=1}^{\infty} a_n = A \qquad (C,1).$$

Thus $\sum_{n=1}^{\infty} (-1)^n = -\frac{1}{2}(C,1)$. For the sequence of partial sums for this series is $-1, 0, -1, 0, -1, 0, \ldots$ and this *sequence* is $(C,1)$ summable to $-\frac{1}{2}$.

The series $1 - 2 + 3 - 4 + 5 - 6 + \cdots$ has the sequence of partial sums $1, -1, 2, -2, 3, -3, \ldots$. As we saw after theorem 2.11B, this sequence is not $(C,1)$ summable. Hence the series $1 - 2 + 3 - 4 + 5 - 6 + \cdots$ is not $(C,1)$ summable.

3.9B. DEFINITION. A method T of summability for series is called regular if every convergent series is T summable to its sum. (That is, if $\sum_{n=1}^{\infty} a_n$ is T summable to A whenever $\sum_{n=1}^{\infty} a_n$ converges to A.)

3.9C. THEOREM. The $(C,1)$ summability method for series is regular.

PROOF: If $\Sigma_{n=1}^{\infty}a_n$ converges to A, then the sequence of partial sums $\{s_n\}_{n=1}^{\infty}$ converges to A. Hence, by 2.11B, $\lim_{n\to\infty}s_n = A(C,1)$. This in turn implies $\Sigma_{n=1}^{\infty}a_n = A(C,1)$, which proves the theorem.

3.9D. We know that a series may be $(C,1)$ summable even if it does not converge. We now prove a theorem giving a simple condition on a series which, together with $(C,1)$ summability of the series, will ensure that the series does converge. We first need a lemma.

LEMMA. Let $\Sigma_{n=1}^{\infty}a_n$ be $(C,1)$ summable and let $t_n = a_1 + 2a_2 + \cdots + na_n$. Then if

$$\lim_{n\to\infty}\frac{t_n}{n} = 0, \tag{1}$$

the series $\Sigma_{n=1}^{\infty}a_n$ converges.

PROOF: We first show that

$$t_n = (n+1)s_n - n\sigma_n \qquad (n \in I) \tag{2}$$

where $\sigma_n = n^{-1}(s_1 + s_2 + \cdots + s_n)$. We have $t_1 = a_1 = s_1 = \sigma_1$ so that (2) is certainly true for $n = 1$. We proceed by induction. Suppose (2) is true for some value of n. We then have

$$\begin{aligned} t_{n+1} &= t_n + (n+1)a_{n+1} = (n+1)s_n - n\sigma_n + (n+1)a_{n+1} \\ &= (n+1)(s_n + a_{n+1}) - n\sigma_n = (n+1)(s_{n+1}) - (s_1 + \cdots + s_n) \\ &= (n+2)(s_{n+1}) - (s_1 + \cdots + s_n + s_{n+1}), \end{aligned}$$

and so

$$t_{n+1} = (n+2)s_{n+1} - (n+1)\sigma_{n+1}.$$

Thus (2) is true for $n+1$ which completes the induction.

Now suppose that $\Sigma_{n=1}^{\infty}a_n = A(C,1)$. Then $\lim_{n\to\infty}\sigma_n = A$. From (2) and (1) we then have

$$s_n = \frac{t_n}{n+1} + \frac{n}{n+1}\sigma_n = \frac{n}{n+1}\left(\frac{t_n}{n} + \sigma_n\right),$$

$$\lim_{n\to\infty} s_n = \lim_{n\to\infty}\frac{n}{n+1}\left(\frac{t_n}{n} + \sigma_n\right) = 1\cdot(0 + A) = A.$$

In particular, $\{s_n\}_{n=1}^{\infty}$ converges, which proves the lemma.

THEOREM. If $\Sigma_{n=1}^{\infty}a_n$ is $(C,1)$ summable and if $\lim_{n\to\infty}na_n = 0$, then $\Sigma_{n=1}^{\infty}a_n$ converges.

PROOF: The sequence $\{na_n\}_{n=1}^{\infty}$ converges to 0 by hypothesis. By 2.11B, $\{na_n\}_{n=1}^{\infty}$ is $(C,1)$ summable to 0. That is,

$$\lim_{n\to\infty}\frac{a_1 + 2a_2 + \cdots + na_n}{n} = 0.$$

But this says precisely that $\lim_{n\to\infty}t_n/n = 0$ where t_n is as in the lemma. The hypotheses of the lemma thus hold and the theorem follows.

3.9E. The only method of summability for series that we have discussed in this section is the $(C,1)$ method. However, from our development it should be clear that any regular method of summability for sequences defines a corresponding regular method for series.

Thus we would say $\sum_{n=1}^{\infty} a_n = A(C,2)$ if the sequence $\{s_n\}_{n=1}^{\infty}$ of partial sums of $\sum_{n=1}^{\infty} a_n$ is $(C,2)$ summable to A. From 2.11D it then follows, for example, that

$$1-2+3-4+5-6+\cdots=\tfrac{1}{4} \qquad (C,2).$$

(For the sequence of partial sums is, in this case, $1, -1, 2, -2, 3, -3, \ldots$.) This $(C,2)$ method for series is regular since $(C,2)$ summability for sequences is regular.

In a later section we consider methods of summability that do not deal with the sequence of partial sums. See Section 9.6.

Exercises 3.9

1. Examine whether the following series are $(C,1)$ summable.
   (a) $1-3+1-3+1-3+\cdots$.
   (b) $1-\frac{1}{2}+\frac{1}{3}-\frac{1}{4}+\cdots$.
   (c) $1+0-1-0+1+0-1-0+\cdots$.
2. Does the series (c) of the preceding exercise have the same $(C,1)$ sum as
$$1-1+1-1+1-1+\cdots?$$
3. Prove that
$$1+1-1+1+1-1+1+1-1+\cdots$$
is not $(C,1)$ summable.
4. Show that a divergent series of positive terms cannot be $(C,1)$ summable.
5. Prove that if $\sum_{n=1}^{\infty} a_n$ is $(C,1)$ summable, then the sequence $\{a_n\}_{n=1}^{\infty}$ is $(C,1)$ summable to 0. (Compare this result with 3.1D.)
6. Prove that if $\sum_{n=1}^{\infty} a_n$ is $(C,1)$ summable, then $\lim_{n\to\infty} s_n/n=0$ where $s_n=a_1+a_2+\cdots+a_n$.

3.10 THE CLASS ℓ^2

Most of the important concepts we have so far introduced (set, sequence, function, series) are involved in the definition of the class ℓ^2.

3.10A. DEFINITION. The class ℓ^2 is the class of all sequences $s=\{s_n\}_{n=1}^{\infty}$ such that $\sum_{n=1}^{\infty} s_n^2<\infty$.

The elements of ℓ^2 are thus sequences. The sequence $0,0,0,\ldots$ is clearly an element of ℓ^2. By 3.7C, the sequence $\{1/n\}_{n=1}^{\infty}$ is an element of ℓ^2. By 3.2C, the sequence $\{1/\sqrt{n}\}_{n=1}^{\infty}$ is not an element of ℓ^2. We shortly show that if $s\in\ell^2$ and $t\in\ell^2$, then $s+t\in\ell^2$.

Here are two famous inequalities.

3.10B. THEOREM. (THE SCHWARZ* INEQUALITY). If $s=\{s_n\}_{n=1}^{\infty}$ and $t=\{t_n\}_{n=1}^{\infty}$ are in ℓ^2, then $\sum_{n=1}^{\infty} s_n t_n$ is absolutely convergent and

$$\left|\sum_{n=1}^{\infty} s_n t_n\right| \leqslant \left(\sum_{n=1}^{\infty} s_n^2\right)^{1/2}\left(\sum_{n=1}^{\infty} t_n^2\right)^{1/2}. \tag{1}$$

*Often called the Cauchy or the Buniakovski inequality.

PROOF: We may assume that there is at least one s_n—say s_N—not equal to 0. Otherwise the theorem is trivial. For fixed $n \geqslant N$ and any $x \in R$ we have

$$\sum_{k=1}^{n} (xs_k + t_k)^2 \geqslant 0.$$

Expanding the parentheses on the left we have

$$x^2 \sum_{k=1}^{n} s_k^2 + 2x \sum_{k=1}^{n} s_k t_k + \sum_{k=1}^{n} t_k^2 \geqslant 0.$$

This can be written $Ax^2 + Bx + C \geqslant 0$ where

$$A = \sum_{k=1}^{n} s_k^2 > 0, \qquad B = 2 \sum_{k=1}^{n} s_k t_k, \qquad C = \sum_{k=1}^{n} t_k^2.$$

[From calculus we know that the minimum value of $Ax^2 + Bx + C$ $(A > 0)$ occurs when $x = -B/2A$ (verify!). This is motivation for the next step in the proof.]

If we set $x = -B/2A$, we have $A(-B/2A)^2 + B(-B/2A) + C \geqslant 0$, or $B^2 \leqslant 4AC$. But this says

$$\left(\sum_{k=1}^{n} s_k t_k \right)^2 \leqslant \left(\sum_{k=1}^{n} s_k^2 \right) \cdot \left(\sum_{k=1}^{n} t_k^2 \right). \tag{2}$$

If we replace s_k, t_k by $|s_k|$, $|t_k|$ in (2), we obtain

$$\sum_{k=1}^{n} |s_k t_k| \leqslant \left(\sum_{k=1}^{n} s_k^2 \right)^{1/2} \cdot \left(\sum_{k=1}^{n} t_k^2 \right)^{1/2} \leqslant \left(\sum_{k=1}^{\infty} s_k^2 \right)^{1/2} \cdot \left(\sum_{k=1}^{\infty} t_k^2 \right)^{1/2}.$$

The sequence of partial sums of $\sum_{k=1}^{\infty} |s_k t_k|$ is thus bounded, and so $\sum_{k=1}^{\infty} |s_k t_k| < \infty$. In particular, $\sum_{k=1}^{\infty} s_k t_k$ converges by 3.4B. If we now let n approach infinity in (2) and use 2.7E, we obtain (1).

3.10C. THEOREM. (THE MINKOWSKI INEQUALITY.) If $s = \{s_n\}_{n=1}^{\infty}$ and $t = \{t_n\}_{n=1}^{\infty}$ are in ℓ^2, then $s + t = \{s_n + t_n\}_{n=1}^{\infty}$ is in ℓ^2 and

$$\left[\sum_{n=1}^{\infty} (s_n + t_n)^2 \right]^{1/2} \leqslant \left[\sum_{n=1}^{\infty} s_n^2 \right]^{1/2} + \left[\sum_{n=1}^{\infty} t_n^2 \right]^{1/2}.$$

PROOF: By hypothesis, the series $\sum_{n=1}^{\infty} s_n^2$ and $\sum_{n=1}^{\infty} t_n^2$ converge. Also, the series $\sum_{n=1}^{\infty} s_n t_n$ converges by 3.10B. Since $(s_n + t_n)^2 = s_n^2 + 2s_n t_n + t_n^2$, it follows from 3.1C that $\sum_{n=1}^{\infty} (s_n + t_n)^2$ converges and

$$\sum_{n=1}^{\infty} (s_n + t_n)^2 = \sum_{n=1}^{\infty} s_n^2 + 2 \sum_{n=1}^{\infty} s_n t_n + \sum_{n=1}^{\infty} t_n^2.$$

Applying the Schwarz inequality to the second sum on the right, we obtain

$$\sum_{n=1}^{\infty} (s_n + t_n)^2 \leqslant \sum_{n=1}^{\infty} s_n^2 + 2 \left(\sum_{n=1}^{\infty} s_n^2 \right)^{1/2} \left(\sum_{n=1}^{\infty} t_n^2 \right)^{1/2} + \sum_{n=1}^{\infty} t_n^2,$$

and so

$$\sum_{n=1}^{\infty} (s_n + t_n)^2 \leqslant \left[\left(\sum_{n=1}^{\infty} s_n^2 \right)^{1/2} + \left(\sum_{n=1}^{\infty} t_n^2 \right)^{1/2} \right]^2.$$

Taking the square root on both sides completes the proof.

The class ℓ^2 is used as an example of a metric space in the next chapter. For this purpose it is useful to introduce the notion of the norm of an element in ℓ^2.

3.10D. DEFINITION. If $s=\{s_n\}_{n=1}^{\infty}$ is an element of ℓ^2, we define $\|s\|_2$, called the norm of s, as

$$\|s\|_2=\left(\sum_{n=1}^{\infty} s_n^2\right)^{1/2}.$$

The norm of a sequence of ℓ^2 is thus a nonnegative number. (Actually then, the norm is a function with domain ℓ^2 and range $[0,\infty)$.) The sequence $0,0,0,\ldots$ has norm 0. Any other sequence in ℓ^2 has positive norm. If $c\in R$ and $s\in\ell^2$, then $\|cs\|_2=|c|\cdot\|s\|_2$ (verify).

3.10E. THEOREM. The norm for sequences in ℓ^2 has the following properties:

$$\|s\|_2\geqslant 0 \qquad (s\in\ell^2), \tag{1}$$

$$\|s\|_2=0 \quad \text{if and only if} \quad s=\{0\}_{n=1}^{\infty}, \tag{2}$$

$$\|cs\|_2=|c|\cdot\|s\|_2 \qquad (c\in R, s\in\ell^2), \tag{3}$$

$$\|s+t\|_2\leqslant\|s\|_2+\|t\|_2 \qquad (s,t\in\ell^2). \tag{4}$$

PROOF: Only (4) has not been verified. But (4) is just a restatement of the Minkowski inequality.

3.10F. Do not read this section unless you know a little bit about vector spaces.

First of all, since $s+t\in\ell^2$ if $s\in\ell^2$ and $t\in\ell^2$, and since $cs\in\ell^2$ if $c\in R$, $s\in\ell^2$, it is clear that ℓ^2 is a vector space over the real numbers.

In n-dimensional Euclidean vector space, a vector $\langle s_1,\ldots,s_n\rangle$ has length $(\sum_{k=1}^{n}s_k^2)^{1/2}$. The dot (scalar) product of two vectors $\langle s_1,\ldots,s_n\rangle$ and $\langle t_1,\ldots,t_n\rangle$ is equal to $\sum_{k=1}^{n}s_kt_k$ and is in absolute value less than the product of the lengths of the two vectors. That is,

$$\left|\sum_{k=1}^{n} s_kt_k\right|\leqslant\left(\sum_{k=1}^{n} s_k^2\right)^{1/2}\cdot\left(\sum_{k=1}^{n} t_k^2\right)^{1/2}. \tag{*}$$

Thus the norm $(\sum_{k=1}^{\infty}s_k^2)^{1/2}$ of a sequence in ℓ^2 is analogous to the length $(\sum_{k=1}^{n}s_k^2)^{1/2}$ of a vector in n-space. The Schwarz inequality corresponds to (*). The Minkowski inequality states that the "length" (norm) of the sum of two vectors (sequences) in ℓ^2 is less than or equal to the sum of their lengths. In n-space the corresponding fact is that a straight line joining two points is no longer than any broken line joining them.

Exercises 3.10

1. Which of the following sequences are in ℓ^2?

(a) $\left\{\dfrac{1}{\log n}\right\}_{n=2}^{\infty}$.

(b) $\left\{\dfrac{1}{e^n}\right\}_{n=1}^{\infty}$.

(c) $\left\{\sin\dfrac{n\pi}{10}\right\}_{n=1}^{\infty}$.

2. Give an example of a sequence $\{s_n\}_{n=1}^{\infty}$ in ℓ^2 such that $\sum_{n=1}^{\infty}|s_n|=\infty$.

3. Prove that if $\{s_n\}_{n=1}^{\infty} \in \ell^2$, then $\lim_{n\to\infty} s_n = 0$.
4. Show that equality can hold in (1) of 3.10B if and only if $\{t_n\}_{n=1}^{\infty}$ is "proportional" to $\{s_n\}_{n=1}^{\infty}$ – that is, if and only if

$$t_n = \lambda s_n \qquad (n \in I)$$

for some $\lambda \in R$.
5. If $\{a_n\}_{n=1}^{\infty} \in \ell^2$, prove that $\sum_{n=1}^{\infty} a_n/n$ converges absolutely.
6. Prove that if $\sum_{n=1}^{\infty} |a_n| < \infty$, then $\{a_n\}_{n=1}^{\infty}$ is in ℓ^2.
7. For each $k \in I$ let e_k be the sequence $0, 0, \ldots, 0, 1, 0, 0, \ldots$, all of whose terms are 0 except the kth term, which is equal to 1. Show that $\|e_k - e_j\|_2 = \sqrt{2}$ if $k \neq j$. What is $\|e_k\|_2$?
(In n-space can you find infinitely many vectors of length 1 such that any two of the vectors are distance $\sqrt{2}$ from each other? Answer: No. In n-space how many such vectors can you find? Answer: n.)
8. If $a_n \geqslant 0 (n \in I)$ and if $\sum_{n=1}^{\infty} a_n$ converges, prove that

$$\sum_{n=1}^{\infty} \frac{\sqrt{a_n}}{n}$$

converges.

3.11. REAL NUMBERS AND DECIMAL EXPANSIONS

The theory of infinite series enables us to treat more carefully the connection between real numbers and "decimal expansions" first mentioned in Section 1.7.

We consider only expansions for numbers in $[0, 1]$. Here is the definition of decimal expansion.

3.11A. DEFINITION. A decimal expansion is an infinite series $\sum_{n=1}^{\infty} a_n/10^n$ where each a_n is an integer, $0 \leqslant a_n \leqslant 9$.

It is, of course, customary to write the decimal expansion $\sum_{n=1}^{\infty} a_n/10^n$ as $.a_1a_2a_3\cdots$. The number a_n is called the nth digit of the decimal expansion. We now show that every decimal expansion "represents" a number in $[0, 1]$.

3.11B THEOREM. Every decimal expansion converges to a number in $[0, 1]$.

PROOF: The decimal expansion $\sum_{n=1}^{\infty} a_n/10^n$ is clearly dominated by the series $\sum_{n=1}^{\infty} 9/10^n$. But $\sum_{n=1}^{\infty} 9/10^n$ is (absolutely) convergent and thus, by 3.6B, so is $\sum_{n=1}^{\infty} a_n/10^n$. In fact,

$$\sum_{n=1}^{\infty} \frac{9}{10^n} = 9 \sum_{n=1}^{\infty} \left(\frac{1}{10}\right)^n = 9\left[\sum_{n=0}^{\infty} \left(\frac{1}{10}\right)^n - 1\right] = 9\left[\frac{1}{1-\frac{1}{10}} - 1\right] = 1.$$

Thus $0 \leqslant \sum_{n=1}^{\infty} a_n/10^n \leqslant \sum_{n=1}^{\infty} 9/10^n \leqslant 1$ and the proof is complete.

3.11C. DEFINITION. If the decimal expansion $\sum_{n=1}^{\infty} a_n/10^n = .a_1a_2a_3\cdots$ converges to $x \in [0, 1]$, we say that $.a_1a_2a_3\cdots$ is a decimal expansion for x (or that x has the decimal expansion $.a_1a_2a_3\cdots$).

Thus 3.11B tells us that every decimal expansion is the decimal expansion for some $x \in [0,1]$. We next show that each $x \in [0,1]$ has at least one decimal expansion.

3.11D. THEOREM. If $x \in [0,1]$, then there is a decimal expansion converging to x.

PROOF: Suppose first that x can be written as $x = k/10^n$ for some $k, n \in I$ where $k < 10^n$. Then $k = a_1 \cdot 10^{n-1} + a_2 \cdot 10^{n-2} + \cdots + a_n$ for some integers a_i. Thus $x = a_1/10 + a_2/10^2 + \cdots + a_n/10^n$ so that $.a_1 a_2 \cdots a_n 000 \cdots$ converges to x.

If x cannot be written as $x = k/10^n$ for any $n \in I$, then x lies in one of the ten open intervals $(0, \frac{1}{10}), (\frac{1}{10}, \frac{2}{10}), \ldots, (\frac{9}{10}, \frac{10}{10})$. If $x \in (m/10, (m+1)/10)$, let $a_1 = m$. Then $x - .a_1 < \frac{1}{10}$. Now x lies in one of the ten open intervals

$$\left(\frac{a_1}{10}, \frac{a_1}{10} + \frac{1}{100}\right), \left(\frac{a_1}{10} + \frac{1}{100}, \frac{a_1}{10} + \frac{2}{100}\right), \ldots, \left(\frac{a_1}{10} + \frac{9}{100}, \frac{a_1}{10} + \frac{10}{100}\right).$$

If

$$x \in \left(\frac{a_1}{10} + \frac{p}{100}, \frac{a_1}{10} + \frac{p+1}{100}\right),$$

Let $a_2 = p$. Then $x - .a_1 a_2 < \frac{1}{100}$. Continuing in this fashion we can define a decimal expansion $.a_1 a_2 a_3 \cdots$ that clearly converges to x. This completes the proof.

Unfortunately, some numbers in $[0,1]$ have more than one decimal expansion. For example, $\frac{1}{2} = 0.5000 \cdots$ *and* $\frac{1}{2} = 0.4999 \cdots$. Indeed, since $1 = 0.999 \cdots$ we have $1/10^n = .000 \cdots 0999 \cdots$ (where n 0's precede the string of 9's). From this and the proof of 3.11B, the reader should have no difficulty in proving the following result.

3.11E. THEOREM. Every decimal expansion that ends in a string of 9's converges to a number of the form $k/10^n$. Conversely, every number in $(0,1]$ of the form $k/10^n$ has a decimal expansion ending in a string of 9's. Thus every number in $(0,1)$ of the form $k/10^n$ has (at least) two decimal expansions, one ending in a string of 9's and another in a string of 0's.

The next theorem enables us to prove that any number in $[0,1]$ not of the form $k/10^n$ has precisely *one* decimal expansion.

3.11F. THEOREM. If the two decimal expansions $.a_1 a_2 a_3 \cdots$ and $.b_1 b_2 b_3 \cdots$ are distinct (that is, if $a_k \neq b_k$ for some $k \in I$), and if neither expansion ends in a string of 9's, then they converge to different sums.

PROOF: Let n be the largest integer such that $a_k = b_k$ $(k \leqslant n)$. Then $a_1 = b_1$, $a_2 = b_2, \ldots, a_n = b_n$, but $a_{n+1} \neq b_{n+1}$. Let us assume, say, that $a_{n+1} > b_{n+1}$. Then $a_{n+1} \geqslant b_{n+1} + 1$ and so

$$.a_{n+1}000 \cdots \geqslant .b_{n+1}000 \cdots + 0.1000 \cdots = .b_{n+1}000 \cdots + 0.0999 \cdots$$
$$= .b_{n+1}999 \cdots > .b_{n+1}b_{n+2}b_{n+3} \cdots,$$

since $.b_1 b_2 \cdots$ does not end in a string of 9's. Certainly, then, $.a_{n+1}a_{n+2}a_{n+3} \cdots > .b_{n+1}b_{n+2}b_{n+3} \cdots$ and so $.a_1 a_2 a_3 \cdots > .b_1 b_2 b_3 \ldots$. This proves the theorem.

3.11G. COROLLARY. If $x \in (0,1)$ is not of the form $k/10^n$, then x has one and only one decimal expansion. If x is of the form $k/10^n$, then x has precisely two decimal expansions.

PROOF: The proof follows directly from 3.11D, 3.11E, and 3.11F. (Verify!)

Exercises 3.11

1. A *repeating* decimal expansion is one of the form
$$.a_1a_2\cdots a_nb_1b_2\cdots b_mb_1b_2\cdots b_mb_1b_2\cdots b_m\cdots.$$
Show that the decimal expansion of a *rational* number is repeating. (*Hint*: If $x=p/q$, calculate the decimal expansion for x by repeated short or long division and consider the number of possible remainders.)
2. Conversely, show that every repeating decimal expansion converges to a rational number.
3. Prove that between any two distinct real numbers there is a rational number.
4. Prove that the open interval $(0,1)$ is equivalent to the open square
$$\{\langle x,y\rangle | 0<x,y<1\}.$$

3.12 NOTES AND ADDITIONAL EXERCISES FOR CHAPTERS 1, 2, AND 3.

I. Some results from set theory.

3.12A. The following result is known as the Schröder-Bernstein theorem.

THEOREM: Let A and B be disjoint nonempty sets. Suppose that A is equivalent to a subset of B and that B is equivalent to a subset of A. Then A and B are equivalent.

PROOF: We begin the proof as follows: By hypothesis, there is a 1-1 function f from A onto a subset B_1 of B, and a 1-1 function g from B onto a subset A_1 of A.

Fix $x\in A$. If $x\in A-A_1$, then x is not the image under g of any point in B. In this case we say that x has no ancestor. However, if $x\in A_1$, then there is a (unique) $y\in B$ such that $x=g(y)$. In this case we say that y is an ancestor of x. Next, consider this y. If $y\in B_1$, then there is a unique $z\in A$ such that $f(z)=y$. We then say that z is an ancestor of y *and* an ancestor of x. On the other hand, if $y\in B-B_1$, then y has no ancestor and y is the only ancestor of x. Note that y, if it exists, is equal to $g^{-1}(x)$, and z, if it exists, is equal to $f^{-1}(y)=f^{-1}[g^{-1}(x)]$. Thus for any $x\in A$, the ancestors of x are the elements of those sets in the sequence

$$g^{-1}(x), f^{-1}\left[g^{-1}(x)\right], g^{-1}\left\{f^{-1}\left[g^{-1}(x)\right]\right\},\cdots \tag{*}$$

that are not empty (and hence consist of a single point of B or A). Of course, if any term of (*) is empty, then so are all the subsequent terms.

So any $x\in A$ has $0,1,2,\cdots$ or possibly infinitely many ancestors. (If the same ancestor occurs more than once in (*), it should be counted as many times as it occurs. Can the same ancestor occur exactly twice?) Let A_I be the set of all points in A which have infinitely many ancestors, and let A_O, A_E be respectively the sets of all points in A that have a (finite) odd or even number of ancestors. Let B_I, B_O, B_E be the corresponding subsets of B.

EXERCISE: Finish the proof.

3.12B. By a partially ordered set we mean a set A together with a relation $\leqslant$ between some but not necessarily all elements of A satisfying

$$x \leqslant x \qquad (x \in A), \tag{1}$$

$$x \leqslant y, y \leqslant x \quad \text{imply} \quad x = y \qquad (x,y \in A), \tag{2}$$

$$x \leqslant y, y \leqslant z \quad \text{imply} \quad x \leqslant z \qquad (x,y,z \in A). \tag{3}$$

More precisely, $\leqslant$ is a subset of $A \times A$ and hence is a set of ordered pairs $\langle x,y \rangle$ where $x,y \in A$. Instead of writing $\langle x,y \rangle \in \leqslant$ we write $x \leqslant y$.

For example, let X be any nonempty set and let A be the family of all subsets of X. If $x,y \in A$, define $x \leqslant y$ to mean $x \subset y$ (remember that x and y are subsets of X). Then the relation $x \leqslant y$ holds for some of the $x,y \in A$ and it is clear that the conditions (1)–(3) are satisfied. Hence A, the family of subsets of X, in "partially ordered by inclusion."

For another example let $A = I = \{1,2,\cdots\}$ and define $x \leqslant y$ to mean that x divides y (that is, that y is an integral multiple of x). Thus $4 \leqslant 24$, $5 \nleqslant 24$. Note that $5 \nleqslant 24$ does not imply $24 \leqslant 5$.

EXERCISE: Show that (1)–(3) are satisfied in this example.

3.12C. Suppose A is a set partially ordered by the relation $\leqslant$. If $B \subset A$, $x \in A$, we say that x is an upper bound for B if

$$b \leqslant x \qquad (b \in B).$$

Thus in the last example, if $B = \{3,6,10\}$, then 120 is an upper bound for B.

Next we define a chain in A as a subset B of A such that any two elements of B are comparable. That is, we say B is a chain if, whenever $x,y \in B$, either $x \leqslant y$ or $y \leqslant x$.

EXERCISE: Let X, Y be nonempty sets. Let A be the set of all 1-1 functions f from nonempty subsets of X into Y. If $f,g \in A$, define $f \leqslant g$ to mean that g is an extension of f.

1. Prove that this definition of $\leqslant$ makes A a partially ordered set.
2. If B is any chain in A, prove that B has an upper bound.

3.12D. Let A be a partially ordered set. If $x \in A$, we say that x is a maximal element of A if there is no element $y \in A$ distinct from x, such that $x \leqslant y$. A famous result of set theory is:

ZORN'S LEMMA. Let A be a nonempty partially ordered set. Suppose that every chain in A has an upper bound. Then A contains a maximal element.

EXERCISE: Use Zorn's lemma to prove the following:

THEOREM: Let X, Y be nonempty sets. Then either X is equivalent to a subset of Y, or Y is equivalent to a subset of X.

3.12E. Here is a description of Russell's famous paradox.

Let A be the set of all sets which are not elements of themselves. That is,

$$A = \{X \mid X \notin X\}.$$

We ask the question, "Is A an element of itself?"

Suppose $A \in A$. Then, since A consists only of sets that are not elements of themselves, we must have $A \notin A$—a contradiction. On the other hand, suppose $A \notin A$. But then, by definition of A, we must have $A \in A$—again a contradiction. So both statements $A \in A$ and $A \not\in A$ lead to contradictions! Something is obviously wrong.

Modern theories of sets attempt to avoid Russell's paradox (and others) by restricting in one way or another the use of the term "set." Only certain collections of objects may be called sets. In one theory it is permissible to talk about the *collection* A of all *sets* which are not elements of themselves. It just turns out that A is not a set. Since the elements of A *are* sets it is then clear that $A \not\in A$ and no contradiction arises. (That is, $A \not\in A$ does not imply $A \in A$.) In another theory it is not necessarily permissible even to *form* the collection of all sets with a certain property. However, it *is* permissible to form the set of all subsets of a given *set*, and (important with respect to Russell's paradox) to form a set by collecting all elements within a given set that have a specified property. Thus according to this theory, given a set Y we could define the set A where

$$A = \{X | X \in Y \quad \text{and} \quad X \not\in X\}.$$

EXERCISE: Show that assuming $A \in A$ leads to a contradiction. Deduce that $A \not\in A$. Prove that $A \not\in Y$. Note that no contradiction arises.

We wish to reassure the reader that although we have been glib about the use of the word "set" (see the first sentence of chapter 1), all examples called sets in this book are indeed sets even according to the more restrictive usage required by the various rigorous theories.

II. On divergent series.

3.12F. Abel wrote in 1828: "Divergent series are the invention of the devil, and it is shameful to base on them any demonstration whatsoever."

Indeed our modern attitudes toward and definitions of convergence and divergence were imposed on mathematics at about that time, largely by Abel and Cauchy. Mathematicians of earlier generations were well aware of the pitfalls involved in using divergent series but nevertheless used them under more or less restrictive conditions. Many of their results are correct when interpreted according to an appropriate summability method. Others of their results are correct as they stand because they involve series that are in fact convergent. Even these correct results were often derived by nonrigorous methods.

For the reader who can manipulate complex numbers we present a typical nonrigorous argument which will produce the formula

$$\frac{1}{1^2} + \frac{1}{3^2} + \frac{1}{5^2} + \frac{1}{7^2} + \cdots = \frac{\pi^2}{8}. \tag{1}$$

(This formula, incidentally, is not easy to derive rigorously. It is usually established either with Fourier series or with the theory of analytic functions of a complex variable.)

Let

$$s = 1 - z + z^2 - z^3 + \cdots$$

where z is a complex number. Then

$$zs = z - z^2 + z^3 - \cdots = 1 - (1 - z + z^2 - \cdots),$$

$$zs = 1 - s,$$

so that

$$s = \frac{1}{1+z}.$$

Hence

$$\frac{1}{1+z} = 1 - z + z^2 - z^3 + \cdots.$$

Set $z = e^{i\theta}$. Then

$$\frac{1}{1+e^{i\theta}} = 1 - e^{i\theta} + e^{2i\theta} - \cdots. \tag{2}$$

Using the formula $e^{i\alpha\theta} = \cos\alpha\theta + i\sin\alpha\theta$, we see that the real part of the right side of (2) is $1\text{-}\cos\theta + \cos 2\theta - \cdots$. The left side of (2) is equal to

$$\frac{1}{1+e^{i\theta}} \cdot \frac{1+e^{-i\theta}}{1+e^{-i\theta}} = \frac{1+\cos\theta - i\sin\theta}{2+2\cos\theta}$$

whose real part is $\frac{1}{2}$. Hence, from (2),

$$\frac{1}{2} = 1 - \cos\theta + \cos 2\theta - \cos 3\theta + \cdots.$$

Integrate term by term from $\theta = 0$ to $\theta = t$ to obtain

$$\frac{t}{2} = t - \sin t + \frac{\sin 2t}{2} - \frac{\sin 3t}{3} + \cdots$$

or

$$\frac{t}{2} = \sin t - \frac{\sin 2t}{2} \frac{\sin 3t}{3} - \cdots.$$

Integrate again from $t = 0$ to $t = x$. Then

$$\frac{x^2}{4} = (1 - \cos x) - \frac{(1-\cos 2x)}{2^2} + \frac{(1-\cos 3x)}{3^2} - \cdots.$$

Now set $x = \pi$. Since $\cos n\pi = 1$ if n is even, $\cos n\pi = -1$ if n is odd, this yields

$$\frac{\pi^2}{4} = 2 - 0 + \frac{2}{3^2} - 0 + \frac{2}{5^2} - \cdots$$

from which (1) follows.

EXERCISE: Criticize the above derivation. What parts, if any, are wrong? What parts do you feel require justification that has not yet been presented?

III. A really general summability theorem.

3.12G. Let $T = (c_{mn})_{m,n=1}^{\infty}$ be an infinite matrix of real numbers. Given a sequence $\{s_n\}_{n=1}^{\infty}$ of real numbers, define a sequence $\{t_m\}_{m=1}^{\infty}$ by

$$t_m = \sum_{n=1}^{\infty} c_{mn} s_n \qquad (m = 1, 2, \cdots). \tag{*}$$

(Thus the column vector of the t_m is the product of the matrix T with the column vector of the s_n.)

We say that $\{s_n\}_{n=1}^{\infty}$ is T summable to $L \in R$ if (the series in (*) all converge and)

$$\lim_{n\to\infty} t_n = L.$$

EXERCISE: Suppose T is the matrix

$$\begin{bmatrix} 1 & 0 & 0 & 0 & \cdots \\ \frac{1}{2} & \frac{1}{2} & 0 & 0 & \cdots \\ \frac{1}{3} & \frac{1}{3} & \frac{1}{3} & 0 & \cdots \\ \frac{1}{4} & \frac{1}{4} & \frac{1}{4} & \frac{1}{4} & \cdots \\ \vdots & & & & \end{bmatrix}$$

Show that T summability is precisely $(C,1)$ summability.

EXERCISE: Show that $(C,2)$ summability is a special case of T summability.

3.12H. We recall from 2.11H that a summability method is said to be regular if it sums every convergent sequence to its limit.

We now state conditions on the c_{mn} that turn out to be both necessary and sufficient for T summability to be regular. These conditions are

A. There exists $M>0$ such that

$$\sum_{n=1}^{\infty} |c_{mn}| \leqslant M \qquad (m=1,2,\cdots).$$

B. $\lim_{m\to\infty} \sum_{n=1} c_{mn}=1.$

C. $\lim_{m\to\infty} c_{mn}=0 \qquad (n=1,2,\cdots).$

Notice that A and B are conditions on the rows of $T=(c_{mn})$, while condition C applies to the columns.

EXERCISE: Fill in the details of the following proof that conditions A, B, C are necessary if T summability is to be regular.

Suppose T is regular. That is, suppose that whenever $\{s_n\}_{n=1}^{\infty}$ converges to L then each series in $(*)$ is convergent, and $\{t_m\}_{m=1}^{\infty}$ converges to L. To show that C holds, fix n and define $\{s_k\}$ as follows:

$$s_k=0 \qquad (k\neq n)$$

$$s_n=1.$$

Then $\{s_k\}$ converges to 0. Calculate the corresponding $\{t_m\}$ and apply regularity.

To show that B holds, consider $s_n=1$ $(n=1,2,\dots)$.

Now we have to show that A must hold. This is the hard part.

First show that

$$\sum_{n=1}^{\infty} |c_{mn}|<\infty \qquad (m=1,2,\dots). \tag{1}$$

To do this, assume

$$\sum_{n=1}^{\infty} |c_{mn}|=\infty$$

for some m. Then there exists $\{\epsilon_n\}$ converging to zero such that

$$\sum_{n=1}^{\infty} \epsilon_n |c_{mn}|=\infty. \tag{2}$$

(Why?) Let $s_n = \epsilon_n$ for those n such that $c_{mn} \geqslant 0$ and let $s_n = -\epsilon_n$ for those n such that $c_{mn} < 0$. For this $\{s_n\}$ show that t_m is given by (2), which contradicts our assumption of regularity. Hence (1) holds.

Let

$$k_m = \sum_{n=1}^{\infty} |c_{mn}| < \infty \qquad (m = 1, 2, \ldots).$$

To show that condition A holds it is enough to show that $\{k_m\}$ is bounded. Again, prove by contradiction.

Assume $\{k_m\}$ is not bounded. Justify all the following assertions. Let n_1 be any positive integer. Then there exists m_1 such that

$$\sum_{n=1}^{n_1-1} |c_{m_1 n}| < 1, \qquad k_{m_1} > 1^2 + 2.$$

Then there exists $n_2 > n_1$ such that

$$\sum_{n=n_2}^{\infty} |c_{m_1 n}| < 1.$$

It follows that

$$\sum_{n=n_1}^{n_2-1} |_{m_1 n}| > 1^2.$$

Now there exists $m_2 > m_1$ such that

$$\sum_{n=1}^{n_2-1} |c_{m_2 n}| < 1, \qquad k_{m_2} > 2^2 + 2.$$

Then there exists $n_3 > n_2$ such that

$$\sum_{n=n_3}^{\infty} |c_{m_2 n}| < 1.$$

It follows that

$$\sum_{n=n_2}^{n_3-1} |c_{m_2 n}| > 2^2.$$

Continuing in this manner, define

$$n_1 < n_2 < \cdots < n_r < \cdots \quad \text{and} \quad m_1 < m_2 < \cdots < m_r < \cdots$$

such that, for each $r = 1, 2, \ldots,$

$$\sum_{n=1}^{n_r-1} |c_{m_r n}| < 1, \qquad \sum_{n=n_r}^{n_{r+1}-1} |c_{m_r n}| > r^2, \qquad \sum_{n=n_{r+1}}^{\infty} |c_{m_r n}| < 1.$$

Now define $\{s_n\}$ as follows:

$$s_n = 0 \qquad (n < n_1).$$

If $n_r \leqslant n < n_{r+1}$, let

$$s_n = \frac{1}{r} \quad \text{if} \quad c_{m_r n} \geqslant 0,$$

$$s_n = \frac{-1}{r} \quad \text{if} \quad c_{m_r n} < 0.$$

Then $\{s_n\}$ converges to zero. Finish the proof by showing that $t_{m_r} > r-2$. This is a contradiction (why?) and the proof is complete.

3.12I. Now we prove that conditions A, B, C are sufficient for T summability to be regular.

Assume A, B, C. Suppose $\{s_n\}$ converges to L. We must prove that $\{t_m\}$ also converges to L. We have

$$t_m - L = \sum_{n=1}^{\infty} c_{mn} s_n - L$$

$$= \sum_{n=1}^{\infty} c_{mn}[s_n - L] + L\left[\sum_{n=1}^{\infty} c_{mn} - 1\right].$$

EXERCISE: Finish the proof.

Consult the book *Divergent Series*, by G. H. Hardy (Oxford, 1949), for more material of this nature.

IV. More about the product of series.

3.12J. EXERCISE: The assumption in 3.5G of absolute convergence for $\sum_{n=0}^{\infty} a_n$ and $\sum_{n=0}^{\infty} b_n$ cannot be weakened simply to convergence. Demonstrate this by taking

$$a_k = b_k = \frac{(-1)^k}{\sqrt{k+1}} \qquad (k = 0, 1, 2, \ldots)$$

and showing that $\sum_{n=0}^{\infty} c_n$ diverges. However, a stronger theorem than 3.5G can be proved.

3.12K. MERTEN'S THEOREM Suppose $\sum_{n=0}^{\infty} a_n$ converges absolutely to A, and $\sum_{n=0}^{\infty} b_n$ converges (not necessarily absolutely) to B. Then $\sum_{n=0}^{\infty} c_n$ converges to $C = AB$, where

$$c_n = \sum_{k=0}^{n} a_k b_{n-k} \qquad (n = 0, 1, 2, \ldots).$$

SKETCH OF PROOF:

CASE I; $B = 0$. It must be shown that $C_n \to 0$ as $n \to \infty$, where $C_n = \sum_{k=0}^{n} c_k$. Let

$$B_n = \sum_{k=0}^{n} b_k, \qquad \alpha = \sum_{k=0}^{\infty} |a_k| < \infty.$$

First show that

$$C_n = a_0 B_n + a_1 B_{n-1} + \cdots + a_n B_0 \qquad (n = 0, 1, 2, \ldots).$$

But $B_n \to 0$ as $n \to \infty$ (why?). Hence given $\epsilon > 0$ there exists $N \in I$ such that

$$|C_n| \leqslant |a_n B_0 + \cdots + a_{n-N} B_N| + \epsilon\alpha \qquad (n \geqslant N).$$

CASE II; $B \neq 0$. Let

$$b_0' = b_0 - B,$$
$$b_n' = b_n \qquad (n \in I).$$

Then $\sum_{n=0}^{\infty} b_n' = 0$. Apply Case I to $\sum_{n=0}^{\infty} a_n$ and $\sum_{n=0}^{\infty} b_n'$.

EXERCISE: Fill in the details and finish the proof.

MISCELLANEOUS EXERCISES

1. Suppose $f: X \to Y$.
   (a) If $A \subset X$, show that $A \subset f^{-1}[f(A)]$.
   (b) If $B \subset Y$, show that $f[f^{-1}(B)] \subset B$.
2. Suppose $f: X \to Y$. Show that f is onto if and only if
$$E = f[f^{-1}(E)]$$
for every $E \subset Y$.
3. Suppose $f: X \to Y$. Show that f is $1-1$ if and only if
$$A = f^{-1}[f(A)]$$
for every $A \subset X$.
4. Let $s_1 = 1$ and
$$s_{n+1} = \sqrt{2 + s_n} \qquad (n \geq 2).$$
Show that $\{s_n\}_{n=1}^{\infty}$ is bounded and nondecreasing. Then compute $\lim_{n\to\infty} s_n$.
5. Prove that $\{nx^n\}_{n=1}^{\infty}$ converges to 0 for every x such that $0 \leq x < 1$.
6. Let $\{s_n\}_{n=1}^{\infty}$ be a sequence of real numbers. Prove that $\{s_n\}_{n=1}^{\infty}$ converges to L if and only if every subsequence of $\{s_n\}_{n=1}^{\infty}$ has a subsequence that converges to L.
7. Let $\{s_n\}_{n=1}^{\infty}$ be a bounded sequence of real numbers. Let A be the set of all numbers L such that $\{s_n\}_{n=1}^{\infty}$ has a subsequence that converges to L. Prove that
$$\limsup_{n\to\infty} s_n = \text{l.u.b.}\, A.$$
8. Let A be a nonempty set and let
$$X_n \subset A, Y_n \subset A \qquad (n \in I).$$
   (a) Prove that
$$\liminf_{n\to\infty} X_n \cup \liminf_{n\to\infty} Y_n \subset \liminf (X_n \cup Y_n)$$
   and that equality need not occur.
   (b) Prove that
$$\limsup_{n\to\infty} X_n \cup \limsup_{n\to\infty} Y_n = \limsup_{n\to\infty} (X_n \cup Y_n).$$
9. Prove $\lim_{n\to\infty} n^{1/n} = 1$. (Begin by setting $s_n = n^{1/n} - 1$ and use the binomial theorem.)
10. Prove that $\sum_{n=1}^{\infty} 1/n^{(1+1/n)}$ diverges.
11. Suppose that $a_n \geq 0$ $(n \in I)$ and that $\sum_{n=1}^{\infty} a_n$ converges. Prove that $\sum_{n=1}^{\infty} \sqrt{a_n a_{n+1}}$ converges.
12. Prove that $\frac{1}{4}$ belongs to the Cantor set but is not an end point of any of the open intervals removed.

13. The following game was invented by S. Mazur. Player A owns the irrationals in $[0,1]$; player B owns the rationals in $[0,1]$. One player (either A or B) starts by choosing a closed interval in $[0,1]$ of length $\leqslant \frac{1}{2}$. The second player now chooses a closed interval of length $\leqslant \frac{1}{3}$ inside the interval already chosen. The first player now chooses an interval of length $\leqslant \frac{1}{4}$ inside the preceding interval, and so on. By Theorem 2.10E there is a unique point x in all of these intervals. If x is irrational, then A wins. If x is rational, then B wins.
Prove that A can always win, independent of B's strategy.

4

LIMITS AND METRIC SPACES

4.1 LIMIT OF A FUNCTION ON THE REAL LINE

In Chapter 2 we define limit of a sequence. We now recall from calculus the definition of limit of a "function of a real variable" on which are based the definitions of continuous function and derivative. Later in the chapter we generalize to a wide class of spaces (called metric spaces) which includes the real line R as a very special case.

Let $a \in R$, and let f be a real-valued function whose domain includes all points in some open interval $(a-h, a+h)$ except possibly the point a itself.

4.1A. DEFINITION. We say that $f(x)$ approaches L (where $L \in R$) as x approaches a if given $\epsilon > 0$, there exists $\delta > 0$ such that

$$|f(x)-L|<\epsilon \qquad (0<|x-a|<\delta).$$

In this case we write $\lim_{x\to a} f(x) = L$ or $f(x) \to L$ as $x \to a$. We sometimes say "f has the limit L at a" instead of "$f(x)$ approaches L as x approaches a."

We emphasize that the point a need not be in the domain of f. You remember from calculus (we hope) that $\lim_{x\to 0}(\sin x/x) = 1$, even though $\sin x/x$ is not defined for $x = 0$. We verify this limit in a later chapter, after we give a rigorous definition of $\sin x$.

Consider Figure 12. In order for $f(x)$ to approach L as x approaches a the following must be true: Given any ϵ parentheses about L there must exist δ parentheses about a such that every arrow which begins inside the δ parentheses (except possibly the arrow, if there is one, that starts at a) must end inside the ϵ parentheses.

Roughly speaking, the following can be seen on the x-y graph of a function f such that $\lim_{x\to a} f(x) = L$: As the x coordinate of a moving point on the graph gets close to a (from either the right or the left), the height $f(x)$ of the point heads toward L. Be sure to think out why this is a geometric interpretation of definition 4.1A. For example, the functions in Figures 13, 14, and 15 all satisfy $\lim_{x\to a} f(x) = L$. (An empty dot, as in Figure 15, indicates a point not on the graph.) On the other hand, the function in Figure 16 has no limit at a. This is because $f(x)$ gets close to 3 when x gets close to a on the left, while $f(x)$ gets close to 1 when x gets close to a on the right. Hence there is no single number L such that $f(x)$ gets close to L when x gets close to a.

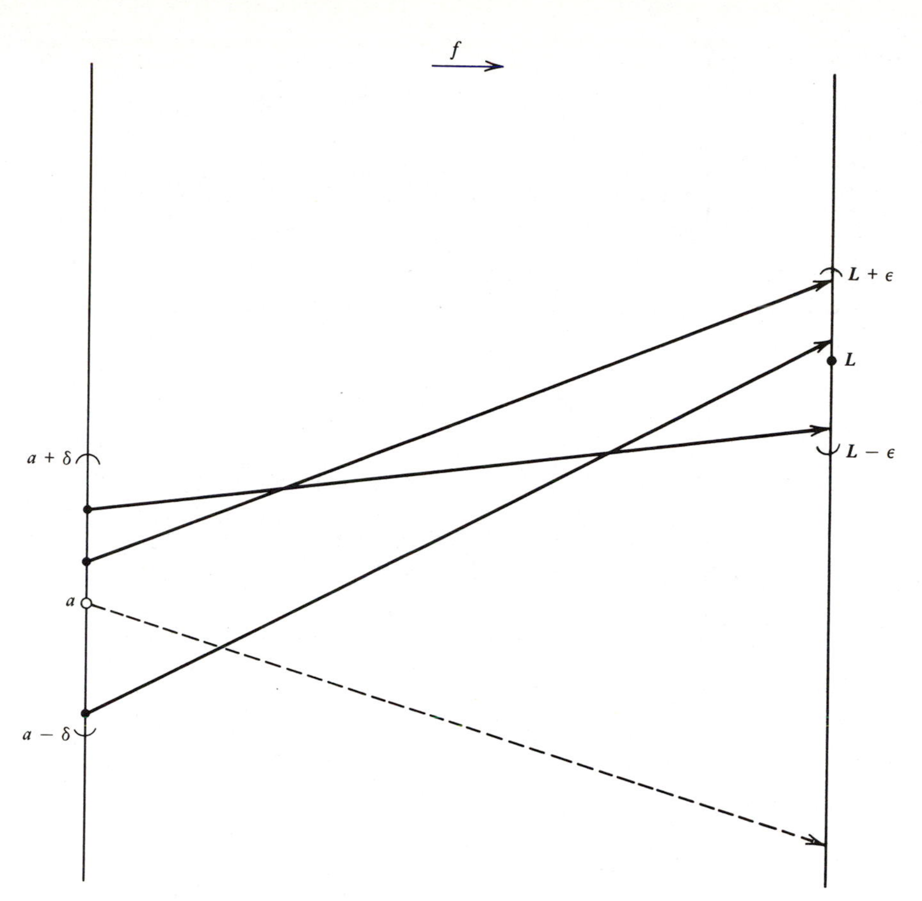

FIGURE 12.

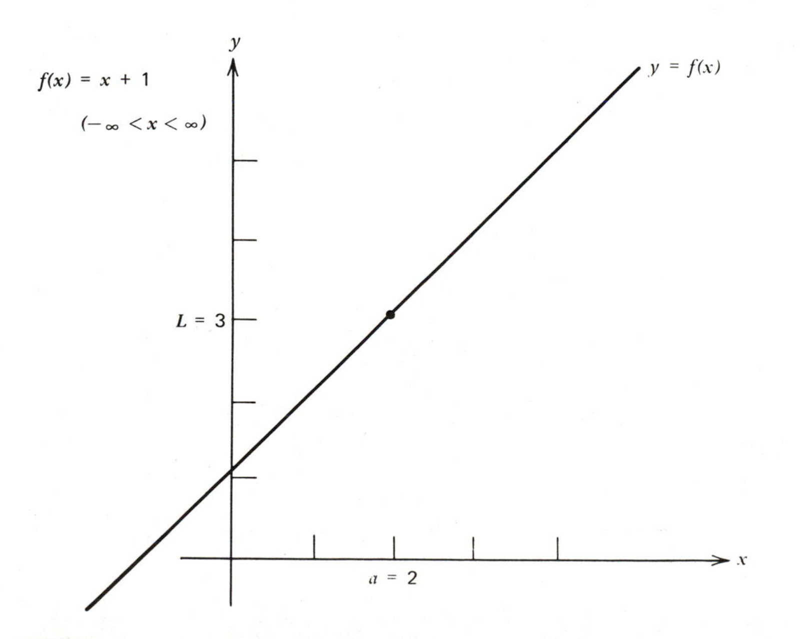

FIGURE 13. Example 1 of $\lim_{x \to a} f(x) = L$

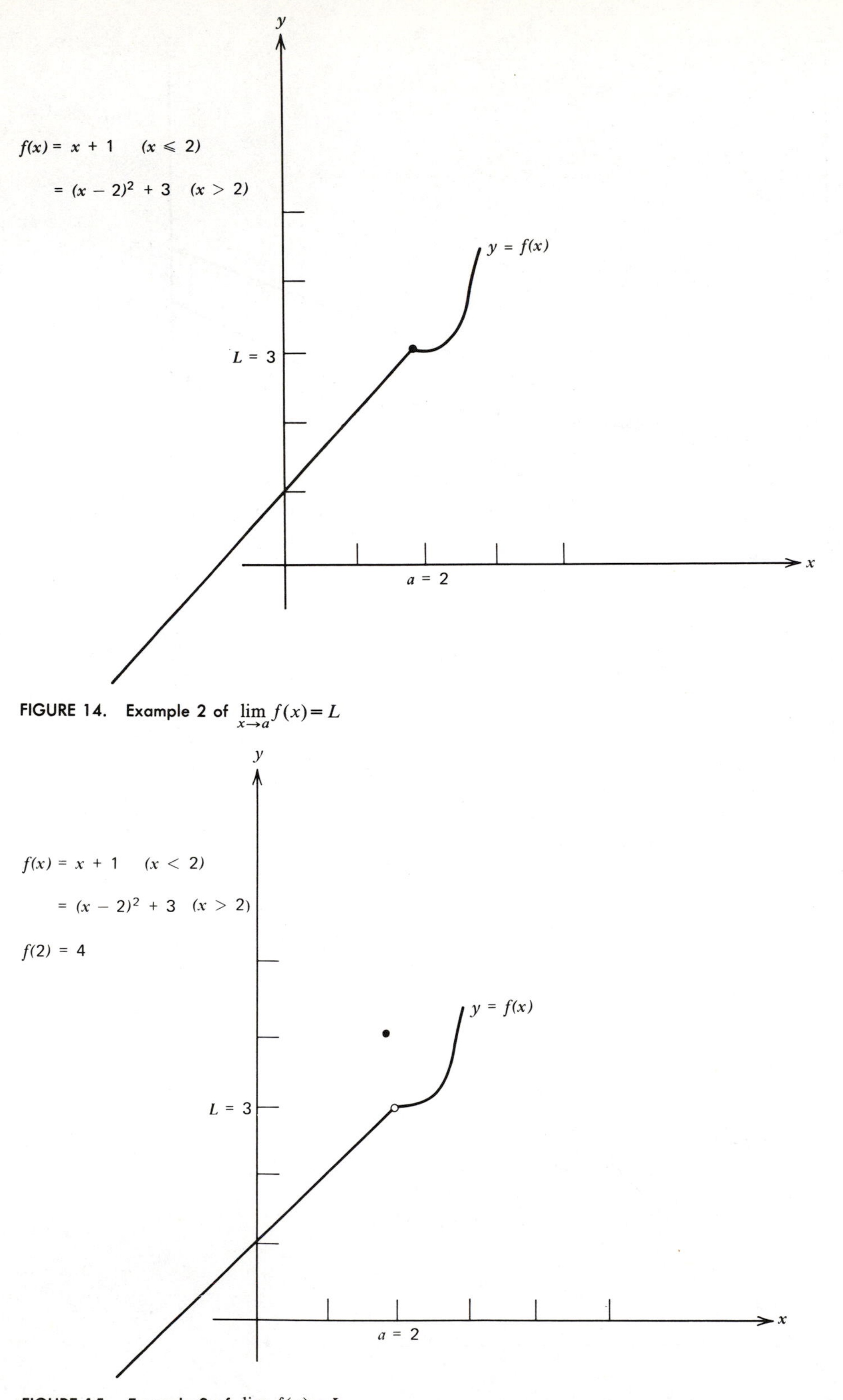

FIGURE 14. Example 2 of $\lim_{x \to a} f(x) = L$

FIGURE 15. Example 3 of $\lim_{x \to a} f(x) = L$

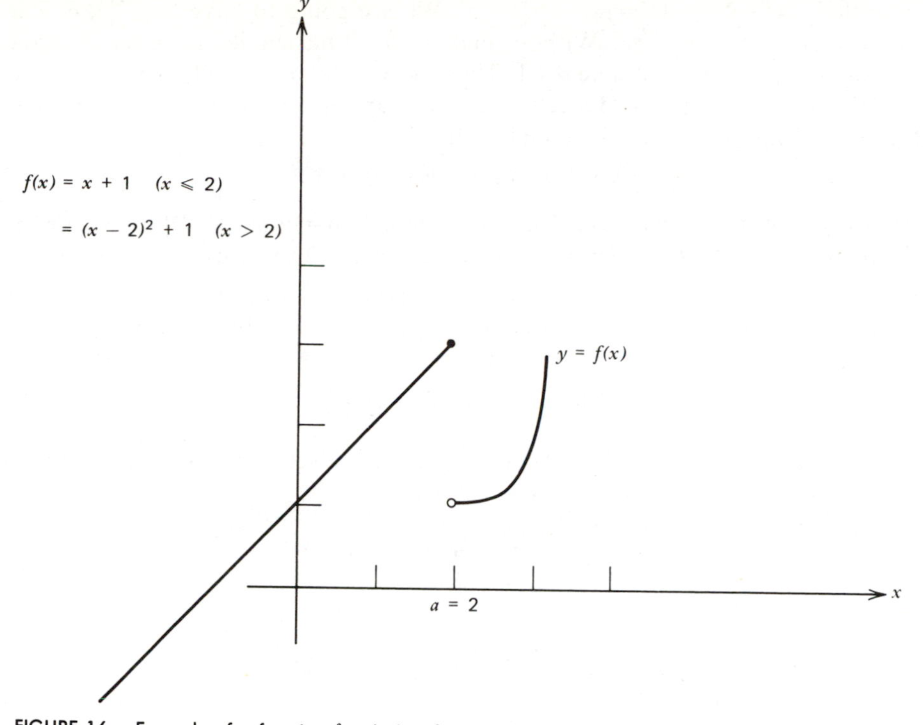

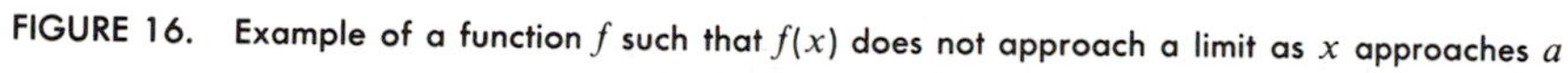
FIGURE 16. Example of a function f such that $f(x)$ does not approach a limit as x approaches a

For a final pictorial example, consider Figure 17. This shows the graph of f where

$$f(x) = \sin\frac{1}{x} \qquad (x \neq 0).$$

Here, as x gets close to $a=0$, the value $f(x)$ oscillates rapidly. Even if we look on only one side of a, it is clear that there is no number L toward which the value $f(x)$ tends. Hence f has no limit at 0.

Now, for some examples with *proofs*.

First, let us prove $\lim_{x\to 3}(x^2+2x)=15$. Here $f(x)=x^2+2x$, $L=15$, $a=3$. Given $\epsilon>0$ we must find $\delta>0$ such that

$$|(x^2+2x)-15|<\epsilon \qquad (0<|x-3|<\delta). \tag{1}$$

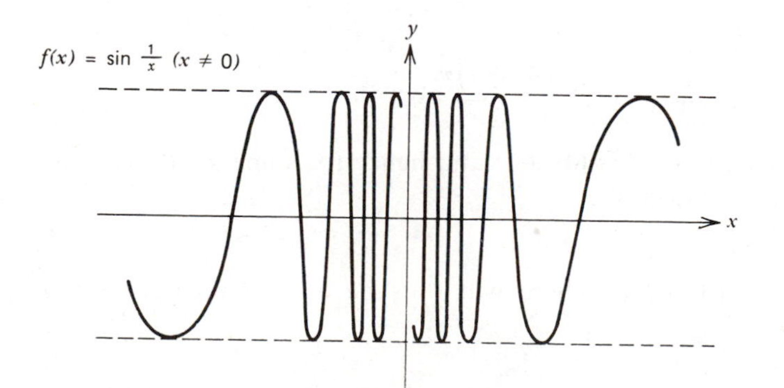

FIGURE 17. The function has no limit at $a=0$

We first note that $|(x^2+2x)-15|=|x-3|\cdot|x+5|$. We are going to have $|x-3|<\delta$. The question is, how big can $|x+5|$ be? Without making our final choice of δ, let us agree that when we do choose it we will take $\delta \leqslant 1$. Then, if $|x-3|<\delta$, we will have $|x-3|<1$. Thus $x\in(2,4)$ and so $x+5\in(7,9)$. Hence $|x+5|<9$ if $|x-3|<\delta<1$, and so $|x-3|\cdot|x+5|<\delta\cdot 9$ if $|x-3|<\delta$ and $\delta\leqslant 1$. Let $\delta=\min(1,\epsilon/9)$. Then

$$|x-3|\cdot|x+5|<9\delta\leqslant\epsilon \qquad (|x-3|<\delta),$$

which implies (1). Given $\epsilon>0$ we have found a δ [namely $\delta=\min(1,\epsilon/9)$] for which (1) holds, and this proves $\lim_{x\to 3}(x^2+2x)=15$. Note that in this example $a=3$ and $|f(x)-L|<\epsilon$ even for $x=a$.

For a second example we will show that $\lim_{x\to 1}\sqrt{x+3}=2$. Here $f(x)=\sqrt{x+3}$, $L=2, a=1$. Given $\epsilon>0$ we must find $\delta>0$ such that

$$|\sqrt{x+3}-2|<\epsilon \qquad (0<|x-1|<\delta). \tag{2}$$

Multiplying the left side by $|(\sqrt{x+3}+2)/(\sqrt{x+3}+2)|$, we see that (2) is equivalent to

$$\frac{|(\sqrt{x+3})^2-2^2|}{|\sqrt{x+3}+2|}<\epsilon \qquad (0<|x-1|<\delta)$$

or

$$\frac{|x-1|}{|\sqrt{x+3}+2|}<\epsilon \qquad (0<|x-1<\delta). \tag{3}$$

If we agree to take $\delta\leqslant 1$, then $|x-1|<\delta$ implies $x\in(0,2)$ and hence $\sqrt{x+3}+2>\sqrt{3}+2$. Thus if $|x-1|<\delta\leqslant 1$, then

$$\frac{|x-1|}{|\sqrt{x+3}+2|}<\frac{\delta}{\sqrt{3}+2}.$$

If we pick $\delta=\min(1,\epsilon(\sqrt{3}+2))$, then $\delta/(\sqrt{3}+2)\leqslant\epsilon$, hence (3) holds, hence (2) holds, and we are done.

For an example in the other direction we shall *prove* what we have already inferred from Figure 17—namely, that $\sin(1/x)$ does not approach a limit as $x\to 0$. For assume the contrary—that is, assume there exists $L\in R$ such that $\lim_{x\to 0}\sin(1/x)=L$. Then for $\epsilon=1$ there would exist $\delta>0$ such that

$$\left|\sin\frac{1}{x}-L\right|<1 \qquad (0<|x|<\delta). \tag{4}$$

Now

$$\sin\left(2n\pi+\frac{\pi}{2}\right)=\sin\frac{(4n+1)\pi}{2}=1$$

for any $n\in I$. Thus $\sin(1/x)=1$ for $x=2/\pi(4n+1)$ and hence for some $x\in(0,\delta)$, since $\lim_{n\to\infty}2/\pi(4n+1)=0$. For this x (4) implies

$$|1-L|<1. \tag{5}$$

Similarly, $\sin(2n\pi+3\pi/2)=-1$ for $n\in I$. There will thus be an $x\in(0,\delta)$ for which $\sin(1/x)=-1$. By (4) again,

$$|-1-L|<1. \tag{6}$$

The reader should be able to deduce a contradiction from (5) and (6). Hence $\lim_{x\to 0}\sin(1/x)$ does not exist.

4.1B. We wish to emphasize the strong analogy between definition 4.1A and definition 2.2A. Indeed, consider the "table of analogues."

TABLE OF ANALOGUES

2.2*A*	4.1*A*
$S=\{s_n\}_{n=1}^{\infty}$	f
n	x
s_n	$f(x)$
L	L
∞	a
ϵ	ϵ
N	δ
$n \geqslant N$	$0<\|x-a\|<\delta$

If we substitute each entry in the right-hand column for the corresponding entry on the left, we change definition 2.2A into definition 4.1A.

However, more than a mechanical process is involved here. Corresponding entries in the table actually "have the same meaning." For example, $S=\{s_n\}_{n=1}^{\infty}$ is the *function* (sequence) involved in definition 2.2A, while f is the function involved in 4.1A. Also, s_n is the value of S at n, while $f(x)$ is the value of f at x. Finally, $n \geqslant N$ means "n is sufficiently close to infinity" (but not, of course, equal to infinity), while $0<|x-a|<\delta$ means "x is sufficiently close to a but not equal to a."

We will now prove a theorem corresponding to 2.7A. The reader should first study the proof and then see how it could be obtained from the proof of 2.7A by mechanical substitution from our table.

4.1C. THEOREM. If* $\lim_{x\to a} f(x)=L$ and $\lim_{x\to a} g(x)=M$, then $f(x)+g(x)$ has a limit as $x\to a$ and, in fact, $\lim_{x\to a}[f(x)+g(x)]=L+M$.

PROOF: Given $\epsilon>0$ we must find $\delta>0$ such that

$$|[f(x)+g(x)]-(L+M)|<\epsilon \qquad (0<|x-a|<\delta). \tag{1}$$

Since $\lim_{x\to a} f(x)=L$, there exists $\delta_1>0$ such that

$$|f(x)-L|<\frac{\epsilon}{2} \qquad (0<|x-a|<\delta_1).$$

Similarly, there exists $\delta_2>0$ such that

$$|g(x)-M|<\frac{\epsilon}{2} \qquad (0<|x-a|<\delta_2).$$

Thus if $\delta=\min(\delta_1,\delta_2)$ and if $0<|x-a|<\delta$, then

$$|f(x)-L|<\frac{\epsilon}{2}, \qquad |g(x)-M|<\frac{\epsilon}{2},$$

and so

$$|[f(x)+g(x)]-(L+M)|=|[f(x)-L]+[g(x)-M]|$$
$$\leqslant |f(x)-L|+|g(x)-M|<\frac{\epsilon}{2}+\frac{\epsilon}{2}=\epsilon.$$

Thus (1) holds for $\delta=\min(\delta_1,\delta_2)$ and the proof is completed.

* In this chapter, whenever we write a hypothesis such as $\lim_{x\to a} f(x)=L$ it is understood that f is a function whose domain contains all points whose distance from a is less than some $h(h>0)$ except perhaps the point a itself.

Using the proofs of 2.7D, 2.7G, and 2.7I as models, the reader should now be able to prove the following theorem.

4.1D. THEOREM. If $\lim_{x\to a} f(x) = L$ and $\lim_{x\to a} g(x) = M$, then

(a) $\lim_{x\to a}[f(x) - g(x)] = L - M$,
(b) $\lim_{x\to a} f(x)\,g(x) = L \cdot M$,

and if $M \neq 0$,

(c) $\lim_{x\to a}[f(x)/g(x)] = L/M$.

We occasionally need to handle limits of the form $\lim_{x\to\infty} f(x)$.

4.1E. DEFINITION. We say that $f(x)$ approaches L as x approaches infinity, if given $\epsilon > 0$, there exists $M \in R$ such that

$$|f(x) - L| < \epsilon \qquad (x > M).$$

In this case we write $\lim_{x\to\infty} f(x) = L$, or $f(x) \to L$ as $x \to \infty$.

Definition 4.1E requires, of course, that the domain of the (real-valued) function f contain some interval of the form (c, ∞). Note the very strong resemblance of 4.1E to 2.2A.

For example, let us prove that $\lim_{x\to\infty}(1/x^2) = 0$. Given $\epsilon > 0$ we must find $M \in R$ such that

$$\left|\frac{1}{x^2} - 0\right| < \epsilon \qquad (x > M). \tag{1}$$

Since (1) is equivalent to

$$\frac{1}{x} < \sqrt{\epsilon} \qquad (x > M),$$

it is clear that (1) will hold if we take $M = 1/\sqrt{\epsilon}$.

It is also useful to consider "one-sided" limits.

4.1F. DEFINITION. We say that $f(x)$ approaches L as x approaches a from the right, if given $\epsilon > 0$, there exists $\delta > 0$ such that

$$|f(x) - L| < \epsilon \qquad (a < x < a + \delta).$$

In this case we write $\lim_{x\to a+} f(x) = L$. (The number L is called the right-hand limit of f at a.)

We say that $f(x)$ approaches M as x approaches a from the left, if given $\epsilon > 0$, there exists $\delta > 0$ such that

$$|f(x) - M| < \epsilon \qquad (a - \delta < x < a).$$

In this case we write $\lim_{x\to a-} f(x) = M$. (The number L is called the left-hand limit of f at a.)

Thus the statement $\lim_{x\to a+} f(x) = L$ involves only values of $f(x)$ for x to the right of a, while $\lim_{x\to a-} f(x) = M$ involves only values of $f(x)$ for x to the left of a. It should be obvious to the reader that $\lim_{x\to a} f(x) = L$ if and only if

$$\lim_{x\to a+} f(x) = \lim_{x\to a-} f(x) = L.$$

On the other hand, both $\lim_{x\to a+} f(x)$ and $\lim_{x\to a-} f(x)$ may exist without being equal to each other. For example, if

$$f(x) = x \qquad (0 \leqslant x < 1),$$
$$f(x) = 3 - x \qquad (1 \leqslant x \leqslant 2),$$

then $\lim_{x \to 1-} f(x) = 1$ while $\lim_{x \to 1+} f(x) = 2$.

It is not difficult to show that theorems analogous to 4.1C and 4.1D also hold for limits of the form $\lim_{x \to \infty} f(x), \lim_{x \to a+} f(x)$, or $\lim_{x \to a-} f(x)$. (See problem 15.)

"One-sided" limits always exist for an important class of functions—namely, monotone functions.

4.1G. DEFINITION. If f is a real-valued function on an interval $J \subset R$, we say that f is nondecreasing of J if

$$f(x) \leqslant f(y) \qquad (x < y; \quad x, y \in J).$$

We say that f is nonincreasing on J if

$$f(x) \geqslant f(y) \qquad (x < y; \quad x, y \in J).$$

We say that f is monotone if f is either nondecreasing or nonincreasing.

Thus definition 4.1G is analogous to definition 2.6A for sequences. As for sequences, we say that a function f on an interval $J \subset R$ is bounded above or bounded below if the range of f is respectively bounded above or bounded below. We then have the following important result analogous to 2.6B.

4.1H. THEOREM. Let f be a nondecreasing function on the bounded open interval (a, b). If f is bounded above on (a, b), then $\lim_{x \to b-} f(x)$ exists. Also, if f is bounded below on (a, b), then $\lim_{x \to a+} f(x)$ exists.

PROOF: If f is bounded above and nondecreasing on (a, b), let

$$M = \underset{x \in (a,b)}{\text{l.u.b.}} f(x).$$

Given $\epsilon > 0$ the number $M - \epsilon$ is thus not an upper bound for the range of f. Hence there exists $y \in (a, b)$ such that $f(y) > M - \epsilon$. Let $\delta = b - y$. Then

$$f(b - \delta) = f(y) > M - \epsilon.$$

Since f is nondecreasing, this implies

$$f(x) > M - \epsilon \qquad (b - \delta < x < b).$$

Hence since $f(x) \leqslant M$ for all $x \in (a, b)$ we have

$$|f(x) - M| < \epsilon \qquad (b - \delta < x < b).$$

This proves that $\lim_{x \to b-} f(x) = M$.

If f is bounded below, a similar argument will show that $\lim_{x \to a+} f(x) = m$ where $m = \text{g.l.b.}_{x \in (a,b)} f(x)$.

If f is nonincreasing on (a, b), the following result may be proved by applying 4.1H to $-f$ (which will be nondecreasing).

4.1I. THEOREM. Let f be a nonincreasing function on the bounded open interval (a, b). If f is bounded below on (a, b), then $\lim_{x \to b-} f(x)$ exists, while if f is bounded above on (a, b) then $\lim_{x \to a+} f(x)$ exists.

We then have the important corollary 4.1J.

4.1J. COROLLARY. If f is a monotone function on the open interval (a,b), and if $c \in (a,b)$, then $\lim_{x\to c+} f(x)$ and $\lim_{x\to c-} f(x)$ both exist.

PROOF: Suppose that f is nondecreasing. Choose $\delta > 0$ such that $(c-\delta, c+\delta)$ (bounded open interval) is contained in (a,b). Then the values of f on the open interval $(c-\delta, c)$ are bounded above by $f(c)$ and hence by 4.1H, $\lim_{x\to c-} f(x)$ exists. Similarly, the values of f on the open interval $(c, c+\delta)$ are bounded below by $f(c)$. Hence again by 4.1H, $\lim_{x\to c+} f(x)$ exists.

If f in nonincreasing we use 4.1I instead of 4.1H. This completes the proof. Note that we did not assume that (a,b) was bounded.

As long as we are on this subject we may as well define strictly increasing function and strictly decreasing function.

4.1K. DEFINITION. The real-valued function f on the interval $J \subset R$ is said to be strictly increasing if

$$f(x) < f(y) \qquad (x < y;\quad x, y \in J).$$

Similarly, f is said to be strictly decreasing if

$$f(x) > f(y) \qquad (x < y;\quad x, y \in J).$$

Thus if f is nonincreasing on J, then f is strictly increasing on J if and only if f is 1–1 on J.

Exercises 4.1

1. (a) If $|x-2| < 1$, prove that $|x^2-4| < 5$.
   (b) If $|x-3| < \frac{1}{10}$, prove that $|x^2-x-6| < 0.51$.
   (c) If $|x+1| < \frac{1}{10}$, prove that $|x^3+1| < 0.331$.
2. Let δ be any number such that $0 < \delta < 1$.
   (a) If $|x-2| < \delta$, prove that $|x^2-4| < 5\delta$.
   (b) If $|x-3| < \delta$, prove that $|x^2-x-6| < 6\delta$.
   (c) If $|x+1| < \delta$, prove that $|x^3+1| < 7\delta$.
   (d) If $|x-2| < \delta$, prove that $|x-2)/(x+3)| < \delta/4$.
3. (a) Let $f(x) = x^2+4x$. Find $\delta > 0$ such that

   $$|f(x)-5| < \tfrac{1}{10} \qquad (0 < |x-1| < \delta).$$

   (b) Prove directly from definition 4.1A that $\lim_{x\to 1}(x^2+4x) = 5$.
4. For each of the functions in Figures 13–17, draw a diagram of the type in Figure 7. Relate the diagrams to definition 4.1A.
5. Prove, using only definition 4.1A, the truth of the following statements.
   (a) $\lim\limits_{x\to -2} x^2+3x = -2$.
   (b) $\lim\limits_{x\to 1} \dfrac{x^2-1}{x-1} = 2$.
   (c) $\lim\limits_{x\to 0} \sqrt{4-x} = 2$.
6. Prove that if $\lim_{x\to a} f(x) = L$ and $\lim_{x\to a} f(x) = M$, then $L = M$. (Compare with theorem 2.3B.)
7. If $\lim_{x\to a} f(x) = L$ and $c \in R$, prove that $\lim_{x\to a} cf(x) = cL$.

8. Prove

$$\lim_{x\to 1}\frac{x^7-2x^5+1}{x^3-3x^2+2}=1.$$

9. If $c\in R$ and $f(x)=c$ for all $x\in R$, prove that $\lim_{x\to a}f(x)=c$, where a is any point in R.
10. Let $[x]$ denote the greatest integer not exceeding x. (For example, $[-4]=-4$, $[-4.1]=-5$, $[15.4]=15$.) Prove that if $n\in I$, then

$$\lim_{x\to n^+}[x]=n, \qquad \lim_{x\to n^-}[x]=n-1.$$

11. Let

$$f(x)=[1-x^2] \qquad (-1\leqslant x\leqslant 1).$$

Does $\lim_{x\to 0}f(x)$ exist? If so, evaluate it.
12. For any $a\in R$ prove that $\lim_{x\to a}x=a$. Then, using theorems from this section, prove that $\lim_{x\to a}P(x)=P(a)$ where P is any polynomial function.
13. If $f(x)\geqslant 0$ for $|x-a|<h$ and if $\lim_{x\to a}f(x)=L$, prove that $L\geqslant 0$.
14. If $\lim_{x\to a}f(x)=L>0$, show that there exists $\delta>0$ such that

$$f(x)>0 \qquad (0<|x-a|<\delta).$$

(*Hint*: Take $\epsilon=L/2$.)
15. If $\lim_{x\to\infty}f(x)=A$ and $\lim_{x\to\infty}g(x)=B$, prove that $\lim_{x\to\infty}[f(x)+g(x)]=A+B$. Do the same with $\lim_{x\to\infty}$ replaced by $\lim_{x\to a+}$ and $\lim_{x\to b-}$.
16. Let f and g be nondecreasing functions on an interval (a,b) and let $h=f-g$. If $c\in(a,b)$ prove that $\lim_{x\to c}h(x)$ and $\lim_{x\to c-}h(x)$ exist.
17. If f is a real-valued function on $(0,\infty)$ and if

$$g(x)=f\left(\frac{1}{x}\right) \qquad (0<x<\infty),$$

prove that $\lim_{x\to\infty}f(x)=L$ if and only if $\lim_{x\to 0+}g(x)=L$.
18. Write out a definition of

$$\lim_{x\to-\infty}f(x)=L.$$

Prove, for this limit, a theorem corresponding to 4.1C.
19. Give an example of a 1–1 function on $(0,\infty)$ that is not monotone.
20. Give an example of a nondecreasing function on $[0,1]$ that is not strictly increasing.
21. Prove that if f is nondecreasing and bounded above on (a,∞), then $\lim_{x\to\infty}f(x)$ exists.
22. Let f be real-valued function on R and suppose $\lim_{x\to a}f(x)=L$. If $\{x_n\}_{n=1}^{\infty}$ is any sequence of real numbers which converges to a, and if $x_n\neq a(n\in I)$, prove that the sequence $\{f(x_n)\}_{n=1}^{\infty}$ converges to L.
23. Conversely, suppose $\lim_{n\to\infty}f(x_n)=L$ for every sequence $\{x_n\}_{n=1}^{\infty}$ such that $x_n\neq a$ $(n\in I)$ and $\lim_{n\to\infty}x_n=a$. Prove that $\lim_{x\to a}f(x)=L$.
24. Suppose only that $\lim_{n\to\infty}f(x_n)$ *exists* for every sequence $\{x_n\}_{n=1}^{\infty}$ such that $x_n\neq a$ and $\lim_{n\to\infty}x_n=a$. Prove that $\lim_{x\to a}f(x)$ exists.
25. Use exercises 22 and 23 to give a new proof of theorem 4.1C.

4.2 METRIC SPACES

4.2A. In the proofs of theorems 2.7A, 2.7D, 2.7G, and 2.7I, and their counterparts 4.1C and 4.1D, the following crucial properties of the absolute-value function were used.

$$|0|=0, \tag{1}$$

$$|a|>0 \qquad (a\in R, a\neq 0), \tag{2}$$

$$|a|=|-a| \qquad (a\in R), \tag{3}$$

$$|a+b|\leqslant |a|+|b| \qquad (a,b\in R). \tag{4}$$

Now, for $x,y\in R$, the geometric interpretation of $|x-y|$ is the distance from x to y. If we define the "distance function" ρ by

$$\rho(x,y)=|x-y| \qquad (x,y\in R),$$

then the properties (1)–(4) have the following consequences for any points $x,y,z\in R$:

$$\rho(x,x)=0. \tag{5}$$

(That is, the distance from a point to itself is 0.)

$$\rho(x,y)>0 \qquad (x\neq y). \tag{6}$$

(The distance between two distinct points is strictly positive.)

$$\rho(x,y)=\rho(y,x). \tag{7}$$

(The distance from x to y is equal to the distance from y to x.)

$$\rho(x,y)\leqslant \rho(x,z)+\rho(z,y) \qquad \text{(triangle inequality).} \tag{8}$$

[This is proved by setting $a=x-z, b=z-y$ in (4). The inequality (8) says that going from x to y directly never takes longer than going from x to z and then to y.]

A satisfactory definition of limit can be constructed, not only for R, but for any set M which has a "distance function" ρ satisfying (5)–(8). A "distance function" is usually called a metric.

4.2B. DEFINITION. Let M be any set. A metric for M is a function ρ with domain $M\times M$ and range contained in $[0,\infty)$ such that

$$\rho(x,x)=0 \qquad (x\in M),$$

$$\rho(x,y)>0 \qquad (x,y\in M, x\neq y),$$

$$\rho(x,y)=\rho(y,x) \qquad (x,y\in M),$$

$$\rho(x,y)\leqslant \rho(x,z)+\rho(z,y) \qquad (x,y,z\in M) \qquad \text{(triangle inequality).}$$

If ρ is a metric for M, then the ordered pair $\langle M,\rho\rangle$ is called a metric space. (In many cases we abuse language slightly and refer to the metric space $\langle M,\rho\rangle$ simply by M. Thus if we say "let M be a metric space," there is always a metric ρ for M lurking in the background.)

A metric for M thus has all the properties (5)–(8) of the distance function $|x-y|$ for R.

4.2C. Here are five examples of metric spaces.

1. The function ρ defined by $\rho(x,y)=|x-y|$ is obviously a metric for the set R of real numbers. We denote the resulting metric space $\langle R,\rho\rangle$ by R^1. We call this metric ρ the absolute value metric.

2. Here is another metric for the set R. Define $d: R \times R \to [0, \infty)$ by
$$d(x,x)=0 \qquad (x \in R),$$
$$d(x,y)=1 \qquad (x,y \in R; \quad x \neq y).$$
That is, the "distance" $d(x,y)$ between any two distinct points $x,y \in R$ is 1. The reader should verify that d is a metric for R. The metric d is called the *discrete metric*. We will henceforth denote the metric space $\langle R,d \rangle$ by R_d. The examples 1 and 2 show that a given set may have more than one metric.
3. Fix $n \in I$. If $x = \langle x_1, \ldots, x_n \rangle$ and $y = \langle y_1, \ldots, y_n \rangle$ are two ordered n-tuples of real numbers, define
$$\rho(x,y) = \left[\sum_{k=1}^{n} (x_k - y_k)^2 \right]^{1/2}.$$
[For $n=2$, $\rho(x,y)$ is thus the usual distance formula for points in the Cartesian plane.] We will show that ρ satisfies the triangle inequality. Thus, if $z = \langle z_1, \ldots, z_n \rangle$, we must show $\rho(x,y) \leqslant \rho(x,z) + \rho(z,y)$. For $k = 1, \ldots, n$ let $a_k = x_k - z_k, b_k = z_k - y_k$. Then
$$\rho(x,z) = \left(\sum_{k=1}^{n} a_k^2 \right)^{1/2},$$
$$\rho(z,y) = \left(\sum_{k=1}^{n} b_k^2 \right)^{1/2},$$
and
$$\rho(x,y) = \left[\sum_{k=1}^{n} (a_k + b_k)^2 \right]^{1/2}.$$
We must thus show that
$$\left[\sum_{k=1}^{n} (a_k + b_k)^2 \right]^{1/2} \leqslant \left(\sum_{k=1}^{n} a_k^2 \right)^{1/2} + \left(\sum_{k=1}^{n} b_k^2 \right)^{1/2}.$$
But this follows from 3.10C. It is trivial to verify that ρ satisfies the other requirements for a metric. We denote by R^n the metric space formed by the set of all n-tuples of real numbers with this metric ρ. The metric space R^n is called Euclidean n-space. (Note that for $n=1$, R^n becomes the R^1 of example 1 since
$$\left[\sum_{k=1}^{1} (x_k - y_k)^2 \right]^{1/2} = |x_1 - y_1|.$$
4. Let ℓ^∞ denote the set of all bounded sequences of real numbers. If $x = \{x_n\}_{n=1}^\infty$ and $y = \{y_n\}_{n=1}^\infty$ are points in ℓ^∞, define
$$\rho(x,y) = \underset{1 \leqslant n < \infty}{\text{l.u.b.}} |x_n - y_n|.$$
For example, if $x = \{1 + 1/n\}_{n=1}^\infty$, $y = \{2 - 1/n\}_{n=1}^\infty$, then
$$\rho(x,y) = \underset{1 \leqslant n < \infty}{\text{l.u.b.}} \left| \left(1 + \frac{1}{n}\right) - \left(2 - \frac{1}{n}\right) \right| = \underset{1 \leqslant n < \infty}{\text{l.u.b.}} \left| -1 + \frac{2}{n} \right| = 1.$$
Again, it is easy to see that ρ satisfies the first three requirements of a metric. To demonstrate the triangle inequality, let $z = \{z_n\}_{n=1}^\infty$ also be a point in ℓ^∞. For any

$k \in I$ we have

$$|x_k - y_k| = |x_k - z_k + z_k - y_k| \leqslant |x_k - z_k| + |z_k - y_k|$$
$$\leqslant \underset{1 \leqslant n < \infty}{\text{l.u.b.}} |x_n - z_n| + \underset{1 \leqslant n < \infty}{\text{l.u.b.}} |z_n - y_n|,$$

and so

$$|x_k - y_k| \leqslant \rho(x,z) + \rho(z,y) \qquad (k \in I).$$

From this it follows that $\text{l.u.b.}_{1 \leqslant k < \infty} |x_k - y_k| \leqslant \rho(x,z) + \rho(z,y)$ (why?), and this is the triangle inequality for ρ.

It is customary to denote this metric space $\langle \ell^\infty, \rho \rangle$ simply by ℓ^∞. (The reason for the ∞ symbol on the ℓ is usually disclosed in the next course in analysis.)

5. For a final example of a metric space, consider the set ℓ^2 from Section 3.10. For $x,y \in \ell^2$ define $\rho(x,y) = \|x - y\|_2$. Then theorem 3.10E shows that ρ is a metric for ℓ^2. For example, using (3) of 3.10E, we have

$$\rho(x,y) = \|x - y\|_2 = \|(-1)(y - x)\|_2$$
$$= |-1| \cdot \|y - x\|_2 = \|y - x\|_2 = \rho(y,x).$$

Also, for $x,y,z \in \ell^2$ we have, using (4) of 3.10E,

$$\rho(x,y) = \|x - y\|_2 = \|x - z + z - y\|_2$$
$$\leqslant \|x - z\|_2 + \|z - y\|_2 = \rho(x,z) + \rho(z,y).$$

We denote the metric space $\langle \ell^2, \rho \rangle$ simply by ℓ^2.

We have thus listed R^1, R_d, R^n, ℓ^∞, and ℓ^2 as examples of metric spaces.

It is important to note that if ρ is a metric for the set M, then ρ defines a metric for any subset of M in an obvious way. For example, $\rho(x,y) = |x - y|$ defines a metric for any closed interval $[a,b]$ of real numbers.

4.2D. In the next section we will make use of the concept of cluster point.

DEFINITION. Let $\langle M, \rho \rangle$ be a metric space and suppose $A \subset M$. The point $a \in M$ is called a cluster point of A in M if, for every $h > 0$, there exists a point $x \in A$ such that $0 < \rho(x,a) < h$.

That is, a is a cluster point of A if there are points of A distinct from a but arbitrarily near a. Note that a need not belong to A.

For example, let $M = R^1$ and let $A = \{1, \frac{1}{2}, \frac{1}{4}, \frac{1}{8}, \ldots\}$. Then 0 is the only cluster point of A in M. See Figure 18.

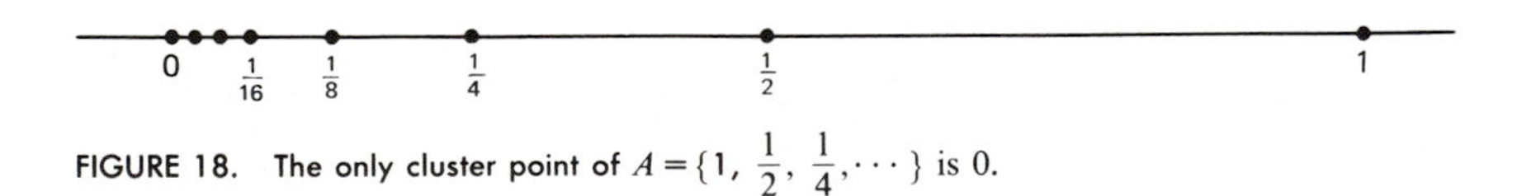

FIGURE 18. The only cluster point of $A = \{1, \frac{1}{2}, \frac{1}{4}, \cdots\}$ is 0.

Exercises 4.2

1. Show that if ρ is a metric for a set M, then so is 2ρ.
2. Show that if ρ and σ are both metrics for a set M, then $\rho + \sigma$ is also a metric for M.
3. Suppose ρ_1 and ρ_2 are metrics for the set M. Prove that $\max(\rho_1, \rho_2)$ is also a metric for M.
4. Let $\langle M, \rho \rangle$ be a metric space. Prove that $\min(1, \rho)$ is also a metric for M.

5. Let

$$\rho(x,y)=\left|\frac{1}{x}-\frac{1}{y}\right| \qquad (x>0,y>0).$$

Prove that ρ is a metric for $(0,\infty)$.

6. Let $\langle M,\rho\rangle$ be a metric space. Prove that

$$|\rho(x,y)-\rho(x,y)|\leqslant\rho(y,z) \qquad (x,y,z\in M).$$

7. Let ℓ^1 be the class of all sequences $\{s_n\}_{n=1}^{\infty}$ of real numbers such that $\Sigma_{n=1}^{\infty}|s_n|<\infty$. Show that if $s=\{s_{n=1}$ and $t=\{t_n\}_{n=1}^{\infty}$ are in ℓ^1, then $\rho(s,t)=\Sigma_{n=1}^{\infty}|s_n-t_n|$ defines a metric for ℓ^1.

8. For $P\langle x_1,y_1\rangle$ and $Q\langle x_2,y_2\rangle$, define

$$\sigma(P,Q)=|x_1-x_2|+|y_1-y_2|.$$

Show that σ is a metric for the set of ordered pairs of real numbers.

Also, if

$$\tau(P,Q)=\max(|x_1-x_2|,|y_1-y_2|),$$

show that τ defines a metric for the same set.

9. Let 0 denote the point $\langle 0,0\rangle$ in R^2. For σ,τ as in Exercise 4, sketch the following subsets of R^2:

$$A=\{P\in R^2|\sigma(0,P)<1\},$$

$$B=\{P\in R^2|\tau(0,P)<1\}.$$

Compare with

$$C=\{P\in R^2|\rho(0,P)<1\},$$

where ρ is the metric for R^2.

10. If P, Q, R are points in R^3 and $\rho(P,R)=\rho(P,Q)+\rho(Q,R)$, what can you say about the relative position of P, Q, R?

Answer the same question with R_d in place of R^3.

11. Let A denote the open interval $(0,1)$. Show that the set of cluster points of A in R^1 is $[0,1]$.

12. If $A=(0,1)$, find the set of cluster points of A in R_d.

4.3 LIMITS IN METRIC SPACES

If we examine definition 4.1A we see that $\lim_{x\to a}f(x)=L$ means that given $\epsilon>0$ there exists $\delta>0$ such that the distance from $f(x)$ to L is less than ϵ provided that the distance from x to a is less than δ (but greater than 0). Now that we have stated this definition in terms of distances, it is not difficult to see how to formulate the corresponding definition for arbitrary metric spaces.

Suppose that $\langle M_1,\rho_1\rangle$ and $\langle M_2,\rho_2\rangle$ are metric spaces, that $a\in M_1$, and that f is a function whose range is contained in M_2 and whose domain contains all $x\in M_1$ such that $\rho_1(a,x)<h$ (for some $h>0$) except possibly $x=a$. We also assume that a is a cluster point of the domain of f.

4.3A. DEFINITION. We say that $f(x)$ approaches L (where $L\in M_2$) as x approaches a if given $\epsilon>0$, there exists $\delta>0$ such that

$$\rho_2(f(x),L)<\epsilon \qquad (0<\rho_1(x,a)<\delta).$$

In this case we write $\lim_{x\to a} f(x)=L$, or $f(x)\to L$ as $x\to a$. If $\langle M_1,\rho_1\rangle=\langle M_2,\rho_2\rangle=R^1$, then $\rho_2(f(x),L)=|f(x)-L|$, $\rho_1(x,a)=|x-a|$, and 4.3A reduces to 4.1A.

In later chapters we very often consider functions f on the metric space $M=[a,b]$ (closed bounded interval with absolute-value metric). For this space the statement

$$\lim_{x\to a} f(x)=L \tag{*}$$

involves only points x to the right of a (since points in R^1 to the left of a are not in M). In 4.1F, L is referred to as the "right-hand limit of f." However, there is no need for us to use this terminology as long as we remember on what space f is defined. A similar remark applies to

$$\lim_{x\to b} f(x)=N.$$

These remarks are also relevant when we define the derivative of a real-valued function on $[a,b]$.

Here is an example illustrating 4.3A. Let $f: \ell^2\to R^1$ be defined as follows: If $x=\{x_n\}_{n=1}^{\infty}\in \ell^2$, let $f(x)=x_1$. That is, the image under f of any sequence in ℓ^2 is the first term of the sequence. Now let $a=\{a_n\}_{n=1}^{\infty}$ be any fixed element of ℓ^2. We will prove that $\lim_{x\to a} f(x)=a_1$. Given $\epsilon>0$ we must find $\delta>0$ such that the distance from $f(x)$ to a_1 (in the metric for R^1) is less than ϵ whenever the distance from x to a (in the metric for ℓ^2) is less than δ but greater than 0. That is, we must find $\delta>0$ such that

$$|f(x)-a_1|<\epsilon \qquad (0<\|x-a\|_2<\delta),$$

or

$$|x_1-a_1|<\epsilon \qquad (0<\|x-a\|_2<\delta). \tag{1}$$

But

$$\|x-a\|_2=\left[\sum_{n=1}^{\infty}(x_n-a_n)^2\right]^{1/2}\geqslant\left[(x_1-a_1)^2\right]^{1/2}=|x_1-a_1|,$$

and so $|x_1-a_1|\leqslant\|x-a\|_2$. If we thus choose $\delta=\epsilon$, then $\|x-a\|_2<\delta=\epsilon$ implies $|x_1-a_1|\leqslant\|x-a\|_2<\epsilon$ and (1) holds. This proves $\lim_{x\to a} f(x)=a_1$. [Note that $a_1=f(a)$ so that we have shown $\lim_{x\to a} f(x)=f(a)$.]

We most often apply definition 4.3A to real-valued functions—that is, when $\langle M_2,\rho_2\rangle=R_1$. The proof of the following theorem is then an exact duplicate of the proofs of 4.1C and 4.1D.

4.3B. THEOREM. Let $\langle M,\rho\rangle$ be a metric space and let a be a point in M. Let f and g be real-valued* functions whose domains are subsets of M. If† $\lim_{x\to a} f(x)=L$ and $\lim_{x\to a} g(x)=N$, then

$$\lim_{x\to a}[f(x)+g(x)]=L+N,$$

$$\lim_{x\to a}[f(x)-g(x)]=L-N,$$

$$\lim_{x\to a}[f(x)\,g(x)]=LN,$$

* Henceforth, whenever we use the phrase "real-valued function," we mean a function with range in R^1. That is, the metric in the range is the absolute-value metric.
†See footnote p. 113.

and, if $N \neq 0$,

$$\lim_{x \to a} \frac{f(x)}{g(x)} = \frac{L}{N}.$$

PROOF: We prove only that $\lim_{x \to a} f(x)g(x) = LN$. (Compare with the second proof of 2.7G.)

Since $\lim_{x \to a} g(x) = N$ we have, for some $\delta_1 > 0$,

$$|g(x) - N| < 1 \qquad (0 < \rho(x,a) < \delta_1).$$

Thus

$$|g(x)| < |N| + 1 = Q \qquad (0 < \rho(x,a) < \delta_1).$$

Now

$$\begin{aligned} f(x)g(x) - LN &= f(x)g(x) - Lg(x) + Lg(x) - LN \\ &= g(x)[f(x) - L] + L[g(x) - N], \end{aligned}$$

$$|f(x)g(x) - LN| \leqslant |g(x)| \cdot |f(x) - L| + |L| \cdot |g(x) - N|.$$

Hence if $0 < \rho(x,a) < \delta_1$,

$$|f(x)g(x) - LN| \leqslant Q \cdot |f(x) - L| + |L| \cdot |g(x) - N|. \tag{1}$$

Given $\epsilon > 0$ there exists $\delta_2 > 0$ such that

$$Q|f(x) - L| < \frac{\epsilon}{2} \qquad (0 < \rho(x,a) < \delta_2), \tag{2}$$

and there exists $\delta_3 > 0$ such that

$$|L||g(x) - N| < \frac{\epsilon}{2} \qquad (0 < \rho(x,a) < \delta_3). \tag{3}$$

If we let $\delta = \min(\delta_1, \delta_2, \delta_3)$, then from (1), (2), and (3) it follows that

$$|f(x)g(x) - LN| < \epsilon \qquad (0 < \rho(x,a) < \delta).$$

This proves $\lim_{x \to a} f(x)g(x) = LN$.

4.3C. A sequence of points in a metric space M is a function from I into M. As with sequences of real numbers, we will use the notation $\{a_n\}_{n=1}^{\infty}$ for a sequence of points M. For such sequences, convergence is defined as in 2.2A and 2.3A.

DEFINITION. Let $\langle M, \rho \rangle$ be a metric space and let $\{s_n\}_{n=1}^{\infty}$ be a sequence of points in M. We say that s_n approaches L (where $L \in M$) as n approaches infinity if given $\epsilon > 0$, there exists $N \in I$ such that

$$\rho(s_n, L) < \epsilon \qquad (n \geqslant N).$$

In this case we write $\lim_{n \to \infty} s_n = L$, or $s_n \to L$ as $n \to \infty$ and say that $\{s_n\}_{n=1}^{\infty}$ is convergent in M to the point L.

Cauchy sequences are defined as in 2.10A.

4.3D. DEFINITION. Let $\langle M, \rho \rangle$ be a metric space and let $\{s_n\}_{n=1}^{\infty}$ be a sequence of points in M. We say that $\{s_n\}_{n=1}^{\infty}$ is a Cauchy sequence if given $\epsilon > 0$, there exists $N \in I$ such that

$$\rho(s_m, s_n) < \epsilon \qquad (m, n \geqslant N).$$

The proof of the following theorem is identical to that of 2.10B.

4.3E. THEOREM. Let $\langle M,\rho\rangle$ be a metric space. If $\{s_n\}_{n=1}^{\infty}$ is a convergent sequence of points of M, then $\{s_n\}_{n=1}^{\infty}$ is Cauchy.

4.3F. Now comes a very important point. For some metric spaces there are Cauchy sequences which are *not* convergent. That is, theorem 2.10D cannot be extended to all metric spaces.

For example, let M be the set of all points $\langle x,y\rangle$ in the Euclidean plane R^2 such that $x^2+y^2<1$, with the R^2 metric used as metric for M. The sequence $A=\{\langle 0,n/n+1\rangle\}_{n=1}^{\infty}$ is a Cauchy sequence of points in M but there is no $L\in M$ such that A is convergent to L! (Draw a picture, then verify.) Hence the sequence of A points of M does not converge in M.

Of course, the sequence A considered as a sequence of points in R^2 does converge to the point $\langle 0,1\rangle$ in R^2. But the fact remains that A is not a convergent sequence in M (according to 4.3C) even though it is Cauchy.

The reader should carefully reexamine the proof of 2.10D to see where properties special to R^1 are used and thus why the proof does not immediately extend to cover all metric spaces as did the proof of 2.10B.

Exercises 4.3

1. Show that a sequence of points in any metric space cannot converge to two distinct limits.
2. For each $n\in I$ let $P_n=\langle x_n,y_n\rangle$ be a point in R^2. Show that $\{P_n\}_{n=1}^{\infty}$ converges to $P=\langle x,y\rangle$ in R^2 if and only if $\{x_n\}_{n=1}^{\infty}$ and $\{y_n\}_{n=1}^{\infty}$ converge in R^1 to x and y, respectively.
3. Let $s=\{1/k\}_{k=1}^{\infty}$. Find a sequence $\{s_n\}_{n=1}^{\infty}$ of points in ℓ^2 such that each s_n is distinct from s and such that $\{s_n\}_{n=1}^{\infty}$ converges to s in ℓ^2.
4. Suppose that ρ and σ are metrics for M such that
$$\lim_{n\to\infty} x_n = x \text{ in } \langle M,\rho\rangle$$
if and only if
$$\lim_{n\to\infty} x_n = x \text{ in } \langle M,\sigma\rangle.$$
(That is, a sequence converges in $\langle M,\rho\rangle$ if and only if it converges in $\langle M,\sigma\rangle$ and the limits are the same.) We then say that ρ and σ are equivalent.

   Prove that the usual metric for R^2, and the metrics τ and σ of Exercise 8 of Section 4.2 are all equivalent to one another.
5. If ρ and σ are metrics for M, and if there exists $k>1$ such that
$$\frac{1}{k}\sigma(x,y)\leqslant\rho(x,y)\leqslant k\sigma(x,y)\qquad (x,y\in M),$$
prove that ρ and σ are equivalent.
6. Show that if $\{s_n\}_{n=1}^{\infty}$ is a Cauchy sequence in a metric space $\langle M,\rho\rangle$, then the sequence of real numbers $\{\rho(s_1,s_n)\}_{n=1}^{\infty}$ is bounded.
7. If $\{x_n\}_{n=1}^{\infty}$ is a Cauchy sequence of points in the metric space M, and if $\{x_n\}_{n=1}^{\infty}$ has a subsequence which converges to $x\in M$, prove that $\{x_n\}_{n=1}^{\infty}$ itself is convergent to x.
8. Show that if $\{x_n\}_{n=1}^{\infty}$ is a convergent sequence in R_d, then there exists $N\in I$ such

that $x_N = x_{N+1} = x_{N+2} \cdots$. (That is, a sequence in R_d is convergent if and only if all the terms of the sequence are the same from some point on.)

9. Show that every Cauchy sequence in R_d *is* convergent.
10. Let $\{x_n\}_{n=1}^{\infty}$ and $\{y_n\}_{n=1}^{\infty}$ be convergent sequences in a metric space $\langle M, \rho \rangle$. Prove that $\{\rho(x_n, y_n)\}_{n=1}^{\infty}$ is convergent in R^1.
11. Let $\{x_n\}_{n=1}^{\infty}$ and $\{y_n\}_{n=1}^{\infty}$ be Cauchy sequences in a metric space $\langle M, \rho \rangle$. Prove that $\{\rho(x_n, y_n)\}_{n=1}^{\infty}$ is Cauchy in R^1.
12. Explain why we do not want definition 4.3A to apply if a is not a cluster point of the domain of f.

5

CONTINUOUS FUNCTIONS ON METRIC SPACES

5.1 FUNCTIONS CONTINUOUS AT A POINT ON THE REAL LINE

Theorems about continuous real-valued functions on a closed bounded interval $[a,b]$ such as, "If f is continuous on $[a,b]$, then f takes on a maximum and a minimum value," and "If f is continuous on $[a,b]$, then f takes on every value between $f(a)$ and $f(b)$," are tools in the proof of the basic theorems in differential and integral calculus. We deduce these theorems as special cases of theorems about continuous functions on metric spaces. However, we first review the concept of continuity in its most elementary form.

Let a be a point in R^1 and suppose f is a real-valued function domain contains all points of some open interval* $(a-h, a+h)$ including a itself.

5.1A. DEFINITION. We say that the function f is continuous at $a \in R^1$ if $\lim_{x \to a} f(x) = f(a)$.

The definition really demands that two conditions be fulfilled in order that f be continuous at a. The first condition is that the $\lim_{x \to a} f(x)$ exist; the second is that this limit be equal to $f(a)$. In particular, if $f(a)$ is not defined, then f cannot be continuous at a. For example, the function f defined by

$$f(x) = \frac{\sin x}{x} \qquad (x \in R^1, x \neq 0)$$

is not defined at $x=0$ and hence is not continuous at $x=0$ even though $\lim_{x \to 0}(\sin x / x)$ exists (and is equal to 1).

However, the function g defined by

$$g(x) = \frac{\sin x}{x} \qquad (x \neq 0),$$

$$g(0) = 1,$$

is continuous at $x=0$ since $\lim_{x \to 0} g(x) = g(0)$.

* $h > 0$, of course.

It is often the case that a function f fails to be continuous at a point a because $\lim_{x\to a} f(x)$ does not exist. Consider, for example, the characteristic function χ of the rational numbers. That is,

$$\chi(x)=1 \qquad (x\in R^1, x \text{ rational}),$$

$$\chi(x)=0 \qquad (x\in R^1, x \text{ irrational}).$$

Then $\chi(a)$ is defined for any $a\in R^1$ but $\lim_{x\to a}\chi(x)$ does not exist for any a. To see this, assume the contrary—that $\lim_{x\to a}\chi(x)=L$ for some $L\in R^1$. Given $\epsilon=\frac{1}{3}$ there would exist $\delta>0$ such that $|\chi(x)-L|<\frac{1}{3}$ if $0<|x-a|<\delta$. But in the interval $(a,a+\delta)$, say, there is both a rational number and an irrational. If $x\in(a,a+\delta)$ is rational, we would have $|1-L|<\frac{1}{3}$, while if $x\in(a,a+\delta)$ is irrational, we would have $|0-L|<\frac{1}{3}$. A contradiction follows easily.

On the other hand, most of the functions that are "easy to write down" turn out to be continuous at all points where they are defined. In Section 4.1, for example, we proved that $\lim_{x\to 3}(x^2+2x)=15$. This shows that the function f defined by

$$f(x)=x^2+2x \qquad (x\in R^1)$$

is continuous at $x=3$. For $f(3)=15$ and $\lim_{x\to 3}f(x)=15$. The next example in Section 4.1 shows that the function g defined by

$$g(x)=\sqrt{x+3} \qquad (0<x<2)$$

is continuous at $x=1$.

From theorems 4.1C and 4.1D we deduce the following important result.

5.1B. THEOREM. If the real-valued functions f and g are continuous at $a\in R^1$, then so are $f+g, f-g$, and fg. If $g(a)\neq 0$, then f/g is also continuous at a.

PROOF: Since f and g are continuous at a we have

$$\lim_{x\to a} f(x)=f(a), \quad \lim_{x\to a} g(x)=g(a).$$

Then, by 4.1C, $\lim_{x\to a}[f(x)+g(x)]=f(a)+g(a)$. In other "words,':

$$\lim_{x\to a}(f+g)(x)=(f+g)(a).$$

This proves that $f+g$ is continuous at a. The remainder of the theorem is proved similarly.

A continuous function of a continuous function is continuous. More precisely,

5.1C. THEOREM. If f and g are real-valued functions, if f is continuous at a, and if g is continuous at $f(a)$, then $g\circ f$ is continuous at a.

PROOF: We must show $\lim_{x\to a} g\circ f(x)=g\circ f(a)$ or,

$$\lim_{x\to a} g[f(x)]=g[f(a)].$$

That is, given $\epsilon>0$ we must find $\delta>0$ such that

$$|g[f(x)]-g[f(a)]|<\epsilon \qquad (0<|x-a|<\delta). \tag{1}$$

Let $b=f(a)$. Now by hypothesis $\lim_{y\to b} g(y)=g(b)$. Hence there exists $\eta>0$ such that

$$|g(y)-g(b)|<\epsilon \qquad (|y-b|<\eta). \tag{2}$$

(Why didn't we have to write $0<|y-b|<\eta$?) But, also by hypothesis,

$$\lim_{x\to a} f(x)=f(a).$$

Thus (using η where we usually use ϵ) there exists δ such that

$$|f(x)-f(a)|<\eta \qquad (|x-a|<\delta),$$

or

$$|f(x)-b|<\eta \qquad (|x-a|<\delta). \tag{3}$$

Thus if $|x-a|<\delta$, then $f(x)$ is within η of b and so we may substitute $f(x)$ for y in (2). Hence

$$|g[f(x)]-g(b)|<\epsilon \qquad (|x-a|<\delta),$$

which implies (1), and the proof is complete. (We give a more elegant proof of this theorem later on.) See Figure 19.

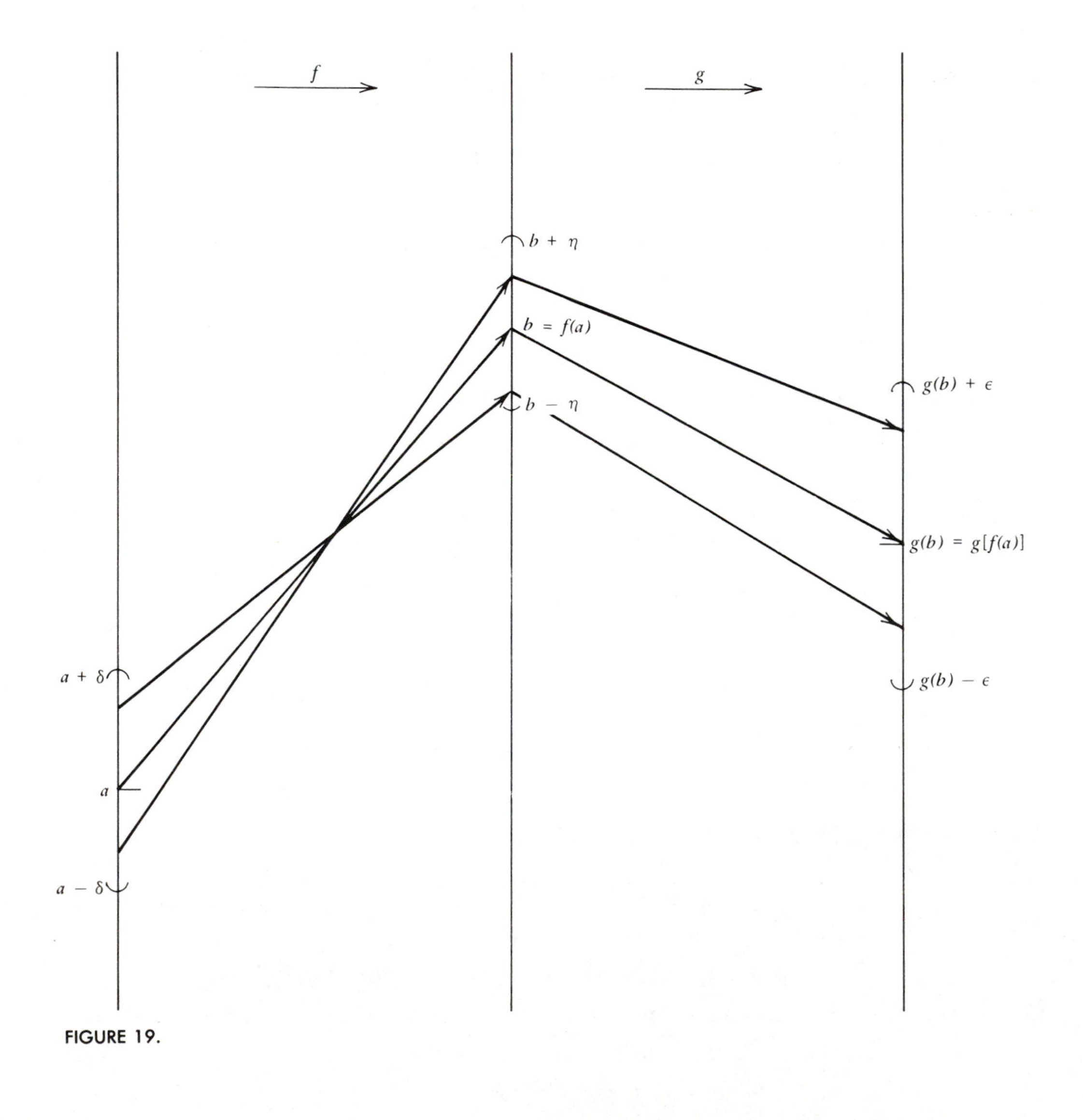

FIGURE 19.

Exercises 5.1

1. Which of the functions in Figures 13–17 are continuous at a?
2. If f is continuous at a, and if $c \in R$, prove that cf is continuous at a.
3. If f is continuous at a and $f(a)>0$, prove that there exists $h>0$ such that

$$f(x)>0 \qquad (a-h<x<a+h).$$

(*Hint*: Set $f(a)=2\epsilon$.)

4. Prove that if f is continuous at $a \in R^1$, then $|f|$ is also continuous at a.
5. If f is continuous at a and $f(a)\neq 0$, prove that there exists $h>0$ such that

$$|f(x)|>0 \qquad (a-h<x<a+h).$$

6. Prove that if both f and g are continuous at a, then $\max(f,g)$ and $\min(f,g)$ are also continuous at a.
7. If

$$f(x)=x \qquad (-\infty<x<\infty),$$

prove that f is continuous at each point in R^1.

8. If $n \in I$ and

$$f(x)=x^n \qquad (-\infty<x<\infty),$$

prove that f is continuous at each point in R^1.

9. Prove that any polynomial function is continuous at each point in R^1.
10. (a) Prove that if

$$g(x)=\sqrt{x} \qquad (0<x<\infty),$$

then g is continuous at each point of $(0,\infty)$.

(b) Prove that if

$$h(x)=\sqrt{1-x^2} \qquad (-1<x<1),$$

then h is continuous at each point of $(-1,1)$. [Use 5.1C. Note that $h=g\circ f$ where g is as in part (a) and

$$f(x)=1-x^2 \qquad (-1<x<1).]$$

5.2 REFORMULATION

We have defined "f is continuous at a" to mean $\lim_{x\to a} f(x)=f(a)$. That is, f is continuous at a if for any $\epsilon>0$ there exists $\delta>0$ such that $|f(x)-f(a)|<\epsilon$ if $0<|x-a|<\delta$. However (as you were asked to observe in the last proof), the inequality $|f(x)-f(a)|<\epsilon$ obviously holds if $x=a$. Thus we need only write $|x-a|<\delta$ instead of $0<|x-a|<\delta$. Here, then, is a reformulation of definition 5.1A.

5.2A. THEOREM. The real-valued function f is continuous at $a \in R^1$ if and only if given $\epsilon>0$ there exists $\delta>0$ such that

$$|f(x)-f(a)|<\epsilon \qquad (|x-a|<\delta).$$

By 5.2A then, f is continuous at a if for any $\epsilon>0$, there exists $\delta>0$ such that, if the distance from x to a is less than δ, then the distance from $f(x)$ to $f(a)$ will be less than ϵ. [This is sometimes put roughly as "if x is close to a, then $f(x)$ is close to $f(a)$."] Theorem 5.2A shows that the definition of continuity is based on the metric in R^1.

Consider Figure 20. In order for f to be continuous at a, given any ϵ parentheses about $f(a)$ there must be δ parentheses about a such that an arrow which begins inside the δ parentheses must end inside the ϵ parentheses.

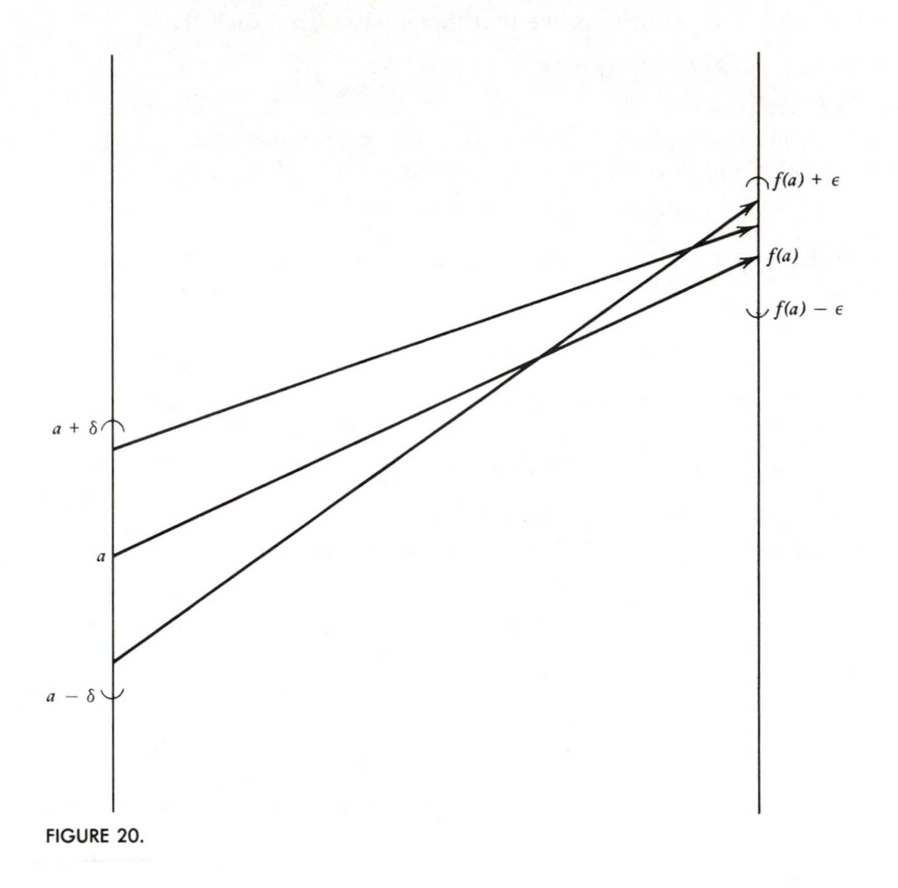

FIGURE 20.

In order to give another reformulation of the definition of continuity we introduce the following definition.

5.2B. DEFINITION. If $a \in R^1$ and $r > 0$, we define $B[a;r]$ to be the set of all $x \in R^1$ whose distance to a is less than r. That is,

$$B[a;r] = \{x \in R^1 \,|\, |x-a| < r\}.$$

We call $B[a;r]$ the open ball of radius r about a.

It is clear that $B[a;r]$ is just a fancy way of denoting the bounded open interval $(a-r, a+r)$. However, in an arbitrary metric space there is no such thing as an interval. But the object $B[a;r]$ does have a counterpart in any metric space, which is the reason we define it in terms of distance.

Theorem 5.2A thus reads "f is continuous at a if and only if given $\epsilon > 0$ there exists $\delta > 0$ such that $f(x) \in B[f(a);\epsilon]$ if $x \in B[a;\delta]$." That is, the entire open ball $B[a;\delta]$ is mapped by f into the open $B[f(a);\epsilon]$.

Thus f is continuous at a if and only if, for any open ball B about $f(a)$, there is an open ball about a which f maps entirely into B. It turns out to be more useful to state this definition in terms of inverse images.

5.2C. THEOREM. The real-valued function f is continuous at $a \in R^1$ if and only if the inverse image* under f of any open ball $B[f(a);\epsilon]$ about $f(a)$ contains an open ball $B[a;\delta]$ about a. (That is, given $\epsilon > 0$ there exists $\delta > 0$ such that

$$f^{-1}(B[f(a);\epsilon]) \supset B[a;\delta].)$$

It is of the utmost importance that the reader fully understand 5.2C before going on.

Our final reformulation of the continuity concept will be in terms of sequences. Observe first that the sequence $\{x_n\}_{n=1}^{\infty}$ converges to a if and only if given $\epsilon > 0$ there exists $N \in I$ such that

$$x_n \in B[a;\epsilon] \qquad (n \geqslant N).$$

That is, given any open ball B about a, all but a finite number of the x_n are in B.

5.2D. THEOREM. The real-valued function f is continuous at $a \in R^1$ if and only if, whenever $\{x_n\}_{n=1}^{\infty}$ is a sequence of real numbers converging to a, then the sequence $\{f(x_n)\}_{n=1}^{\infty}$ converges to $f(a)$. That is, f is continuous at a if and only if

$$\lim_{n\to\infty} x_n = a \quad \text{implies} \quad \lim_{n\to\infty} f(x_n) = f(a). \tag{*}$$

PROOF: Let us first assume that f is continuous at a and prove that (*) holds. Let $\{x_n\}_{n=1}^{\infty}$ be any sequence of real numbers converging to a. [Then $f(x_n)$ will be defined for n sufficiently large.] We must show that $\lim_{n\to\infty} f(x_n) = f(a)$—that is, given $\epsilon > 0$ we must find $N \in I$ such that

$$f(x_n) \in B[f(a);\epsilon] \qquad (n \geqslant N). \tag{1}$$

But since f is continuous at a there exists $\delta > 0$ such that

$$f(x) \in B[f(a);\epsilon] \qquad (x \in B[a;\delta]). \tag{2}$$

Furthermore, since $\lim_{n\to\infty} x_n = a$, there exists $N \in I$ such that

$$x_n \in B[a;\delta] \qquad (n \geqslant N). \tag{3}$$

For this N, (1) follows from (2) and (3).

Conversely, suppose (*) holds. We must prove that f is continuous at a. Assume the contrary. Then, by 5.2C for some $\epsilon > 0$ the inverse image under f of $B = B[f(a);\epsilon]$ contains no open ball about a. In particular, $f^{-1}(B)$ does not contain $B[a;1/n]$ for any $n \in I$. Thus for each $n \in I$, there is a point $x_n \in B[a;1/n]$ such that $f(x_n) \notin B$. That is,

$$|x_n - a| < \frac{1}{n} \quad \text{but} \quad |f(x_n) - f(a)| \geqslant \epsilon.$$

This clearly contradicts (*), so f must be continuous at a.

5.2E. We use 5.2D to give an easy proof of 5.1C.

Suppose, then, that the hypotheses of 5.1C hold. By 5.1D, all we need show is that

$$\lim_{n\to\infty} g[f(x_n)] = g[f(a)] \tag{1}$$

where $\{x_n\}_{n=1}^{\infty}$ is any sequence of real numbers such that

$$\lim_{n\to\infty} x_n = a. \tag{2}$$

Since f is continuous at a, (2) and 5.2D imply

$$\lim_{n\to\infty} f(x_n) = f(a). \tag{3}$$

But since g is continuous at $f(a)$, (3) and 5.2D imply (1) and the proof is complete.

* See 1.3C.

Exercises 5.2

1. Use 5.2D to prove 5.1B.
2. Use 5.2C to prove 5.1C.
3. Use 5.2D to prove that if f is continuous at $a \in R^1$, then $|f|$ is also continuous at a.

5.3 FUNCTIONS CONTINUOUS ON A METRIC SPACE

All the formulations of the continuity of a real-valued function at a point in R^1 were based on the metric for R^1. It is therefore easy to extend the concept of continuity to functions from any metric space into another. We first define "open ball" for a metric space.

5.3A. DEFINITION. Let $\langle M, \rho \rangle$ be a metric space. If $a \in M$ and $r > 0$, then $B[a;r]$ is defined to be the set of all points in M whose distance to a is less than r. That is

$$B[a;r] = \{x \in M \mid \rho(x,a) < r\}.$$

We call $B[a,r]$ the open ball of radius r about a.

For example, the open ball of radius 1 about the origin in Euclidean 3-space is the set of all points $\langle x,y,z \rangle$ such that $x^2 + y^2 + z^2 < 1$. This example shows why we use the term "ball."

If M is the closed interval $[0,1]$ with the absolute value metric, then $B[\frac{1}{4}; \frac{1}{2}]$ is the interval $[0, \frac{3}{4})$. (Points in R^1 to the left of 0 are not in M.)

If $M = R_d$, the real line with discrete metric, and if a is any point in R_d, then $B[a;1] = \{a\}$. For the only point in R_d whose distance to a is less than 1 is a itself. On the other hand, $B[a;2] = R_d$.

We now define continuity. Let $\langle M_1, \rho_1 \rangle$ and $\langle M_2, \rho_2 \rangle$ be metric spaces, let $a \in M_1$, and let f be any function whose range is contained in M_2 and whose domain contains some open ball $B[a;h]$ $(h > 0)$.

5.3B. DEFINITION. The function f is continuous at $a \in M_1$ if $\lim_{x \to a} f(x) = f(a)$ (where limit is defined in 4.3A).

The proof of the following theorem consists merely of translating the proofs of 5.2A, 5.2C, and 5.2D into metric-space notation. Suppose f is as in the paragraph preceding 5.3B.

5.3C. THEOREM. The function f is continuous at $a \in M_1$ if and only if any one (and hence all) of the following conditions hold.

(a) Given $\epsilon > 0$ there exists $\delta > 0$ such that

$$\rho_2[f(x), f(a)] < \epsilon \qquad (\rho_1(x,a) < \delta).$$

(b) The inverse image under f of any open ball $B[f(a);\epsilon]$ about $f(a)$ contains an open ball $B[a;\delta]$ about a.

(c) Whenever $\{x_n\}_{n=1}^{\infty}$ is a sequence of points in M_1 converging to a, then the sequence $\{f(x_n)\}_{n=1}^{\infty}$ of points in M_2 converges to $f(a)$.

If a is not a cluster point of the domain of f (see the paragraph preceding 4.3A), then $\lim_{x \to a} f(x)$ is not defined. However, the properties (a), (b), and (c) of 5.3C *do* make sense for such a point a and are equivalent to one another. We may take any one of these properties as the definition of the continuity of f at a. (It will follow that f must be continuous at a. Verify.)

The analogue of 5.1C reads as follows.

5.3D. THEOREM. Let $\langle M_1,\rho_1\rangle$, $\langle M_2,\rho_2\rangle$, $\langle M_3,\rho_3\rangle$ be metric spaces and let $f: M_1 \to M_2$, $g: M_2 \to M_3$.* If f is continuous at $a \in M_1$ and g is continuous at $f(a) \in M_2$, then $g \circ f$ is continuous at a.

PROOF: By (c) of 5.3C all we need show is that

$$\lim_{n\to\infty} g[f(x_n)] = g[f(a)]$$

whenever $\{x_n\}_{n=1}^{\infty}$ is a sequence in M_1 such that

$$\lim_{n\to\infty} x_n = a.$$

The proof then proceeds exactly like the proof of 5.2E.

For real-valued functions on metric spaces there is a generalization of 5.1B. The following theorem may be easily deduced from 4.3B (in the same way that 5.1B was deduced from 4.1C and 4.1D).

5.3E. THEOREM. Let M be a metric space, and let f and g be real-valued functions which are continuous at $a \in M$. Then $f+g$, $f-g$, and fg are also continuous at a. Furthermore, if $g(a) \neq 0$, then f/g is continuous at a.

We emphasize that, so far, only continuity *at a point* has been defined. The continuity of a function f at a point a is a local property—that is, continuity of f at a depends only on "what goes on near a."

Now we will define what we mean by a function continuous on a whole metric space.

5.3F. DEFINITION. Let M_1 and M_2 be metric spaces and let $f: M_1 \to M_2$. We say that f is a continuous function from M_1 into M_2 (or, more simply, f is continuous on M_1) if f is continuous at each point in M_1.

5.3G. THEOREM. If f and g are real-valued continuous functions on a metric space M, then so are $f+g$, $f-g$, and fg. Furthermore, if $g(x) \neq 0$ $(x \in M)$, then f/g is also continuous on M.

PROOF: The proof follows directly from 5.3F and 5.3E.

Any polynomial function f [that is, $f(x) = a_0x^n + a_1x^{n-1} + \cdots + a_n$] is thus a continuous function on R^1. For constant functions are continuous on R^1 and so is the function $g(x) = x$. The function f can be written as a sum of products of these kinds of functions and is thus, by 5.3G, continuous.

The function h defined by $h(x) = (1+x^3)/(1+x^2)$ can be written f/g where f and g are polynomials. Since $g(x)$ is never zero, it follows that h is continuous on R^1.

Here is a more curious illustration. Let f be any function from the metric space R_d into a metric space M. We have already observed that for any $a \in R_d$, the open ball $B[a;1]$ contains only the point a. Thus for any $\epsilon > 0$, the inverse image under f of $B[f(a);\epsilon]$

* For simplicity of statement we are assuming that the domains of f and g are all of M_1 and M_2 respectively.

certainly contains $B[a;1]$. By (b) of 5.3C, this shows that f is continuous at a. Since a was an arbitrary point in R_d we have

5.3H. CURIOSITY. Every function from R_d (into a metric space) is continuous on R_d.

Exercises 5.3

1. Give an example of a function which is continuous on R^1 and whose range is
   (a) $(0, \infty)$ (b) $[0, \infty)$
   (c) $(0,1)$ (d) $[0,1]$.
   You may assume that e^x, $\log x$, $\sin x$, and so on, are continuous where they look continuous.
2. Let f be the function from R^2 onto R^1 defined by
   $$f(\langle x,y\rangle)=x \qquad (\langle x,y\rangle \in R^2).$$
   Show that f is continuous on R^2.
3. If $f: R^2 \to R^2$ is defined by
   $$f(\langle x,y\rangle)=\langle y,x\rangle \qquad (\langle x,y\rangle \in R^2),$$
   show that f is continuous on R^2.
4. If $f: R^1 \to R^1$, $g: R^1 \to R^1$, if f and g are both continuous on R^1, and if
   $$h(\langle x,y\rangle)=\langle f(x),g(y)\rangle \qquad (\langle x,y\rangle \in R^2),$$
   prove that h is continuous on R^2.
5. Define $f: \ell^2 \to \ell^2$ as follows. If $s \in \ell^2$ is the sequence $s_1, s_2, \ldots$, let $f(s)$ be the sequence $0, s_1, s_2, \ldots$. Show that f is continuous on ℓ^2.
6. Let M be a metric space. Suppose $f: M \to R_d$ and that f is 1–1. Show that if f is continuous at $a \in M$, then $\{a\}$ is an open ball in M.
7. True or false: If f is a 1–1 continuous function from a metric space M_1 into a metric space M_2, and if B is an open ball in M_1, then $f(B)$ is an open ball in M_2.
8. Let A be a nonempty set. Find a metric ρ for A for which there exist $r_1, r_2 \in R$ with $0 < r_1 < r_2$ and such that
   $$B[a;r_1] = B[a;r_2]$$
   for every $a \in A$.
9. Let M_1, M_2 be metric spaces and suppose $f: M_1 \to M_2$. Prove that f is continuous if and only if f sends convergent sequences in M_1 to convergent sequences in M_2.
10. For any rational number r in $(0,1)$, write $r = p/q$ where p and q are integers with no common factor and $q > 0$. Then define $f(r) = 1/q$. Define $f(x) = 0$ for all irrational numbers x in $(0,1)$. Thus $f: (0,1) \to [0,1]$.
    (a) Prove that f is *not* continuous at any rational.
    (b) Prove that f *is* continuous at each irrational. (*Hint*: Show that for any $\epsilon > 0$ there are only a finite number of rational numbers p/q in $(0,1)$ such that $1/q \geqslant \epsilon$.)
    (c) Show that f can be extended to a function g on R^1 such that g *is* continuous at each irrational but is *not* continuous at any rational.

5.4 OPEN SETS

In order to formulate properties of continuous functions on a metric space M we need to attach names to various kinds of subsets of M such as open, closed, bounded, totally bounded, compact, and so on. We begin by defining open set.

5.4A. DEFINITION. Let M be a metric space. We say that the subset G of M is an open subset of M (or, more simply, that G is open) if for every $x \in G$, there exists a number $r > 0$ such that the entire open ball $B[x;r]$ is contained in G.

As an intuitive example consider the set A of all points in the plane R^2 inside an ellipse. (Draw a picture.) If $P \in A$, we can draw a circle with center P which lies entirely inside the ellipse. The set B of points inside this circle is then an open ball (in R^2) which lies entirely in A. This shows that A is open in R^2.

Next, let us prove that for an arbitrary metric space $\langle M, \rho \rangle$, any open ball $\mathfrak{B} = B[a;s]$ is itself an open set. (This will justify the use of the word "open" in "open ball.") If $x \in \mathfrak{B}$, we must find $r > 0$ such that $B[x;r] \subset \mathfrak{B}$. Let $t = \rho(x,a)$ and let r be any positive number less than $s - t$. (Why is $s - t$ positive?) If $y \in B[x;r]$, then $\rho(a,y) \leqslant \rho(a,x) + \rho(x,y)$. But $\rho(a,x) = t$, and $\rho(x,y) < r$ since $y \in B[x;r]$. Thus $\rho(a,y) < t + r < t + s - t = s$. Hence $y \in B[a;s] = \mathfrak{B}$, which proves $B[x;r] \subset \mathfrak{B}$, and we are done. See Figure 21.

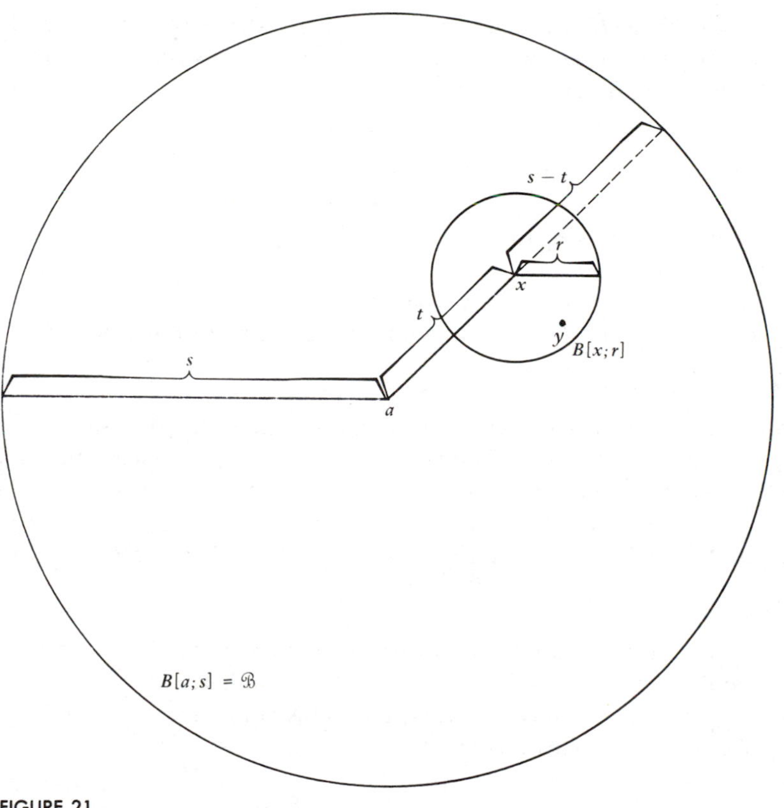

FIGURE 21.

As a third example, consider R_d. If $a \in R_d$, then $\{a\} = B[a;1]$ and hence $\{a\}$ is an open set in R_d. That is, any set with only one point in it is open in R_d.

On the other hand, if $a \in R^1$, then $\{a\}$ is *not* an open set in R^1. For every open ball in R^1 is a nonempty open interval and, certainly, $\{a\}$ contains no such interval.

The last two paragraphs show that whether a set A is open or not depends on what metric space is under consideration. As another illustration of this important point we note that the "half-open" interval $[0, \frac{1}{2})$ is not an open subset of R^1. However $[0, \frac{1}{2})$ *is* an open subset of the metric space $[0,1]$. Indeed $[0,\frac{1}{2})$ is precisely the open ball $B[0;\frac{1}{2}]$ in the metric space $[0,1]$.

5.4B. THEOREM. In any metric space $\langle M,\rho\rangle$ both M and the empty set $\varnothing$ are open sets.

PROOF: If $x\in M$, then (by definition of $B[x;r]$) every open ball $B[x;r]$ is contained in M. Hence M is open. The empty set $\varnothing$ is open simply because there are no x in $\varnothing$ and hence every $x\in\varnothing$ satisfies the condition of 5.4A.

We can put any number of open sets together and obtain a new open set. That is, the union of finitely many, countably many, or even uncountably many open sets is again an open set. Here is the proof.

5.4C. THEOREM. Let $\mathcal{F}$ be any nonempty family of open subsets of a metric space M. Then $\bigcup_{G\in\mathcal{F}} G$ is also an open subset of M.

PROOF: Let $H=\bigcup_{G\in\mathcal{F}} G$. If $x\in H$, we must show that there is an open ball $B[x;r]$ contained in H. But if $x\in H$, then $x\in G$ for some $G\in\mathcal{F}$. Since G is open there is some $B[x;r]$ with $B[x;r]\subset G$. But $G\subset H$ and so $B[x;r]\subset H$, which is what we wished to show.

An interesting consequence of 5.4C is the following.

5.4D. THEOREM. Every subset of R_d is open.

PROOF: In the third example following 5.4A we showed that all one-point subsets of R_d are open. But any subset G of R_d is obviously a union of such sets. By 5.4C, then, G is open.

It is not true, however, that the *intersection* of an *infinite* number of open sets in a metric space is always open. In R^1, for example, if I_n denotes the open interval $(-1/n,1/n)$, then $\bigcap_{n=1}^{\infty} I_n$ contains only 0 and is therefore not open. However,

5.4E. THEOREM. If G_1 and G_2 are open subsets of the metric space M, then $G_1\cap G_2$ is also open.

PROOF: If $x\in G_1\cap G_2$, we must find an open ball $B[x;r]$ contained in $G_1\cap G_2$. Since $x\in G_1$ and G_1 is open, there is a ball $B[x;r_1]$ with $B[x;r_1]\subset G_1$. Similarly, there is a ball $B[x;r_2]$ with $B[x;r_2]\subset G_2$. Thus if $r=\min(r_1,r_2)$, then $B[x;r]$ is contained in G_1 *and* G_2 and thus $B[x;r]\subset G_1\cap G_2$. This completes the proof.

From 5.4E it follows easily by induction that the intersection of any *finite* number of open sets is open.

5.4F. It is useful to know precisely what the open sets in R^1 look like. From 5.4C we know that if $I_1,I_2,\dots$ are open intervals, then $\bigcup_{n=1}^{\infty} I_n$ is an open set in R^1. We now prove the converse.

THEOREM. Every open subset G of R^1 can be written $G=\bigcup I_n$, where $I_1,I_2,\dots$ are a finite number or a countable number of open intervals which are mutually disjoint. (That is, $I_m\cap I_n=\varnothing$ if $m\neq n$.)

PROOF: If $x \in G$, then there is an open interval (open ball) B containing x such that $B \subset G$. Let I_x denote the largest open interval containing x such that $I_x \subset G$.* [I_x may be an unbounded interval, for example, (a, ∞).] Then $G = \cup_{x \in G} I_x$. Now if $x \in G, y \in G$, then either $I_x = I_y$ or $I_x \cap I_y = \varnothing$. For if $I_x \neq I_y$ and $I_x \cap I_y \neq \varnothing$, then $I_x \cup I_y$ would be an open interval contained in G which is larger than I_x. This contradicts the definition of I_x. Finally, each I_x contains a rational number. Since disjoint intervals cannot contain the same rational and since there are only countably many rationals, there cannot be uncountably many mutually disjoint intervals I_x. The theorem now follows.

We can use the notion of open set to give a necessary and sufficient condition that a function on a metric space be continuous. The following theorem is fundamental.

5.4G. THEOREM. Let $\langle M_1, \rho_1 \rangle$ and $\langle M_2, \rho_2 \rangle$ be metric spaces and let $f: M_1 \to M_2$. Then f is continuous on M_1 if and only if $f^{-1}(G)$ is open in M_1 whenever G is open in M_2. (Briefly, f is continuous if and only if the inverse image of every open set is open.)

PROOF: Suppose first that f is continuous on M_1. We wish to show that if G is open in M_2, then $f^{-1}(G)$ is open in M_1. Thus if $x \in f^{-1}(G)$, we must find an open ball $B[x;r]$ contained in $f^{-1}(G)$. Now since $x \in f^{-1}(G)$ then $y = f(x) \in G$. Hence there is an open ball $B[y;s]$ contained in G (since G is open in M_2). By (b) of 5.3C, $f^{-1}(B[y;s])$ contains some $B[x;r]$. Hence $f^{-1}(G) \supset f^{-1}(B[y;s]) \supset B[x;r]$ which is what we wished to show.

Now suppose $f^{-1}(G)$ is open in M_1 whenever G is open in M_2. To show that f is continuous on M_1 it is sufficient to show that f is continuous at an arbitrary point $a \in M_1$. Let $\mathfrak{B} = B[f(a); \epsilon]$ be any ball about $f(a)$. Then $\mathfrak{B}$ is open in M_2 and so, by assumption, $f^{-1}(\mathfrak{B})$ is open in M_1. Since $a \in f^{-1}(\mathfrak{B})$ and $f^{-1}(\mathfrak{B})$ is open, there is an open ball $B[a;\delta]$ contained in $f^{-1}(\mathfrak{B})$. But then, by (b) of 5.3C, f is continuous at a. This completes the proof.

Exercises 5.4

1. This concerns the proof of 5.4F. Show that if G is an open subset of R^1 and if $x \in G$, then there *is* such a thing as the largest open interval I_x containing x such that $I_x \subset G$.
2. Use your intuition to decide which of the following subsets of R^2 are open.
   (a) $\{\langle x,y \rangle \mid x + y = 1\}$.
   (b) $\{\langle x,y \rangle \mid x + y > 1\}$.
   (c) $\{\langle x,y \rangle \mid x \text{ and } y \text{ rational}\}$.
   (d) $R^2 - \{\langle 0,0 \rangle\}$. (That is, R^2 with the origin removed.)
3. Let x_1, x_2 be distinct points in a metric space M. Find disjoint open sets G_1 and G_2 such that $x_1 \in G_1$ and $x_2 \in G_2$.
4. Let E be the set of positive real numbers. Find $f^{-1}(E)$ for each of the following functions f.
   (a) $f(x) = \sin x \quad (-\infty < x < \infty)$.
   (b) $f(x) = x^2 \quad (-\infty < x < \infty)$,
   (c) $f(x) = 0 \quad (x < 0)$,
   $\quad\;\; = 1 \quad (x \geqslant 0)$.
5. Prove that if f is any continuous real-valued function on R^1, then $f^{-1}(E)$ is open in R^1 (where E is as in the preceding exercise).
6. Let f and g be continuous real-valued functions on the metric space M. Let A be the set of all $x \in M$ such that $f(x) < g(x)$. Prove that A is open.

* See Exercise 1.

7. Let A be the set of all sequences $\{s_n\}_{n=1}^{\infty}$ in ℓ^2 such that $\Sigma_{n=1}^{\infty} s_n^2 < 1$. Prove that A is an open subset of ℓ^2.
8. Let G be an open subset of R^1. Prove that χ_G (the characteristic function of G) is continuous at each point of G.
9. Give an example of subsets A and B of R^2 such that all three of the following conditions hold.
   (a) Neither A nor B is open;
   (b) $A \cap B = \varnothing$;
   (c) $A \cup B$ is open.
10. Do the preceding exercise with R^2 replaced by R^1.
11. If A and B are open subsets of R^1, prove that $A \times B$ is an open subset of R^2.

5.5 CLOSED SETS

5.5A. DEFINITION. Let E be a subset of the metric space M. A point $x \in M$ is called a limit point* of E if there is a sequence $\{x_n\}_{n=1}^{\infty}$ of points of E which converges to x. The set $\bar{E}$ of all limit points of E is called the closure of E.

It follows immediately that any point x of E is a limit point of E. For the sequence $x, x, x, \ldots$ of points of E converges to x. Thus if $x \in E$, then $x \in \bar{E}$. In other words,

5.5B. COROLLARY. If E is any subset of the metric space M, then $E \subset \bar{E}$.

It very often happens, however, that $E \neq \bar{E}$ (that is, that E does not contain all its limit points). For example, let E denote the open interval $(0,1)$ considered as a subset of R^1. Then 0 is a limit point of E, since the sequence $\{1/n\}_{n=1}^{\infty}$ of points of E converges to 0. But 0, although a limit point of E, does not lie in E.

On the other hand, the closed interval $[0,1]$ does contain all its limit points (verify).

For a third example, consider Figure 22. If E is the set of points $\langle x,y \rangle$ in R^2 such that $x^2 + y^2 < 1$, then $\bar{E}$ is the set of $\langle x,y \rangle$ such that $x^2 + y^2 \leqslant 1$. This is because every point $P_1 = \langle x_1, y_1 \rangle$ such that $x_1^2 + y_1^2 = 1$ is a limit point of E, while any point $P_2 = \langle x_2, y_2 \rangle$ such that $x_2^2 + y_2^2 > 1$ is, clearly, not a limit point of E.

We now define a closed subset of M as a subset that contains all its limit points.

5.5C. DEFINITION. Let E be a subset of the metric space M. We say that E is a closed subset of M if $E = \bar{E}$.

In view of 5.5B, to show that a subset E of M is closed it is enough to show that $\bar{E} \subset E$.

Before proceeding to examples we give another formulation of the concept of limit point.

5.5D. THEOREM. Let E be a subset of the metric space M. Then the point $x \in M$ is a limit point of E if and only if every open ball $B[x;r]$ about x contains at least one point of E.

PROOF: Suppose x is a limit point of E. Then there is a sequence $\{x_n\}_{n=1}^{\infty}$ of points of E that converges to x. If $B[x;r]$ is any open ball about x, then $B[x;r]$ contains x_n for any n such that $\rho(x_n, x) < r$. Hence $B[x;r]$ contains a point of E.

*Some authors use the term *limit point* to mean what we call *cluster point*.

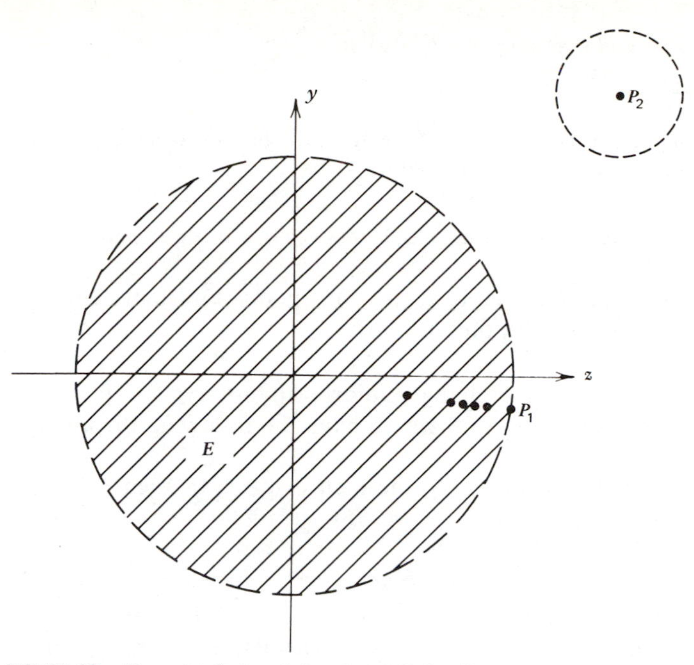

FIGURE 22. The point P_1 is a limit point of E, but P_2 is not.

Conversely, let $x \in M$ and suppose every $B[x;r]$ contains a point of E. Then for $n \in I$, the open ball $B[x;1/n]$ contains a point $x_n \in E$. The sequence $\{x_n\}_{n=1}^{\infty}$ obviously converges to x [since $\rho(x,x_n) < 1/n$], and hence x is a limit point of E. The proof is complete.

Theorem 5.5D says roughly that $x \in M$ is a limit point of $E \subset M$ if and only if there are points of E arbitrarily close to x.

If $E \subset R^2$ (the plane), then x is a limit point of E if and only if inside every circle about x there is a point of E. It is intuitively clear, then, that if L is a straight line in R^2 then no point outside of L can be a limit point of L. Hence L is a closed subset of R^2. Similarly, a plane in R^3 is a closed subset of R^3.

For *any* metric space M, if $x \in M$, then $\{x\}$ is a closed subset of M. For the only sequence of points in $\{x\}$ is $x, x, x, \ldots$, and hence x itself is the only limit point of $\{x\}$. Thus $\{x\}$ contains all its limit points and is therefore closed. Thus if $a \in R_d$, then the set $\{a\}$ is both open and closed in R_d.

This shows, as the saying goes, that "sets are not like doors." A set may be simultaneously open and closed!

In the other direction, a set may be neither open nor closed! For example, the half-open interval $[0,1)$ is neither a closed subset nor an open subset of R^1.

If we take the closure of any subset E of a metric space M we obtain a closed set.

5.5E. THEOREM. If E is any subset of a metric space M, then $\overline{E}$ is closed. That is, $\overline{E} = \overline{\overline{E}}$.

PROOF: Since $\overline{E} \subset \overline{\overline{E}}$ (5.5B) we need only prove $\overline{E} \supset \overline{\overline{E}}$. Take any $x \in \overline{\overline{E}}$. To show that $x \in \overline{E}$ it is enough (by 5.5D) to show that any open ball $B[x;r]$ contains a point of E. Since $x \in \overline{\overline{E}}$, the ball $B[x;r]$ contains a point $y \in \overline{E}$ (again by 5.5D). Let $s = \rho(x,y)$ and choose any positive number t with $t < r - s$. Since $y \in \overline{E}$ the ball $B[y;t]$ contains, by

5.5D, a point $z \in E$. But $\rho(x,y)=s, \rho(y,z)<t<r-s$, and so

$$\rho(x,z) \leqslant \rho(x,y)+\rho(y,z)<s+r-s=r.$$

Hence $z \in B[x;r]$. Thus $B[x;r]$ contains a point of E, which is what we wished to show.
Corresponding to 5.4E we also have the following result

5.5F. THEOREM. In any metric space $\langle M,\rho\rangle$ the sets M and $\varnothing$ are both closed.

PROOF: It should be obvious that M contains all its limit points and that $\varnothing$ has no limit points (and hence contains all its limit points).

In theorems 5.4C and 5.4E we must interchange union and intersection to obtain correct results for closed sets. Corresponding to 5.4E we have the following theorem.

5.5G. THEOREM. If F_1 and F_2 are closed subsets of the metric space M, then $F_1 \cup F_2$ is also closed.

PROOF: Let $x \in \overline{F_1 \cup F_2}$. Then there is a sequence $\{x_n\}_{n=1}^{\infty}$ of points of $F_1 \cup F_2$ converging to x. But $\{x_n\}_{n=1}^{\infty}$ must have a subsequence consisting wholly of points in F_1 or a subsequence consisting of points in F_2. Since any subsequence of $\{x_n\}_{n=1}^{\infty}$ must converge to x, this shows that either $x \in \overline{F}_1 = F_1$ or $x \in \overline{F}_2 = F_2$. Thus $x \in F_1 \cup F_2$. Hence $F_1 \cup F_2 \supset \overline{F_1 \cup F_2}$, and the proof is complete.

The *union* of an *infinite* number of closed sets need not be closed. For example, $\cup_{n=2}^{\infty}[1/n, 1-1/n]=(0,1)$, which is not closed in R^1. Indeed, any set can be written as the union of closed sets since one-point sets are always closed.
On the other hand, the intersection of any number of closed sets is closed. (Compare 5.4C.)

5.5H. THEOREM. If $\mathcal{F}$ is any family of closed subsets of a metric space M, then $\cap_{F \in \mathcal{F}} F$ is also closed.

PROOF: Let $x \in \overline{\cap_{F \in \mathcal{F}} F}$. Then any ball $B[x;r]$ contains a point $y \in \cap_{F \in \mathcal{F}} F$. Thus for any $F \in \mathcal{F}$, the ball $B[x;r]$ contains a point of F—namely, y. Hence $x \in \overline{F} = F$. Thus x lies in every $F \in \mathcal{F}$ and so $x \in \cap_{F \in \mathcal{F}} F$. This proves $\cap_{F \in \mathcal{F}} F \supset \overline{\cap_{F \in \mathcal{F}} F}$ and thus $\cap_{F \in \mathcal{F}} F$ is closed.

Now we come to the extremely important relationship between open sets and closed sets—namely, that a set is open if and only if its complement is closed.

5.5I. THEOREM. Let G be an open subset of the metric space M. Then $G' = M - G$ is closed. Conversely, if F is a closed subset of M, then $F' = M - F$ is open.

PROOF: Suppose first that G is open. If $x \in G$, then there is a ball $B = B[x;r]$ which lies entirely in G. Hence B contains no point of G'. By 5.5D (with $E = G'$) the point x cannot be a limit point of G'. Thus no point in G is a limit point of G', so G' contains all its limit points and is thus closed.
Now suppose F is closed. If $y \in F'$, there must be a ball $B[y;r]$ which contains no point of F. For otherwise y would be a limit point of F. We would then have $y \in F$ (since

F is closed), which contradicts $y \in F'$. Thus for every $y \in F'$ there is a ball $B[y;r]$ lying entirely in F'. Hence F' is open.

Theorem 5.5I enables us to prove theorems on closed sets from theorems on open sets. For example, let us deduce 5.5G from 5.4E.

Suppose F_1 and F_2 are closed. Then, by 5.5I, F_1' and F_2' are open. But then, by 5.4E, $F_1' \cap F_2'$ is open. Now we use 1.2H to show that $F_1' \cap F_2' = (F_1 \cup F_2)'$ so that $(F_1 \cup F_2)'$ is open. By 5.5I once more, $F_1 \cup F_2$ [the complement of $(F_1 \cup F_2)'$] is closed. This proves 5.5G.

Similarly, 5.5H may be deduced from 5.4C. (However, to do this we must first show that if $\mathfrak{F}$ is any family of sets, then $\cup_{F \in \mathfrak{F}} F' = (\cap_{F \in \mathfrak{F}} F)'$, even if $\mathfrak{F}$ consists of infinitely many sets. The proof of this is essentially the same as that of 1.2H. We leave all this to the reader.)

We can now formulate continuity in terms of closed sets (compare 5.4G).

5.5J. THEOREM. Let $\langle M_1, \rho_1 \rangle$ and $\langle M_2, \rho_2 \rangle$ be metric spaces, and let $f: M_1 \to M_2$. Then f is continuous on M_1 if and only if $f^{-1}(F)$ is a closed subset of M_1 whenever F is a closed subset of M_2.

PROOF: Suppose first that f is continuous on M_1. If $F \subset M_2$ is a closed set, then, by 5.5I, F' is open. By 5.4G, $f^{-1}(F')$ is open in M_1. But since $F \cup F' = M_2$ we have, by 1.3E, $f^{-1}(F) \cup f^{-1}(F') = f^{-1}(M_2)$. That is, $f^{-1}(F) \cup f^{-1}(F') = M_1$. Hence $f^{-1}(F)$ is the complement (relative to M_1) of $f^{-1}(F')$. Since $f^{-1}(F')$ is open, then $f^{-1}(F)$ is closed, which is what we wished to show. The converse part of the proof is left to the reader.

If the reader understands the equivalence of the various formulations of continuity, he should have no difficulty in proving the following theorem.

5.5K. THEOREM. Let f be a 1–1 function from a metric space M_1 onto a metric space M_2. (That is, f is a 1–1 correspondence between M_1 and M_2.) Then if f has any one of the following properties, it has them all.

(a) Both f and f^{-1} are continuous (on M_1 and M_2, respectively).
(b) The set $G \subset M_1$ is open if and only if its image $f(G) \subset M_2$ is open.
(c) The set $F \subset M_1$ is closed if and only if its image $f(F)$ is closed.

5.5L. DEFINITION. If f has any one (and hence all) of the properties in 5.5K, we call f a homeomorphism from M_1 onto M_2. If a homeomorphism from M_1 onto M_2 exists, we say that M_1 and M_2 are homeomorphic.

The metric spaces $[0,1]$ and $[0,2]$ (with absolute value metric) are thus homeomorphic. For if $f(x) = 2x$, then f is a homeomorphism of $[0,1]$ onto $[0,2]$.

If $f(x) = \log x$, then f is a homeomorphism of $(0,\infty)$ onto R^1 (verify).

See if you can prove that $(0,1)$ and $[0,1]$ are not homeomorphic.

The final concept in this section is that of "dense subset."

5.5M. DEFINITION. Let M be a metric space. The subset A of M is said to be dense in M if $\overline{A} = M$. (That is, A is dense in M if every point in M is a limit point of A.)

For example, the set A of rationals in dense in R^1. For, by 3.11D, every irrational is the limit of a sequence of rationals.

On the other hand, R_d has no dense subset (except R_d itself). For if $A \subset R_d$, then $\bar{A} = A$ (by 5.4D and 5.5I). Hence, if $A \neq R_d$ then $\bar{A} \neq R_d$, and so A is not dense in R_d.

Exercises 5.5

1. Use your intuition to decide which of the sets of exercise 2, Section 5.4, are closed in R^2.
2. For each of five distinct metric spaces give an example of a subset which is *neither* open *nor* closed.
3. Prove that any finite subset of a metric space M is closed.
4. Let A and B be subsets of a metric space M. If $A \subset B$, prove that $\bar{A} \subset \bar{B}$.
5. (a) True or false? If A and B are subsets of R^1 and if $\bar{A} \subset \bar{B}$, then $A \subset B$.
   (b) The same question with R^1 replaced by R_d.
6. If $a \in R^1$, prove that $[a, \infty)$ is a closed subset of R^1.
7. Let F, G be subsets of a metric space M such that F is closed and G is open in M. Show that $F - G$ is closed and $G - F$ is open in M.
8. If $0 < r < s$ and a is a point in the metric space M, show that the set
$$\{x \in M \mid r < d(x,a) < s\}$$
   is open in M.
9. Let A, B be subsets of the metric space M. Prove that
$$\overline{A \cup B} = \bar{A} \cup \bar{B}.$$
   Also, prove that
$$\overline{A \cap B} \subset \bar{A} \cap \bar{B},$$
   and give an example to show that equality need not occur.
10. Let f be a continuous real-valued function on the metric space M. Let A be the set of all $x \in M$ such that $f(x) \geqslant 0$. Prove that A is closed.
11. Let f be a continuous real-valued function on the metric space M. Let B be the set of all $x \in M$ such that $f(x) = 0$. Prove that B is closed.
12. If A and B are closed subsets of R^1, prove that $A \times B$ is a closed subset of R^2.
13. Give an example of a sequence $A_1, A_2, \ldots$ of nonempty closed subsets of R^1 such that both of the following conditions hold:
   (a) $A_1 \supset A_2 \supset A_3 \supset \cdots$.
   (b) $\bigcap_{n=1}^{\infty} A_n = \varnothing$.
14. Let
$$f(x) = \frac{x}{1+|x|} \qquad (-\infty < x < \infty).$$
   Prove that f is a homeomorphism of R^1 onto $(-1, 1)$.
15. Show that R^1 and R_d are not homeomorphic.
16. Let M_1, M_2, M_3 be metric spaces. If M_1 and M_2 are homeomorphic, and if M_2 and M_3 are homeomorphic, prove that M_1 and M_3 are homeomorphic.
17. Prove that $(0, \infty)$ (with absolute-value metric) is homeomorphic to $(0, 1)$.
18. Give an example of a countable subset of R^2 which is dense in R^2.
19. Give an example of a countable subset of ℓ^2 which is dense in ℓ^2. (This is a difficult one.)

20. Let M be a metric space and let $A \subset B \subset M$. If A is dense in B and if B is dense in M, prove that A is dense in M.
21. Give an example of a set E such that both E and its complement are dense in R^1. Can E be closed?

5.6 DISCONTINUOUS FUNCTIONS ON R^1

As an interesting digression from our discussion of metric spaces, we are going to investigate the set of points at which a given real-valued function on R^1 is discontinuous (discontinuous = not continuous).

After the proof of 5.5G we noted that a countable union of closed subsets of R^1 need not be closed.

5.6A. DEFINITION. The subset D of R^1 is said to be of type F_σ if $D = \bigcup_{n=1}^{\infty} F_n$ where each F_n is a closed subset of R^1.

Thus if F is closed, then F is of type F_σ since we can write $F = \bigcup_{n=1}^{\infty} F_n$ where $F_1 = F$ and $F_2 = F_3 = \cdots = \varnothing$.

Any open interval (a,b) is also of type F_σ since (a,b) is the union of the (countably many) closed intervals $[a+1/n, b-1/n]$ for $n \in I$ with $2/n < b-a$.

What we now wish to show is that if $f: R^1 \to R^1$, and D is the set of points of R^1 at which f is *not* continuous, then D is of type F_σ. This, however, requires a little machinery.

5.6B. DEFINITION. Let $f: R^1 \to R^1$. If J is any bounded open interval in R^1, we define $\omega[f;J]$ (called the oscillation of f over J) as*

$$\omega[f;J] = \underset{x \in J}{\text{l.u.b.}}\, f(x) - \underset{x \in J}{\text{g.l.b.}}\, f(x).$$

Then if $a \in R^1$, we define $\omega[f;a]$ (called the oscillation of f at a) to be

$$\omega[f;a] = \text{g.l.b.}\,\omega[f;J]$$

where the g.l.b. is taken over all bounded open intervals J containing a.

Clearly $\omega[f;J] \geqslant 0$ for any interval J, and hence $\omega[f;a] \geqslant 0$ for any point a. The number $\omega[f;J]$ measures, roughly, the distance from the "lowest part" to the "highest part" of the graph of f on J. It is intuitively clear that if f is continuous at a, and J is a "small" interval containing a, then $\omega[f;J]$ must be "small."

We leave the proof of the next theorem to the reader.

5.6C. THEOREM. If $f: R^1 \to R^1$ and $a \in R^1$, then the following statements hold:

(1) If f is continuous at a, then $\omega[f;a] = 0$,
(2) If f is not continuous at a, then $\omega[f;a] > 0$.

Now we start using closed sets.

5.6D. THEOREM. Let $f: R^1 \to R^1$. For any $r > 0$ let E_r be the set of all $a \in R^1$ such that $\omega[f;a] \geqslant 1/r$. Then E_r is closed.

* For simplicity, we assume throughout this section that f is bounded. That is, we assume that the range of f is a bounded subset of R^1. This is to avoid infinite values for $\omega[f;J]$. All the results of this section hold also for unbounded f.

PROOF: Let x be any limit point of E_r. We must show that $x \in E_r$. That is, we must show that $\omega[f;x] \geqslant 1/r$. To do this, it is sufficient to show that if J is a bounded open interval containing x, then $\omega[f;J] \geqslant 1/r$ (since $\omega[f;x]$ is the g.l.b. of such $\omega[f;J]$). But by 5.5D the open interval (open ball) J must contain a point y of E_r (since J contains the limit point x of E_r). But then $\omega[f;J] \geqslant \omega[f;y] \geqslant 1/r$ and the proof is complete.

The result we have been looking for now follows.

5.6E. THEOREM. Let $f: R^1 \to R^1$ and let D be the set of points in R^1 at which f is not continuous. Then D is of type F_σ.

PROOF: If $x \in D$, then, by 5.6C, $\omega[f;x] > 0$. For some $n \in I$, then, we must have $\omega[f;x] \geqslant 1/n$. This proves that $D \subset \bigcup_{n=1}^{\infty} E_{1/n}$ where $E_{1/n}$ is as in 5.6D. Conversely, if $x \in \bigcup_{n=1}^{\infty} E_{1/n}$, then $\omega[f;x] > 0$ and so $x \in D$. Thus $D = \bigcup_{n=1}^{\infty} E_{1/n}$. But by 5.6D each $E_{1/n}$ is closed. Thus D is a countable union of closed sets, which is what we wished to show.

In exercise 10 of Section 5.1 we gave an example of a function which was continuous at each irrational but discontinuous at each rational.

We are going to show that there is *no* function which is continuous at each rational but discontinuous at each irrational. To do this it is enough, by 5.6E, to show that the set of all irrationals is *not* of type F_σ. This involves introducing the notion of category, which has great importance in higher analysis.

5.6F. DEFINITION. The subset A of R^1 is said to be nowhere dense (in R^1) if $\overline{A}$ contains no (nonempty) open interval.

Thus the closed set F is nowhere dense if F itself contains no open interval. For example, the set I of positive integers is nowhere dense. The Cantor set K is another example of a closed set that is nowhere dense. (For K is closed since its complement is the union of open intervals. Furthermore, K is nowhere dense since, according to 1.6D, a chunk of every open interval will be removed in the geometric construction of K.)

5.6G. DEFINITION. The subset D of R^1 is said to be of the first category if $D = \bigcup_{n=1}^{\infty} E_n$ where each E_n is nowhere dense in R^1. If D is not of the first category, we say that D is of the second category.

It follows immediately that any countable set D is of the first category since D is the countable union of one-point sets, and any one-point set is (closed and) nowhere dense. In particular, the set of rationals is of the first category. Furthermore

5.6H. THEOREM. If A and B are sets of the first category, then $A \cup B$ is also of the first category.

PROOF: If $A = \bigcup_{n=1}^{\infty} H_n$ and $B = \bigcup_{n=1}^{\infty} E_n$ where each E_n and each H_n is nowhere dense, then $A \cup B$ is the union of all the E_n's and H_n's (of which there are a countable number). Hence $A \cup B$ is the first category.

On the other hand, the whole space R^1 is not of the first category. This important result is known as the Baire category theorem (for R^1).

5.6I. THEOREM. The set R^1 is of the second category.

PROOF: Suppose the contrary. Then $R^1 = \cup_{n=1}^{\infty} F_n$ where each F_n is nowhere dense. We may assume that the F_n are closed. Otherwise we could consider $\bar{F}_n$ in place of F_n since $R^1 = \cup_{n=1}^{\infty} \bar{F}_n$ and the $\bar{F}_n$ are *closed* and nowhere dense. Take any x_1 not in F_1. Since F_1 is closed there is an open interval I_1 about x_1 which does not intersect F_1. Let J_1 be a closed interval with $0 < \text{length } J_1 < 1$ such that $J_1 \subset I_1$. Then $J_1 \cap F_1 = \varnothing$. Now F_2 is nowhere dense and thus does not contain all of the interior* of J_1. Take any x_2 in the interior of J_1 such that $x_2 \not\in F_2$. Then there is an open interval I_2 about x_2 which does not intersect F_2 such that $I_2 \subset J_1$. Let J_2 be a closed interval with $0 < \text{length } J_2 < \frac{1}{2}$ such that $J_2 \subset I_2$. Then $J_2 \cap F_2 = \varnothing$. Continuing in this fashion we may construct a sequence of nonempty closed intervals $J_1 \supset J_2 \supset J_3 \supset \cdots$ such that $0 < \text{ length } J_n < 1/n$ and $J_n \cap F_n = \varnothing$. By theorem 2.10E, there is a point $y \in R^1$ contained in $\cap_{n=1}^{\infty} J_n$. But for each n, y is in J_n and hence y is not in F_n. Hence $y \not\in \cup_{n=1}^{\infty} F_n$. This is a contradiction since $\cup_{n=1}^{\infty} F_n$ is by assumption equal to R^1. The contradiction proves that R^1 must be of the second category.

5.6J. COROLLARY. The set of all irrationals is of the second category.

PROOF: The set of rationals is of the first category, as we have already observed. If the set of all irrationals were of the first category, then, by 5.6H, R^1 would be of the first category, contradicting 5.6I. Thus the irrationals must be of the second category.

5.6K. COROLLARY. The set of all irrationals is not of type F_σ.

PROOF: Let A denote the set of all irrationals. If A is of type F_σ, then $A = \cup_{n=1}^{\infty} F_n$ where each F_n is closed. But each F_n contains only irrationals. Hence F_n contains no nonempty open interval. Thus each F_n is closed and nowhere dense. This implies A is of the first category, contradicting 5.6J.

We thus obtain the result we have been seeking.

5.6L. THEOREM. There is no real-valued function f on R^1 which is continuous at each rational but discontinuous at each irrational.

PROOF: The proof follows directly from 5.6E and 5.6K.

Exercises 5.6

1. If E is nowhere dense in R^1, prove that any subset of E is nowhere dense in R^1.
2. If $E_1, E_2, \ldots$ are a countable number of subsets of R^1, and if each E_n is of the first category, prove that $\cup_{n=1}^{\infty} E_n$ is of the first category.
3. Prove that any nonempty open interval in R^1 is of the second category.
4. Let G be an open subset of R^1. Prove that G is dense in R^1 if and only if G' (the complement of G) is nowhere dense.
5. Let $G_1, G_2, \ldots$ be a sequence of open subsets of R^1 each of which is dense in R^1. Prove that $\cap_{n=1}^{\infty} G_n$ is dense in R^1. (*Hint*: Use the preceding two exercises.)
6. Let χ be the function on R^1 defined by

$$\chi(x) = 0 \text{ if } x \text{ is rational,}$$

$$\chi(x) = 1 \text{ if } x \text{ is irrational.}$$

* By the interior of a closed interval $[a,b]$ we mean (a,b).

(That is, χ is the characteristic function of the set of irrationals.) Fill in the details in the proof of the following theorem.

THEOREM. There is no sequence $\{f_n\}_{n=1}^{\infty}$ of functions continuous on R^1 such that

$$\lim_{n\to\infty} f_n(x)=\chi(x) \qquad (-\infty<x<\infty). \tag{*}$$

PROOF: Suppose the contrary—that is, suppose (*) holds for some sequence $\{f_n\}_{n=1}^{\infty}$ of continuous functions.

(a) For each $n\in I$ let $E_n=\{x|f_n(x)\geqslant\frac{1}{2}\}$. Then E_n is closed (why?).
(b) For each $N\in I$ let $F_N=E_N\cap E_{N+1}\cap E_{N+2}\cdots=\cap_{n=N}^{\infty}E_n$. Then F_N is closed (why?).
(c) If x is irrational, then $\lim_{n\to\infty}f_n(x)=\chi(x)=1$ and so there exists $N\in I$ such that $x\in F_N$ (why?).
(d) If $x\in F_N$ for some N, then x is irrational (why?).
(e) Thus $\cup_{N=1}^{\infty}F_N$ is precisely the set of irrationals. But this implies that the set of irrationals is of type F_σ (why?). This contradicts 5.6K, and the contradiction proves the theorem.

5.7 THE DISTANCE FROM A POINT TO A SET

It is fruitful to extend the notion of distance between points to the notion of distance from a point to a set.

5.7A. DEFINITION. Let $\langle M,\rho\rangle$ be a metric space. Let A be a nonempty subset of M and let x be a point of M. We define $\rho(x,A)$, the distance from x to A, as

$$\rho(x,A)=\text{g.l.b.}\{\rho(x,y)|y\in A\}.$$

For example, if $M=R^1$, $A=(0,1)$, and $x=2$, then

$$\rho(x,A)=\underset{0<y<1}{\text{g.l.b.}}|2-y|=1.$$

Note that although $\rho(x,A)=1$, there is no $y\in A$ such that $\rho(x,y)=1$.

Thus if there is a closest point z in A to x (or more than one closest point), then $\rho(x,A)=\rho(x,z)$. However, as in the example, there may be no closest point.

Here is a connection with closed sets.

5.7B. THEOREM. Let A be a nonempty subset of the metric space M and let x be a point of M. Then

$$\rho(x,A)=0$$

if and only if $x\in\overline{A}$.

PROOF: First suppose $\rho(x,A)=0$. Then 0 is the greatest lower bound of the set $\{\rho(x,y)|y\in A\}$. Thus if $\epsilon>0$, then ϵ is not a lower bound for the set so there exists $y\in A$ with $\rho(x,y)<\epsilon$. This shows that every $B[x;\epsilon]$ contains a point of A. By theorem 5.5E, x is a limit point of A, and so $x\in\overline{A}$.

Conversely, suppose $x \in \bar{A}$. Then there exists a sequence $\{x_n\}_{n=1}^{\infty}$ in A such that $\lim_{n\to\infty} x_n = x$. But

$$\rho(x,A) \leqslant \rho(x,x_n) \qquad (n \in I).$$

Letting $n\to\infty$ we obtain $\rho(x,A) \leqslant 0$, so that $\rho(x,A)=0$. This completes the proof.

If A is closed, we have the following corollary.

5.7C. COROLLARY. Let A be a nonempty closed subset of the metric space M and let x be a point of M. Then $x \in A$ if and only if $\rho(x,A)=0$. Hence $x \in A'$ (the complement of A) if and only if $\rho(x,A)>0$.

Viewed as a function, the distance $\rho(x,A)$ has interesting properties and applications. (See the exercises.)

5.7D. THEOREM. Let A be a nonempty subset of the metric space M. Define

$$f(x)=\rho(x,A) \qquad (x \in M).$$

Then f is continuous on M.

PROOF: Let x be a point of M. Given $\epsilon>0$ let x_1 be any point of M such that $\rho(x,x_1)<\delta=\epsilon/2$. Since

$$f(x)=\rho(x,A)=\underset{y\in A}{\text{g.l.b.}}\,\rho(x,y),$$

there exists $y_1 \in A$ such that $f(x)+\epsilon/2>\rho(x,y_1)$. Then

$$\rho(x_1,y_1) \leqslant \rho(x_1,x)+\rho(x,y_1)<\delta+f(x)+\frac{\epsilon}{2}=f(x)+\epsilon.$$

Hence

$$f(x_1)=\rho(x_1,A)=\underset{y\in A}{\text{g.l.b.}}\,\rho(x_1,y) \leqslant \rho(x_1,y_1)<f(x)+\epsilon,$$

so that $f(x_1)-f(x)<\epsilon$. But, reversing the roles of x and x_1 will show $f(x)-f(x_1)<\epsilon$. We have thus shown that

$$|f(x)-f(x_1)|<\epsilon \qquad (\rho(x,x_1)<\delta)$$

which proves that f is continuous at x. Since x was an arbitrary point of M, the proof is complete.

Exercises 5.7

1. Let A be a nonempty subset of a metric space. If $\epsilon \geqslant 0$, show that the set of $x \in M$ such that $\rho(x,A) \geqslant \epsilon$ is closed.
2. Show that every open subset of a metric space is the union of countably many closed sets.
3. Let F_1, F_2 be disjoint closed subsets of a metric space M. Prove that there exists disjoint open sets G_1, G_2 such that $G_1 \supset F_1, G_2 \supset F_2$.

6

CONNECTEDNESS, COMPLETENESS, AND COMPACTNESS

6.1 MORE ABOUT OPEN SETS

6.1A. As we noted in the previous chapter, if we denote the metric space $[0,1]$ (with absolute value metric) by A, then the interval $[0,\frac{1}{2})$ is an open subset of A even though $[0,\frac{1}{2})$ is not open subset of R^1. Thus whether a given set is open or not depends on the metric space of which it is considered a subset.

Indeed, if $\langle M,\rho\rangle$ is any metric space and $A \subset M$, then, by 5.4B, A is always an open set in the metric space $\langle A,\rho\rangle$ even though A may not be open in $\langle M,\rho\rangle$.

Before we take up new concepts in this chapter, we wish to investigate this phenomenon more closely.

Let $\langle M,\rho\rangle$ be any metric space and let A be any nonempty subset of M. Then $\langle A,\rho\rangle$ is also a metric space. Now if $a \in A$ we must distinguish between open balls in A about a and open balls in M about a. For example, if $\langle M,\rho\rangle = R^1$ and $\langle A,\rho\rangle = [0,1]$, then the open ball $B[0;\frac{1}{2}]$ in R^1 is the interval $(-\frac{1}{2},+\frac{1}{2})$, while the open ball $B[0;\frac{1}{2}]$ in $A=[0,1]$ is the interval $[0,\frac{1}{2})$. (That is, in $A=[0,1]$ the set of all points of A whose distance to 0 is less than $\frac{1}{2}$ is the interval $[0,\frac{1}{2})$.) Consequently, let us introduce the following notations: If $a \in A$, let

$$B_A[a;r] = \{x \in A \mid \rho(a,x) < r\},$$
$$B_M[a;r] = \{x \in M \mid \rho(a,x) < r\}.$$

Then it is clear that

$$B_A[a;r] = A \cap B_M[a;r]. \qquad (*)$$

We can now throw some light on the question of what is open in what.

6.1B. THEOREM. Let $\langle M,\rho\rangle$ be a metric space and let A be a proper subset of M. Then the subset G_A of A is an open subset of $\langle A,\rho\rangle$ if and only if there exists an open subset G_M of $\langle M,\rho\rangle$ such that $G_A = A \cap G_M$. That is, a set is open in $\langle A,\rho\rangle$ if and only if it is the intersection with A of a set that is open in $\langle M,\rho\rangle$.

PROOF: Suppose first that G_A is open in A. Then for each $a \in G_A$ there exists $r_a > 0$ such that $B_A[a; r_a] \subset G_A$. Define G_M as

$$G_M = \bigcup_{a \in G_A} B_M[a; r_a].$$

Then G_M is open in M since G is the union of open balls of M (5.4C). Also, from (*) of 6.1A it follows that $G_M \cap A = G_A$.

Conversely, suppose G_M is open in M and let $G_A = A \cap G_M$. We wish to prove that G_A is open in A. If $a \in G_A$, then $a \in G_M$. Since G_M is open in M there is an open ball $B_M[a; r]$ contained in G_M. But then $B_M[a; r] \cap A \subset G_M \cap A$, which says that $B_A[a; r] \subset G_A$. For each $a \in G_A$, then, we have shown that there is an open ball $B_A[a; r]$ contained in G_A. This proves that G_A is open in A.

For example, if $M = R^1$ and $A = [0, 1]$, the set $G_A = [0, \frac{1}{2})$ is open in A. But $G_A = A \cap (-\infty, \frac{1}{2})$, and $(-\infty, \frac{1}{2})$ is open in M. Thus, the G_M of 6.1B can be taken to be $(-\infty, \frac{1}{2})$.

The importance of 6.1B is that it enables us to give more than one formulation of connectedness.

Exercises 6.1

1. Give an example of a metric space M and a nonempty proper subset $A \subset M$ with the property that every open subset of A is also an open subset of M.
2. Let $A = [0, 1]$. Which of the following subsets of A are open subsets of A?
   (a) $(\frac{1}{2}, 1]$,
   (b) $(\frac{1}{2}, 1)$,
   (c) $[\frac{1}{2}, 1)$.
   Which of (a), (b), and (c) are open subsets of R^1? Which are open subsets of R^2? (Regard R^1 as a subset of R^2.)
3. Is there any subset of R^1 that is an open subset of R^2?
4. Let A be an open subset of the metric space M. Prove that if $B \subset A$, then B is open in A if and only if B is open in M.

6.2 CONNECTED SETS

Our intuition tells us that, whatever the definition of "connected set" turns out to be, the interval $[0, 1]$ should be called a connected subset of R^1 while the union $[0, 1] \cup [2, 3]$ should not.

The usual way to define connectedness, however, does not immediately appeal to the intuition. It is usually not trivial to prove that a given set is connected even though it looks "connected."

We will list two equivalent properties and then define a set to be connected if it has either (and hence both) of these properties. We recall first that in a metric space $\langle M, \rho \rangle$ the sets M and $\varnothing$ are both open and closed. If these are the only subsets of M that are open and closed, we will (ultimately) call M connected.

6.2A. THEOREM. Let $\langle M, \rho \rangle$ be a metric space and let A be a subset of M. Then if A has either one of the following properties, it has the other.

(a) It is impossible to find nonempty subsets A_1, A_2 of M such that $A = A_1 \cup A_2, \overline{A}_1 \cap A_2 = \varnothing, A_1 \cap \overline{A}_2 = \varnothing$. (Here $\overline{A}_i$ means the closure of A_i in $\langle M, \rho \rangle$.)

(b) When $\langle A,\rho\rangle$ is itself regarded as a metric space, then there is no set except A and $\varnothing$ which is both open and closed in $\langle A,\rho\rangle$.

PROOF: We will prove that (a) implies (b). Suppose, then, that (a) holds. If (b) were false, then there would be a nonempty proper subset A_1 of A such that A_1 is both open and closed in A. But then, by 5.5I, $A_2 = A - A_1$ would also be both open and closed in A.

We now show that $A_1 \cap \overline{A_2} = \varnothing$. Suppose $x \in M$ is a point in $\overline{A_2}$. Then x is the limit of a sequence of points in A_2. If x were in A_1, then x would be in A. Since x is the limit of a sequence of points in A_2, this implies $x \in A_2$, since A_2 is closed in A. But then $A_1 \cap A_2 \neq \varnothing$ contradicting $A_2 = A - A_1$. Hence if $x \in \overline{A_2}$, then $x \notin A_1$, so that $A_1 \cap \overline{A_2} = \varnothing$. Similarly $\overline{A_1} \cap A_2 = \varnothing$. Clearly $A = A_1 \cup A_2$. But this contradicts (a), and so (b) must be true if (a) is.

Now we will prove that (b) implies (a). Suppose that (b) is true. If (a) were false, then there would exist nonempty subsets A_1, A_2 of M such that $A = A_1 \cup A_2$ and $\overline{A_1} \cap A_2 = \varnothing = A_1 \cap \overline{A_2}$. Let $G = M - \overline{A_2}$. Then G is an open subset of M (by 5.5I). Since A_1 is disjoint from $\overline{A_2}$ we have $A_1 \subset G$. This proves that $G \cap A = A_1$. But then, by 6.1B, A_1 is an open subset of A. Similar reasoning will show that A_2 is an open subset of A. Hence $A_1 = A - A_2$ is a closed subset of A. We have thus produced a nonempty proper subset of A (namely A_1) that is both open and closed in A. This contradicts (b). Thus (a) must be true if (b) is.

6.2B. DEFINITION. Let $\langle M,\rho\rangle$ be a metric space and let A be a subset of M. If A has either (and hence both) of the properties (a) and (b) of theorem 6.2A, we say that A is connected.

It should be clear from (b) of 6.2A that whether or not A is connected is determined purely by A and ρ and has nothing to do with M. That is, A will be connected when regarded as a subset of $\langle M,\rho\rangle$ if and only if A is connected when regarded as a subset of $\langle A,\rho\rangle$. (In this respect, then, "connected" is very different from "open.") However, in some proofs it is useful to consider A as a subset in a larger space.

We can now show that $A = [0,1] \cup [2,3]$ is not a connected subset of R^1. For $[0,1]$ is both open and closed in A. More generally,

6.2C. THEOREM. The subset A of R^1 is connected if and only if whenever $a \in A, b \in A$ with $a < b$, then $c \in A$ for any c such that $a < c < b$. (That is, whenever $a \in A, b \in A, a < b$, then $(a,b) \subset A$.)

PROOF: First suppose $A \subset R^1$ and that there exist $a \in A, b \in A$ with $a < b$ and a number $c \in R^1 - A$ with $a < c < b$. We will show that A is not connected. Indeed, if we let $A_1 = A \cap (-\infty, c), A_2 = A \cap (c, \infty)$, then $A = A_1 \cup A_2$. If $x \in \overline{A_1}$, then x is the limit of a sequence of numbers in $(-\infty, c)$. By 2.7D, then $x \leqslant c$. Hence $x \notin A_2$. This proves that $\overline{A_1} \cap A_2 = \varnothing$. Similarly $A_1 \cap \overline{A_2} = \varnothing$, and hence A is not connected.

Now suppose A is not connected. Then there exist nonempty sets A_1, A_2 such that $A = A_1 \cup A_2, \overline{A_1} \cap A_2 = \varnothing = A_1 \cap \overline{A_2}$. Choose any points $a_1 \in A_1, a_2 \in A_2$. Then $a_1 \neq a_2$ and we may assume $a_1 < a_2$. We will show that $A \not\supset (a_1, a_2)$. Now let B be the set of all $x \in A_1$ such that $a_1 \leqslant x \leqslant a_2$. That is, $B = A_1 \cap [a_1, a_2]$. Then B is a bounded nonempty set of real numbers and thus, by 1.7D, has a least upper bound $\bar{a}$. Now $\bar{a} \in \overline{B}$ (why?) and hence, since $B \subset A_1$, we have $\bar{a} \in \overline{A_1}$. Hence $\bar{a} \notin A_2$ since $\overline{A_1} \cap A_2 = \varnothing$. But $\bar{a} \in [a_1, a_2]$ and so $\bar{a} < a_2$.

Now either $\bar{a} \in A$ or $\bar{a} \notin A$. If $\bar{a} \notin A$, then $\bar{a} \neq a_1$ and so $\bar{a} > a_1$. Thus $a_1 < \bar{a} < a_2$ and

$\bar{a} \notin A$ so that $A \not\supset (a_1, a_2)$. On the other hand, if $\bar{a} \in A$, then $\bar{a} \in A_1$ (since $\bar{a} \notin A_2$). Hence $\bar{a} \notin \bar{A}_2$ and so $\bar{a}$ is not a limit point of A_2. Hence there is a number c with $\bar{a} < c < a_2$ such that $c \notin A_2$. But $c \notin A_1$ by definition of $\bar{a}$. Hence $c \notin A$. Since $a_1 < c < a_2$ this proves $A \not\supset (a_1, a_2)$. In either case, then, $A \not\supset (a_1, a_2)$ which is what we wished to show.

Consider next the subset B of R^2 consisting of the graph of $y = \sin(1/x)$ $(0 < x \leqslant 1)$ together with the closed interval on the y axis from $\langle 0, -1 \rangle$ to $\langle 0, 1 \rangle$. Surprisingly enough, it may be shown that B is a connected set in R^2. We will not give details. This example shows that untrained intuition may not always be helpful in connectedness problems.

From the following theorem we can deduce a result useful in calculus.

6.2D. THEOREM. Let f be a continuous function from a metric space M_1 into a metric space M_2. If M_1 (the domain of f) is connected, then the range of f is also connected.

PROOF: Let $A = f(M_1)$ so that $f: M_1 \Rightarrow A$. If A were *not* connected, then there would exist a nonempty proper subset B of A such that B is both open and closed in A. But then, by 5.4G and 5.5J, $f^{-1}(B)$ would be a nonempty proper subset of M_1 that is both open and closed in M_1. This would contradict the hypothesis that M_1 is connected. Hence A is connected and the proof is complete.

The special case of 6.2D in which M_1 is a closed bounded interval $[a, b]$ and $M_2 = R^1$ yields the corollary we have been looking for.

6.2E. COROLLARY. If f is a continuous real-valued function on the closed bounded interval $[a, b]$, then f takes on every value between $f(a)$ and $f(b)$.

PROOF: By 6.2C the interval $[a, b]$ is connected. By 6.2D the range of f is thus connected. The corollary then follows from 6.2C.

Here is another application of connectedness. We will show that $[0, 1)$ and $(0, 1)$ are not homeomorphic. Assume the contrary. Then there exists a 1–1 function f from $[0, 1)$ onto $(0, 1)$ such that both f and f^{-1} are continuous. Let $a = f(0)$. Then the restriction of f to $(0, 1)$ is a continuous function from $(0, 1)$ onto $(0, a) \cup (a, 1)$. But this contradicts 6.2D since $(0, 1)$ is connected while $(0, a) \cup (a, 1)$ is not. See Figure 23.

We next give an interesting reformulation of connectedness.

6.2F. THEOREM. Let M be a metric space. Then M is connected if and only if every continuous characteristic function on M is constant. That is, M is connected if and only if the function identically 0 and the function identically 1 are the only characteristic functions on M that are continuous on M.

PROOF: Let M be any metric space and let χ be the characteristic function of $A \subset M$. Then $A = \chi^{-1}(1)$. Thus if χ is continuous, then A is closed (by 5.5J) since A is the inverse image under χ of the closed subset $\{1\}$ of R^1. Similarly, if χ is continuous, then $M - A = \chi^{-1}(0)$ is closed, and so A is also open. Thus, if χ is a continuous characteristic function on M, then A is both open and closed in M. But if M is connected, then either $A = M$ or $A = \varnothing$, and in either case χ is constant.

The proof of the converse is left to the reader.

Theorem 6.2F can be used to give a snappy proof of 6.2D. Indeed, suppose f is a continuous function from the connected metric space M_1 into a metric space M_2. To

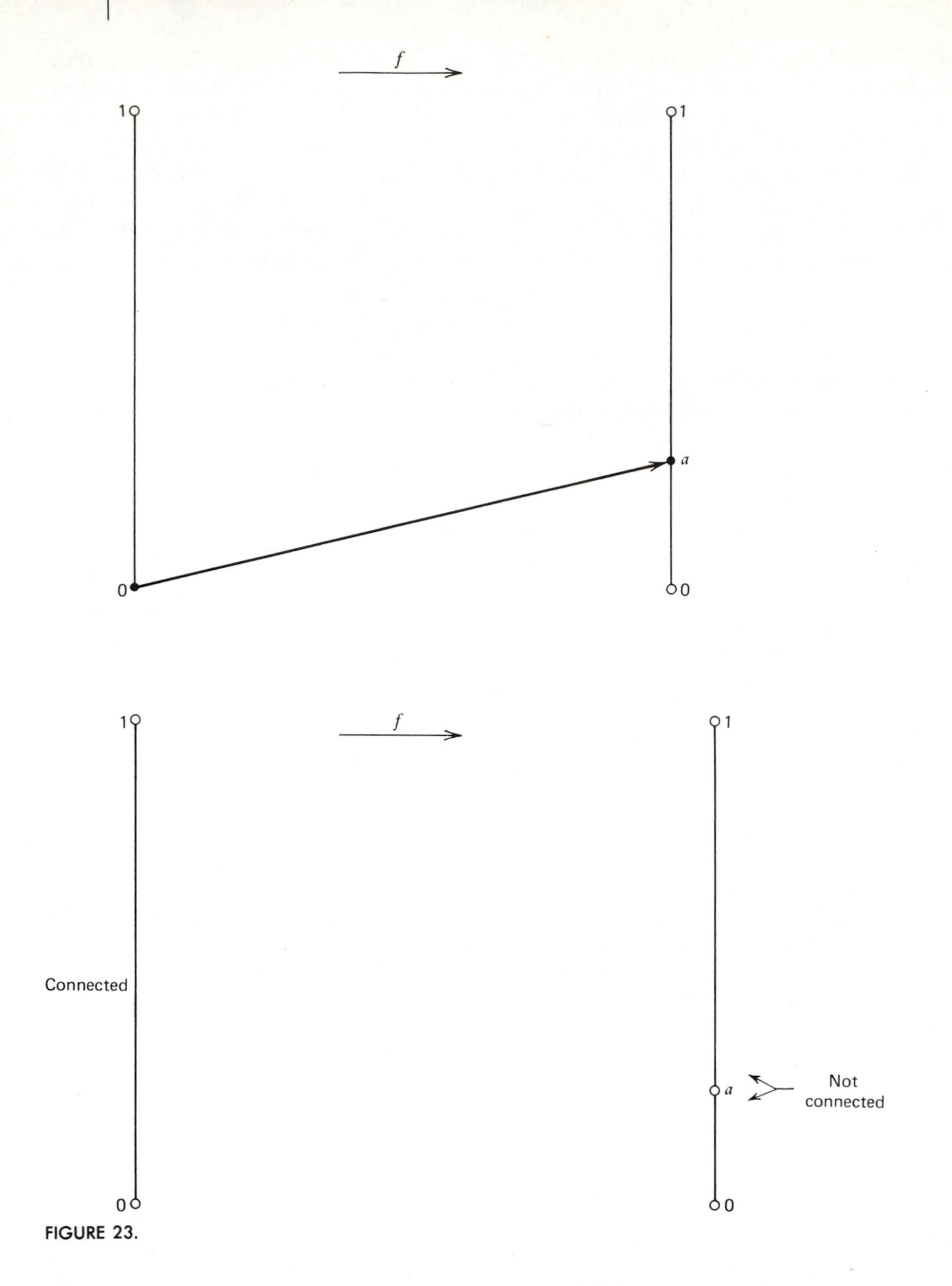

FIGURE 23.

show that $f(M_1)$ is connected it is sufficient, by 6.2F, to show that any continuous characteristic function χ on $f(M_1)$ is constant. But since χ and f are both continuous, then $\chi \circ f$ is a continuous characteristic function on M_1 (use 5.3D), and hence $\chi \circ f$ is constant by 6.2F. It follows that χ is constant.

Another interesting application of 6.2F is the following.

6.2G. THEOREM. If A_1 and A_2 are connected subsets of a metric space M, and if $A_1 \cap A_2 \neq \varnothing$, then $A_1 \cup A_2$ is also connected.

PROOF: Let χ be a continuous characteristic function on $A_1 \cup A_2$. If $x_0 \in A_1 \cap A_2$, then by 6.2F

$$\chi(x)=\chi(x_0) \qquad (x \in A_1)$$

since A_1 is connected, and

$$\chi(x)=\chi(x_0) \qquad (x \in A_2)$$

since A_2 is connected. Hence χ is identically equal to $\chi(x_0)$ and is thus constant. By 6.2F, $A_1 \cup A_2$ is connected and the proof is complete.

Exercises 6.2

1. Prove that if f is a nonconstant real-valued continuous function on R^1, then the range of f is not countable.
2. Prove that there is no continuous real-valued function f on R^1 such that

$$f(x) \text{ is irrational if } x \text{ is rational}$$

and

$$f(x) \text{ is rational if } x \text{ is irrational.}$$

3. Prove that the interval $[0,1]$ is not a connected subset of R_d.
4. True or false? If A and C are connected subsets of the metric space M, and if

$$A \subset B \subset C,$$

then B is connected.
5. If A is a connected subset of the metric space M, prove that $\overline{A}$ is connected. (*Hint*: Use 6.2F.)
6. If A is a connected subset of the metric space M, and if $A \subset B \subset \overline{A}$, prove that B is connected.
7. Prove that the set of all points on a minus sign (considered as a subset of R^2) is not homeomorphic to the set of all points on a plus sign. (*Hint*: First decide whether or not the center point of the plus sign can be the image under a homeomorphism of an end point of the minus sign.)
8. Fill in the details in the following proof that every number $c \geqslant 0$ has a square root.
   (a) Let $f(x)=x^2$ $(0 \leqslant x < \infty)$. Then f is continuous on $[0,\infty)$, and 0 is in the range of f.
   (b) If $c \geqslant 0$, then $c \leqslant (1+c)^2$ and $(1+c)^2$ is in the range of f.
   (c) Therefore, c is in the range of f (why?). Hence $f(x_1)=x_1^2=c$ for some x_1. The number x_1 is the square root of c.

6.3 BOUNDED SETS AND TOTALLY BOUNDED SETS

6.3A. DEFINITION. Let $\langle M,\rho \rangle$ be a metric space. We say that the subset A of M is bounded if there exists a positive number L such that

$$\rho(x,y) \leqslant L \qquad (x,y \in A).$$

If A is bounded, we define the diameter of A (denoted diam A) as

$$\operatorname{diam} A = \underset{\substack{x \in A \\ y \in A}}{\text{l.u.b.}}\, \rho(x,y).$$

If A is not bounded, we write $\operatorname{diam} A = \infty$.

Thus our defintion of bounded set in an arbitrary metric space is consistent with the definition of bounded set of real numbers in 1.7A. A subset A of R^1 is bounded if and only if A is contained in some interval of finite length. Similarly it is easy to see that a subset of R^2 (respectively R^3) is bounded if and only if it is contained in some square (respectively cube) whose edge has finite length.

The interval $(0,\infty)$ is not a bounded subset of R^1. However, $(0,\infty)$ *is* a bounded subset of R_d, since $\rho(x,y) \leqslant 1$ for any $x,y \in R_d$. Indeed the diameter of any subset A of R_d is equal to 1, provided that A contains at least two points.

6.3B. Another example which will be of interest later is the following. For each $k \in I$ let e_k denote the sequence all of whose terms are equal to 0 except the kth term, which is equal to 1. Thus e_3, for example, is the sequence $0,0,1,0,0,0,\ldots$. Then $e_k \in \ell^2$. Let $E \subset \ell^2$ be the set of all the e_k for $k \in I$—

$$E = \{e_1, e_2, e_3, \ldots\}.$$

If $j \neq k$, then $\rho(e_j, e_k) = \|e_j - e_k\|_2 = \sqrt{2}$. Hence E is bounded and $\operatorname{diam} E = \sqrt{2}$.

The last example shows that a subset of ℓ^2 can be bounded and still be pretty "big" in the sense of having infinitely many points none of which is "close" to any of the others. For the theory of general metric spaces the concept of "totally bounded" turns out to be more useful than "bounded."

6.3C. DEFINITION. Let $\langle M,\rho\rangle$ be a metric space. The subset A of M is said to be totally bounded if, given $\epsilon > 0$, there exist a finite number of subsets $A_1, A_2, \ldots, A_n$ of M such that $\operatorname{diam} A_k < \epsilon$ $(k = 1, \ldots, n)$ and such that $A \subset \bigcup_{k=1}^{n} A_k$.

(It is clear that the phrase "$A \subset \bigcup_{k=1}^{n} A_k$" may be replaced by "$A = \bigcup_{k=1}^{n} A_k$.")

If a set A is contained in the union of sets $A_1, A_2, \ldots$, we sometimes say that the A_k cover A. Thus $A \subset M$ is totally bounded if and only if, for every $\epsilon > 0$, A can be covered by a finite number of subsets of M whose diameters are all less than ϵ.

Some authors use the term "precompact" instead of "totally bounded". "Totally bounded" is a stronger restriction than "bounded."

6.3D. THEOREM. If the subset A of the metric space $\langle M,\rho\rangle$ is totally bounded, then A is bounded.

PROOF: If A is totally bounded, then there exist nonempty subsets $A_1, A_2, \ldots, A_n$ of M such that $\operatorname{diam} A_k < 1$ $(k = 1, \ldots, n)$ and $A \subset \bigcup_{k=1}^{n} A_k$. For each $k = 1, \ldots, n$, let a_k be any point in A_k. Then let $D = \rho(a_1, a_2) + \rho(a_2, a_3) + \cdots + \rho(a_{n-1}, a_n)$. Now for any points $x,y \in A$, we have $x \in A_i, y \in A_j$ for some i and j (since the A_k cover A). We may assume that $i \leqslant j$. But then $\rho(x,y) \leqslant \rho(x,a_i) + [\rho(a_i, a_{i+1}) + \cdots + \rho(a_{j-1}, a_j)] + \rho(a_j, y)$. Since $\operatorname{diam} A_i < 1$ we have $\rho(x, a_i) < 1$. Similarly $\rho(a_j, y) < 1$. Hence

$$\rho(x,y) < 1 + D + 1 = D + 2 \qquad (x,y \in A)$$

and thus A is bounded.

In R^1 it turns out that bounded and totally bounded mean the same thing. In fact, if $A \subset R^1$ is bounded, then $A \subset [-L, L]$ for some $L > 0$. Given $\epsilon > 0$, A is certainly covered by

$$\left[-L, -L + \frac{\epsilon}{2}\right], \left[-L + \frac{\epsilon}{2}, -L + 2\frac{\epsilon}{2}\right], \ldots, \left[-L + (n-1)\frac{\epsilon}{2}, -L + n\frac{\epsilon}{2}\right]$$

where n is any positive integer such that $n(\epsilon/2) \geqslant 2L$. Thus, in R^1, a set is totally bounded if and only if it is bounded.

The same result holds true for R^n. The proof (for $n=2$) is reserved for an exercise.

On the other hand, in R_d "bounded" and "totally bounded" are not at all equivalent. For we have seen that every subset of R_d is bounded. However, if $B \subset R_d$ and $\operatorname{diam} B < \frac{1}{2}$, then B can contain at most one point. A finite number of subsets of R_d each of which has diameter $< \frac{1}{2}$ can therefore cover only a finite subset of R_d. Thus

6.3E. COROLLARY. A subset A of R_d is totally bounded if and only if A contains only a finite number of points.

We will soon show that the subset E of ℓ^2 defined in 6.3B is bounded but is not totally bounded.

In summary then, in any metric space a totally bounded set is bounded. However, in some metric spaces there are bounded sets which are not totally bounded.

We now give two important reformulations of "totally bounded."

6.3F. DEFINITION. Let A be a subset of the metric space M. The subset B of A is said to be ϵ-dense in A (where $\epsilon > 0$) if for every $x \in A$ there exists $y \in B$ such that $\rho(x,y) < \epsilon$. (That is, B is ϵ-dense in A if each point of A is within distance ϵ from some point of B.)

6.3G. THEOREM. The subset of the metric space $\langle M, \rho \rangle$ is totally bounded if and only if, for every $\epsilon > 0$, A contains a finite subset $\{x_1, \ldots, x_n\}$ which is ϵ-dense in A.

PROOF: Fix $\epsilon > 0$. If A is totally bounded, then $A = \cup_{i=1}^{n} A_i$ where $\operatorname{diam} A_i < \epsilon$. We may assume $A_i \neq \varnothing$. If $a_i \in A_i (i = 1, \ldots, n)$, then $\{a_1, \ldots, a_n\}$ is ϵ-dense in A. Hence, if A is totally bounded, then A has a finite ϵ-dense subset.

Conversely, if $\{x_1, \ldots, x_n\}$ is $\epsilon/3$-dense in A, then $B[x_i; \epsilon/3], \ldots, B[x_n; \epsilon/3]$ form a covering of A by sets of diameter $< \epsilon$. The theorem follows.

We now present the most important property of totally bounded sets.

6.3H. THEOREM. Let $\langle M, \rho \rangle$ be a metric space. The subset A of M totally bounded if and only if every sequence of points of A contains a Cauchy subsequence.

PROOF: Suppose A is totally bounded. Let $\{x_n\}_{n=1}^{\infty}$ be a sequence of points of A. We wish to show that $\{x_n\}_{n=1}^{\infty}$ has a Cauchy sequence. The set A can be covered by a finite number of subsets of A of diameter < 1. One of these sets, call it A_1, must contain x_n for infinitely many values of n (why?). Choose any $n_1 \in I$ such that $x_{n_1} \in A_1$. Now A_1 is clearly totally bounded and hence can be covered by a finite number of subsets of A_1 of diameter $< \frac{1}{2}$. One of the sets, call it A_2, must contain x_n for infinitely many n. Let n_2 be any integer greater than n_1 such that $x_{n_2} \in A_2$. Since $A_2 \subset A_1$ we also have $x_{n_2} \in A_1$. Continuing in this fashion we obtain, for any $k \in I$, a subset A_k of A_{k-1} with $\operatorname{diam} A_k < 1/k$, and a term $x_{n_k} \in A_k$ of the sequence $\{x_n\}_{n=1}^{\infty}$. Since

$$x_{n_k}, x_{n_{k+1}}, x_{n_{k+2}}, \ldots$$

all lie in A_k, and since $\operatorname{diam} A_k < 1/k$, it follows that $\{x_{n_k}\}_{k=1}^{\infty}$ is a Cauchy subsequence of $\{x_n\}_{n=1}^{\infty}$.

Conversely, suppose every sequence $\{x_n\}_{n=1}^{\infty}$ of points of some subset A of M contains a Cauchy subsequence. We wish to show that A is totally bounded. Suppose the contrary. Then by 6.5G there exists some $\epsilon>0$ such that A contains no finite ϵ-dense subset. Thus if $x_1 \in A$, then the set $\{x_1\}$ is not ϵ-dense in A, so there exists $x_2 \in A$ such that $\rho(x_1, x_2) \geqslant \epsilon$. But then $\{x_1, x_2\}$ is not ϵ-dense in A and so there exists $x_3 \in A$ such that $\rho(x_1, x_3) \geqslant \epsilon$ and $\rho(x_2, x_3) \geqslant \epsilon$. Continuing in this fashion we may construct a sequence $\{x_n\}_{n=1}^{\infty}$ of points of A such that $\rho(x_j, x_k) \geqslant \epsilon$ for any $j, k \in I, (j \neq k)$. But then $\{x_n\}_{n=1}^{\infty}$ has no Cauchy subsequence, which contradicts our hypothesis. This contradiction shows that A must be totally bounded and the proof is complete.

From 6.3H it follows immediately that the subset E of ℓ^2 defined in 6.3B is not totally bounded. For, since $\rho(e_j, e_k) = \sqrt{2}$ if $j \neq k$, the sequence $e_1, e_2, \ldots$ has no Cauchy subsequence.

Exercises 6.3

1. Prove that every bounded subset of R^2 is totally bounded.
2. Give an example of a bounded subset of ℓ^∞ which is not totally bounded.
3. Give an example of an infinite subset of ℓ^2 which is totally bounded.
4. Prove that every finite subset of a metric space M is totally bounded.
5. If $\langle M, \rho \rangle$ is totally bounded and $A \subset M$, prove that $\langle A, \rho \rangle$ is totally bounded.
6. Let B be a subset of the metric space M. Prove that B is dense in M if and only if B is ϵ-dense in M for every $\epsilon > 0$.
7. Let A be an infinite bounded subset of R^1. Prove that there is at least one cluster point of A in R^1. (*Hint*: Suppose $A \subset J_1$ where J_1 is a closed bounded interval. If J_1 is divided in half, then at least one of the halves must contain infinitely many points of A. Call this half J_2. Continue this process.) This result is known as the Bolzano-Weierstrass theorem.
8. Give another proof of the Bolzano-Weierstrass theorem, beginning as follows: Let A be an infinite bounded subset of R^1. Let $\{a_n\}_{n=1}^{\infty}$ be a sequence of distinct points of A. Then $\{a_n\}_{n=1}^{\infty}$ contains a Cauchy subsequence (why?). Now finish the proof.

6.4 COMPLETE METRIC SPACES

In 2.10D we saw that in the metric space R^1 every Cauchy sequence of points in R^1 converges to a point in R^1. In 4.3F we noted that there are metric spaces $\langle M, \rho \rangle$ in which not all Cauchy sequences of points of M converge to a point in M.

6.4A. DEFINITION. We say that the metric space M is complete if every Cauchy sequence of points in M converges to a point in M.

Thus R^1 is complete by 2.10D. In the exercises you will be asked to show that R^2 and R_d are also complete.

6.4B. We now show that ℓ^2 is complete. If $s^{(1)}, s^{(2)}, \ldots$ is a Cauchy sequence of points in ℓ^2, we must find $s \in \ell^2$ such that $s^{(n)} \to s$ as $n \to \infty$. Since each $s^{(n)}$ is itself a sequence, the notation will be a little complicated. Denote the kth term of the sequence $s^{(n)}$ by $s_k^{(n)}$ so that

$$s^{(n)} = \{s_k^{(n)}\}_{k=1}^{\infty} \quad \text{and} \quad \|s^{(n)}\|_2^2 = \sum_{k=1}^{\infty} s_k^{(n)^2}.$$

Since $s^{(1)}, s^{(2)}, \ldots$ is a Cauchy sequence in ℓ^2, given $\epsilon > 0$ there exists $N \in I$ such that $\rho[s^{(n)}, s^{(m)}] < \epsilon$ if $n, m \geqslant N$. That is,

$$\|s^{(n)} - s^{(m)}\|_2 < \epsilon \qquad (n, m \geqslant N), \tag{1}$$

which implies

$$\|s^{(n)} - s^{(N)}\|_2 < \epsilon \qquad (n \geqslant N).$$

Thus, if $n \geqslant N$,

$$\|s^{(n)}\|_2 = \|[s^{(n)} - s^{(N)}] + s^{(N)}\|_2 < \epsilon + \|s^{(N)}\|_2.$$

Thus, for some $A > 0$,

$$\|s^{(n)}\|_2 \leqslant A \qquad (n \geqslant N). \tag{2}$$

Now, for any $k \in I$, we have from (1)

$$|s_k^{(n)} - s_k^{(m)}| \leqslant \|s^{(n)} - s^{(m)}\|_2 < \epsilon \qquad (n, m \geqslant N).$$

Hence (for fixed k) the sequence $\{s_k^{(n)}\}_{n=1}^{\infty}$ is a Cauchy sequence in R^1 and so, by 2.10D, converges to a number $s_k \in R^1$. Let s denote the sequence $\{s_k\}_{k=1}^{\infty}$. First we will show that $s \in \ell^2$. From (2) we have

$$\sum_{k=1}^{\infty} s_k^{(n)^2} \leqslant A^2 \qquad (n \geqslant N).$$

Hence for any integer $L \in I$,

$$\sum_{k=1}^{L} s_k^{(n)^2} \leqslant A^2 \qquad (n \geqslant N). \tag{3}$$

But for $k = 1, 2, \ldots, L$, we have $s_k^{(n)} \to s_k$ as $n \to \infty$. Hence letting $n \to \infty$ in (3) and using 2.7A and 2.7E, we have

$$\sum_{k=1}^{L} s_k^2 \leqslant A^2 \qquad (L = 1, 2, \ldots).$$

It follows that

$$\sum_{k=1}^{\infty} s_k^2 \leqslant A^2,$$

which proves that $s = \{s_k\}_{k=1}^{\infty}$ is in ℓ^2. From (1) we have

$$\sum_{k=1}^{\infty} (s_k^{(n)} - s_k^{(m)})^2 < \epsilon^2 \qquad (n, m \geqslant N).$$

Hence for $L \in I$,

$$\sum_{k=1}^{L} (s_k^{(n)} - s_k^{(m)})^2 < \epsilon^2 \qquad (n, m \geqslant N).$$

Letting $m \to \infty$ (and using $\lim_{m \to \infty} s_k^{(m)} = s_k$, 2.7A, and 2.7E), we have

$$\sum_{k=1}^{L} (s_k^{(n)} - s_k)^2 \leqslant \epsilon^2 \qquad (n \geqslant N; L \in I),$$

and so

$$\sum_{k=1}^{\infty} (s_k^{(n)} - s_k)^2 \leqslant \epsilon^2 \qquad (n \geqslant N).$$

But this says that $\rho(s^{(n)},s)=\|s^{(n)}-s\|_2 \leqslant \epsilon$ if $n \geqslant N$, which proves that $s^{(1)},s^{(2)},\ldots$ converges in ℓ^2 to the point s. This completes the proof.

6.4C. THEOREM. If $\langle M,\rho\rangle$ is a complete metric space and A is a closed subset of M, then $\langle A,\rho\rangle$ is also complete.

PROOF: Let $\{x_n\}_{n=1}^{\infty}$ be a Cauchy sequence of points in $\langle A,\rho\rangle$. We must show that $\{x_n\}_{n=1}^{\infty}$ converges to a point in A. Since $A \subset M$, $\{x_n\}_{n=1}^{\infty}$ is a Cauchy sequence of points of M. Thus since M is complete, $\{x_n\}_{n=1}^{\infty}$ must converge to some $x \in M$. But x is a limit point of A because x is the limit of a sequence of points in A. Hence $x \in A$ because A is closed, and the proof is complete.

Thus the metric space $[0,1]$ (with absolute value metric) is complete. For $[0,1]$ is a closed subset of R^1.

Here is a generalization of the nested interval theorem 2.10E.

6.4D. THEOREM. Let $\langle M,\rho\rangle$ be a complete metric space. For each $n \in I$ let F_n be a non-empty, closed bounded subset of M such that

(a) $F_1 \supset F_2 \supset \cdots \supset F_n \supset F_{n+1} \supset \cdots,$

and

(b) $\operatorname{diam} F_n \to 0$ as $n \to \infty$.

Then $\cap_{n=1}^{\infty} F_n$ contains precisely one point.

PROOF: For each $n \in I$, let a_n be any point in F_n. Then, by (a),

$$a_n, a_{n+1}, a_{n+2}, \ldots \quad \text{all lie in } F_n. \tag{1}$$

Given $\epsilon > 0$ there exists, by (b), an integer $N \in I$ such that $\operatorname{diam} F_N < \epsilon$. Now a_N, $a_{N+1}, a_{N+2}, \ldots$ all lie in F_N. For $m,n \geqslant N$ we then have $\rho(a_n,a_m) \leqslant \operatorname{diam} F_N < \epsilon$. This proves that $\{a_n\}_{n=1}^{\infty}$ is a Cauchy sequence. Since M is complete there exists $a \in M$ such that $\lim_{n\to\infty} a_n = a$. For any $n \in I$, the statement (1) then shows that a is a limit point of the closed set F_n and hence $a \in F_n$. Thus $a \in \cap_{n=1}^{\infty} F_n$. If $b \in M, b \neq a$, then $\rho(a,b) > \operatorname{diam} F_n$ for n sufficiently large. Hence b cannot be in $\cap_{n=1}^{\infty} F_n$. This completes the proof.

6.4E. We are now going to discuss a class of functions called contractions. Although their usefulness will not be immediately apparent, they turn out to have important applications. In a later chapter we use a result on contractions to prove an existence theorem for differential equations.

To simplify notation in this discussion, if $T: M \to M$ and if $x \in M$, we will write Tx instead of $T(x)$. We will also write T^2 instead of $T \circ T$, T^3 instead of $T \circ T^2$, etc.

DEFINITION. Let $\langle M,\rho\rangle$ be a metric space. If $T: M \to M$, we say that T is a contraction on M if there exists $\alpha \in R$ with $0 \leqslant \alpha < 1$ such that

$$\rho(Tx,Ty) \leqslant \alpha\rho(x,y) \qquad (x,y \in M).$$

(We emphasize that the number α must be independent of x and y.)

Thus T is a contraction if the distance from Tx to Ty is not greater than α times the distance from x to y. We see that applying T to each of two points "contracts" the distance between them.

The reader should verify that if T is a contraction on M then T is continuous on M.

Here is an easy example of a contraction. If $u=\{u_n\}_{n=1}^{\infty}\in \ell^2$, let $Tu=\{u_n/2\}_{n=1}^{\infty}$. Then T is a contraction on ℓ^2. For if $v=\{v_n\}_{n=1}^{\infty}$ is any other point in ℓ^2, then

$$\rho(Tu,Tv)=\|Tu-Tv\|_2=\left[\sum_{n=1}^{\infty}\left(\frac{u_n}{2}-\frac{v_n}{2}\right)^2\right]^{1/2}=\tfrac{1}{2}\|u-v\|_2$$
$$=\tfrac{1}{2}\rho(u,v).$$

Thus in this example, α may be taken to be $\frac{1}{2}$. For this T it is obvious that there is one and only one sequence $s\in\ell^2$ such that $Ts=s$—namely, the sequence $0,0,0,\ldots$. This illustrates the following theorem, which is called the Picard or the Banach fixed-point theorem.

6.4F. THEOREM. Let $\langle M,\rho\rangle$ be a complete metric space. If T is a contraction on M, then there is one and only one point x in M such that $Tx=x$. (This is often stated as "T has precisely one fixed point.")

PROOF: Suppose $x,y\in M$. We have $\rho(Tx,Ty)\leqslant\alpha\rho(x,y)$ for some $\alpha, 0\leqslant\alpha<1$. Then $\rho(T^2x,T^2y)\leqslant\alpha\rho(Tx,Ty)\leqslant\alpha^2\rho(x,y)$. Indeed, for any $n\in I$, it is easy to show that

$$\rho(T^nx,T^ny)\leqslant\alpha^n\rho(x,y)\qquad(x,y\in M). \tag{1}$$

Now choose any $x_0\in M$. Let $x_1=Tx_0, x_2=Tx_1,\ldots,x_{n+1}=Tx_n$. Then $x_2=T^2x_0$ and, for any $n\in I, x_n=T^nx_0$. We will first show that $\{x_n\}_{n=1}^{\infty}$ is a Cauchy sequence. For if $m,n\in I$ (and $m>n$, say, so that $m=n+p$) we have

$$\rho(x_n,x_m)=\rho(x_n,x_{n+p})\leqslant\rho(x_n,x_{n+1})+\rho(x_{n+1},x_{n+2})+\cdots+\rho(x_{n+p-1},x_{n+p})$$
$$=\rho(T^nx_0,T^nx_1)+\rho(T^{n+1}x_0,T^{n+1}x_1)+\cdots+\rho(T^{n+p-1}x_0,T^{n+p-1}x_1).$$

Thus by (1)

$$\rho(x_n,x_m)\leqslant\alpha^n\rho(x_0,x_1)+\alpha^{n+1}\rho(x_0,x_1)+\cdots+\alpha^{n+p-1}\rho(x_0,x_1)$$
$$\leqslant\alpha^n\rho(x_0,x_1)(1+\alpha+\alpha^2+\cdots),$$

and hence

$$\rho(x_n,x_m)\leqslant\frac{\alpha^n\rho(x_0,x_1)}{1-\alpha}.$$

Since $\lim_{n\to\infty}\alpha^n=0$, it follows easily that $\rho(x_n,x_m)$ can be made arbitrarily small by taking n (and hence m) sufficiently large. Thus $\{x_n\}_{n=1}^{\infty}$ is Cauchy. Since (by hypothesis) M is complete, there exists $x\in M$ such that $\lim_{n\to\infty}x_n=x$. Therefore $\lim_{n\to\infty}Tx_n=Tx$ (why?). But $Tx_n=x_{n+1}$ and so $\{Tx_n\}_{n=1}^{\infty}$, being a subsequence of $\{x_n\}_{n=1}^{\infty}$, must converge to $\lim_{n\to\infty}x_n=x$. It follows that $Tx=x$, so that x is a fixed point. All that remains to be shown is that if $y\in M, y\neq x$, then y cannot be a fixed point. For suppose the contrary. Then $Ty=y$ and so (since $Tx=x$), $\rho(x,y)=\rho(Tx,Ty)\leqslant\alpha\rho(x,y)$. Since $\rho(x,y)\neq0$ this implies $1\leqslant\alpha$ which is a contradiction. Hence $Ty\neq y$ and the proof is complete.

Exercises 6.4

1. Prove that R_d is complete.
2. Prove that the interval $(0, 1)$ with absolute value metric is not a complete metric space. Prove that $(0, 1)$ with the metric of R_d *is* a complete metric space.
3. Prove that R^2 is complete.
4. Prove that ℓ^∞ is complete. (Model your proof after the proof that ℓ^2 is complete.)
5. If
$$T(x)=x^2 \qquad (0<x\leqslant \tfrac{1}{3}),$$
prove that T is a contraction on $(0, \frac{1}{3}]$, but that T has no fixed point.
6. If $T:[0,1]\to[0,1]$ and if there is a real number α with $0\leqslant\alpha<1$ such that
$$|T'(x)|\leqslant\alpha \qquad (0\leqslant x\leqslant 1),$$
where T' is the derivative of T, prove that T is a contraction on $[0, 1]$.
7. Let M be a metric space that is both totally bounded and complete. Prove that every sequence of points of M has a subsequence that converges to a point of M.
8. Let $M=[0,\infty)$ with the absolute value metric $\rho(x,y)=|x-y|$. Let
$$f(x)=\frac{1}{1+x^2} \qquad (0\leqslant x<\infty).$$
Show that $f: M\to M$, that
$$\rho[f(x),f(y)]<\rho(x,y) \qquad (x,y\in M),$$
but that f has no fixed point

6.5 COMPACT METRIC SPACES

It is because the closed bounded interval $[a,b]$ is compact that many of the theorems about continuous functions on $[a,b]$ hold. We now begin a general discussion of compact metric spaces.

6.5A. DEFINITION. The metric space $\langle M,\rho\rangle$ is said to be compact if $\langle M,\rho\rangle$ is both complete and totally bounded.

For example, the metric space $[a,b]$ (with absolute value metric) is totally bounded and, by 6.4C, is complete. Hence $[a,b]$ is compact. The space R^1 is complete but not totally bounded. Hence R^1 is not compact. The metric space $(0,1)$ (with absolute value metric) is totally bounded but not complete and hence is not compact.

From 6.3E we see that an infinite subset of R_d cannot be compact. We leave it to the reader to show that every finite subset of R_d actually is compact. (See Exercise 2.)

A very useful reformulation of compactness is continued in the following.

6.5B. THEOREM. The metric space $\langle M,\rho\rangle$ is compact if and only if every sequence of points in M has a subsequence converging to a point in M.

PROOF: Suppose first that M is compact and that $\{x_n\}_{n=1}^{\infty}$ is any sequence of points in M. Then, by 6.3H, since M is totally bounded, the sequence $\{x_n\}_{n=1}^{\infty}$ has a Cauchy subsequence $\{x_{n_k}\}_{k=1}^{\infty}$. But $\{x_{n_k}\}_{k=1}^{\infty}$ converges to a point in M since M is complete. Thus if M is compact, then every sequence in M contains a convergent subsequence.

Conversely, suppose every sequence in M has a convergent subsequence. Then, by 6.3H, M is totally bounded. To show that M is complete we must show that every Cauchy sequence $\{x_n\}_{n=1}^{\infty}$ in M converges to a point of M. By assumption, $\{x_n\}_{n=1}^{\infty}$ has a subsequence $\{x_{n_k}\}_{k=1}^{\infty}$ which converges to a point x in M. Since $\{x_n\}_{n=1}^{\infty}$ is Cauchy, it is then not difficult to show that $\{x_n\}_{n=1}^{\infty}$ itself converges to x. Hence M is complete and so M is compact. This completes the proof.

Here is a useful corollary.

6.5C. COROLLARY. If A is a closed subset of the compact metric space $\langle M,\rho\rangle$ then the metric space $\langle A,\rho\rangle$ is also compact.

PROOF: Any sequence $\{x_n\}_{n=1}^{\infty}$ of points of A is a sequence of points of M and hence, by 6.5B, has a subsequence converging to a point x in M. But then x is a limit point of A and so $x\in A$ (since A is closed). Thus any sequence in A has a subsequence converging to a point in A. By 6.5B, A is compact.

In the other direction we have the following.

6.5D. THEOREM. Let A be a subset of the metric space $\langle M,\rho\rangle$. If $\langle A,\rho\rangle$ is compact, then A is a closed subset of $\langle M,\rho\rangle$.

PROOF: Let $x\in M$ be any limit point of A. Then there is a sequence $\{x_n\}_{n=1}^{\infty}$ in A converging to x. But $\{x_n\}_{n=1}^{\infty}$ is a Cauchy sequence in A and so, since A is complete, $\{x_n\}_{n=1}^{\infty}$ converges to a point in A. This point must be x and so $x\in A$. Thus A contains all its limit points and so A is closed. (Note that the proof uses only the completeness of A. However, it is in the context of compactness, rather than completeness, that the result will be useful.)

6.5E. If M is any set, the family $\mathfrak{F}$ of subsets A of M is said to form a covering of M if $M\subset\bigcup_{A\in\mathfrak{F}}A$. We will be primarily interested in *open* coverings—that is, coverings $\mathfrak{F}$ of a metric space M where each $A\in\mathfrak{F}$ is an open subset of M.

For example, the family of open intervals $(1/n,1-1/n)$ for $n=3,4,5,\ldots$ is an open covering of the metric space $(0,1)$ (with absolute value metric). Note that there are infinitely many open sets in this covering of $(0,1)$ and that no finite number of these sets form a covering!

On the other hand, the reader should experiment with some open coverings of $[0,1]$. We will find that if $\mathfrak{F}$ is any open covering of $[0,1]$, then some finite number of sets in $\mathfrak{F}$ will also form a covering. A proof of this will be given shortly.

6.5F. DEFINITION. A metric space M is said to have the Heine-Borel property if, whenever $\mathfrak{F}$ is an open covering of M, then there exist a finite number of sets $G_1,\ldots,G_n\in\mathfrak{F}$ such that $\{G_1,\ldots,G_n\}$ is still a covering of M.

The Heine-Borel property is sometimes stated as "every open covering of M admits a finite subcovering." From the first example in 6.5E we see that the metric space $(0,1)$ does not have the Heine-Borel property. The next two theorems show the reason the Heine-Borel property is discussed in this section.

6.5G. THEOREM. If M is a compact metric space, then M has the Heine-Borel property.*

PROOF: Suppose the contrary. Then, for some open covering $\mathcal{F}$, no finite number of sets in $\mathcal{F}$ form a covering of M. Now M is totally bounded and hence may be written as the union of a finite number of bounded subsets each of whose diameter is less than 1. But then one of these subsets, call it A_1, cannot be covered by a finite number of sets in $\mathcal{F}$. (Otherwise, all of M could be covered by a finite number of sets in $\mathcal{F}$.) But, since $\operatorname{diam} \bar{A}_1 = \operatorname{diam} A_1$ (verify), $\bar{A}_1$ is a closed subset of M (5.5E) whose diameter is less than 1 and which cannot be covered by a finite number of sets in $\mathcal{F}$. Since $\bar{A}_1$ is itself totally bounded, the same reasoning shows the existence of a subset A_2 of $\bar{A}_1$ with $\operatorname{diam} A_2 < \frac{1}{2}$ and such that A_2 cannot be covered by a finite number of sets in $\mathcal{F}$. Thus $\bar{A}_2 \subset \bar{A}_1$, $\operatorname{diam} \bar{A}_2 < \frac{1}{2}$, and $\bar{A}_2$ cannot be covered by a finite number of sets in $\mathcal{F}$. Continuing in this fashion we can show, for any $n \in I$, the existence of $\bar{A}_n \subset M$ such that $\bar{A}_1 \supset \bar{A}_2 \supset \cdots \supset \bar{A}_n \supset \cdots$, $\operatorname{diam} \bar{A}_n < 1/n$, and such that no finite number of sets in $\mathcal{F}$ form a covering of any $\bar{A}_n$. By 6.4D, there is precisely one point x in $\cap_{n=1}^{\infty} \bar{A}_n$. Now, since $\mathcal{F}$ is a covering of M, there is a set G in $\mathcal{F}$ such that $x \in G$. But G is open (since $\mathcal{F}$ is an open covering) and so there exists an open ball $B[x;r] \subset G$ for some $r > 0$. But if $N \in I$ satisfies $1/N < r$, then $\operatorname{diam} \bar{A}_N < 1/N < r$. Since $x \in \bar{A}_N$ we have $\bar{A}_N \subset B[x;r] \subset G$. Hence G alone covers $\bar{A}_N$. But this is a contradiction since no finite number of sets in $\mathcal{F}$ were supposed to form a covering of $\bar{A}_N$. The contradiction proves the theorem.

The converse of 6.5G is also true.

6.5H. THEOREM. If the metric space M has the Heine-Borel property, then M is compact.

PROOF: Let M be a metric space with the Heine-Borel property, and let $\{x_n\}_{n=1}^{\infty}$ be any sequence of points of M. In order to prove that M is compact it is enough, by 6.5B, to show that $\{x_n\}_{n=1}^{\infty}$ has a subsequence which converges to a point of M.

Suppose first that about each point x in M there were an open ball B_x which contained x_n for only finitely many values of n. The family of all such B_x would then be an open covering of M. By hypothesis, then, M could be covered by a finite number of these B_x. This is clearly impossible. For, since each B_x contains x_n for only a finite number of values of n, the union of any finite number of the B_x could not contain all the x_n.

Hence there must be some point x in M such that *every* open ball about x contains x_n for infinitely many values of n. Thus there exists $n_1 \in I$ such that $x_{n_1} \in B[x;1]$; there exists $n_2 > n_1$ such that $x_{n_2} \in B[x;\frac{1}{2}]$; indeed, for any $k \in I$ there exists $n_k > n_{k-1}$ such that $x_{n_k} \in B[x;1/k]$. This subsequence $\{x_{n_k}\}_{k=1}^{\infty}$ of $\{x_n\}_{n=1}^{\infty}$ clearly converges to the point x in M, and the proof is complete.

We list one more condition equivalent to compactness.

6.5I. DEFINITION. A family $\mathcal{F}$ of subsets of a set M is said to have the finite intersection property if the intersection of any finite number of sets in $\mathcal{F}$ is never empty. (That is, if whenever $F_1, F_2, \ldots, F_n \in \mathcal{F}$ then $F_1 \cap F_2 \cap \cdots \cap F_n \neq \varnothing$.)

* The Heine-Borel property is often taken as the definition of compactness, since the Heine-Borel property does not depend directly on the metric but rather on the notion of open set which is available in a class of spaces more general than metric spaces.

Thus the family of all closed intervals $[-1/n, 1/n]$ has the finite intersection property. So does the family of open intervals $(0, 1/n)$.

6.5J. THEOREM. The metric space M is compact if and only if, whenever $\mathcal{F}$ is a family of closed subsets of M with the finite intersection property, then $\cap_{F \in \mathcal{F}} F \neq \varnothing$.

PROOF: Suppose first that M is a compact metric space and that $\mathcal{F}$ is a family of closed subsets of M with the finite intersection property. We must show that

$$\bigcap_{F \in \mathcal{F}} F \neq \varnothing. \tag{1}$$

For each $F \in \mathcal{F}$ let $G = F' = M - F$, and let $\mathcal{G}$ be the family of all such open sets G. If $F_1, F_2, \ldots, F_n$ are any finite number of sets in $\mathcal{F}$, then

$$F_1 \cap F_2 \cap \cdots \cap F_n = M - (G_1 \cup G_2 \cup \cdots \cup G_n) \tag{2}$$

by (1) of 1.2H. Similarly,

$$\bigcap_{F \in \mathcal{F}} F = M - \bigcup_{G \in \mathcal{G}} G. \tag{3}$$

By hypothesis, the left side of (2) is not empty, and so $M - (G_1 \cup \cdots \cup G_n) \neq \varnothing$. Hence no finite number of sets in $\mathcal{G}$ form a covering of M. Since M is compact, it follows from 6.5G that $\mathcal{G}$ itself is not a covering of M. Hence the right side of (3) is not empty. This proves (1), which is what we wished to show.

We leave the proof of the converse to the reader.

Exercises 6.5

1. Use 6.5A to prove 6.5C.
2. Prove that every finite subset of any metric space is compact.
3. Prove that a subset A of R^2 is compact if and only if A is closed and bounded.
4. If A and B are compact subsets of R^1, prove that $A \times B$ is a compact subset of R^2.
5. Let f be a continuous real-valued function on $[a,b]$. Prove that the graph of f is a compact subset of R^2.
6. For each x in $(0,1)$ let I_x denote the open interval $(x/2, (x+1)/2)$. Show that the family $\mathcal{G}$ of all such I_x is an open covering of $(0,1)$ which admits no finite subcovering of $(0,1)$.
7. Add two appropriate sets to the family $\mathcal{G}$ of the preceding exercise to form an open covering $\mathcal{H}$ of $[0,1]$. Show that $\mathcal{H}$ does admit a finite subcovering of $[0,1]$.
8. Give an example of a connected subset of R^1 that is not compact.
9. Prove that a connected subset of R_d is compact.
10. Give an example of a closed bounded subset of ℓ^2 which is not compact.
11. Prove that the metric space M is compact if and only if every infinite subset of M has a cluster point in M.

6.6 CONTINUOUS FUNCTIONS ON COMPACT METRIC SPACES

Because of the tremendous importance of the following theorem we give two proofs, each based on a different criterion for compactness.

6.6A. THEOREM. Let f be a continuous function from the compact metric space M_1 into the metric space M_2. Then the range $f(M_1)$ of f is also compact.

PROOF 1: Let $\mathcal{G}$ be any open covering of $f(M_1)$. For each $G \in \mathcal{G}$ the set $f^{-1}(G)$ is, by 5.4G, an open subset of M_1. The family of all such sets $f^{-1}(G)$ for $G \in \mathcal{G}$ is therefore an open covering of M_1. Since M_1 is compact, a finite number of these sets—say $f^{-1}(G_1),\ldots,f^{-1}(G_n)$—also form a covering of M_1. But then $G_1,\ldots,G_n$ form a covering of $f(M_1)$. Thus there are a finite number of sets in $\mathcal{G}$ which form a covering of $f(M_1)$, and so $f(M_1)$ is compact.

PROOF 2: Let $\{y_n\}_{n=1}^{\infty}$ be any sequence of points in $f(M_1)$. For each $n \in I$ choose $x_n \in M_1$ such that $f(x_n)=y_n$. By 6.5B, $\{x_n\}_{n=1}^{\infty}$ has a subsequence $\{x_{n_k}\}_{k=1}^{\infty}$ which converges to a point x in M_1. Since f is continuous on M_1, it then follows from 5.3C that $\lim_{k\to\infty} f(x_{n_k})=f(x)$. That is, $\{y_{n_k}\}_{k=1}^{\infty}$ converges to $f(x) \in f(M_1)$. Thus any sequence $\{y_n\}_{n=1}^{\infty}$ in $f(M_1)$ has a convergent subsequence and so, by 6.5B, $f(M_1)$ is compact.

Since compact spaces are bounded the following corollary is immediate.

6.6B. COROLLARY. Let f be a continuous function from the compact metric space M_1 into the metric space M_2. Then the range $f(M_1)$ of f is a bounded subset of M_2.

At this time it is convenient to introduce the notion of "bounded function."

6.6C. DEFINITION. Let f be a function from a set A into a metric space M. We say that the function f is bounded if its range $f(A)$ is a bounded subset of M.

Thus 6.6B states that a continuous function on a compact metric space M_1 (into a metric space M_2) must be bounded. When $M_2 = R^1$ and M_1 is the closed bounded interval $[a,b]$ we thus have corollary 6.6D.

6.6D. COROLLARY. If the real-valued function f is continuous on a closed bounded interval in R^1, then f must be bounded.

The reader should construct examples to show that 6.6D is no longer true if either of the words "closed" or "bounded" is removed from the hypothesis.

6.6E. If f is a real-valued function on a set A, we say (quite naturally) that f attains a maximum value at $a \in A$ if

$$f(a) \geqslant f(x) \qquad (x \in A).$$

For example, if

$$f(x)=x^2 \qquad (-1 \leqslant x \leqslant 1),$$

then f attains a maximum at $x=1$ and $x=-1$. If

$$g(x)=x^2 \qquad (-\infty < x < \infty),$$

then g clearly does not attain a maximum value. Indeed if a real-valued function f on a set A is not bounded above, then f cannot attain a maximum at any point of A.

On the other hand, the examples

$$f(x)=x \qquad (0 \leqslant x < 1),$$

$$f(x)=\frac{x-1}{x} \qquad (1 \leqslant x < \infty),$$

show that even a bounded continuous real-valued function need not attain a maximum at any point in its domain. See Figure 24.

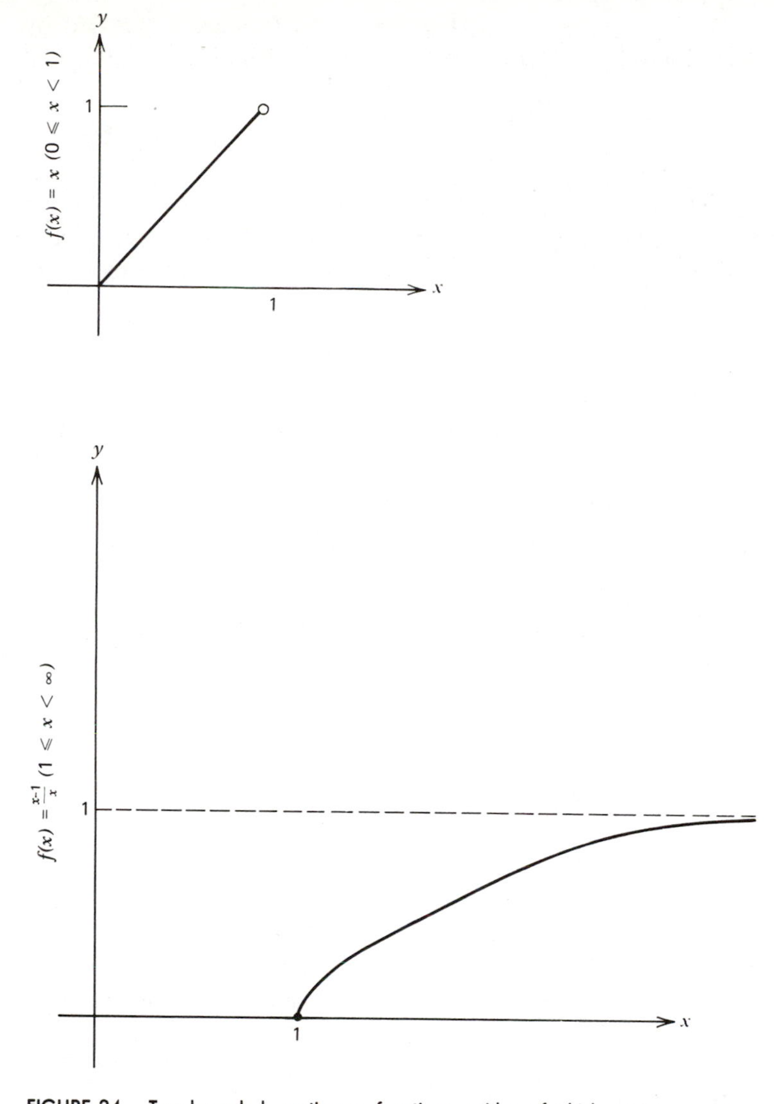

FIGURE 24. Two bounded, continuous functions, neither of which attains a maximum

We leave it to the reader to formulate the corresponding definition and examples for minimum values.

Here is where compactness enters into the picture.

6.6F. THEOREM. If the real-valued function f is continuous on the compact metric space M, then f attains a maximum value at some point of M. Also, f attains a minimum value at some point of M.

PROOF 1: By 6.6D, the function f must be bounded. Let $L = \text{l.u.b.}_{x \in M} f(x)$. By definition of least upper bound, the number L must be a limit point of $f(M)$. But by 6.6A we know that $f(M)$ is compact. Hence, by 6.5D, $f(M)$ is a closed subset of R^1. Thus the number L [which is a limit point of $f(M)$] must lie in $f(M)$. That is, $L = f(a)$ for some $a \in M$. It is then clear that f attains a maximum at the point a.

The assertion about a minimum value may be proved in a similar manner.

PROOF 2: Again let $L = \text{l.u.b.}_{x \in M} f(x)$. If $L \notin f(M)$, then the function g defined by $g(x) = L - f(x)$ $(x \in M)$ is continuous on M and g never takes the value 0. By 5.3G, $1/g$ is continuous on M and thus, by 6.6D, $1/g$ must be bounded. Thus

$$\frac{1}{g(x)} = \frac{1}{L - f(x)} \leqslant N$$

for some $N \geqslant 0$ and all $x \in M$. Hence $f(x) \leqslant L - 1/N$ for all $x \in M$. But this says that $L - 1/N$ is an upper bound for $f(M)$, which is a contradiction, since L was supposed to be the least upper bound. Thus we must have $L \in f(M)$, and the conclusion follows as in proof 1.

For our treatment of calculus we need the following consequence of 6.6F.

6.6G. COROLLARY. If the real-valued function f is continuous on the closed bounded interval $[a,b]$, then f attains a maximum and a minimum value at points of $[a,b]$.

Exercises 6.6

1. Show that if $f: A \to R^1$ and f attains a maximum value at $a \in A$, then
$$f(a) = \underset{x \in A}{\text{l.u.b.}}\, f(x).$$
2. Show that if
$$f(x) = \frac{1}{1+x^2} \qquad (-\infty < x < \infty),$$
then f attains a maximum value but does not attain a minimum value.
3. Give an example of a continuous bounded function on $(-\infty, \infty)$ that attains neither a maximum nor a minimum value.
4. Give an example of a real-valued continuous function on $[0,1)$ that attains a minimum value but does not attain a maximum.
5. If f is a continuous real-valued function on the compact connected metric space M, prove that f takes on every value between its minimum value and its maximum value.

6.7 CONTINUITY OF THE INVERSE FUNCTION

6.7A. We raise the following question: If the function f is continuous and 1–1, is the inverse function f^{-1} necessarily continuous? It is not hard to show that the answer is no.

For a first example consider $f: R_d \Rightarrow R^1$ defined by

$$f(x) = x \qquad (-\infty < x < \infty).$$

That is, f maps every real number onto itself. However, in the domain of f we have the discrete metric, while in the range of f we have the absolute value metric. Now f is obviously 1–1. Moreover f is continuous, since every function on R_d is continuous. However, f^{-1} is *not* continuous, for if f^{-1} were continuous, then f would be a homeomorphism of R_d onto R^1 (5.5L). But then, by (b) of 5.5K, $f(G)$ would be an open set in R^1 whenever G is an open set in R_d. Since $f(G) = G$, this would imply that if G is open in R_d then G is open in R^1. But this is clearly false, since every subset of R_d is open (5.4D), but not every subset of R^1 is open.

For a more intuitive example consider

$$g(x) = \langle \cos x, \sin x \rangle \qquad (0 \leqslant x < 2\pi).$$

Then g is a 1–1 continuous function from the interval $J = [0, 2\pi)$ onto the circumference C of the unit circle in R^2. But g^{-1} then maps the circumference onto the interval and hence cannot be continuous, since a circumference must be "broken" in order to be spread out on an interval. Indeed, we see that g^{-1} sends the point $\langle 1, 0 \rangle$ onto $0 \in J$, but sends all points of C near but below $\langle 1, 0 \rangle$ onto points near 2π in J.

However, if the domain of the 1–1 continuous function f is compact, then f^{-1} will be continuous.

6.7B. THEOREM. If f is a 1–1 continuous function from the compact metric space M_1 onto the metric space M_2, then f^{-1} is continuous (on M_2), and hence f is a homeomorphism of M_1 onto M_2.

PROOF: By 5.5J, to show that f^{-1} is continuous we must show that if F is any closed subset of M_1, then the inverse image of F under f^{-1} is closed in M_2. But the inverse image of F under f^{-1} is precisely $f(F)$ (verify). Thus we need to show that if F is closed in M_1, then $f(F)$ is closed in M_2. But if F is closed in M_1, then, by 6.5C, F itself is compact. By 6.6A we then know that $f(F)$ is compact. Hence, by 6.5D, $f(F)$ is closed in M_2, which is what we wished to show.

As an application of 6.7B let us show that $\sqrt{x}$ is a continuous function of x. More precisely, if g is defined by

$$g(x) = \sqrt{x} \qquad (0 \leqslant x < \infty),$$

we will prove that g is continuous on $[0, \infty)$. For any $N \in I$ the function f defined by

$$f(x) = x^2 \qquad (0 \leqslant x \leqslant N)$$

is a 1–1 continuous function from the compact space $[0, N]$ onto $[0, N^2]$. Hence, by 6.7B, f^{-1} is continuous on $[0, N^2]$. But f^{-1} is precisely the restriction of g to $[0, N^2]$. It follows that g must be continuous on $[0, \infty)$.

Exercises 6.7

1. Use your intuition on this one. Let f be the function which sends a point on a flat map of the world onto the corresponding point on a globe.
   (a) Is f continuous?
   (b) Is f^{-1} continuous?
2. Prove that $\sqrt[n]{x}$ is a continuous function for any $n \in I$.
3. Show that if f is a $1-1$ continuous function from a metric space M into R_d then f^{-1} is continuous and hence f is a homeomorphism.
4. Show that if f is a 1–1 continuous function from R^1 into R^1, then f is a homeomorphism.

6.8 UNIFORM CONTINUITY

6.8A. We have defined the real-valued function f as continuous at the point $a \in R^1$ if given $\epsilon > 0$, there exists $\delta > 0$ such that

$$|f(x) - f(a)| < \epsilon \qquad (|x - a| < \delta).$$

Now, in general, the number δ depends not only on ϵ but also on which point a is under consideration. For example, let

$$g(x)=x^2 \qquad (-\infty<x<\infty).$$

Then with $\epsilon=2$, the statement

$$|g(x)-g(a)|<2 \qquad (|x-a|<\tfrac{1}{2}) \tag{1}$$

is true if $a=1$. [For then $g(x)-g(a)=x^2-a^2=x^2-1$. If $|x-a|<\frac{1}{2}$, then $\frac{1}{2}<x<\frac{3}{2}$ and so $-\frac{3}{4}<g(x)-g(a)<\frac{5}{4}$.] However, the statement (1) is false for $a=10$. For, when $a=10$, we have $g(x)-g(a)=x^2-100$. If $x=10\frac{1}{4}$, then $|x-a|<\frac{1}{2}$ but $g(x)-g(a)=(10\frac{1}{4})^2-10^2$ $=5\frac{1}{16}$ and so $|g(x)-g(a)|>2$.

Thus (even though g is continuous at the point 10 as well as at the point 1) the number $\delta=\frac{1}{2}$ is usable at $a=1$ but not at $a=10$ as a δ corresponding to $\epsilon=2$.

It is not difficult to show, in fact, that there is no one $\delta>0$ such that the statement

$$|g(x)-g(a)|<2 \qquad (|x-a|<\delta) \tag{2}$$

is true for *all* $a\in R^1$. For $g(x)-g(a)=(x-a)(x+a)$. Suppose there were a δ for which (2) held for all a. We would then have for $a>0$ and $x=a+\delta/2$,

$$|g(x)-g(a)|=|x-a|\cdot|x+a|=\frac{\delta}{2}\cdot\left|2a+\frac{\delta}{2}\right|<2.$$

This would imply

$$a\delta<2$$

for all $a>0$, and this is clearly false. Hence, for this function g, corresponding to $\epsilon=2$ there is no δ that will "work" for all a simultaneously. (Nevertheless, g is continuous at each $a\in R^1$.)

If a continuous function is such that, given ϵ, we can always choose δ so that δ depends only on ϵ but *not* on a, then we say that the function is uniformly continuous. We now make this precise for functions on metric spaces.

6.8B. DEFINITION. Let $\langle M_1,\rho_1\rangle$ and $\langle M_2,\rho_2\rangle$ be metric spaces. If $f: M_1\to M_2$, we say that f is uniformly continuous on M_1 if given $\epsilon>0$, there exists $\delta>0$ such that

$$\rho_2[f(x),f(a)]<\epsilon \qquad [\rho_1(x,a)<\delta; a\in M_1].$$

In the special case $\langle M_1,\rho_1\rangle=\langle M_2,\rho_2\rangle=R^1$ we thus have "the real-valued function f is uniformly continuous on R^1 if given $\epsilon>0$ there exists $\delta>0$ such that

$$|f(x)-f(a)|<\epsilon \qquad (|x-a|<\delta;\ -\infty<a<\infty)."$$

What must be emphasized here is that, given ϵ, the δ must be such so that the statement

$$|f(x)-f(a)|<\epsilon \qquad (|x-a|<\delta)$$

is true for all a simultaneously. The function g of 6.8A is therefore not *uniformly* continuous on R^1 even though g is continuous on R^1.

Thus not every function which is continuous on a metric space M is uniformly continuous on M. On the other hand, it is clear from 6.8B that if f is uniformly continuous on M then f is continuous at each $a\in M$ and hence f is continuous on M.

We will now show that if M is compact, then the continuity of f on M implies the uniform continuity of f on M. (Thus on a *compact* metric space, a function is continuous if and only if it is uniformly continuous.)

6.8C. THEOREM. Let $\langle M_1,\rho_1\rangle$ be a compact metric space. If f is a continuous function from M_1 into a metric space $\langle M_2,\rho_2\rangle$, then f is uniformly continuous on M_1.

PROOF: By hypothesis, f is continuous at each $a\in M_1$. Thus for each $a\in M_1$, given $\epsilon>0$ there exists $r>0$ (depending on a) such that

$$\rho_2[f(x),f(a)]<\frac{\epsilon}{2} \qquad [\rho_1(x,a)<r]. \tag{1}$$

The family of open balls $B[a;r/2]$ for all $a\in M_1$ is an open covering of M_1. Since M_1 is compact, there are a finite number of these balls—say $B[a_1;r_1/2],\ldots,B[a_n;r_n/2]$—which form a covering of M_1. Now let $\delta=\min(r_1/2,\ldots,r_n/2)$. For any $a\in M_1$ we have $a\in B[a_j;r_j/2]$ for some $j=1,\ldots,n$, and so $\rho(a,a_j)<r_j/2$. Now if $\rho_1(x,a)<\delta$, then $\rho_1(x,a)<r_j/2$ and hence $\rho_1(x,a_j)<r_j$. By (1) (with a_j,r_j in place of a,r) we then have

$$\rho_2[f(x),f(a_j)]<\frac{\epsilon}{2}$$

and

$$\rho_2[f(a),f(a_j)]<\frac{\epsilon}{2},$$

from which it follows that $\rho_2[f(x),f(a)]<\epsilon$. Thus for $\delta=\min(r_1/2,\ldots,r_n/2)$, we have shown that, for all $a\in M_1$,

$$\rho_2[f(x),f(a)]<\epsilon \qquad [\rho_1(x,a)<\delta].$$

This proves that f is uniformly continuous on M_1.

6.8D. COROLLARY. If the real-valued function f is continuous on the closed bounded interval $[a,b]$, then f is uniformly continuous on $[a,b]$.

6.8E. The function f defined by

$$f(x)=\frac{1}{x} \qquad (0<x\leqslant 1)$$

is continuous on $(0,1]$. However, there is no way to define $f(0)$ so that the resulting extension of f is continuous on $[0,1]$ (why?).

For another such example, the function g defined by

$$g(x)=\sin\frac{1}{x} \qquad (x\neq 0)$$

is continuous and bounded on $R^1-\{0\}$. But g cannot be extended to a function continuous on all of R^1. (For $\lim_{x\to 0} g(x)$ does not exist.)

We pose the following problem: If the function f is continuous on a dense subset (5.5M) of a metric space M, when can f be extended to a function continuous on all of M? The partial answer given here involves uniform continuity.

6.8F. THEOREM. Let $\langle M_1,\rho_1\rangle$ be a metric space and let A be a dense subset of M_1. If f is a *uniformly* continuous function from $\langle A,\rho_1\rangle$ into a *complete* metric space $\langle M_2,\rho_2\rangle$, then f can be extended to a uniformly continuous function F from M_1 into M_2.

PROOF: First we will prove that if $\{x_n\}_{n=1}^{\infty}$ is a Cauchy sequence of points in A, then $\{f(x_n)\}_{n=1}^{\infty}$ is Cauchy in M_2. Indeed, given $\epsilon>0$ choose $\delta>0$ such that

$$\rho_2[f(x),f(y)]<\epsilon \qquad [\rho_1(x,y)<\delta;x,y\in A] \tag{1}$$

(We can find such a δ since f is uniformly continuous on A.) Now, since $\{x_n\}_{n=1}^{\infty}$ is Cauchy in A there exists $N \in I$ such that

$$\rho_1(x_m, x_n) < \delta \qquad (m, n \geqslant N). \tag{2}$$

From (1) and (2) it follows that

$$\rho_2[f(x_m), f(x_n)] < \epsilon \qquad (m, n \geqslant N),$$

which proves that $\{f(x_n)\}_{n=1}^{\infty}$ is Cauchy in M_2.

Now if $x \in A$, define $F(x)$ to be equal to $f(x)$. If $x \in M_1$ but $x \notin A$, then, by hypothesis, x is a limit point of A. Hence $\lim_{n\to\infty} x_n = x$ for some sequence $\{x_n\}_{n=1}^{\infty}$ of points of A. The sequence $\{x_n\}_{n=1}^{\infty}$ is convergent in M_1 and hence is a Cauchy sequence of points of A. According to the first paragraph, $\{f(x_n)\}_{n=1}^{\infty}$ is then Cauchy in M_2. Since M_2 is complete, $\lim_{n\to\infty} f(x_n)$ exists. We define

$$F(x) = \lim_{n\to\infty} f(x_n).$$

We leave it to the reader to show that if $\{y_n\}_{n=1}^{\infty}$ is any other sequence in A which converges to x, then $\lim_{n\to\infty} f(y_n) = \lim_{n\to\infty} f(x_n)$. This will show that the definition of $F(x)$ does not depend on the choice of the sequence $\{x_n\}_{n=1}^{\infty}$.

We have thus defined $F(x)$ for all $x \in M_1$, and F is clearly an extension of f. It remains to show that F is uniformly continuous on M_1. Given $\epsilon > 0$ choose δ_1 such that

$$\rho_2[f(x), f(y)] < \frac{\epsilon}{3} \qquad [\rho_1(x,y) < \delta_1; x, y \in A]. \tag{3}$$

If $a, b \in M_1$, choose $x, y \in A$ such that $\rho_1(x,a) < \delta_1/3, \rho_1(y,b) < \delta_1/3$ and such that

$$\rho_2[F(x), F(a)] < \frac{\epsilon}{3} \tag{4}$$

$$\rho_2[F(y), F(b)] < \frac{\epsilon}{3}. \tag{5}$$

[This is possible by the very definition of $F(a)$ and $F(b)$.] But then, if $\rho_1(a,b) < \delta_1/3$, we have

$$\rho_1(x,y) < \rho_1(x,a) + \rho_1(a,b) + \rho_1(b,y) < \frac{\delta_1}{3} + \frac{\delta_1}{3} + \frac{\delta_1}{3} = \delta_1.$$

From (3) we then have [since $f(x) = F(x)$ and $f(y) = F(y)$]

$$\rho_2[F(x), F(y)] < \frac{\epsilon}{3}. \tag{6}$$

We conclude from (4), (5), and (6) that

$$\rho_2[F(a), F(b)] < \epsilon$$

provided only that $\rho_1(a,b) < \delta/3$. This shows that F is uniformly continuous on M_1, and the proof is complete.

From 6.8F we see that neither f nor g in 6.8E is uniformly continuous.

Exercises 6.8

1. Given $\epsilon > 0$ find $\delta > 0$ such that

$$|\sin x - \sin a| < \epsilon \qquad (|x - a| < \delta; -\infty < a < \infty).$$

[*Hint*: Apply the theorem (or law) of the mean to $f(x) = \sin x$.] Deduce that the sine function is uniformly continuous on $(-\infty, \infty)$.

2. Suppose that f is a real-valued function on $[a,b]$ and that the absolute value of the slope of every secant line on the graph of f is $\leqslant 1$. Prove that f is uniformly continuous on $[a,b]$.
3. Which of the following functions are uniformly continuous on the indicated domain?
   (a) $f(x)=x^3$ $(0\leqslant x\leqslant 1)$.
   (b) $f(x)=x^3$ $(0\leqslant x<\infty)$.
   (c) $f(x)=\sin x^2$ $(0\leqslant x<\infty)$.
   (d) $f(x)=1/(1+x^2)$ $(0\leqslant x<\infty)$.
4. Let f be uniformly continuous real-valued function on $(0,1)$. Prove that $\lim_{x\to 0+}f(x)$ exists.
5. Suppose f is a continuous real-valued function on R^1 and that

$$\lim_{x\to\infty} f(x)=0=\lim_{x\to-\infty} f(x). \tag{*}$$

   Prove that f is uniformly continuous on R^1. [*Hint*: Use (*) and 6.8D.]
6. Prove that every function from R_d into a metric space is uniformly continuous.
7. Let M be a metric space, $x_0\in M$, and

$$f(x)=\rho(x,x_0) \qquad (x\in M).$$

   Prove that f is uniformly continuous on M.
8. Prove that a uniformly continuous function sends Cauchy sequences to Cauchy sequences.

6.9 NOTES AND ADDITIONAL EXERCISES FOR CHAPTERS 4, 5, AND 6.

I. A theory that allows infinite limits.

6.9A. We have by design allowed only finite values for limits of sequences of real numbers. This is appropriate, we think, in an introduction to the subject in order not to compound the difficulty of the limit concept.

There are, however, certain advantages to a theory which allows ∞ and $-\infty$ as limits, and we will indicate here how to proceed with this alternate approach. Since we have already developed in detail the "theory of finite limits," the exposition will not be long.

First we will define the set known as the extended real numbers. This set, denoted R^*, consists of all real numbers together with the symbols ∞ and $-\infty$. We postulate

$$a<\infty \qquad (a\in R),$$
$$-\infty<a \qquad (a\in R),$$

so that the ordering of R is extended to R^*.

EXERCISE: Let $f(\infty)=1$, $f(-\infty)=-1$, and

$$f(x)=\frac{x}{1+|x|} \qquad (x\in R),$$

and define

$$\rho(x,y)=|f(x)-f(y)| \qquad (x,y\in R^*).$$

Prove that ρ is a metric for R^*.

6.9B. We make the following definitions to extend addition and subtraction from R to R^*;

$$a+\infty=\infty+a=\infty \qquad (a\in R),$$
$$a+(-\infty)=(-\infty)+a=-\infty \qquad (a\in R),$$
$$\infty+\infty=\infty,$$
$$(-\infty)+(-\infty)=-\infty,$$
$$-(-\infty)=\infty,$$

and

$$a-b=a+(-b) \qquad (a,b\in R^*),$$

except that we do not define $\infty-\infty$ or $\infty+(-\infty)$. For multiplication we define

$$a(\infty)=\infty(a)=\infty \qquad (0<a\leqslant\infty),$$
$$a(-\infty)=(-\infty)a=-\infty \qquad (0<a\leqslant\infty),$$
$$a(\infty)=(\infty)a=-\infty \qquad (-\infty\leqslant a<0),$$
$$a(-\infty)=(-\infty)a=\infty \qquad (-\infty\leqslant a<0).$$

We do not define $0(\infty)$, $0(-\infty)$, $(\infty)0$, or $(-\infty)0$.

EXERCISE: Show that addition and multiplication in R^* are commutative and associative. Does the distributive law

$$a(b+c)=ab+ac \qquad (a,b,c\in R^*)$$

hold provided that both sides are defined?

6.9C. Next we define limits. Let $\{s_n\}_{n=1}^{\infty}$ be a sequence of elements of R^*.

If $L\in R$, we say that $\{s_n\}_{n=1}^{\infty}$ has the limit L and write $\lim_{n\to\infty}s_n=L$ if the condition in 2.2A holds. Note that in this case s_n can be infinite for only a finite number of n.

We say that $\{s_n\}_{n=1}^{\infty}$ has the limit ∞ and write $\lim_{n\to\infty}s_n=\infty$ if the condition in 2.4A holds. In this case s_n can be $-\infty$ for only a finite number of n. However, s_n can be ∞ for any number of n—possibly all n.

We treat $-\infty$ as a limit in similar fashion.

We avoid in this section the terminology convergent and divergent.

EXERCISE 1. Prove that every monotone sequence of elements of R^* has a limit. (This simple statement replaces 2.6B, 2.6D, and 2.6E.)

EXERCISE 2. Prove that the limit of a sum (of sequences in R^*) is equal to the sum of the limits, provided that the sum of the limits is defined. Do the same for products.

EXERCISE 3. Prove that $\lim_{n\to\infty}\rho(n,\infty)=0$, where ρ is the metric for R^* defined in 6.9A. Note that according to 4.3C, this may be written as

$$n\to\infty \quad \text{as} \quad n\to\infty.$$

EXERCISE 4. Let $\{s_n\}_{n=1}^{\infty}$ be a sequence in R^*. Show that $\{s_n\}_{n=1}^{\infty}$ has the limit s in R^* if and only if

$$\lim_{n\to\infty}\rho(s_n,s)=0.$$

6.9D. Now we take up the supremum (sup) and infimum (inf) which play for R^* the role played by the least upper bound and greatest lower bound for R.

Let E be a nonempty subset of R^*. If $E \subset R$, and E is bounded above as in 1.7A, we define

$$\sup E = \text{l.u.b.}\, E.$$

If $\infty \in E$ or if $E \subset R$, but E is not bounded above, we define

$$\sup E = \infty.$$

Similarly, we define

$$\inf E = \text{g.l.b.}\, E$$

if $E \subset R$ and E is bounded below, and

$$\inf E = -\infty$$

if $-\infty \in E$ or $E \subset R$, but E is not bounded below.

Thus $\sup E$ and $\inf E$ are defined for all nonempty subsets of R^*. This leads to definitions of lim sup and lim inf that are less complicated than those in 2.9.

Let $\{s_n\}_{n=1}^{\infty}$ be a sequence of elements in R^*. Let

$$M_n = \sup\{s_n, s_{n+1}, \cdots\} \qquad (n = 1, 2, \cdots)$$

so that $\{M_n\}_{n=1}^{\infty}$ is nonincreasing and hence has a limit. We define

$$\limsup_{n\to\infty} s_n = \lim_{n\to\infty} M_n.$$

Thus in our treatment of limits in R^*, there is no need to separate the definition into two parts as in 2.9A and 2.9B. Notice also that $\limsup_{n\to\infty} s_n$ is actually the lim(it) of a sup(remum). That is,

$$\limsup_{n\to\infty} s_n = \lim_{n\to\infty} \left[\sup\{s_n, s_{n+1}, \cdots\}\right].$$

Similarly, we define

$$\liminf_{n\to\infty} s_n = \lim_{n\to\infty} \left[\inf\{s_n, s_{n+1}, \cdots\}\right].$$

Thus just as in R, every sequence in R^* has a lim sup and a lim inf.

EXERCISE 1. Let $\{s_n\}$ be a sequence in R^*. Show that $\{s_n\}$ has a limit if and only if $\liminf_{n\to\infty} s_n = \limsup_{n\to\infty} s_n$.

EXERCISE 2. If $\limsup_{n\to\infty} s_n = L \in R^*$, show that $\{s_n\}$ has a subsequence whose limit is L.

EXERCISE 3. Let $\{s_n\}$ be a sequence in R^*. Then $\{s_n\}$ has a subsequence which has a limit. (Compare with 2.9M.)

This ends our treatment of infinite limits.

In the remainder of the book (except in some of the additional exercises) we will return to our usage as introduced in sections 2.1 through 2.12—namely,

$\lim_{n\to\infty} s_n$ must be finite,

$\{s_n\}$ converges means $\{s_n\}$ has a limit,

$\{s_n\}$ diverges means $\{s_n\}$ does not converge.

II. Discontinuities of monotonic functions.

6.9E. It is convenient to introduce the notation $f(c+)$ defined by

$$f(c+) = \lim_{x \to c^+} f(x)$$

provided that the one-sided limit exists. Similarly,

$$f(c-) = \lim_{x \to c^-} f(x).$$

We see from 4.1 that if f is a nondecreasing function on $[0,1]$, then $f(c+), f(c-)$ exist for every $c \in (0,1)$ (as do $f(0+)$ and $f(1-)$). Thus if $0 < c < 1$, then f will be continuous at c if and only if

$$f(c-) = f(c) = f(c+).$$

Now, clearly,

$$f(c-) \leqslant f(c) \quad \text{and} \quad f(c) \leqslant f(c+)$$

since f is nondecreasing. Hence if f is not continuous at c, we must have $f(c-) < f(c+)$. In this case, the number

$$f(c+) - f(c-)$$

is called the jump of f at c, and we say that f has a jump discontinuity at c. If f is not continuous at 0, then the jump of f at 0 is $f(0+) - f(0)$. Similarly, the jump of f at 1 (if there is a jump) is $f(1) - f(1-)$. The reason for the word "jump" is evident from the graph of the following simple example:

$$\begin{aligned} g(x) &= 0 \qquad (0 \leqslant x < c), \\ g(c) &= \tfrac{1}{2}, \\ g(x) &= 1 \qquad (c < x \leqslant 1). \end{aligned}$$

Thus if a nondecreasing function f on $[0,1]$ has a discontinuity at $c \in [0,1]$, this discontinuity must be a jump discontinuity (in contrast to the kind of discontinuity that the "highly oscillating" function h has at 0 where

$$\begin{aligned} h(x) &= \sin \frac{1}{x} \qquad (0 < x \leqslant 1), \\ h(0) &= 0). \end{aligned}$$

It is, perhaps, unexpected that a nondecreasing function f on $[0,1]$ can have infinitely many discontinuities. Consider the following:

$$\begin{aligned} f(x) &= 0 \qquad (0 \leqslant x < \tfrac{1}{2}), \\ f(x) &= \tfrac{1}{2} \qquad (\tfrac{1}{2} \leqslant x < \tfrac{3}{4}), \\ f(x) &= \tfrac{3}{4} \qquad (\tfrac{3}{4} \leqslant x < \tfrac{7}{8}), \end{aligned}$$

and so on. We set $f(1) = 1$.

EXERCISE: Show that for $n = 1, 2, \cdots$, this function has a jump of $1/(2^n)$ at the point $1 - 1/(2^n)$.

6.9F. The preceding example shows that a nondecreasing function on $[0,1]$ may have countably many discontinuities. It may not have uncountably many.

THEOREM: Let f be a nondecreasing function on $[0,1]$. Then f has at most countably many discontinuities.

Begin the proof by asking how many discontinuities f can have where the jump is greater than 1. How many can it have where the jump is greater than $\frac{1}{2}$?

EXERCISE: Finish the proof.

6.9G. It is interesting to note that the set of points of discontinuity of a nondecreasing function can be any countable set. For example, here is a nondecreasing function on $[0,1]$ which is discontinuous at every rational in $(0,1)$.

Let $r_1, r_2, \cdots$ be an enumeration of all the rationals in $(0,1)$. For each $n=1,2,\cdots$ define the function t_n by

$$t_n(x)=0 \qquad (0 \leqslant x < r_n)$$

$$t_n(x)=\frac{1}{n^2} \qquad (r_n \leqslant x \leqslant 1),$$

so that t_n has a jump of $(1)/(n^2)$ at the point r_n. Let

$$f(x)=\sum_{n=1}^{\infty} t_n(x) \qquad (0 \leqslant x \leqslant 1).$$

EXERCISE: Prove that f is nondecreasing on $[0,1]$, and that f is discontinuous at every r_n. (It can also be proved that f is continuous at every irrational. We will take this up after we study uniform covergence in Chapter 9.)

III. Cluster points, isolated points, and discrete spaces.

6.9H. Let M be a metric space. Recall that if $A \subset M$, we call the point $x \in M$ a cluster point of A if every open ball about x contains a point of A distinct from x. It follows that in any open ball about a cluster point x of A there must be infinitely many points of A (why?).

Hence the point $x \in M$ is a cluster point of M if every open ball about x contains infinitely many points of M.

On the other hand, if the point $x \in M$ is not a cluster point of M, then there is an open ball about x which contains no other point of M. In this case we say that x is an isolated point of M.

So every point of a metric space is either a cluster point or an isolated point. For example, if

$$N=\{1,\tfrac{1}{2},\tfrac{1}{3},\cdots\} \cup \{0\}$$

with the absolute value metric, then 0 is the only cluster point of N—all other points are isolated.

If x is an isolated point of the metric space M, then $\{x\}$ is an open ball—hence an open set. If M consists entirely of isolated points, then every one-point subset of M is open—so every subset of M is open. We call such a metric space a discrete space.

For example, the space R_d is discrete since, clearly, every point is isolated. Note, however, that a metric space $\langle M, \rho \rangle$ can be discrete even though ρ is not the discrete metric (that is, the metric taking only the values 0 and 1). Indeed the space

$$\{1, \tfrac{1}{2}, \tfrac{1}{3}, \tfrac{1}{4}, \cdots\},$$

with absolute value metric, is discrete.

In each of the following exercises, M is a metric space.

EXERCISE 1. Suppose $\bigcap_{G \in \mathcal{F}} G$ is open whenever $\mathcal{F}$ is a family of open subsets of M. Prove that M is discrete.

EXERCISE 2. Suppose that every subset of M is either open or closed. Prove that M has at most one cluster point.

EXERCISE 3. If M is infinite (that is, M has infinitely many points), prove that M contains an open subset G such that G and G' (the complement of G) are both infinite.

EXERCISE 4. If M is infinite, show that M has an infinite subset A which, with respect to the metric for M, is a discrete space.

EXERCISE 5. Suppose that for every open subset G of M, the closure $\overline{G}$ of G is also open . Prove that M is discrete.

MISCELLANEOUS EXERCISES

1. Let $\langle M, \rho \rangle$ be a metric space. Show that

$$\frac{\rho}{1+\rho}$$

   is also a metric for M.
2. Show that every metric space is homeomorphic to some bounded metric space.
3. Give an example of subsets A, B of R^1 such that none of the four sets $A \cap \overline{B}$, $\overline{A} \cap B$, $\overline{A \cap B}$, $\overline{A} \cap \overline{B}$ is equal to any of the others.
4. Show that there is a nonempty open subset of R^2 which cannot be written as a finite or countable disjoint union of open balls in R^2. (Thus theorem 5.4F fails in R^2.)
5. Let A be a subset of the metric space M. Prove that each of the following statements implies the others.
   (a) A is dense in M.
   (b) $B[a, r] \cap A \neq \varnothing$ for every $r > 0, a \in M$.
   (c) $A \cap G \neq \varnothing$ for every nonempty open subset G of M.
6. Let M be a metric space. We have defined the open ball $B[a; r]$ as

$$B[a; r] = \{x \in M \mid \rho(a, x) < r\}.$$

   Define the closed ball $B^c[a; r]$ as

$$B^c[a; r] = \{x \in M \mid \rho(a, x) \leqslant r\}.$$

   (a) Show that $B^c[a; r]$ is a closed set.
   (b) Show by example that $\overline{B[a; r]}$ is not necessarily equal to $B^c[a; r]$. That is, the closure of the open ball may not be equal to the closed ball.

(c) Give an example of a metric space in which there is an open ball that is a closed set but is not a closed ball. (Use a subset of R^2 which contains $\langle 1,0\rangle$, $\langle -1,0\rangle$, and an appropriate interval on the y axis.)

7. Let M be a metric space which has a countable dense subset. If $A \subset M$, prove that A has a countable dense subset.
8. Suppose $f: R^1 \to R^1$. Prove that f is continuous at $a \in R^1$ if and only if whenever $\{x_n\}_{n=1}^{\infty}$ converges to a, the sequence $\{f(x_n)\}_{n=1}^{\infty}$ has a subsequence converging to $f(a)$.
9. Let M_1, M_2 be metric spaces and suppose $f: M_1 \to M_2$. Prove that f is continuous if and only if
$$f(\bar{A}) \subset \overline{f(A)}$$
for every $A \subset M_1$.
10. Give an example of a continuous function f from a metric space M into a metric space N such that
$$f^{-1}(\bar{A}) \neq \overline{f^{-1}(A)}$$
for some $A \subset N$.
11. Let A be a subset of the metric space M. Define
$$f(x) = \rho(x, A) \qquad (x \in M).$$
Prove that f is uniformly continuous on M.
12. Let A be a connected metric space with at least two points. Show that there exists a continuous real-valued function on A that is not constant. Use this to prove that A is uncountable.
13. A map of the state of Iowa falls on the ground (flat) somewhere within the state. Prove that there is a point on the map that is directly above the corresponding point on the ground.
14. Let f be a continuous function from $[0,1]$ into $[0,1]$. Prove that f has a fixed point.
15. Let A be an infinite subset of the compact metric space M. If the space A is discrete, prove that A is not closed in M.
16. Let M be a compact metric space and $f: M \to M$. Assume
$$\rho[f(x), f(y)] < \rho(x,y) \qquad (x, y \in M; x \neq y).$$
Prove that there exists $x \in M$ such that $f(x) = x$. (Start by showing that $\rho[x, f(x)]$ attains a minimum on M.)
17. Let M be a compact metric space in which the closure of every open ball $B[a;r]$ is the closed ball $B^c[a;r]$. Prove that every open ball is connected.

SKETCH OF PROOF: Prove by contradiction. Suppose $B = B[a;r]$ is not connected. Then $B = C \cup D$ where C and D are nonempty and open in B. We may assume $a \in C$. Let
$$f(x) = \rho(a, x) \qquad (x \in \bar{B} - C).$$
If $\underset{x \in \bar{B} - C}{\text{g.l.b.}} f(x) = s$, there is a point $d \in D$ such that $f(d) = s$. The point d is a limit point of $B[a;s] \subset C$. Fill in the details and finish the proof.

18. Let M be a compact metric space. Suppose $f: M \to M$ is such that
$$\rho(x,y) \leqslant \rho[f(x), f(y)] \qquad (x, y \in M).$$
Prove that f is onto and that
$$\rho(x,y) = \rho[f(x), f(y)] \qquad (x, y \in M).$$

Thus f preserves distances. (Such an f is called an isometry.)

SKETCH OF PROOF: For $x \in M$ let $x_1 = f(x), x_2 = f(x_1), \cdots, x_{n+1} = f(x_n), \cdots$. Similarly, for $y \neq x$, let $y_1 = f(y), y_2 = f(y_1)$, and so on.

Use Theorem 6.5B to show that given ϵ, there exists $k \in I$ such that both

$$\rho(x, x_k) < \epsilon \quad \text{and} \quad \rho(y, y_k) < \epsilon.$$

Thus the range of f is dense in M. Next suppose that $\rho[f(x), f(y)] > \rho(x,y)$. Choose ϵ such that $\rho[f(x), f(y)] > \rho(x,y) + 2\epsilon$. Deduce a contradiction. Thus $\rho[f(x), f(y)] = \rho(x,y)$ for all $x, y \in M$. Finally, show that f is onto M.

7

CALCULUS

7.1 SETS OF MEASURE ZERO

In the next section we define the Riemann integral—the integral considered in elementary calculus courses. We will see that a bounded function f has a Riemann integral provided f is continuous at "almost every" point. The precise meaning of "almost every" will be defined in terms of the following concept of set of measure zero.

If J is an interval of real numbers, we denote the length of J by $|J|$.

7.1A. DEFINITION. The subset E of R^1 is said to be of measure zero if for each $\epsilon > 0$, there exists a finite or countable number of open intervals $I_1, I_2, \ldots$ such that $E \subset \cup_n I_n$ and $\Sigma_n |I_n| < \epsilon$.

Thus E is of measure zero if, given $\epsilon > 0$, E can be covered by a union of open intervals whose lengths add up to be less than ϵ. It is obvious, then, that a set consisting of one point has measure zero.

The following result is very useful.

7.1B. THEOREM. If each of the subsets $E_1, E_2, \ldots$ of R^1 is of measure zero, then $\cup_{n=1}^{\infty} E_n$ is also of measure zero.

PROOF: Fix $\epsilon > 0$. Since E_n has measure zero, for each $n \in I$ there exists a finite or countable number of open intervals that cover E_n and whose lengths add up to less than $\epsilon/2^n$. The union of all such open intervals (for all $n \in I$) then covers $\cup_{n=1}^{\infty} E_n$, and the lengths of all these (countably many*) intervals add up to $< \epsilon/2 + \epsilon/2^2 + \cdots = \epsilon$. Hence $\cup_{n=1}^{\infty} E_n$ has measure zero.

Since one-point sets are of measure zero we deduce the following corollary.

7.1C. COROLLARY. Every countable subset of R^1 has measure zero.

* We are using 1.5F.

In fact, there are even uncountable subsets of R^1 that have measure zero. In Chapter 11 we will be able to show that the Cantor set of 1.6D (which is uncountable) is of measure zero. On the other hand, a nonempty open interval (no matter how small) is never of measure zero. (See Exercise 3.)

We now make "almost every" precise.

7.1D. DEFINITION. A statement is said to hold at almost every point of $[a,b]$ (or almost everywhere in $[a,b]$) if the set of points of $[a,b]$ at which the statement does not hold is of measure zero.

Thus "f is continuous at almost every point of $[a,b]$" means the same as "if E is the set of points of $[a,b]$ at which f is not continuous, then E is of measure zero." We could also say "f is continuous almost everywhere in $[a,b]$."

Exercises 7.1

1. If A is not of measure zero, if $B \subset A$, and if B is of measure zero, prove that $A - B$ is not of measure zero.
2. If $a < b$, prove that $[a,b]$ cannot be covered by a *finite* number of open intervals whose lengths add up to less than $b-a$.
   Use the Heine-Borel property of $[a,b]$ to deduce that $[a,b]$ is not of measure zero.
3. If $a < b$, prove that (a,b) is not of measure zero.
4. (a) Show that the set of all rational numbers is of measure zero.
   (b) Show that the set of all irrational numbers is not of measure zero.
5. True or false? If f is continuous on $[0,1]$, and if $g(x) = f(x)$ at almost every $x \in [0,1]$, then g is continuous almost everywhere in $[0,1]$.

7.2 DEFINITION OF THE RIEMANN INTEGRAL

Throughout the remainder of this chapter we consider only real-valued functions.

7.2A. DEFINITION. Let $\mathcal{J}$ be any bounded interval of real numbers, and let f be a bounded (real-valued) function on $\mathcal{J}$. We define $M[f;\mathcal{J}]$, $m[f;\mathcal{J}]$ and $\omega[f;\mathcal{J}]$ as

$$M[f;\mathcal{J}] = \underset{x\in\mathcal{J}}{\text{l.u.b.}}\, f(x),$$

$$m[f;\mathcal{J}] = \underset{x\in\mathcal{J}}{\text{g.l.b.}}\, f(x),$$

$$\omega[f;\mathcal{J}] = M[f;\mathcal{J}] - m[f;\mathcal{J}].$$

(Thus $\omega[f;\mathcal{J}]$ is exactly the same as in definition 5.6B save that we now do not require that $\mathcal{J}$ be open.) If a is a point of $\mathcal{J}$, we define $\omega[f;a]$ as

$$\omega[f;a] = \text{g.l.b.}\,\omega[f;J]$$

where the g.l.b. is taken over all open subintervals J of $\mathcal{J}$ such that $a \in J$.* (This is also consistent with 5.6B.)

7.2B. DEFINITION. By a subdivision of the closed bounded interval $[a,b]$ we mean a finite subset $\{x_0, x_1, \ldots, x_n\}$ of $[a,b]$ such that $a = x_0 < x_1 < \cdots < x_n = b$. If σ and τ are two subdivisions of $[a,b]$, we say that τ is a refinement of σ if $\sigma \subset \tau$. (That is, τ is a

* Remember, for example, that $[0, \frac{1}{2})$ is an open subinterval of $[0,1]$.

refinement of σ means that the subdivision τ is obtained from the subdivision σ by adding more "points of subdivision.")

If $\sigma=\{x_0,x_1,\ldots,x_n\}$ is a subdivision of $[a,b]$, then the closed intervals $I_1=[x_0,x_1], I_2=[x_1,x_2],\ldots,I_n=[x_{n-1},x_n]$ are called the component intervals of σ.

7.2C. DEFINITION. Let f be a bounded function on the closed bounded interval $[a,b]$ and let σ be any subdivision of $[a,b]$. We define $U[f;\sigma]$, called the upper sum for f corresponding to σ, as*

$$U[f;\sigma]=\sum_{k=1}^{n} M[f;I_k]\cdot|I_k|$$

where $I_1,\ldots,I_n$ are the component intervals of σ. Similarly, the lower sum $L[f;\sigma]$ is defined as

$$L[f;\sigma]=\sum_{k=1}^{n} m[f;I_k]\cdot|I_k|.$$

Obviously, $U[f;\sigma]\geqslant L[f;\sigma]$. Note that if f is continuous and nonnegative valued on $[a,b]$, then $U[f;\sigma]$ is the sum of the areas of n rectangles each of which has one of the I_k as base and whose height is equal to $\max_{x\in I_k} f(x)$. That is, $U[f;\sigma]$ is the sum of the areas of the "circumscribed rectangles" as pictured in claculus texts. Similarly, $L[f;\sigma]$ is the sum of the areas of the "inscribed rectangles." This geometric interpretation makes the following result quite plausible (at least for continuous functions).

7.2D. LEMMA. Let f be a bounded function on $[a,b]$. Then every upper sum for f is greater than or equal to every lower sum for f. That is, if σ and τ are any two subdivisions of $[a,b]$, then $U[f;\sigma]\geqslant L[f;\tau]$.

PROOF: We will show that if σ^* is any refinement of σ, then

$$U[f;\sigma]\geqslant U[f;\sigma^*]. \tag{1}$$

It is enough to prove this in the case where σ^* is obtained from σ by adding only one point of subdivision. (For we can then apply induction.) Thus we may suppose that σ has component intervals $I_1,\ldots,I_k,\ldots,I_n$ and σ^* has component intervals $I_1,\ldots,I_k^*,I_k^{**},\ldots,I_n$ where $I_k=I_k^*\cup I_k^{**}$ and $|I_k|=|I_k^*|+|I_k^{**}|$. Since $I_k^*\subset I_k$ we have $M[f;I_k^*]\leqslant M[f;I_k]$. Similarly, $M[f;I_k^{**}]\leqslant M[f;I_k]$. Thus

$$\begin{aligned}U[f;\sigma^*]&=\sum_{\substack{j=1\\ j\neq k}}^{n} M[f;I_j]\cdot|I_j|+M[f;I_k^*]\cdot|I_k^*|+M[f;I_k^{**}]\cdot|I_k^{**}|\\ &\leqslant \sum_{\substack{j=1\\ j\neq k}}^{n} M[f;I_j]\cdot|I_j|+M[f;I_k](|I_k^*|+|I_k^{**}|)=U[f;\sigma],\end{aligned}$$

which proves (1). Similarly, if τ^* is any refinement of τ, it may be shown that

$$L[f;\tau]\leqslant L[f;\tau^*]. \tag{2}$$

But, since $\sigma\cup\tau$ is a refinement of both σ and τ, it follows from (1) and (2) that

$$U[f;\sigma]\geqslant U[f;\sigma\cup\tau]\geqslant L[f;\sigma\cup\tau]\geqslant L[f;\tau].$$

This proves the lemma.

* For any bounded interval J, $|J|$ denotes the length of J.

7.2E. From the preceding lemma it follows that

$$\text{g.l.b.}\, U[f;\sigma] \geqslant \text{l.u.b.}\, L[f;\sigma], \tag{1}$$

where the g.l.b. and l.u.b. are both taken over all subdivisions σ of $[a,b]$. For, if τ is any subdivision of $[a,b]$, then the lemma shows that $L[f;\tau]$ is a lower bound for the set of all upper sums $U[f;\sigma]$. Hence

$$L[f;\tau] \leqslant \text{g.l.b.}\, U[f;\sigma],$$

for every subdivision τ. But this says that g.l.b. $U[f;\sigma]$ is an upper bound for the set of all lower sums $L[f;\tau]$. Hence

$$\underset{\tau}{\text{l.u.b.}}\, L[f;\tau] \leqslant \underset{\sigma}{\text{g.l.b.}}\, U[f;\sigma],$$

which is equivalent to (1). The inequality (1) gives us an important relation between the upper and lower integrals of a function.

DEFINITION. Let f be a bounded function on the closed bounded interval $[a,b]$ We define

$$\overline{\int_a^b} f(x)\,dx,$$

called the upper integral of f over $[a,b]$, as

$$\overline{\int_a^b} f(x)\,dx = \text{g.l.b.}\, U[f;\sigma],$$

where the g.l.b. is taken over all subdivisions σ of $[a,b]$. Similarly, we define

$$\underline{\int_a^b} f(x)\,dx,$$

called the lower integral of f over $[a,b]$, as

$$\underline{\int_a^b} f(x)\,dx = \text{l.u.b.}\, L[f;\sigma].$$

For simplicity we sometimes denote the upper and lower integrals of f by

$$\overline{\int_a^b} f \quad \text{and} \quad \underline{\int_a^b} f.$$

Note that we are attaching no meaning to dx alone. From inequality (1) we see that

$$\underline{\int_a^b} f \leqslant \overline{\int_a^b} f. \tag{2}$$

We will presently show that for continuous functions f (as well as some other functions)

$$\underline{\int_a^b} f \quad \text{and} \quad \overline{\int_a^b} f$$

are equal. However, there exist f for which

$$\underline{\int_a^b} f < \overline{\int_a^b} f.$$

For example, if χ is the characteristic function of the rational numbers in $[0,1]$, then, for any interval $J \subset [0,1]$

$$M[\chi;J] = 1, \qquad m[\chi;J] = 0.$$

Hence, for any subdivision σ we have $U[\chi;\sigma]=1, L[\chi;\sigma]=0$. It follows that

$$\underline{\int_0^1}\chi=0 \quad \text{but} \quad \overline{\int_0^1}\chi=1.$$

7.2F. DEFINITION. If f is a bounded function on the closed bounded interval $[a,b]$, we say that f is Riemann integrable on $[a,b]$ if

$$\underline{\int_a^b} f = \overline{\int_a^b} f.$$

In this case we define $\displaystyle\int_a^b f(x)\,dx \left(\text{or} \int_a^b f\right)$ as

$$\int_a^b f = \underline{\int_a^b} f = \overline{\int_a^b} f.$$

We denote by $\mathcal{R}[a,b]$ the class of all functions f which are Riemann integrable on $[a,b]$.

Thus the function χ in 7.2E is not Riemann integrable on $[0,1]$. On the other hand, it is clear that any constant function on a closed bounded interval $[a,b]$ *is* Riemann integrable on $[a,b]$. In the next section we show exactly which functions belong to $\mathcal{R}[a,b]$. The following theorem will be useful.

7.2G. THEOREM. Let f be a bounded function on the closed bounded interval $[a,b]$. Then $f\in\mathcal{R}[a,b]$ if and only if, for each $\epsilon>0$, there exists a subdivision σ of $[a,b]$ such that

$$U[f;\sigma] < L[f;\sigma] + \epsilon. \tag{1}$$

PROOF: Suppose first that given $\epsilon>0$ there exists σ such that (1) holds. Then, since

$$\overline{\int_a^b} f \leqslant U[f;\sigma] \quad \text{and} \quad \underline{\int_a^b} f \geqslant L[f;\sigma],$$

we have

$$\overline{\int_a^b} f < \underline{\int_a^b} f + \epsilon.$$

Since ϵ was arbitrary, it follows that

$$\overline{\int_a^b} f \leqslant \underline{\int_a^b} f$$

and hence, by (2) of 7.2E, that

$$\underline{\int_a^b} f = \overline{\int_a^b} f.$$

This proves $f\in\mathcal{R}[a,b]$.

Conversely, suppose $f\in\mathcal{R}[a,b]$. Then

$$\overline{\int_a^b} f = \underset{\sigma}{\text{g.l.b.}}\, U[f;\sigma] = \underset{\tau}{\text{l.u.b.}}\, L[f;\tau] = \underline{\int_a^b} f.$$

Given $\epsilon > 0$ we may (by definition of g.l.b.) choose a subdivision σ such that

$$\overline{\int_a^b} f + \frac{\epsilon}{2} > U[f;\sigma].$$

Similarly, we may choose a subdivision τ such that

$$\underline{\int_a^b} f - \frac{\epsilon}{2} < L[f;\tau].$$

Hence,

$$L[f;\tau] + \frac{\epsilon}{2} > U[f;\sigma] - \frac{\epsilon}{2}.$$

By (1) and (2) of 7.2D we then have

$$L[f;\sigma\cup\tau] + \frac{\epsilon}{2} > U[f;\sigma\cup\tau] - \frac{\epsilon}{2}.$$

This is equivalent to (1) (with $\sigma \cup \tau$ in place of σ).

Exercises 7.2

1. Let $f(x) = x$ $(0 \leqslant x \leqslant 1)$. Let σ be the subdivision $\{0, \frac{1}{3}, \frac{2}{3}, 1\}$ of $[0,1]$. Compute $U[f;\sigma]$ and $L[f;\sigma]$.
2. For each $n \in I$ let σ_n be the subdivision $\{0, 1/n, 2/n, \ldots, n/n\}$ of $[0,1]$. Compute
$$\lim_{n\to\infty} U[f;\sigma_n]$$
for the function f of the preceding exercise.
3. For σ_n as in the preceding exercise, compute
$$\lim_{n\to\infty} L[f;\sigma_n]$$
for the function $f(x) = x^2$ $(0 \leqslant x \leqslant 1)$. [You will need the formula
$$1^2 + 2^2 + 3^2 + \cdots + n^2 = \frac{n(n+1)(2n+1)}{6} \qquad (n \in I).]$$
4. If $f \in \mathfrak{R}[0,1]$, if $\sigma_n = \{0, 1/n, 2/n, \ldots, n/n\}$, and if
$$\lim_{n\to\infty} U[f;\sigma_n] = \lim_{n\to\infty} L[f;\sigma_n] = A,$$
prove that $\int_0^1 f(x)\,dx = A$.
5. For $f(x) = \sin x$ $(0 \leqslant x \leqslant \pi/2)$ and $\sigma_n = \{0, \pi/2n, 2\pi/2n, \ldots, n\pi/2n\}$ compute $U[f;\sigma_n]$. Use the identity from 3.8D (with $x = \pi/2n$) to show that
$$U[f;\sigma_n] = \frac{\pi/4n}{\sin(\pi/4n)}\left[\cos\frac{\pi}{4n} - \cos\left(\frac{\pi}{2} + \frac{\pi}{4n}\right)\right].$$
Then prove that
$$\lim_{n\to\infty} U[f;\sigma_n] = \cos 0 - \cos\frac{\pi}{2} = 1.$$
6. Let f be a continuous function on $[a,b]$. Fill in the details in the following proof that $f \in \mathfrak{R}[a,b]$.
   (a) The function f is uniformly continuous on $[a,b]$.
   (b) Given $\epsilon > 0$ there exists $\delta > 0$ such that
$$|f(x) - f(y)| < \frac{\epsilon}{b-a} \qquad (|x-y| < \delta;\, x,y \in [a,b]).$$

(c) For this $\epsilon > 0$ there exists a subdivision σ of $[a,b]$ such that, if I_k is any component interval of σ, then

$$M[f;I_k] - m[f;I_k] < \frac{\epsilon}{b-a}.$$

(d) For this σ,

$$U[f;\sigma] - L[f;\sigma] < \epsilon.$$

(e) Therefore $f \in \mathcal{R}[a,b]$.

7. If f is continuous on $[0,1]$, if $\sigma_n = \{0, 1/n, 2/n, \ldots, n/n\}$, and if x_k^* is any point in the interval $[(k-1)/n, k/n]$ $(k = 1, \ldots, n)$, show that

$$L[f,\sigma_n] \leqslant \frac{1}{n} \sum_{k=1}^{n} f(x_k^*) \leqslant U[f,\sigma_n].$$

Then show that

$$\lim_{n\to\infty} \frac{1}{n} \sum_{k=1}^{n} f(x_k^*) = \int_0^1 f.$$

(This result is used many times in elementary calculus.) (*Hint*: Use the uniform continuity of f to show that

$$U[f;\sigma_n] - L[f;\sigma_n] < \epsilon$$

for large n. Conclude that

$$U[f;\sigma_n] - \int_0^1 f < \epsilon$$

for large n and hence that

$$\lim_{n\to\infty} U[f;\sigma_n] = \int_0^1 f.$$

Similarly, show

$$\lim_{n\to\infty} L[f;\sigma_n] = \int_0^1 f.)$$

8. If f is continuous on $[0,1]$, prove that

$$\lim_{n\to\infty} \frac{1}{n} \sum_{k=1}^{n} f\left(\frac{k}{n}\right) = \int_0^1 f.$$

9. Evaluate the following limits.

(a) $\displaystyle \lim_{n\to\infty} \frac{1}{n}\left[\left(\frac{1}{n}\right)^2 + \left(\frac{2}{n}\right)^2 + \cdots + \left(\frac{n}{n}\right)^2\right].$

(b) $\displaystyle \lim_{n\to\infty} \frac{1}{n}\left(\sin\frac{\pi}{n} + \sin\frac{2\pi}{n} + \cdots + \sin\frac{n\pi}{n}\right).$

(c) $\displaystyle \lim_{n\to\infty} \frac{1}{n}\left(e^{3/n} + e^{6/n} + \cdots + e^{3n/n}\right).$

10. Prove that $\int_a^a f = 0$ for any function f on the "interval" $[a,a]$.

11. True or false? If $f \in \mathcal{R}[a,b]$, then

$$\int_a^b f(x)\,dx = \int_a^b f(u)\,du = \int_a^b f(g)\,dg.$$

7.3. EXISTENCE OF THE RIEMANN INTEGRAL

7.3A. THEOREM. Let f be a bounded function on the closed bounded interval $[a,b]$. Then $f \in \mathcal{R}[a,b]$ if and only if f is continuous at almost every point in $[a,b]$.

PROOF: Suppose first that $f \in \mathcal{R}[a,b]$. We wish to show that the set E of points in $[a,b]$ at which f is not continuous is of measure zero. Now, by 5.6C, $x \in E$ if and only if $\omega[f;x]>0$. Hence, $E=\cup_{m=1}^{\infty} E_m$ where E_m is the set of all $x \in [a,b]$ such that $\omega[f;x] \geqslant 1/m$. To prove that E is of measure zero it is sufficient, by 7.1B, to show that each E_m is of measure zero. This we will now do.

Fix m. Since $f \in \mathcal{R}[a,b]$, given $\epsilon>0$ there exists by 7.2G a subdivision σ of $[a,b]$ such that $U[f;\sigma]-L[f;\sigma]<\epsilon/2m$. Thus if $I_1,\ldots,I_n$ are the (closed) component intervals of σ, we have

$$\sum_{k=1}^{n} \omega[f;I_k]\cdot|I_k| = \sum_{k=1}^{n} M[f;I_k]\cdot|I_k| - \sum_{k=1}^{n} m[f;I_k]\cdot|I_k| = U[f;\sigma]-L[f;\sigma],$$

and hence,

$$\sum_{k=1}^{n} \omega[f;I_k]\cdot|I_k| < \frac{\epsilon}{2m}. \tag{1}$$

Now $E_m = E_m^* \cup E_m^{**}$ where E_m^* is the set of points of E_m that are points of the subdivision σ, and $E_m^{**}=E-E_m^*$. Obviously, $E_m^* \subset J_1 \cup \cdots \cup J_p$ where the J_i are open subintervals such that $|J_1|+\cdots+|J_p|<\epsilon/2$ (since there are only a finite number of points of subdivsion). But if $x \in E_m^{**}$, then x is an interior point of some I_k. Hence, $\omega[f;I_k] \geqslant \omega[f;x] \geqslant 1/m$. If we denote by

$$I_{k_1},\ldots,I_{k_r}$$

those of the component intervals of σ that contain a point of E_m^{**} (in their interior), we have

$$\frac{1}{m}\left(|I_{k_1}|+\cdots+|I_{k_r}|\right) \leqslant \omega[f;I_{k_1}]\cdot|I_{k_1}|+\cdots+\omega[f;I_{k_r}]\cdot|I_{k_r}|.$$

Hence, by (1),

$$\frac{1}{m}\left(|I_{k_1}|+\cdots+|I_{k_r}|\right) < \frac{\epsilon}{2m},$$

$$|I_{k_1}|+\cdots+|I_{k_r}| < \frac{\epsilon}{2}.$$

Since E_m^{**} is covered by the interiors of $I_{k_1},\ldots,I_{k_r}$, and since E_m^* is covered by $J_1,\ldots,J_p$, it follows that $E_m = E_m^* \cup E_m^{**}$ is of measure zero, which is what we wished to show.

To prove the converse we need a lemma.

LEMMA. If $\omega[f;x]<a$ for each x in a closed bounded interval J, then there is a subdivision τ of J such that

$$U[f;\tau]-L[f;\tau]<a|J|. \tag{2}$$

PROOF: For each $x \in J$ there is an open subinterval I_x containing x such that $\omega[f;\bar{I}_x]<a$. Since J is compact, a finite number of these I_x will cover J (by 6.5G). Let τ be the set of end points of these I_x. If $I_1,I_2,\ldots,I_n$ are the component intervals of τ, we have $\omega[f;I_k]<a\,(k=1,\ldots,n)$, and (2) follows easily.

Now let us assume that f is continuous at almost every point of $[a,b]$. We wish to show that $f \in \mathfrak{R}[a,b]$. Given $\epsilon > 0$ choose $m \in I$ such that $(b-a)/m < \epsilon/2$. If E_m is defined as in the first part of the proof, then, by hypothesis, E_m is of measure zero. Hence, $E_m \subset \cup_{n=1}^{\infty} I_n$ where each I_n is an *open* subinterval of $[a,b]$ and

$$\sum_{n=1}^{\infty} |I_n| < \frac{\epsilon}{2\omega[f;[a,b]]}.^*$$

But E_m is closed in R^1 by 5.6D. Hence, E_m is a closed subset of $[a,b]$ and is thus compact. Therefore a finite number of the I_n—say $I_{n_1}, \ldots, I_{n_k}$—cover E_m. Now

$$[a,b] - (I_{n_1} \cup \cdots \cup I_{n_k})$$

is a union of closed intervals $J_1, \ldots, J_p$. That is,

$$[a,b] = I_{n_1} \cup \cdots \cup I_{n_k} \cup J_1 \cup \cdots \cup J_p.$$

Since no interval $J_i (i = 1, \ldots, p)$ contains a point of E_m, there exists by the lemma a subdivision τ_i of J_i such that $U[f;\tau_i] - L[f;\tau_i] < |J_i|/m$. Now define the subdivision σ of $[a,b]$ as $\sigma = \tau_1 \cup \cdots \cup \tau_p$. Then the component intervals of σ are the component intervals of $\tau_1, \ldots, \tau_p$ together with $\bar{I}_{n_1}, \ldots, \bar{I}_{n_k}$. Hence,

$$\begin{aligned} U[f;\sigma] - L[f;\sigma] &= \sum_{i=1}^{p} \{U[f;\tau_i] - L[f;\tau_i]\} + \sum_{l=1}^{k} \left\{M\left[f;\bar{I}_{n_l}\right] - m\left[f;\bar{I}_{n_l}\right]\right\} |\bar{I}_{n_l}| \\ &< \frac{1}{m} \sum_{i=1}^{p} |J_i| + \sum_{l=1}^{k} \omega\left[f;\bar{I}_{n_l}\right] \cdot |I_{n_l}| \\ &\leqslant \frac{b-a}{m} + \omega[f;[a,b]] \sum_{l=1}^{k} |I_{n_l}| \\ &< \frac{\epsilon}{2} + \omega[f;[a,b]] \cdot \frac{\epsilon}{2\omega[f;[a,b]]} = \epsilon. \end{aligned}$$

By 7.2G we have $f \in \mathfrak{R}[a,b]$, and the proof is complete.

Exercises 7.3

1. Which of the following functions f are in $\mathfrak{R}[0,1]$?
   (a) The characteristic function of the set $\{0, \frac{1}{10}, \frac{2}{10}, \frac{3}{10}, \ldots, 1\}$.
   (b) $f(x = \sin(1/x) \qquad (0 < x \leqslant 1)$,
   $f(0) = 7$.
   (c) The function f of exercise 10, Section 5.3.
   (d) The characteristic function of a set $E \subset [0,1]$ such that E and $[0,1] - E$ are both dense in $[0,1]$.
2. Compute $\omega[f;x]$ for all $x \in [0,1]$ for each of the functions f of the preceding exercise.
3. If $f \in \mathfrak{R}[a,b]$, prove that $|f| \in \mathfrak{R}[a,b]$.
4. True or false? If $f \in \mathfrak{R}[a,b]$ and if $f(x) = g(x)$ except for a countable number of points $x \in [a,b]$, then $g \in \mathfrak{R}[a,b]$.
5. True or false? If $f \in \mathfrak{R}[a,b]$, and if $f(x) = g(x)$ except for a finite number of points $x \in [a,b]$, then $g \in \mathfrak{R}[a,b]$.

* We may assume that $\omega[f;[a,b]] > 0$

7.4 PROPERTIES OF THE RIEMANN INTEGRAL

All the results in this section are used in proving fundamental calculus theorems as well as in the solution of standard calculus problems. The reader should interpret geometrically each of these results by thinking of $\int_a^b f$ (for continuous nonnegative-valued f) as the area under the curve $y=f(x)$ from $x=a$ to $x=b$.

7.4A. THEOREM. If $f\in\mathcal{R}[a,b]$ and $a<c<b$, then $f\in\mathcal{R}[a,c]$,* $f\in\mathcal{R}[c,b]$, and

$$\int_a^b f=\int_a^c f+\int_c^b f.$$

PROOF: By 7.3A the set E of points in $[a,b]$ at which f is not continuous is of measure zero. Obviously, then, $E\cap[a,c]$ is of measure zero and so (by 7.3A) $f\in\mathcal{R}[a,c]$. Similarly, $f\in\mathcal{R}[c,b]$.

If σ is any subdivision of $[a,c]$ and τ is any subdivision of $[c,b]$, then $\sigma\cup\tau$ is a subdivision of $[a,b]$ whose component intervals are those of σ together with those of τ. Hence,

$$L[f;\sigma]+L[f;\tau]=L[f;\sigma\cup\tau]\leqslant \underline{\int_a^b} f$$

and so

$$L[f;\sigma]+L[f;\tau]\leqslant\int_a^b f.$$

By taking the least upper bound on the left over all σ (keeping τ fixed for the moment) we obtain

$$\int_a^c f+L[f;\tau]\leqslant\int_a^b f. \tag{1}$$

Now taking the least upper bound over all τ we have

$$\int_a^c f+\int_c^b f\leqslant\int_a^b f.$$

Going back to the original σ and τ we also have

$$U[f;\sigma]+U[f;\tau]=U[f;\sigma\cup\tau]\geqslant\overline{\int_a^b} f$$

so that

$$U[f;\sigma]+U[f;\tau]\geqslant\int_a^b f.$$

Taking greatest lower bounds as in the first part of the proof we obtain

$$\int_a^c f+\int_c^b f\geqslant\int_a^b f. \tag{2}$$

The theorem follows from (1) and (2).

* More precisely, the restriction of f to $[a,c]$ is in $\mathcal{R}[a,c]$.

Any freshman student will instinctively write

$$\int_0^1 3x^2\,dx \quad \text{as} \quad 3\int_0^1 x^2\,dx.$$

Here is the justification.

7.4B. THEOREM. If $f \in \mathcal{R}[a,b]$ and λ is any real number, then $\lambda f \in \mathcal{R}[a,b]$ and

$$\int_a^b \lambda f = \lambda \int_a^b f.$$

PROOF: If $\lambda = 0$, the theorem is obvious. Suppose $\lambda > 0$. Since λf is continuous at every point where f is continuous, it is clear that $\lambda f \in \mathcal{R}[a,b]$. Since $\lambda > 0$, if J is any interval contained in $[a,b]$, then

$$M[\lambda f; J] = \lambda M[f; J]$$

(verify), and so, for any subdivision σ of $[a,b]$,

$$U[\lambda f; \sigma] = \lambda U[f; \sigma].$$

It follows easily, on taking the g.l.b. of both sides (over all σ) that

$$\int_a^b \lambda f = \lambda \int_a^b f \qquad (\lambda > 0). \tag{1}$$

Hence, the theorem is proved for $\lambda > 0$.

Now for any J we also have

$$M[-f; J] = -m[f; J].$$

Hence,

$$\int_a^b (-f) = \underset{\sigma}{\text{g.l.b.}}\, U[-f; \sigma] = \underset{\sigma}{\text{g.l.b.}} \{-L[f; \sigma]\} = -\underset{\sigma}{\text{l.u.b.}}\, L[f; \sigma] = -\int_a^b f.$$

That is,

$$\int_a^b (-f) = -\int_a^b f. \tag{2}$$

If $\mu < 0$, then $\lambda = -\mu > 0$ and so, by (2) and (1) (since $\mu f \in \mathcal{R}[a,b]$)

$$\int_a^b \mu f = \int_a^b -(\lambda f) = -\int_a^b \lambda f = -\lambda \int_a^b f = \mu \int_a^b f.$$

This completes the proof.

From theorem 7.4C it follows that

$$\int_0^\pi (\sqrt{1+x} + x^2)\,dx = \int_0^\pi \sqrt{1+x}\,dx + \int_0^\pi x^2\,dx.$$

7.4C. THEOREM. If $f \in \mathcal{R}[a,b], g \in \mathcal{R}[a,b]$, then $f+g \in \mathcal{R}[a,b]$ and

$$\int_a^b (f+g) = \int_a^b f + \int_a^b g.$$

PROOF: By 7.3A, the sets E_f and E_g of points at which f and g, respectively, are not continuous are both of measure zero. Hence, by 7.1B, the set $E_f \cup E_g$ is of measure zero. But if $x \in [a,b]-(E_f \cup E_g)$, then f, g, and hence $f+g$ are continuous at x. Thus $f+g$ is continuous at almost every point in $[a,b]$, and so $f+g \in \mathfrak{R}[a,b]$.

If J is any interval contained in $[a,b]$, and if $y \in J$, we have $f(y)+g(y) \leqslant M[f;J]+M[g;J]$. Hence, $M[f+g;J] \leqslant M[f;J]+M[g;J]$.

For any subdivision σ, then, we have

$$\int_a^b (f+g) \leqslant U[f+g;\sigma] \leqslant U[f;\sigma]+U[g;\sigma]. \tag{1}$$

But given $\epsilon > 0$ there is, by 7.2G, a subdivision σ_1 of $[a,b]$ such that

$$U[f;\sigma_1] < L[f;\sigma_1]+\frac{\epsilon}{2} \leqslant \int_a^b f + \frac{\epsilon}{2}.$$

Also, there is a subdivision σ_2 of $[a,b]$ such that

$$U[g;\sigma_2] < L[g;\sigma_2]+\frac{\epsilon}{2} \leqslant \int_a^b g + \frac{\epsilon}{2}.$$

If $\sigma = \sigma_1 \cup \sigma_2$, then, by (1) of 7.2D,

$$U[f;\sigma] < \int_a^b f + \frac{\epsilon}{2},$$

$$U[g;\sigma] < \int_a^b g + \frac{\epsilon}{2}.$$

From (1) we then have

$$\int_a^b (f+g) < \int_a^b f + \int_a^b g + \epsilon.$$

Since ϵ was arbitrary, this proves

$$\int_a^b (f+g) \leqslant \int_a^b f + \int_a^b g. \tag{2}$$

Since f and g were any Riemann integrable functions we can substitute $-f$, $-g$ for f,g in (2). Hence,

$$\int_a^b (-f-g) \leqslant \int_a^b (-f) + \int_a^b (-g).$$

Using 7.4B, we have

$$-\int_a^b (f+g) \leqslant -\left(\int_a^b f + \int_a^b g\right). \tag{3}$$

Now multiply both sides of (3) by -1. This reverses the inequality, and so

$$\int_a^b (f+g) \geqslant \int_a^b f + \int_a^b g. \tag{4}$$

The theorem follows from (2) and (4).

We leave the proof of the next result to the reader.

7.4D. LEMMA. If $f \in \mathcal{R}[a,b]$ and if

$$f(x) \geqslant 0$$

almost everywhere ($a \leqslant x \leqslant b$), then

$$\int_a^b f \geqslant 0.$$

7.4E. COROLLARY. If $f \in \mathcal{R}[a,b], g \in \mathcal{R}[a,b]$, and if

$$f(x) \leqslant g(x)$$

almost everywhere ($a \leqslant x \leqslant b$), then

$$\int_a^b f \leqslant \int_a^b g.$$

PROOF: By 7.4B and 7.4C the functions $-f$ and $g-f$ are Riemann integrable. Since $g(x)-f(x) \geqslant 0$ almost everywhere, we have, by 7.4D, 7.4C, and 7.4B,

$$0 \leqslant \int_a^b (g-f) = \int_a^b [g+(-f)] = \int_a^b g + \int_a^b (-f) = \int_a^b g - \int_a^b f.$$

This proves the corollary.

7.4F. COROLLARY. If $f \in \mathcal{R}[a,b]$, then $|f| \in \mathcal{R}[a,b]$ and

$$\left| \int_a^b f \right| \leqslant \int_a^b |f|.$$

PROOF: Since $|f|$ will be continuous at every point where f is continuous (exercise 4 of Section 5.1), it is clear by 7.3A that $|f| \in \mathcal{R}[a,b]$. Now, since $f(x) \leqslant |f(x)| = |f|(x)$ for all $x \in [a,b]$, 7.4E implies

$$\int_a^b f \leqslant \int_a^b |f|. \tag{1}$$

Since $-f(x) \leqslant |f(x)|$ for all $x \in [a,b]$, 7.4E implies

$$-\int_a^b f \leqslant \int_a^b |f|. \tag{2}$$

The corollary follows from (1) and (2).

7.4G. If $b < a$, we *define*

$$\int_a^b f \quad \text{to be} \quad -\int_b^a f,$$

provided that $f \in \mathcal{R}[b,a]$. It is then not difficult to show that results such as

$$\int_a^c f + \int_c^b f = \int_a^b f$$

hold, regardless of the order of the points a, b, c.

Exercises 7.4

1. Using only results from Sections 7.3 and 7.4 evaluate $\int_0^1 (2x^2-3x+5)dx$.
2. Do the same for $\int_1^3 (2x-3)dx$.
3. Let $J_1, J_2, \ldots, J_n$ be open intervals in $[a,b]$. Show that $\chi = \chi_{J_1 \cup \cdots \cup J_n}$ (the characteristic function of $J_1 \cup \cdots \cup J_n$) is in $\mathcal{R}[a,b]$. Then show that

$$\chi(x) \leqslant \chi_{J_1}(x) + \chi_{J_2}(x) + \cdots + \chi_{J_n}(x) \qquad (a \leqslant x \leqslant b).$$

Deduce that

$$\int_a^b \chi \leqslant |J_1| + |J_2| + \cdots + |J_n|.$$

4. If f is continuous on $[a,b]$ and if

$$F(x) = \int_a^x f(t)\,dt \qquad (a \leqslant x \leqslant b),$$

prove that F is continuous on $[a,b]$.

5. (a) If $0 \leqslant x \leqslant 1$, show that

$$\frac{x^2}{\sqrt{2}} \leqslant \frac{x^2}{\sqrt{1+x}} \leqslant x^2.$$

(b) Prove that

$$\frac{1}{3\sqrt{2}} \leqslant \int_0^1 \frac{x^2}{\sqrt{1+x}}\,dx \leqslant \frac{1}{3}$$

6. Prove that

$$\frac{2\pi^2}{9} \leqslant \int_{\pi/6}^{\pi/2} \frac{2x}{\sin x}\,dx \leqslant \frac{4\pi^2}{9}.$$

7. If f is continuous on the closed bounded interval $[a,b]$, if

$$f(x) \geqslant 0 \qquad (a \leqslant x \leqslant b),$$

and if $f(c) > 0$ for some $c \in [a,b]$, prove that

$$\int_a^b f(x)\,dx > 0.$$

8. If f is continuous on $[a,b]$, if

$$f(x) \geqslant 0 \qquad (a \leqslant x \leqslant b),$$

and if

$$\int_a^b f(x)\,dx = 0,$$

prove that f is identically zero on $[a,b]$.

7.5 DERIVATIVES

As we saw in Section 7.3, the definition of integral has nothing to do with derivatives. The fact that the integral

$$\int_a^b f$$

can be evaluated by "anti-differentiating f and plugging in b and a" is a *theorem* and not in any sense a definition. We now begin our development of the theory of derivatives, which will ultimately yield this theorem.

7.5A. DEFINITION. Let f be a real-valued function on an interval $J \subset R^1$. If $c \in J$, we say that f has a derivative at c if

$$\lim_{x \to c} \frac{f(x)-f(c)}{x-c} \tag{1}$$

exists.* If this limit exists, we denote it by $f'(c)$.

It is clear that

$$\lim_{x \to c} \frac{f(x)-f(c)}{x-c}$$

means the same as

$$\lim_{h \to 0} \frac{f(c+h)-f(c)}{h}.$$

Instead of "f has a derivative at c" we sometimes say, more simply, that "$f'(c)$ exists."

Thus if E is the set of points c in J at which $f'(c)$ exists (and if $E \neq \varnothing$), then f' is itself a real-valued function on E. It is entirely possible, of course, that E *is* empty. We will soon see that if $f'(c)$ exists, then f is continuous at c. Thus if f is not continuous at any point in J, then f cannot have a derivative at any point in J. It is also possible that $f'(c)$ does not exist even though f *is* continuous at c. For example, if

$$f(x)=|x| \qquad (-\infty < x < \infty),$$

then

$$\frac{f(x)-f(0)}{x-0}=1 \quad \text{if} \quad x>0,$$

while

$$\frac{f(x)-f(0)}{x-0}=-1 \quad \text{if} \quad x<0.$$

* If $J=[a,b]$ and $c=a$ or $c=b$, the limit (1) is a one-sided limit and gives what is sometimes called a one-sided derivative (see 4.1F). Thus if f is a real-valued function on $[a,b]$, then

$$f'(a)=\lim_{x \to a+} \frac{f(x)-f(a)}{x-a}$$

while

$$f'(b)=\lim_{x \to b-} \frac{f(x)-f(b)}{x-b}.$$

It is then sometimes said that f has a right-hand derivative at a and a left-hand derivative at b.

Hence,

$$\lim_{x \to 0} \frac{f(x)-f(0)}{x-0}$$

does not exist. Thus f does not have a derivative at 0 even though f is continuous at 0. (In Chapter 9 we show that there exists a function that is continuous at each point of $[0,1]$ but does not have a derivative at any point in $[0,1]$.)

The function

$$g(x)=x^2 \qquad (-\infty<x<\infty)$$

has a derivative at each point of R^1. For, if $c \in R^1$ and $x \neq c$,

$$\frac{g(x)-g(c)}{x-c}=\frac{x^2-c^2}{x-c}=x+c,$$

and thus

$$\lim_{x \to c} \frac{g(x)-g(c)}{x-c}=\lim_{x \to c}(x+c)=2c.$$

Hence,

$$g'(c)=2c \qquad (-\infty<c<\infty),$$

which probably is no surprise to anybody.

Now we prove that differentiability (having a derivative) at a point implies continuity at that point.

7.5B. THEOREM. If the real-valued function f has a derivative at the point $c \in R^1$, then f is continuous at c.

PROOF: For $x \neq c$ we have

$$f(x)-f(c)=\left[\frac{f(x)-f(c)}{x-c}\right](x-c).$$

Since

$$\lim_{x \to c} \frac{f(x)-f(c)}{x-c}=f'(c) \quad \text{and} \quad \lim_{x \to c}(x-c)=0,$$

theorem 4.1D implies

$$\lim_{x \to c}[f(x)-f(c)]=f'(c) \cdot 0=0.$$

Hence, since $f(x)=f(c)+[f(x)-f(c)]$ we have, by 4.1C,

$$\lim_{x \to c} f(x)=f(c)+0=f(c).$$

This proves the theorem.

We now record the familiar results on taking derivatives of sums, products, and so on.

7.5C. THEOREM. If f and g both have derivatives at $c \in R^1$, then so do $f+g, f-g, fg$, and

$$(f+g)'(c)=f'(c)+g'(c),$$
$$(f-g)'(c)=f'(c)-g'(c),$$
$$(fg)'(c)=f'(c)\,g(c)+f(c)\,g'(c).$$

Furthermore, if $g(c)\neq 0$, then f/g has a derivative at c and

$$\left(\frac{f}{g}\right)'(c)=\frac{g(c)f'(c)-f(c)g'(c)}{[g(c)]^2}.$$

PROOF: We prove only the part concerning fg. If $h=fg$, then, for $x\neq c, h(x)-h(c)=f(x)g(x)-f(c)g(c)=f(x)g(x)-f(c)g(x)+f(c)g(x)-f(c)g(c)$, and so

$$\frac{h(x)-h(c)}{x-c}=\frac{f(x)-f(c)}{x-c}\cdot g(x)+f(c)\cdot\frac{g(x)-g(c)}{x-c}.$$

Since

$$\lim_{x\to c}\frac{f(x)-f(c)}{x-c}=f'(c),\qquad \lim_{x\to c}\frac{g(x)-g(c)}{x-c}=g'(c),$$

and (by 7.5B) $\lim_{x\to c}g(x)=g(c)$, we see, using 4.1D and 4.1C, that h has a derivative at c and

$$h'(c)=\lim_{x\to c}\frac{h(x)-h(c)}{x-c}=f'(c)g(c)+f(c)g'(c),$$

which is what we wished to show.

Here is a statement and proof of the well-known "chain rule."

7.5D THEOREM. Suppose g has a derivative at c and that f has a derivative at $g(c)$. Then $\phi=f\circ g$ has a derivative at c and

$$\phi'(c)=f'[g(c)]g'(c).$$

We first prove a lemma.

LEMMA. If f has a derivative at c, then there exists a function F such that

$$F \quad \text{is continuous at} \quad 0, \tag{1}$$

and

$$f(c+h)=f(c)+hF(h) \tag{2}$$

for all sufficiently small h.

Conversely, if there exists a function F satisfying (1) and (2), then f has a derivative at c.

If such a function F exists, then $F(0)=f'(c)$.

PROOF: Suppose $f'(c)$ exists. Let

$$F(h)=\frac{f(c+h)-f(c)}{h}\qquad (h\neq 0)$$

(provided $c+h$ is in the domain of f) and

$$F(0)=f'(c).$$

It is then clear that F satisfies (1) and (2). Moreover, $F(0)=f'(c)$ by definition of F.

Conversely, suppose for some f that a function F exists which satisfies (1) and (2). From (2) we have

$$\frac{f(c+h)-f(c)}{h}=F(h)\qquad (h\neq 0).$$

By (1), when h tends to zero the right side has the limit $F(0)$. Hence, so does the left side. This shows that $f'(c)$ exists and that $f'(c)=F(0)$, and so the proof of the lemma is complete.

Now we can prove the chain rule.

PROOF OF THEOREM 7.5D. Since g has a derivative at c, the lemma shows the existence of G such that G is continuous at 0, $G(0)=g'(c)$, and

$$g(c+h)=g(c)+hG(h)$$

for small h. Similarly, since f has a derivative at $g(c)$, there exists F such that F is continuous at 0, $F(0)=f'[g(c)]$, and

$$f[g(c)+k]=f[g(c)]+kF(k) \tag{3}$$

for small k.

Let

$$k=g(c+h)-g(c)=hG(h)$$

(which will be sufficiently small if h is). Then

$$f[g(c)+k]=f[g(c+h)] \tag{4}$$

and

$$kF(k)=hG(h)F[hG(h)]. \tag{5}$$

Substituting (4) and (5) into (3) we obtain

$$f[g(c+h)]=f[g(c)]+hG(h)F[hG(h)].$$

Since $\phi=f\circ g$ this says

$$\phi(c+h)=\phi(c)+h\Phi(h) \tag{6}$$

where $\Phi(h)=G(h)F[hG(h)]$. Now Φ is continuous at 0 since F and G are continuous at 0. In view of (6) the lemma implies that $\phi'(c)$ exists and that

$$\phi'(c)=\Phi(0)=F(0)G(0)=f'[g(c)]g'(c),$$

which is what we wished to prove.

The following result on the relationship between derivatives of inverse functions is important in later applications. Remember that if f is a 1–1 function on $[a,b]$, then $\varphi[f(x)]=x$ $(a\leqslant x\leqslant b)$ where φ is the inverse function for f.

7.5E. THEOREM. Let f be a 1–1 real-valued function on an interval J. Let φ be the inverse function for f. If f is continuous at $c\in J$, and if φ has a derivative at $d=f(c)$ with $\varphi'(d)\neq 0$, then $f'(c)$ exists and

$$f'(c)=\frac{1}{\varphi'(d)}.$$

PROOF: For $h\neq 0$ let $k(h)=f(c+h)-f(c)$. [Since f is 1–1, we know $k(h)\neq 0$ if $h\neq 0$.] Then $d+k(h)=f(c)+k(h)=f(c+h)$. Hence,

$$\varphi[d+k(h)]=\varphi[f(c+h)]=c+h.$$

We have

$$\frac{f(c+h)-f(c)}{h}=\frac{[d+k(h)]-d}{c+h-c}=\frac{k(h)}{\varphi[d+k(h)]-\varphi(d)}$$

$$=\frac{1}{\dfrac{\varphi[d+k(h)]-\varphi(d)}{k(h)}}.$$

But $\lim_{h\to 0}k(h)=0$ since, by hypothesis, f is continuous at c. Thus as h approaches 0, the right side of (1) approaches the limit $1/\varphi'(d)$. Hence,

$$\lim_{h\to 0}\frac{f(c+h)-f(c)}{h}=\frac{1}{\varphi'(d)},$$

which is what we wished to prove.

For example, if $f(x)=\sqrt{x}$ $(0\leqslant x<\infty)$, then $\varphi(x)=x^2$ $(0\leqslant x<\infty)$. By 7.5E we know that $f'(3)$ exists and

$$f'(3)=\frac{1}{\varphi'(\sqrt{3}\,)}.$$

We have already shown that $\varphi'(x)=2x$ $(0\leqslant x<\infty)$. Hence, $f'(3)=1/2\sqrt{3}$. (This agrees with the well-known formula $f'(x)=1/2\sqrt{x}$ $(0<x<\infty)$.)

7.5F. In elementary calculus courses derivatives are usually introduced with a geometric interpretation. If f is a real-valued function on an interval J, then f defines a graph—namely, the subset of R^2 consisting of all points $\langle x,y\rangle$ in R^2 such that $x\in J$ and $y=f(x)$. This curve is usually denoted simply by

$$y=f(x). \tag{1}$$

The curve (1) is then said to have a tangent at $c\in J$ if f has a derivative at c. The slope of the tangent at c is then defined to be $f'(c)$. The customary notation for the slope of the tangent at c is $dy/dx|_{x=c}$. That is,

$$\left.\frac{dy}{dx}\right|_{x=c}=f'(c).$$

When f has a derivative at all points of J [that is, when (1) has a tangent at all points of J] this is usually written simply as

$$\frac{dy}{dx}=f'(x).$$

The dy/dx notation permits the results of the last two theorems to be put in a more striking form.

For example, consider the chain rule which says (roughly) that if $\varphi=g\circ f$, then

$$\varphi'(x)=g'[f(x)]f'(x). \tag{2}$$

Now if $y=f(x)$ is a curve in the x-y plane and if $u=g(y)$ is a curve in the y-u plane, then

$$\frac{dy}{dx}=f'(x),\qquad \frac{du}{dy}=g'(y)=g'[f(x)].$$

But then $u=g(y)=g[f(x)]=\varphi(x)$ and so $du/dx=\varphi'(x)$. Hence, (2) can be written

$$\frac{du}{dx}=\frac{du}{dy}\cdot\frac{dy}{dx}. \tag{3}$$

This should give a motive for the use of the term "chain rule."

Now consider 7.5E. This states roughly that if f is 1–1 and

$$y=f(x)$$

[so that $x=\varphi(y)$], then

$$f'(x)=\frac{1}{\varphi'(y)}. \tag{4}$$

But $f'(x)=dy/dx$ and $\varphi'(y)=dx/dy$. Hence, (4) can be written

$$\frac{dy}{dx}=\frac{1}{dx/dy}. \tag{5}$$

Both (3) and (5) would be trivial if dx, dy, du were themselves well-defined quantities that could be dealt with by the laws of algebra. However, we have not given any meaning to dx, dy, and so on (and you will not find out in this text what they are). The fact that (3) and (5) are true is itself a good reason for the dy/dx notation.

In subsequent sections the reader should keep in mind the geometric interpretation of the sign of the derivative. If $f'(c)>0$, the curve (1) is "ascending" at c; if $f'(c)<0$, the curve is "descending" at c; while if $f'(c)=0$, the curve has a horizontal tangent at c.

Also remember that if $f'(c)$ exists, then the curve (1) is "smooth" at $x=c$.

7.5G. Suppose $f'(x)$ exists for every x in some interval J. If $c\in J$ and if

$$\lim_{x\to c}\frac{f'(x)-f'(c)}{x-c} \tag{1}$$

exists, we say that f has a second derivative at c. We then denote the limit (1) by $f''(c)$—that is,

$$f''(c)=\lim_{x\to c}\frac{f'(x)-f'(c)}{x-c}.$$

Similarly, the nth derivative of f at c is defined as

$$f^{(n)}(c)=\lim_{x\to c}\frac{f^{(n-1)}(x)-f^{(n-1)}(c)}{x-c},$$

provided that $f^{(n-1)}(x)$ exists for all x in an interval containing c *and* provided that the limit exists. From 7.5B it then follows that if $f^{(n)}(c)$ exists, then $f^{(n-1)}$ is continuous at c.

We assume that the reader is familiar with the geometric significance of the second derivative f''—namely, that the graph

$$y=f(x)$$

is concave up at those points where $f''(x)>0$, and concave down at those points where $f''(x)<0$.

Exercises 7.5

1. Prove that the derivative of a constant function on $[a,b]$ is the identically zero function on $[a,b]$.
2. If $f(x)$ has a derivative at c, if $b \in R^1$, and if $g(x)=bf(x)$ for all x in an interval containing c, show that

$$g'(c)=bf'(c).$$

3. Find $f'(x)$ by the chain rule if
   (a) $f(x)=\sin x^2 \qquad (-\infty<x<\infty)$.
   (b) $f(x)=\sqrt{x+\sqrt{x+\sqrt{x}}} \qquad (0 \leqslant x<\infty)$.
4. If $n \in I$ and

$$f(x)=x^n \qquad (-\infty<x<\infty),$$

use definition 7.5A to show that

$$f'(x)=nx^{n-1} \qquad (-\infty<x<\infty).$$

5. (a) If n is a negative integer and

$$f(x)=x^n \qquad (-\infty<x<\infty; x \neq 0),$$

show that

$$f'(x)=nx^{n-1} \qquad (-\infty<x<\infty; x \neq 0).$$

(b) If n is a nonzero rational number and

$$f(x)=x^n \qquad (-\infty<x<\infty),$$

show that

$$f'(x)=nx^{n-1} \qquad (-\infty<x<\infty).$$

(*Hint*: If $n=p/q$, write $f(x)$ as $(x^p)^{1/q}$ so that $[f(x)]^q=x^p$.)

6. Suppose that f and g have derivatives of all orders at c and that $h=fg$.
   (a) Prove that

$$h''(c)=f''(c)g(c)+2f'(c)g'(c)+f(c)g''(c).$$

(b) More generally prove that, for any $n \in I$,

$$h^{(n)}(c)=\sum_{k=0}^{n} \frac{n!}{k!(n-k)!} f^{(k)}(c)\, g^{(n-k)}(c).$$

(This formula is known as the Leibniz rule.)

7. If f is a function on $[a,b]$ and $f'(c)>0$ where $a<c<b$, prove that there is an x with $c<x<b$ such that $f(x)>f(c)$.
8. Let

$$f(x)=x \quad \text{if} \quad x \text{ is rational},$$
$$f(x)=\sin x \quad \text{if} \quad x \text{ is irrational}.$$

Prove that $f'(0)=1$.

9. True or false? If f is a function on $[a,b]$, if $c \in [a,b]$, and if $f'(c)>0$, then f is strictly increasing on some open subinterval of $[a,b]$ containing c.
10. If f is a real-valued function on $[a,b]$ and if f has a right-hand derivative at $c \in [a,b]$, prove that f is continuous on the right at c.

7.6 ROLLE'S THEOREM

Maximum and minimum problems are a very important part of elementary calculus. For our purposes, however, we need only the following fraction of maximum-minimum theory.

7.6A. THEOREM. Let f be a continuous real-valued function on the closed bounded interval $[a,b]$. If the maximum value for f is attained at c where $a<c<b$, and if $f'(c)$ exists, then $f'(c)=0$.

PROOF: Suppose the contrary—that is, suppose $f'(c)\neq 0$. If $f'(c)>0$, then

$$\lim_{x\to c}\frac{f(x)-f(c)}{x-c}>0 \quad \text{and so} \quad \frac{f(x)-f(c)}{x-c}>0$$

for $0<|x-c|<\delta_1$ where δ_1 is a suitable positive number.* If $x\in(c,c+\delta_1)$, then $x-c>0$ and hence, $f(x)-f(c)>0$. This contradicts the hypothesis that f attains a maximum at c. If $f'(c)<0$, then

$$\frac{f(x)-f(c)}{x-c}<0$$

for $0<|x-c|<\delta_2$. If $x\in(c-\delta_2,c)$, then $x-c<0$ and hence, $f(x)-f(c)>0$, which is again a contradiction. Hence, $f'(c)=0$.

An obvious modification of the proof will establish the following.

7.6B. THEOREM. Theorem 7.6A remains true with "maximum value" replaced by "minimum value."

If the curve

$$y=f(x) \qquad (a\leqslant x\leqslant b)$$

has its end points on the x-axis, and if the curve is smooth, it is intuitively clear that at some point on the curve there will be a horizontal tangent. This result, made precise, is called Rolle's theorem, which we need in our drive toward the fundamental theorem of calculus.

7.6C. ROLLE'S THEOREM. Let f be a continuous real-valued function on the closed bounded interval $[a,b]$, with $f(a)=f(b)=0$. If $f'(x)$ exists for all x in (a,b), then there is some point $c\in(a,b)$ where $f'(c)=0$.

PROOF: If f is identically zero on $[a,b]$, the conclusion is obvious. If $f(x)>0$ for some $x\in(a,b)$, the maximum value of f on $[a,b]$ (remember f attains a maximum value at some point of $[a,b]$ by 6.6G) will not be attained at a or b since $f(a)=f(b)=0$. Hence, f will attain its maximum value at some $c\in(a,b)$ and the theorem follows from 7.6A. If $f(x)<0$ for some x in (a,b), the theorem follows from 7.6B. This completes the proof.

We wish to emphasize the fact that the proof of Rolle's theorem depends on the theorem that a continuous function on a closed bounded interval attains a maximum and a minimum value.

* See exercise 14 of Section 4.1.

Note that in 7.6C we do not require that f' exist on $[a,b]$ but only on (a,b). Thus if

$$f(x)=\sqrt{1-x^2} \qquad (-1\leqslant x\leqslant 1),$$

then f obeys the hypotheses of 7.6C with $a=-1, b=1$. [Here $f'(x)=-x/\sqrt{1-x^2}$ for $-1<x<1$, while f does not have a derivative at -1 or 1.] For this f we see that the c in 7.6C can be taken as 0. That is, $f'(c)=f'(0)=0$.

It is also important to observe that the hypotheses of 7.6C cannot be weakened. For example, if

$$g(x)=1-|x| \qquad (-1\leqslant x\leqslant 1),$$

then $g(-1)=g(1)=0$ and g is continuous on $[-1,1]$. Also, $g'(x)$ exists for all x in $(-1,1)$ *except* $x=0$. Thus g obeys all the hypotheses of 7.6 except that g fails to have a derivative at 0. For this g there is no c in $(-1,1)$ for which $g'(c)=0$. This shows that the conclusion of Rolle's theorem need not hold if we weaken the last hypothesis.

7.6D. Using the techniques involved in the proof of Rolle's theorem we can prove a very interesting property of derivatives.

First note that if f is defined by

$$f(x)=x^2\sin\frac{1}{x} \qquad (x\neq 0),$$

$$f(0)=0,$$

then f has a derivative at each point of $(-\infty,\infty)$. For, by theorems 7.5D and 7.5C we have*

$$f'(x)=-\cos\frac{1}{x}+2x\sin\frac{1}{x} \qquad (x\neq 0). \tag{1}$$

To show that $f'(0)$ exists we have, for $x\neq 0$,

$$\frac{f(x)-f(0)}{x-0}=x\sin\frac{1}{x},$$

$$\left|\frac{f(x)-f(0)}{x-0}\right|\leqslant |x|,$$

and it follows that

$$\lim_{x\to 0}\frac{f(x)-f(0)}{x-0}=0.$$

Thus

$$f'(0)=0.$$

We have shown that $f'(x)$ exists for all x. But note that f' is not continuous at 0. For, because of the $\cos(1/x)$ term in (1), $\lim_{x\to 0}f'(x)$ does not exist. This example shows that a function can have a derivative at each point of an interval but that the derivative (function) need not be continuous on the interval.

* In Chapter 8 we *prove* that if $y=\sin x$, then $dy/dx=\cos x$.

Nevertheless, derivatives share one important property with continuous functions. Namely, if $f'(x)$ exists for all x in $[a,b]$, then the image of $[a,b]$ under f' must be connected, even though f' may not be continuous. (Compare the following theorem with 6.2E.)

7.6E. THEOREM. If f has a derivative at every point of $[a,b]$, then f' takes on every value between $f'(a)$ and $f'(b)$.

PROOF: It is sufficient to consider the case in which $f'(a)<f'(b)$. Then if $f'(a)<\gamma<f'(b)$, we must show that there exists $c\in(a,b)$ such that $f'(c)=\gamma$. Let

$$g(x)=f(x)-\gamma x \qquad (a\leqslant x\leqslant b)$$

so that

$$g'(x)=f'(x)-\gamma \qquad (a\leqslant x\leqslant b).$$

Then, since $g'(x)$ exists for all $x\in[a,b]$, theorem 7.5B shows that g is continuous on $[a,b]$. Hence, by 6.6G, g takes on a minimum value at some point $c\in[a,b]$. But $g'(a)=f'(a)-\gamma<0$ and so g cannot attain its minimum value at a (why?). Similarly, since $g'(b)=f'(b)-\gamma>0$, g cannot attain its minimum value at b. Thus $a<c<b$. Theorem 7.6B then shows that $g'(c)=0$. Since $g'(c)=f'(c)-\gamma$, this proves $f'(c)=\gamma$, which is what we wished to show.

The property of f in 6.2E or f' in 7.6E is called the Darboux property.

Exercises 7.6

1. Prove that there is no value of k such that the equation
$$x^3-3x+k=0$$
has two distinct roots in $[0,1]$.
2. Which of the following functions obey the hypotheses of Rolle's theorem over the interval indicated?
(a) $f(x)=\sin x\ (0\leqslant x\leqslant\pi)$.
(b) $f(x)=\sqrt{x}\,(x-1)\ \ (0\leqslant x\leqslant 1)$.
(c) $f(x)=\sin\frac{1}{x}\left(-\frac{1}{\pi}\leqslant x\leqslant\frac{1}{\pi};x\neq 0\right)$,
$f(0)=0$.
(d) $f(x)=x^2\ (0\leqslant x\leqslant 1)$.
3. Find a suitable point c of Rolle's theorem for
$$f(x)=(x-a)(b-x) \qquad (a\leqslant x\leqslant b).$$
4. Prove that if
$$\frac{a_0}{n+1}+\frac{a_1}{n}+\cdots+\frac{a_{n-1}}{2}+a_n=0,$$
then the equation
$$a_0x^n+a_1x^{n-1}+\cdots+a_{n-1}x+a_n=0$$
has at least one root between 0 and 1. (*Hint*: Consider
$$f(x)=\frac{a_0x^{n+1}}{n+1}+\frac{a_1x^n}{n}+\cdots+\frac{a_{n-1}x^2}{2}+a_nx.)$$

5. Let

$$f(x)=0 \quad (-1 \leqslant x \leqslant 0),$$
$$f(x)=1 \quad (0<x \leqslant 1).$$

Is there a function g such that

$$g'(x)=f(x) \qquad (-1 \leqslant x \leqslant 1)?$$

7.7 THE LAW OF THE MEAN

For a "smooth" curve

$$y=f(x) \qquad (a \leqslant x \leqslant b)$$

it looks evident that at some point $c \in (a,b)$ the slope of the tangent $f'(c)$ will be equal to the slope of the chord joining the end points of the curve. That is, for some $c \in (a,b)$,

$$f'(c)=\frac{f(b)-f(a)}{b-a}.$$

This result, made precise, is called the law of the mean (or mean-value theorem).

Note that Rolle's theorem states exactly the same thing in the special case where $f(a)=f(b)=0$. Thus the law of the mean is a "rotated" version of Rolle's theorem. The idea of the proof is to subtract from f the function g whose graph

$$y=g(x) \qquad (a \leqslant x \leqslant b)$$

is the chord joining the end points of f. [Thus

$$g(x)=f(a)+\frac{f(b)-f(a)}{b-a}(x-a)$$

for $a \leqslant x \leqslant b$.] Since $f-g$ then takes the value 0 at a and b, we can apply Rolle's theorem to $f-g$. Here are the details.

7.7A. THEOREM (LAW OF THE MEAN). If f is a continuous function on the closed bounded interval $[a,b]$, and if $f'(x)$ exists for all x in (a,b), then there exists c in (a,b) such that

$$f'(c)=\frac{f(b)-f(a)}{b-a}.$$

PROOF: Let h be defined as

$$h(x)=f(x)-f(a)-\frac{f(b)-f(a)}{b-a}(x-a) \qquad (a \leqslant x \leqslant b).$$

Then $h(a)=0=h(b)$, and h obeys the other hypotheses of Rolle's theorem as well. Hence, $h'(c)=0$ for some $c \in (a,b)$. But

$$h'(c)=f'(c)-\frac{f(b)-f(a)}{b-a},$$

which establishes the desired result.

An important application of the law of the mean is the following.

7.7B. THEOREM. If f is a continuous real-valued function on the interval J, and if $f'(x)>0$ for all x in J except possibly the end points of J (if there are any), then f is strictly increasing on J (and hence, f is 1–1).

PROOF: If $a,b \in J$ with $a<b$, we have, by 7.7A,

$$f(b)-f(a)=f'(c)(b-a)$$

for some c between a and b. But $f'(c)>0$ by hypothesis, and hence, $f(b)-f(a)>0$. That is, if $a<b$, then $f(a)<f(b)$, which is what we wished to show.

A useful generalization of the law of the mean can be motivated by considering a smooth curve in parametric representation

$$x=g(t), \qquad y=f(t) \qquad (a \leqslant t \leqslant b).$$

The slope of the chord joining the end points of the curve is

$$\frac{f(b)-f(a)}{g(b)-g(a)}.$$

The slope of the tangent to the curve at $t=c$ is $f'(c)/g'(c)$. The generalized law of the mean asserts that there will always be a value of c in (a,b) for which the slope of the chord is equal to the slope of the tangent at c. We now make this precise.

7.7C. THEOREM. Let f and g be continuous functions on the closed bounded interval $[a,b]$ with $g(a) \neq g(b)$. If both f and g have a derivative at each point of (a,b), and $f'(t)$ and $g'(t)$ are not both equal to zero for any $t \in (a,b)$, then there exists a point $c \in (a,b)$ such that

$$\frac{f'(c)}{g'(c)}=\frac{f(b)-f(a)}{g(b)-g(a)}.$$

PROOF: Let

$$h(x)=f(x)-f(a)-\frac{f(b)-f(a)}{g(b)-g(a)}\left[g(x)-g(a)\right].$$

Then $h(a)=0$, $h(b)=0$ and h obeys the other hypotheses of Rolle's theorem. Hence, $h'(c)=0$ for some $c \in (a,b)$. That is,

$$f'(c)-\frac{f(b)-f(a)}{g(b)-g(a)}\cdot g'(c)=0.$$

If $g'(c)$ were zero, then $f'(c)$ would also be zero, contradicting our hypothesis. Hence, $g'(c) \neq 0$, and the theorem follows.

Exercises 7.7

1. Which of the following functions obey the hypotheses of the law of the mean? For those functions to which the law of the mean applies, calculate a suitable point c.

   (a) $f(x)=\dfrac{x}{x-1} \quad (0 \leqslant x \leqslant 2).$

   (b) $f(x)=\dfrac{x}{x-1} \quad (2 \leqslant x \leqslant 4).$

   (c) $f(x)=Ax+B \quad (a \leqslant x \leqslant b).$

   (d) $f(x)=1-x^{2/3} \quad (-1 \leqslant x \leqslant 1).$

2. Calculate a value of c for which
$$\frac{f(b)-f(a)}{g(b)-g(a)}=\frac{f'(c)}{g'(c)}$$
for each of the following pairs of functions.
(a) $f(x)=x, g(x)=x^2$ $(0 \leqslant x \leqslant 1)$.
(b) $f(x)=\sin x, g(x)=\cos x$ $(-\pi/2 \leqslant x \leqslant 0)$.
3. Show that if $f'(x)$ and $g'(x)$ exist for all x in $[a,b]$, and $g'(x)\neq 0$ $(a \leqslant x \leqslant b)$, then f and g satisfy the hypotheses of 7.7C.
4. If $f'(x)=0$ for all x in (a,b), prove that f is constant on (a,b).
5. If f is continuous on $[a,b]$, if $f'(x)$ exists for $a<x<b$, and if $\lim_{x\to b-} f'(x)=A$, prove that the (left-hand) derivative $f'(b)$ exists and is equal to A.
6. Suppose
$$f'(x)>0 \qquad (a \leqslant x \leqslant b).$$
If φ is the inverse function for f, show that φ is continuous on $[f(a),f(b)]$.

7.8 FUNDAMENTAL THEOREMS OF CALCULUS

We begin by asking the following question. What must we know about a function f on an interval $[a,b]$ in order to be sure that f is the derivative of some function F on $[a,b]$? We see from 7.6E that even such a relatively "nice" function as that given by
$$f(x)=0 \quad (-1 \leqslant x<0),$$
$$f(x)=1 \quad (0 \leqslant x \leqslant 1)$$
is not the derivative of any function on $[-1,1]$. What we now show is that if f is continuous on $[a,b]$, then there *will* exist a function F on $[a,b]$ such that $F'(x)=f(x)$ for all $x\in[a,b]$. [Thus continuity on $[a,b]$ is a *sufficient* condition for a function to be a derivative on $[a,b]$. Note, however, that continuity is not a *necessary* condition. For we showed in Section 7.6 that the derivative of φ is not continuous, where
$$\varphi(x)=x^2\sin\frac{1}{x} \qquad (x\neq 0),$$
$$\varphi(0)=0.]$$

7.8A. THEOREM. If f is continuous on the closed bounded interval $[a,b]$, and if
$$F(x)=\int_a^x f(t)\,dt \qquad (a \leqslant x \leqslant b),$$
then $F'(x)=f(x)$ $(a \leqslant x \leqslant b)$.

PROOF: For any fixed $x\in[a,b]$ we have, if $h\neq 0$ and $x+h\in[a,b]$,
$$F(x+h)-F(x)=\int_a^{x+h} f(t)\,dt-\int_a^x f(t)\,dt.$$
Then, by 7.4A,
$$F(x+h)-F(x)=\int_x^{x+h} f(t)\,dt. \tag{1}$$

Since f is continuous on the closed bounded interval $[x,x+h]$, by 6.6G there are points of $[x,x+h]$ at which f attains a maximum value M and a minimum value m. (We are tacitly assuming that h is positive. If $h<0$, we would write $[x+h,x]$ instead of $[x,x+h]$, and make similar adjustments in the details to follow.) That is,

$$m \leqslant f(t) \leqslant M \qquad (x \leqslant t \leqslant x+h),$$

and $f(t_1)=m, f(t_2)=M$ for some t_1, t_2 in $[x,x+h]$. By 7.4E we then have

$$\int_x^{x+h} m\,dt \leqslant \int_x^{x+h} f(t)\,dt \leqslant \int_x^{x+h} M\,dt,$$

$$mh \leqslant \int_x^{x+h} f(t)\,dt \leqslant Mh,$$

and finally

$$m \leqslant \theta \leqslant M,$$

where

$$\theta = \frac{1}{h}\int_x^{x+h} f(t)\,dt.$$

$$\left(\text{How do we know that } \int_x^{x+h} m\,dt = mh?\right)$$

By 6.2D there must be a point $c(h)$ in $[x,x+h]$ such that $f[c(h)]=\theta$. Thus we have shown that if $h>0$, there exists $c(h)$ in $[x,x+h]$ such that

$$f[c(h)] = \frac{1}{h}\int_x^{x+h} f(t)\,dt.$$

From (1) we then have

$$\frac{F(x+h)-F(x)}{h} = f[c(h)]. \tag{2}$$

But, clearly, $\lim_{h\to 0} c(h) = x$ [since $x \leqslant c(h) \leqslant x+h$]. Thus as $h \to 0$, the right side of (2) has the limit $f(x)$, since f is continuous at x. Hence, the left side of (2) approaches $f(x)$ as $h \to 0$, and we have

$$F'(x) = \lim_{h\to 0} \frac{F(x+h)-F(x)}{h} = f(x),$$

which is what we wished to show.

Theorem 7.8A is the first result we have encountered that links the concepts of integral and derivative. It shows not only that when f is continuous there will be a function F such that $F'=f$, but also that F can be expressed in terms of an integral. Theorem 7.8A is sometimes called the fundamental theorem of calculus. However, since theorem 7.8E also sometimes goes by that name we will call 7.8A the first fundamental theorem of calculus.

We can improve 7.8A by assuming only the Riemann integrability of f and assuming the continuity of f at the point x. That is,

7.8B. THEOREM. If $f \in \mathcal{R}[a,b]$, if

$$F(x) = \int_a^x f(t)\,dt \qquad (a \leqslant x \leqslant b),$$

and if f is continuous at $x_0 \in [a,b]$, then $F'(x_0) = f(x_0)$.

PROOF: For $h>0$ let I_h denote the interval $[x_0, x_0+h]$. Then, with $\omega[f;I_h]$ as in 7.2A we have

$$f(x_0)-\omega[f;I_h] \leqslant f(t) \leqslant f(x_0)+\omega[f;I_h] \qquad (t \in I_h).$$

Hence, by 7.4E,

$$h[f(x_0)-\omega[f;I_h]] \leqslant \int_{x_0}^{x_0+h} f(t)\,dt \leqslant h[f(x_0)+\omega[f;I_h]].$$

Dividing by h we have

$$f(x_0)-\omega[f;I_h] \leqslant \frac{F(x_0+h)-F(x_0)}{h} \leqslant f(x_0)+\omega[f;I_h]. \tag{1}$$

[Equation (1) can be established for $h<0$ in identical fashion.] But, since by hypothesis f is continuous at x_0, we have

$$\lim_{h\to 0} \omega[f;I_h]=0. \tag{2}$$

The theorem then follows from (1) and (2).

It should be clear that 7.8A is a consequence of 7.8B.

We now head toward the second fundamental theorem of calculus that justifies evaluating an integral

$$\int_a^b f$$

by "anti-differentiating the integrand and plugging in a and b." We first need two theorems that are of vast importance.

7.8C THEOREM. If $f'(x)=0$ for every x in the closed bounded interval $[a,b]$, then f is constant on $[a,b]$—that is,

$$f(x)=C \qquad (a \leqslant x \leqslant b)$$

for some $C \in R$.

PROOF: For any x with $a<x\leqslant b$, theorem 7.5B shows that f is continuous on the interval $[a,x]$. By the law of the mean 7.7A, there exists $c \in (a,x)$ such that

$$f'(c)=\frac{f(x)-f(a)}{x-a}.$$

But, by hypothesis, $f'(c)=0$. Thus $f(x)=f(a)$ for all x in $[a,b]$, and the theorem is proved [with $C=f(a)$].

Here is an immediate consequence.

7.8D. THEOREM. If $f'(x)=g'(x)$ for all x in the closed bounded interval $[a,b]$, then $f-g$ is constant—that is,

$$f(x)=g(x)+C \qquad (a \leqslant x \leqslant b)$$

for some $C \in R$.

PROOF: By 7.5C, $(f-g)'(x)=f'(x)-g'(x)$ for all x in $[a,b]$. Hence, by the hypothesis, $(f-g)'(x)=0$ for all $x \in [a,b]$. The theorem follows from 7.8C.

7.8E. THEOREM (SECOND FUNDAMENTAL THEOREM OF CALCULUS). If f is a continuous function on the closed bounded interval $[a,b]$, and if

$$\Phi'(x)=f(x) \qquad (a \leqslant x \leqslant b), \tag{1}$$

then

$$\int_a^b f(x)\,dx=\Phi(b)-\Phi(a).$$

PROOF: If

$$F(x)=\int_a^x f(t)\,dt,$$

then, by 7.8A,

$$F'(x)=f(x) \qquad (a \leqslant x \leqslant b). \tag{2}$$

From (1) and (2) we see that $F'(x)=\Phi'(x)$ for all x in $[a,b]$. Hence, by 7.8D,

$$F(x)=\Phi(x)+C \qquad (a \leqslant x \leqslant b)$$

for some $C \in R$. Hence, $F(b)-F(a)=[\Phi(b)+C]-[\Phi(a)+C]=\Phi(b)-\Phi(a)$. But

$$F(a)=\int_a^a f(t)\,dt=0.$$

Thus $F(b)=\Phi(b)-\Phi(a)$. Since

$$F(b)=\int_a^b f(t)\,dt,$$

the theorem is proved.

A computation such as

$$\int_1^2 x^2\,dx=\frac{x^3}{3}\bigg|_1^2=\frac{2^3}{3}-\frac{1^3}{3} \tag{$*$}$$

is thus justified by 7.8E. For, if $f(x)=x^2$ $(1 \leqslant x \leqslant 2)$, then f is continuous on $[1,2]$. Moreover, if $\Phi(x)=x^3/3$ $(1 \leqslant x \leqslant 2)$, then $\Phi'(x)=f(x)$ $(1 \leqslant x \leqslant 2)$. Hence, by 7.8E

$$\int_1^2 f(x)=\Phi(2)-\Phi(1),$$

which is equivalent to $(*)$.

7.8F. In elementary calculus, integrals are often evaluated by the method of substitution (or "change of variable"). For example, the integral

$$\int_0^2 \sqrt{4-x^2}\;dx$$

can be reduced to

$$4\int_0^{\pi/2} \cos^2 u\,du$$

by letting $x=2\sin u$ so that, formally, $dx=2\cos u\,du$.

In general, the integral

$$\int_A^B f(x)\,dx$$

is equal to

$$\int_a^b f[\varphi(u)]\varphi'(u)\,du$$

where $\varphi(a)=A, \varphi(b)=B$, provided that f and φ satisfy certain conditions which we will specify. However, we cannot *prove* this by saying "if $x=\varphi(u)$, then $dx=\varphi'(u)du$" since we have not defined dx, nor have we shown that the dx in dx/du has anything to do with the dx in

$$\int_A^B f(x)\,dx.$$

The fact that *formal* substitution of $x=\varphi(u), dx=\varphi'(u)du$ in

$$\int_A^B f(x)\,dx$$

yields an equal integral

$$\int_a^b f[\varphi(u)]\varphi'(u)\,du$$

shows one reason why the $\int_A^B f(x)\,dx$ notation is a good one. But, let us repeat, this substitution is not justified by anything we have as yet proved.

Before we come to the justification we should observe one thing. If φ is a continuous function on the closed bounded interval $[a,b]$, then the image of $[a,b]$ under φ is compact (by 6.6A) and connected (by 6.2D). Hence, the image $\varphi([a,b])$ of $[a,b]$ is itself a closed bounded interval.

7.8G. THEOREM. Let φ be a real-valued function on the closed bounded interval $[a,b]$ such that φ' is continuous on $[a,b]$. Let $A=\varphi(a)$, $B=\varphi(b)$. Then, if f is continuous on $\varphi([a,b])$, we have

$$\int_A^B f(x)\,dx=\int_a^b f[\varphi(u)]\varphi'(u)\,du.$$

PROOF: Since f is a continuous function on the closed bounded interval $\varphi([a,b])$ there is, by 7.8A, a function F such that

$$F'(x)=f(x) \qquad x\in\varphi([a,b]).$$

Let $G(u)=F[\varphi(u)]$ for $a\leqslant u\leqslant b$. Then, by the chain rule,

$$G'(u)=F'[\varphi(u)]\varphi'(u)=f[\varphi(u)]\varphi'(u) \qquad (a\leqslant u\leqslant b).$$

Using 7.8E we have

$$\begin{aligned}\int_a^b f[\varphi(u)]\varphi'(u)\,du&=\int_a^b G'(u)\,du=G(b)-G(a)\\&=F[\varphi(b)]-F[\varphi(a)]=F(B)-F(A)\\&=\int_A^B F'(x)\,dx=\int_A^B f(x)\,dx.\end{aligned}$$

This proves the theorem.

For example, let f be any continuous function on $[0,1]$. Then with $\varphi(u)=\sin u$ we have $\varphi(0)=0, \varphi(\pi/2)=1$. Since $\varphi'(u)=\cos u$ is continuous (believe this for now) on $[0,\pi/2]$ and since f is continuous on $\varphi([0,\pi/2])=[0,1]$ we have, by 7.8G,

$$\int_0^1 f(x)\,dx=\int_0^{\pi/2} f(\sin u)\cos u\,du.$$

Note also that $\varphi(\pi)=0$, $\varphi(9\pi/2)=1$. Since φ' is continuous on $[\pi,9\pi/2]$ we would also have

$$\int_0^1 f(x)\,dx=\int_\pi^{9\pi/2} f(\sin u)\cos u\,du,$$

provided that f is continuous on $[-1,1]$ which is the image of $[\pi,9\pi/2]$ under φ. Thus with $f(x)=\sqrt{x}$,

$$\int_0^1 \sqrt{x}\,dx=\int_0^{\pi/2}\sqrt{\sin u}\,\cos u\,du$$

is true. However

$$\int_0^1 \sqrt{x}\,dx=\int_\pi^{9\pi/2}\sqrt{\sin u}\,\cos u\,du$$

is nonsense, since $\sqrt{x}$ is not defined for $-1<x<0$. On the other hand, both statements

$$\int_0^1 x^2\,dx=\int_0^{\pi/2}\sin^2 u\cos u\,du$$

and

$$\int_0^1 x^2\,dx=\int_\pi^{9\pi/2}\sin^2 u\cos u\,du$$

are valid and follow from 7.8G.

Exercises 7.8

1. If

$$f(x)=\int_0^x \sqrt{t+t^6}\,dt \qquad (x>0),$$

find $f'(2)$.

2. State all theorems on integration used in the following computation:

$$\int_0^2 (3x^2-5)\,dx=3\int_0^2 x^2\,dx-5\int_0^2 dx=3\left(\frac{2^3}{3}-0\right)-5(2-0).$$

3. If f is continuous on $(-\infty,\infty)$ and if

$$F(x)=\int_0^x f(t)\,dt \qquad (-\infty<x<\infty),$$

prove that

$$F'(x)=f(x) \qquad (-\infty<x<\infty).$$

4. If f is continuous on $[a,b]$, if

$$f(x)>0 \qquad (a\leqslant x\leqslant b),$$

and if

$$F(x)=\int_a^x f(t)\,dt \qquad (a \leqslant x \leqslant b),$$

prove that F is strictly increasing on $[a,b]$.

5. If f is continuous on $[a,b]$, prove that there exists $c \in (a,b)$ such that

$$\int_a^b f(x)\,dx = f(c)(b-a).$$

$\left(\textit{Hint; Apply the law of the mean to } F(x)=\int_a^x f(t)\,dt.\right)$ This theorem is sometimes called the first mean-value theorem for integrals.

6. If f and g are continuous on $[a,b]$, and if $g(t) \geqslant 0$ $(a \leqslant t \leqslant b)$, prove there exists $c \in (a,b)$ such that

$$\int_a^b f(x)g(x)\,dx = f(c)\int_a^b g(x)\,dx.$$

This theorem is sometimes called the second mean-value theorem for integrals. The special case $g=1$ yields the preceding theorem.

7. If f' and g' are continuous on $[a,b]$, prove the familiar integration by parts formula:

$$\int_a^b f(x)g'(x)\,dx = f(b)g(b)-f(a)g(a)-\int_a^b f'(x)g(x)\,dx.$$

$\left(\textit{Hint: Apply 7.8E to } \int_a^b (fg)'(x)\,dx.\right)$

8. Prove that if $f \in \mathcal{R}[a,b]$ and if

$$F'(x)=f(x) \qquad (a \leqslant x \leqslant b),$$

then

$$F(b)-F(a)=\int_a^b f(x)\,dx.$$

(This generalizes 7.8E. *Hint*: Given $\epsilon > 0$ write

$$F(b)-F(a)=\sum_{k=1}^{n}\left[F(x_k)-F(x_{k-1})\right]$$

for a subdivision $\sigma=\{x_0,x_1,\ldots,x_n\}$ as in 7.2G. Then apply 7.7A.)

7.9 IMPROPER INTEGRALS

The definition of the integral

$$\int_a^\infty f(x)\,dx$$

does not follow from Section 7.2 since the interval $[a,\infty)$ is not bounded. Such an integral is called an improper integral. The theory of this type of integral resembles to a great extent the theory of infinite series. For this reason we do not give as many details as usual.

We can define

$$\int_a^\infty f(x)\,dx$$

as follows: If $f \in \mathcal{R}[a,s]$ for every $s > a$, then

$$\int_a^\infty f(x)\,dx$$

is defined to be the ordered pair $\langle f, F\rangle$ where

$$F(s) = \int_a^s f(x)\,dx \qquad (a \leqslant s < \infty).$$

The analogy with definition 3.1A is strong. The function f corresponds to the sequence $\{a_k\}_{k=1}^\infty$ while the "partial integral"

$$F(s) = \int_a^s f$$

corresponds to the partial sum $s_n = \sum_{k=1}^n a_k$.

We then say that $\int_a^\infty f$ is convergent to A if $\lim_{s\to\infty} F(s) = A$. In this case we write $\int_a^\infty f = A$. If $\int_a^\infty f$ does not converge, we say that $\int_a^\infty f$ is divergent.

The integral $\int_1^\infty (1/x^2)\,dx$ is thus convergent. For if

$$F(s) = \int_1^s \frac{1}{x^2}\,dx,$$

then $F(s) = 1 - 1/s$ and hence, $\lim_{s\to\infty} F(s) = 1$. Thus

$$\int_1^\infty \frac{1}{x^2}\,dx = 1.$$

On the other hand, the integral

$$\int_1^\infty \frac{1}{\sqrt{x}}\,dx$$

diverges since

$$F(s) = \int_1^s \frac{1}{\sqrt{x}}\,dx = 2(\sqrt{s} - 1).$$

If

$$\int_a^\infty f \quad \text{and} \quad \int_a^\infty g$$

both converge, then so does $\int_a^\infty (f \pm g)$, and

$$\int_a^\infty (f \pm g) = \int_a^\infty f \pm \int_a^\infty g.$$

This may be proved by the method of 3.1C. Similarly, if $\int_a^\infty f$ converges and $c \in R$, then $\int_a^\infty cf$ converges and

$$\int_a^\infty cf = c\int_a^\infty f.$$

If $f \in \mathcal{R}[a,s]$ for every $s > a$ and if $\int_a^\infty |f(x)|dx$ converges, we say that $\int_a^\infty f(x)dx$ converges absolutely. If

$$F(s) = \int_a^s |f(x)|dx$$

and if F is bounded (above) on $[a, \infty)$, then $\lim_{s \to \infty} F(s)$ exists by exercise 21 of Section 4.1 and hence $\int_a^\infty f(x)dx$ converges absolutely.

If $\int_a^\infty f(x)dx$ converges absolutely, if $g \in \mathcal{R}[a,s]$ for every $s > a$, and if

$$|g(x)| \leqslant |f(x)| \qquad (a \leqslant x < \infty),$$

then $\int_a^\infty g(x)dx$ converges absolutely. For

$$G(s) = \int_a^s |g(x)|dx \leqslant \int_a^s |f(x)|dx \leqslant \int_a^\infty |f(x)|dx.$$

Hence, G is bounded above on $[a, \infty)$ and the absolute convergence of $\int_a^\infty g(x)dx$ follows from the preceding paragraph.

This last result allows us to prove that if $\int_a^\infty |f(x)|dx$ converges absolutely, then $\int_a^\infty f(x)dx$ converges. For, since

$$0 \leqslant f(x) + |f(x)| \leqslant 2|f(x)| \qquad (a \leqslant x < \infty),$$

and since $\int_a^\infty 2|f(x)|dx$ converges (by assumption), it follows that

$$\int_a^\infty \{f(x) + |f(x)|\}dx$$

converges *absolutely*. Since $f(x) + |f(x)| \geqslant 0$ this means simply that

$$\int_a^\infty \{f(x) + |f(x)|\}dx$$

converges. But, since $\int_a^\infty |f(x)|dx$ converges, it follows (by subtraction) that

$$\int_a^\infty f(x)dx$$

itself converges, which is what we wished to show.

If $\int_a^\infty f(x)dx$ converges but does not converge absolutely, we say that $\int_a^\infty f(x)dx$ converges conditionally. A classical example of a conditionally convergent improper integral is

$$\int_\pi^\infty \frac{\sin x}{x} dx.$$

To show that $\int_\pi^\infty (\sin x)/x\,dx$ converges, we have for any $s > \pi$ (using integration by parts)

$$\int_\pi^s \frac{\sin x}{x} dx = \frac{1}{\pi} - \frac{\cos s}{s} + \int_\pi^s \frac{\cos x}{x^2} dx. \tag{1}$$

Now

$$\frac{|\cos x|}{x^2} \leqslant \frac{1}{x^2} \qquad (\pi \leqslant x < \infty).$$

Since $\int_\pi^\infty (1/x^2)dx$ converges (absolutely) it follows that $\int_\pi^\infty (\cos x)/x^2 dx$ converges absolutely and hence converges. Thus as $s \to \infty$ all terms on the right of (1) approach limits. Thus

$$\lim_{s\to\infty} \int_\pi^s \frac{\sin x}{x} dx$$

exists, which proves that $\int_\pi^\infty (\sin x)/x\, dx$ converges. [This also shows that

$$\int_\pi^\infty \frac{\sin x}{x} dx = \int_\pi^\infty \frac{\cos x}{x^2} dx + \frac{1}{\pi}$$

even though the second integral is absolutely convergent while the first integral (as we will now show) is not.]

Now we show that $\int_\pi^\infty (\sin x)/x\, dx$ does not converge absolutely. For any $N \in I$ we have

$$\int_\pi^{N\pi} \frac{|\sin x|}{x} dx = \sum_{n=1}^{N-1} \int_{n\pi}^{(n+1)\pi} \frac{|\sin x|}{x} dx \geqslant \sum_{n=1}^{N-1} \frac{1}{(n+1)\pi} \int_{n\pi}^{(n+1)\pi} |\sin x|\, dx$$

$$= \frac{1}{\pi} \sum_{n=1}^{N-1} \frac{1}{n+1} \int_0^\pi |\sin(u+n\pi)|\, du.$$

Now

$$\sin(u+n\pi) = \sin u \cos n\pi + \cos u \sin n\pi = \sin u \cos n\pi.$$

Since $\cos n\pi = \pm 1$ this shows that $|\sin(u+n\pi)| = |\sin u|$. Hence, if $0 \leqslant u \leqslant \pi$, then $|\sin(u+n\pi)| = \sin u$. Thus

$$\int_\pi^{N\pi} \frac{|\sin x|}{x} dx \geqslant \frac{1}{\pi} \sum_{n=1}^{N-1} \frac{1}{n+1} \int_0^\pi \sin u\, du = \frac{2}{\pi} \sum_{n=1}^{N-1} \frac{1}{n+1} = \frac{2}{\pi} \sum_{k=2}^{N} \frac{1}{k}. \tag{2}$$

By 3.2C the right side of (2) can be made as large as we please by taking N sufficiently large. This and (2) show that

$$\lim_{s\to\infty} \int_\pi^s \frac{|\sin x|}{x} dx$$

cannot exist. Hence, $\int_\pi^\infty (\sin x)/x\, dx$ does not converge absolutely.

The same method will prove the following important result.

7.9A. THEOREM. The improper integral

$$\int_1^\infty \frac{1}{x} dx$$

diverges.

PROOF: For any integer N we have

$$\int_1^N \frac{1}{x}dx = \sum_{n=1}^{N-1} \int_n^{n+1} \frac{1}{x}dx \geq \sum_{n=1}^{N-1} \frac{1}{n+1} \int_n^{n+1} 1\cdot dx = \sum_{n=1}^{N-1} \frac{1}{n+1} = \sum_{k=2}^{N} \frac{1}{k}. \tag{1}$$

Again, since as $N \to \infty$ the right side of (1) diverges to infinity, we see that

$$\lim_{s\to\infty} \int_1^s \frac{1}{x}dx$$

does not exist. This proves the theorem.

We have just used the divergence of an infinite series to establish the divergence of an improper integral. It is more usual to use an integral in the investigation of a series. This is known as the integral test for series.

7.9B. THEOREM. Let f be a nonincreasing function on $[1, \infty)$ such that $f(x) \geq 0$ $(1 \leq x < \infty)$. Then $\Sigma_{n=1}^{\infty} f(n)$ will converge if $\int_1^{\infty} f(x)dx$ converges, and $\Sigma_{n=1}^{\infty} f(n)$ will diverge if $\int_1^{\infty} f(x)dx$ diverges.

PROOF: For any $n \in I$ we have

$$f(n) \geq f(x) \geq f(n+1) \qquad (n \leq x \leq n+1)$$

since f is nonincreasing. Integrating from n to $n+1$ we then have

$$\int_n^{n+1} f(n)dx \geq \int_n^{n+1} f(x)dx \geq \int_n^{n+1} f(n+1)dx$$

or

$$f(n) \geq \int_n^{n+1} f(x)dx \geq f(n+1).$$

Thus for $N \in I$ we have

$$\sum_{n=1}^{N-1} f(n) \geq \int_1^N f(x)dx \geq \sum_{n=1}^{N-1} f(n+1) = \sum_{k=2}^{N} f(k). \tag{1}$$

If $\int_1^{\infty} f(x)dx$ converges to A, then, by (1),

$$\sum_{k=2}^{N} f(k) \leq \int_1^N f(x)dx \leq A.$$

The partial sums of $\Sigma_{k=2}^{\infty} f(k)$ are thus bounded above. By 3.2A, the series $\Sigma_{k=2}^{\infty} f(k)$ converges and hence, $\Sigma_{k=1}^{\infty} f(k)$ converges.

If $\int_1^{\infty} f(x)dx$ diverges, the divergence of $\Sigma_{n=1}^{\infty} f(n)$ may be established in similar fashion, using the left-hand inequality in (1). This will complete the proof.

(There is an obvious modification of 7.9B for functions on $[a, \infty)$.)

For example, we may reestablish the convergence of $\Sigma_{n=1}^{\infty}(1/n^2)$ by use of 7.9B. For, if $f(x) = 1/x^2$ $(1 \leq x < \infty)$, then

$$\sum_{n=1}^{\infty} \frac{1}{n^2} = \sum_{n=1}^{\infty} f(n).$$

Since

$$\int_1^{\infty} f(x)dx = \int_1^{\infty} \frac{1}{x^2}dx$$

is convergent, it follows that $\Sigma_{n=1}^{\infty} f(n)$ converges.

If $g(x) = 1/(x\log x)$, then g is nonegative and nondecreasing on $[3, \infty)$. Since $G'(x) = g(x)$ where $G(x) = \log\log x$, then

$$\int_3^s g(x)dx = \log\log s - \log\log 3,$$

and hence, $\int_3^{\infty} g(x)$ diverges. It follows from 7.9B that $\Sigma_{n=1}^{\infty}[1/(n\log n)]$ diverges. (All properties of $\log x$ used here are proved in the next chapter.)

7.9C. An integral of the form $\int_{-\infty}^{a} f(x)dx$ may be treated by the same methods as those used on integrals of the form $\int_b^{\infty} f(x)dx$. Thus we say that $\int_{-\infty}^{a} f(x)dx$ converges to A if

$$\lim_{s\to\infty} \int_{-s}^{a} f(x)dx = A.$$

The change of variable $x = -u$ will change a $\int_{-\infty}^{a}$ problem into a $\int_b^{\infty}$ problem. For example, consider

$$\int_{-\infty}^{-2} \frac{1}{1-x}dx.$$

For any $s > 2$ we have

$$\int_{-s}^{-2} \frac{1}{1-x}dx = \int_s^2 \frac{1}{1+u}(-1)du = \int_2^s \frac{1}{1+u}du.$$

Since $1/(1+u) \geqslant 1/2u$ for $2 \leqslant u < \infty$, it follows from 7.9A that

$$\lim_{s\to\infty} \int_2^s \frac{1}{1+u}du$$

does not exist. Hence, $\lim_{s\to\infty} \int_{-s}^{-2} 1/(1-x)dx$ does not exist, which proves that $\int_{-\infty}^{-2} 1/(1-x)dx$ does not converge. In this problem the divergence of

$$\int_{-\infty}^{-2} \frac{1}{1-x}dx$$

was deduced from the divergence of $\int_2^{\infty} 1/(1+u)du$.

Integrals of the type discussed in this section are sometimes called improper integrals of the first kind, in contrast to a second kind of improper integral, which we discuss in the next section.

Exercises 7.9

1. Which of the following integrals are convergent?

(a) $\int_1^{\infty} \frac{1}{x^{2/3}}dx.$ (b) $\int_1^{\infty} \frac{1}{x^{4/3}}dx.$

(c) $\int_1^{\infty} \frac{x}{1+x^2}\,dx$.

(d) $\int_1^{\infty} \frac{43x^2}{1+2x^2+12x^4}\,dx$.

(e) $\int_1^{\infty} x\cos x\,dx$.

(f) $\int_1^{\infty} \frac{1}{(1+x^3)^{1/2}}\,dx$.

(g) $\int_1^{\infty} \frac{1}{(1+x^3)^{1/3}}\,dx$.

2. Show that

$$\int_0^{\infty} \frac{x}{(1+x)^3}\,dx = \frac{1}{2}\int_0^{\infty} \frac{1}{(1+x)^2}\,dx.$$

3. Show that

$$\int_1^{\infty} \frac{x^{1/2}}{(1+x)^2}\,dx = \frac{1}{2} + \frac{\pi}{4}.$$

4. True or false? If f is continuous on $[1,\infty)$ and if $\int_1^{\infty} f(x)dx$ converges, then $\lim_{x\to\infty} f(x)=0$. (Answer: false.)

5. If $\int_1^{\infty} f(x)dx$ converges and if $\lim_{x\to\infty} f(x)=L$, prove that $L=0$.

6. Give an example of a continuous function f such that

$$f(x) \geqslant 0 \qquad (1 \leqslant x < \infty),$$

and such that

$$\sum_{n=1}^{\infty} f(n) \qquad \text{converges}$$

but

$$\int_1^{\infty} f(x)dx \qquad \text{diverges.}$$

7. Give an example of a continuous function f such that

$$f(x) \geqslant 0 \qquad (1 \leqslant x < \infty)$$

and such that

$$\int_1^{\infty} f(x)dx \qquad \text{converges}$$

but

$$\sum_{n=1}^{\infty} f(n) \qquad \text{diverges.}$$

8. Prove the following analogue of 3.7D. If $f(x) \geqslant 0$ $(1 \leqslant x < \infty)$, if f is nonincreasing on $[1,\infty)$, and if

$$\int_1^{\infty} f(x)dx \qquad \text{converges,}$$

then $\lim_{x\to\infty} xf(x)=0$. (*Hint*: Use 3.7D.)

9. Let f be a continuous function on $[a,\infty)$ such that, if

$$F(x)=\int_a^x f(t)\,dt \qquad (a\leqslant x<\infty),$$

then F is bounded on $[a,\infty)$. Let g be a function on $[a,\infty)$ such that g' is continuous on $[a,\infty), g'(t)\leqslant 0$ for $a\leqslant t<\infty$, and such that $\lim_{t\to\infty} g(t)=0$. Prove that

$$\int_a^\infty f(t)\,g(t)\,dt \qquad \text{converges.}$$

(*Hint*: Integrate by parts.) Compare with 3.8C.

10. Use the preceding exercise to show that

$$\int_3^\infty \frac{\sin t}{\log t}\,dt \qquad \text{converges.}$$

11. Show that $\int_1^\infty \cos u^2\,du$ is convergent.

7.10 IMPROPER INTEGRALS (CONTINUED)

The definition of

$$\int_0^1 \frac{1}{\sqrt{x}}\,dx$$

does not follow from Section 7.2 because the function f defined by

$$f(x)=\frac{1}{\sqrt{x}} \qquad (0<x\leqslant 1)$$

is not bounded. Note, however, that f is bounded (and continuous) on $[\epsilon,1]$ for every $\epsilon>0$. This suggests treating

$$\int_0^1 \frac{1}{\sqrt{x}}\,dx \quad \text{as the} \quad \lim_{\epsilon\to 0+}\int_\epsilon^1 \frac{1}{\sqrt{x}}\,dx$$

(which turns out to be equal to 2).

In general, if $f\in\mathcal{R}[a+\epsilon,b]$ for all ϵ such that $0<\epsilon<b-a$, but $f\not\in\mathcal{R}[a,b]$, we define $\int_a^b f(x)\,dx$ as the ordered pair $\langle f,F\rangle$ where

$$F(\epsilon)=\int_{a+\epsilon}^b f(x)\,dx \qquad (0<\epsilon<b-a).$$

We say that $\int_a^b f$ converges to A if $\lim_{\epsilon\to 0+}F(\epsilon)=A$. We say that $\int_a^b f$ diverges if $\int_a^b f$ does not converge. The integral $\int_a^b f$ is called an improper integral of the second kind.

Thus

$$\int_0^1 \frac{1}{\sqrt{x}}\,dx \qquad \text{converges}$$

while

$$\int_1^3 \frac{1}{(x-1)^2}\,dx \qquad \text{diverges.}$$

Properties such as absolute convergence and conditional convergence for improper integrals of the second kind are defined in the same way as for improper integrals of the first kind, and results on these properties carry over without difficulty to improper integrals of the second kind. For example, if $\int_a^b f(x)dx$ is an absolutely convergent improper integral, and if $|g(x)| \leqslant |f(x)|$ $(a<x\leqslant b)$, then $\int_a^b g(x)dx$ converges absolutely. Thus

$$\int_0^1 \frac{\sin x}{\sqrt{x}}\,dx \qquad \text{converges absolutely,}$$

since

$$\frac{|\sin x|}{\sqrt{x}} \leqslant \frac{1}{\sqrt{x}} \qquad (0<x\leqslant 1).$$

It is often useful to convert an improper integral of the second kind by a change of variable into an improper integral of the first kind. For example:

7.10A. THEOREM. The improper integral

$$\int_0^1 \frac{1}{x}\,dx$$

diverges.

PROOF: For $0<\epsilon<1$ let

$$F(\epsilon)=\int_\epsilon^1 \frac{1}{x}\,dx.$$

If $\varphi(u)=1/u$ $(1\leqslant u\leqslant 1/\epsilon)$, then $\varphi'(u)=-1/u^2du$ $(1\leqslant u\leqslant 1/\epsilon)$. Hence, by 7.8G,

$$F(\epsilon)=\int_{1/\epsilon}^1 u\left(\frac{-1}{u^2}\right)du=\int_1^{1/\epsilon}\frac{1}{u}\,du.$$

By 7.9A,

$$\lim_{\epsilon\to 0+}\int_1^{1/\epsilon}\frac{1}{u}\,du$$

does not exist. Hence, $\lim_{\epsilon\to 0+} F(\epsilon)$ does not exist, and the theorem follows.

Thus far in this section we have treated only integrals $\int_a^b f$ where f is "bad" near a. Corresponding theory holds in the case where f is "bad" near b. Thus if

$$f\in\mathfrak{R}[a,b-\epsilon]$$

for all ϵ such that $0<\epsilon<b-a$, and if

$$\lim_{\epsilon \to 0+} \int_a^{b-\epsilon} f(x)dx$$

exists, we again say that $\int_a^b f(x)dx$ is a convergent improper integral.

7.10B. THEOREM. The improper integral

$$\int_0^1 \frac{1}{\sqrt{1-x^2}} dx$$

is convergent.

PROOF: The integral is improper since $1/\sqrt{1-x^2}$ is not bounded on $[0,1]$. We first show that the improper integral

$$\int_0^1 \frac{1}{\sqrt{1-x}} dx$$

is convergent (and hence, absolutely convergent, since $\sqrt{1-x} \geqslant 0$). If $0<\epsilon<1$, we have

$$\int_0^{1-\epsilon} \frac{1}{\sqrt{1-x}} dx = 2-2\sqrt{\epsilon}$$

and so

$$\lim_{\epsilon \to 0+} \int_0^{1-\epsilon} \frac{1}{\sqrt{1-x}} dx = 2.$$

Hence, $\int_0^1 1/\sqrt{1-x}\, dx$ converges absolutely.

But, for $0 \leqslant x < 1$,

$$\frac{1}{\sqrt{1-x^2}} = \frac{1}{\sqrt{1+x}} \cdot \frac{1}{\sqrt{1-x}} \leqslant \frac{1}{\sqrt{1-x}}.$$

Hence, $\int_0^1 1/\sqrt{1-x^2}\, dx$ is absolutely convergent, which implies the desired conclusion.

7.10C. The integral

$$\int_0^\infty \frac{1}{x^2+\sqrt{x}} dx$$

does not fall into any one of the categories we have thus far described since it is an integral over $(0,\infty)$ *and* $1/(x^2+\sqrt{x}\,)$ is not bounded for x near 0. However, we will agree to call $\int_0^\infty 1/(x^2+\sqrt{x}\,)dx$ a *convergent* improper integral since we can break it up into

$$J_1 = \int_0^1 \frac{1}{x^2+\sqrt{x}} dx \quad \text{and} \quad J_2 = \int_1^\infty \frac{1}{x^2+\sqrt{x}} dx.$$

Now J_1 is a convergent integral of the second kind [since $1/(x^2+\sqrt{x}\,) \leqslant 1/\sqrt{x}$ for

$0 < x \leqslant 1$] and J_2 is a convergent improper integral of the first kind [since $1/(x^2 + \sqrt{x}\,) \leqslant 1/x^2$ for $1 \leqslant x < \infty$].

In general if an integral J can be broken up in this way into two or more improper integrals $J_1, \ldots, J_n$ of the first or second kinds, and if *each* $J_k (k = 1, \ldots, n)$ is convergent, we will say that J is a convergent improper integral. However, if one or more of the J_k is divergent, we will say that J is a divergent improper integral.

Thus $\int_0^\infty (1/x^2)dx$ is a divergent improper integral since

$$\int_0^1 \frac{1}{x^2} dx \quad \text{and} \quad \int_1^\infty \frac{1}{x^2} dx$$

are improper integrals (of the second and first kinds, respectively) one of which is divergent.

Similarly,

$$\int_{-\infty}^{\infty} \frac{1+x}{1+x^2} dx$$

is a divergent improper integral since both

$$\int_0^\infty \frac{1+x}{1+x^2} dx \quad \text{and} \quad \int_{-\infty}^0 \frac{1+x}{1+x^2} dx$$

are divergent improper integrals of the second kind. (Note $(1+x)/(1+x^2) \geqslant 1/x$ for $1 \leqslant x \leqslant \infty$.)

7.10D. As we have just observed, the integral

$$\int_{-\infty}^{\infty} \frac{1+x}{1+x^2} dx$$

diverges since

$$\int_0^s \frac{1+x}{1+x^2} dx \tag{1}$$

does not approach a limit as $s \to \infty$. Similarly,

$$\int_{-s}^0 \frac{1+x}{1+x^2} dx \tag{2}$$

does not approach a limit as $s \to \infty$. However, the sum of (1) and (2) does approach a limit as $s \to \infty$. For the sum of (1) and (2) is

$$\int_{-s}^0 \frac{1+x}{1+x^2} dx + \int_0^s \frac{1+x}{1+x^2} dx = \int_0^s \frac{1-u}{1+u^2} du + \int_0^s \frac{1+u}{1+u^2} du$$

$$= 2\int_0^s \frac{1}{1+u^2} du,$$

and

$$\lim_{s \to \infty} 2\int_0^s \frac{1}{1+u^2} du$$

does exist. The sum of (1) and (2) may also be written $\int_{-s}^{s}(1+x)/(1+x^2)dx$. Hence, we have shown that

$$\lim_{s\to\infty}\int_{-s}^{s}\frac{1+x}{1+x^2}dx$$

exists, even though $\int_{-\infty}^{\infty}(1+x)/(1+x^2)dx$ diverges. We call

$$\lim_{s\to\infty}\int_{-s}^{s}\frac{1+x}{1+x^2}dx$$

the Cauchy principal value of $\int_{-\infty}^{\infty}(1+x)/(1+x^2)dx$.

In general, the Cauchy principal value of $\int_{-\infty}^{\infty}f(x)dx$ is defined to be

$$\lim_{s\to\infty}\int_{-s}^{s}f(x)dx,$$

if this limit exists. We denote Cauchy principal value by C.P.V. Thus $\int_{-\infty}^{\infty}x\,dx$ diverges but

$$\text{C.P.V.}\int_{-\infty}^{\infty}x\,dx=\lim_{s\to\infty}\int_{-s}^{s}x\,dx=\lim_{s\to\infty}0=0.$$

In an exercise you will be asked to show that if $\int_{-\infty}^{\infty}f(x)dx$ converges to A, then C.P.V.$\int_{-\infty}^{\infty}f(x)dx=A$ also. As we have seen, however, the Cauchy principal value of $\int_{-\infty}^{\infty}f(x)dx$ may exist even if the interval diverges.

Exercises 7.10

1. Which of the following improper integrals are convergent?

(a) $\int_0^1 \frac{1}{x^{2/3}}dx$.

(b) $\int_0^1 \frac{1}{x^{4/3}}dx$.

(c) $\int_0^2 \frac{x}{(16-x^4)^{1/3}}dx$.

(d) $\int_{a-1}^{a+1} \frac{1}{(x-a)^{1/3}}dx$.

(e) $\int_0^1 \frac{\log(1/x)}{\sqrt{x}}dx$.

(f) $\int_0^1 \frac{\sin x}{x^{3/2}}dx$.

(g) $\int_0^{\infty} \frac{t^{-1/2}}{1+t}\text{dt}$.

2. Prove that if $s<1$, then

$$\int_a^b (x-a)^{-s}dx=\frac{(b-a)^{1-s}}{1-s}.$$

Prove that if $s\geqslant 1$, then the integral diverges.

3. Prove that

$$\int_0^{\infty} \frac{x^{s-1}}{1+x}\,dx$$

is convergent if and only if $0<s<1$.

4. Let

$$F(x)=\int_0^x \frac{\sin t}{t^{3/2}}\,dt \qquad (0<x<\infty).$$

Prove that the maximum value of $F(x)$ is attained when $x=\pi$.

5. For which of the following integrals does the Cauchy principal value exist?

(a) $\int_{-\infty}^{\infty} \sin t\,dt.$

(b) $\int_{-\infty}^{\infty} |\sin t|\,dt.$

(c) $\int_{-\infty}^{\infty} \frac{1}{1+t^2}\,dt.$

Do any of the above integrals converge?

6. If f is continuous on $(-\infty,\infty)$ and if $\int_{-\infty}^{\infty} f(x)\,dx$ converges to A, prove that

$$\text{C.P.V.} \int_{-\infty}^{\infty} f(x)\,dx=A.$$

7. If f is continuous on $[0,1]$, prove that

$$\int_0^1 \frac{f(x)}{\sqrt{1-x^2}}\,dx$$

is convergent.

Then prove that

$$\int_0^1 \frac{f(x)}{\sqrt{1-x^2}}\,dx=\int_0^{\pi/2} f(\sin u)\,du.$$

Is the integral on the right improper?

8. If f is "well behaved" on $[a,b]$ except near the point $c\in(a,b)$, we define the Cauchy principal value of $\int_a^b f$ as

$$\lim_{\epsilon\to 0+}\left(\int_a^{c-\epsilon} f+\int_{c+\epsilon}^{b} f\right).$$

(a) Show that $\text{C.P.V.} \int_{-1}^{1} \frac{1}{x}\,dx=0.$

(b) Show that $\text{C.P.V.} \int_{-1}^{1} \frac{1}{|x|}\,dx$ does not exist.

8

THE ELEMENTARY FUNCTIONS. TAYLOR SERIES

We wish to define two major families of elementary functions. The first consists of the exponential function, the log function, the hyperbolic functions sinh, cosh, tanh, and so on, and the inverse hyperbolic functions $\sinh^{-1}$ and $\tanh^{-1}$. The second family consists of the trigonometric functions sin, cos, tan, and so on, and the inverse functions $\sin^{-1}$ and $\tan^{-1}$.

Our method of defining all these functions will probably be new to you (although, of course, they turn out to be the same functions you worked with in calculus). The method, although mercilessly analytic, has several advantages. One advantage is that the definition of the functions in the second family parallels the definition of the functions in the first. Second, we do not have to rely on any unverified theorems of geometry in our definition of the trigonometric functions.

8.1 HYPERBOLIC FUNCTIONS

Let

$$U(x)=\int_0^x \frac{1}{\sqrt{1+t^2}}\,dt \qquad (-\infty<x<\infty).$$

We first note that

$$U(-x)=-U(x) \qquad (-\infty<x<\infty). \tag{1}$$

For

$$U(-x)=\int_0^{-x} \frac{1}{\sqrt{1+t^2}}\,dt=\int_0^{x} \frac{1}{\sqrt{1+(-u)^2}}(-1)\,du=-U(x).$$

Thus since $U(x)\geqslant 0$ if $x\geqslant 0$, we have $U(x)\leqslant 0$ if $x\leqslant 0$ and, for example, $U(-4)=-U(4)$. Next, since

$$\frac{1}{\sqrt{1+t^2}}\geqslant\frac{1}{\sqrt{2t^2}}=\frac{1}{\sqrt{2}\,t}$$

for $1 \leqslant t < \infty$, it follows from 7.9A that $\int_1^{\infty} 1/\sqrt{1+t^2}\,dt$ is divergent. Hence, $\int_0^{\infty} 1/\sqrt{1+t^2}\,dt$ diverges, which shows that U is not bounded above. Equation (1) then shows that U is not bounded below, either.

By 7.8A,

$$U'(x)=\frac{1}{\sqrt{1+x^2}} \qquad (-\infty<x<\infty). \tag{2}$$

Hence, by 7.5B, U is continuous on $(-\infty,\infty)$. Since U is bounded neither above nor below, theorem 6.2D implies that the range of U must be $(-\infty,\infty)$. Also, (2) and 7.7B show that U is 1–1 on $(-\infty,\infty)$.

Thus U is a 1–1 continuous function from R^1 onto R^1. If we denote the inverse function of U by S, then S is a 1–1 function from R^1 onto R^1. Moreover, since by 6.7B the restriction of U to any compact interval is a homeomorphism on that interval, it follows that S is continuous at each point of R^1 (see exercise 4 of Section 6.7).

If $S(a)=b$, then $U(b)=a$. By 7.5E and (2),

$$S'(a)=\frac{1}{U'(b)}=\sqrt{1+b^2}\ .$$

Since $b=S(a)$ we thus have

$$S'(a)=\sqrt{1+[S(a)]^2} \qquad (-\infty<a<\infty). \tag{3}$$

Hence, by differentiation,

$$S''(a)=\frac{S(a)S'(a)}{\sqrt{1+[S(a)]^2}}\ . \tag{4}$$

But (3) and (4) then imply

$$S''(a)=S(a) \qquad (-\infty<a<\infty). \tag{5}$$

That is, S is its own second derivative. Finally define the function C by*

$$C(x)=\sqrt{1+[S(x)]^2} \qquad (-\infty<x<\infty). \tag{6}$$

Then, by (3),

$$C(x)=S'(x) \qquad (-\infty<x<\infty). \tag{7}$$

Hence, by (5)

$$C'(x)=S(x) \qquad (-\infty<x<\infty). \tag{8}$$

As you may have guessed by now, U usually goes by the name of $\sinh^{-1}$, $S(x)$ is usually denoted $\sinh x$ (the hyperbolic sine of x) and $C(x)$ is usually denoted $\cosh x$ (the hyperbolic cosine of x). Thus (7) and (8) state the familiar fact that sinh and cosh are derivatives of one another. From (6) we deduce the identity

$$\cosh^2 x-\sinh^2 x=1,$$

which in turn implies $\cosh x \geqslant 1$.

Since $S=U^{-1}$ and

$$U(0)=\int_0^0 \frac{1}{\sqrt{1+t^2}}\,dt=0,$$

* Remember $\sqrt{\ }$ means nonnegative square root.

we have $S(0)=0$. That is, $\sinh 0=0$. From (6) we see then that $\cosh 0=1$.

Also, from (1) it follows that if $U(x)=y$, then $U(-x)=-y$. Hence, $S(y)=x=-(-x)=-S(-y)$. That is,

$$\sinh(-y)=-\sinh y \qquad (-\infty<y<\infty). \tag{9}$$

From (6) and (9) we conclude

$$\cosh(-y)=\cosh y \qquad (-\infty<y<\infty). \tag{10}$$

We have thus defined $\sinh^{-1}$, sinh, cosh. The other hyperbolic functions can now be defined in terms of these. For example, define $\tanh x$ as $(\sinh x)/(\cosh x)$.

Exercises 8.1

1. Use only the results from Section 8.1 to work the following exercises.
   (a) Prove that $\tanh(-x)=-\tanh x (-\infty<x<\infty)$.
   (b) Prove that $C''(x)=C(x)$ $(-\infty<x<\infty)$.
   (c) Show that S is strictly increasing on $(-\infty,\infty)$.
   (d) Show that the graph of S is concave up for $0\leqslant x<\infty$ and concave down for $-\infty<x\leqslant 0$.
   (e) Sketch the graph of S.
2. Prove that tanh is strictly increasing on $(-\infty,\infty)$.

8.2 THE EXPONENTIAL FUNCTION

We define the function E by

$$E(x)=C(x)+S(x) \qquad (-\infty<x<\infty). \tag{11}^*$$

Hence, E is continuous on $(-\infty,\infty)$ and $E(0)=C(0)+S(0)=\cosh 0+\sinh 0=1+0=1$. Also, $E(-x)=C(-x)+S(-x)$ and so, by (9) and (10),

$$E(-x)=C(x)-S(x) \qquad (-\infty<x<\infty). \tag{12}$$

From (11) and (12) we have $E(x)E(-x)=[C(x)]^2-[S(x)]^2=\cosh^2 x-\sinh^2 x=1$. Hence,

$$E(x)E(-x)=1 \qquad (-\infty<x<\infty). \tag{13}$$

This shows that $E(x)$ is never 0. Hence, by 6.2D, either E takes on only positive values or E takes on only negative values. The latter alternative is not possible since $E(0)=1$. Hence,

$$E(x)>0 \qquad (-\infty<x<\infty). \tag{14}$$

Since $C(x)\geqslant 1$ for all x we see from (11) that $E(x)>S(x)$ for all x. Thus since S is not bounded above, neither is E. That is, E takes on arbitrarily large positive values. By (13) E must then take on arbitrarily small positive values. [For, by (13), if $E(x)=M$ (large), then $E(-x)=1/M$ (small).] Theorem 6.2D and (14) then show that the range of E is precisely $(0,\infty)$.

From (11) we have $E'(x)=C'(x)+S'(x)=S(x)+C(x)$. Thus

$$E'(x)=E(x) \qquad (-\infty<x<\infty). \tag{15}$$

From (14) we thus see that $E'(x)>0$ for all x and hence, by 7.7B, that E is 1–1 on

* In these sections only we continue the numbering of equations from one section into the next.

$(-\infty, \infty)$. It follows (exercise 4 of Section 6.7) that E is a homeomorphism of $(-\infty, \infty)$ onto $(0, \infty)$.

Finally, let us prove that

$$E(x+a)=E(x)E(a) \qquad [a, x \in (-\infty, \infty)]. \tag{16}$$

Fix a and let $F(x)=E(x+a)E(-x)$. Using 7.5C and 7.5D we have, on differentiation with respect to x,

$$F'(x)=E(x+a)[-E'(-x)]+E'(x+a)E(-x) \qquad (-\infty<x<\infty).$$

Using (15) we see that $F'(x)=0$ for all x. Hence, by 7.8C, F is constant. That is, $F(x)=F(0)=E(a)$ for all x. Thus

$$E(x+a)E(-x)=E(a) \qquad (-\infty<x<\infty).$$

This and (13) prove (16).

It is, of course, customary to denote $E(x)$ by e^x. Thus $e^0=E(0)=1$. Also, (15) states the familiar fact that e^x is its own derivative, while (16) states the fundamental rule of exponents $e^{x+a}=e^x \cdot e^a$.

[To justify the use of the letter e here, we must show that e^1 as defined here is equal to the e mentioned in (2) of 2.6C. We do this in the next section.]

Exercises 8.2

Use only results from Section 8.1 and 8.2 to work the following exercises.

1. (a) Prove $\lim_{x \to -\infty} e^x = 0$.
   (b) Show that the graph of $y=e^x$ is concave up for all x.
   (c) Sketch
$$y=e^x \qquad (-\infty<x<\infty).$$
2. (a) Prove that $e^{2x}=(e^x)^2 (-\infty<x<\infty)$.
   (b) Prove that $e^x/e^y=e^{x-y} (-\infty<x,y<\infty)$.
3. Show that
$$C(x)=\frac{E(x)+E(-x)}{2}.$$
   [That is, $\cosh x=(e^x+e^{-x})/2$.] Also, show that
$$\sinh x=\frac{e^x-e^{-x}}{2}.$$
4. Prove the following identities.
   (a) $\sinh(x+y)=\sinh x \cosh y+\cosh x \sinh y$.
   (b) $\cosh(x+y)=\cosh x \cosh y+\sinh x \sinh y$.
   (c) $\sinh 2x=2 \sinh x \cosh x$.
   (d) $\cosh 2x=\cosh^2 x+\sinh^2 x$.
   (e) $2[\sinh(x/2)]^2=\cosh x-1$.
5. (a) Show that
$$\tanh x=\frac{e^x-e^{-x}}{e^x+e^{-x}} \qquad (-\infty<x<\infty).$$
   (b) What is the range of tanh?
   (c) If w is the inverse function for tanh, show that
$$w'(x)=\frac{1}{1-x^2}$$
   for all x in the domain of w.

8.3 THE LOGARITHMIC FUNCTION. DEFINITION OF x^a

In the last section we found that E is a homeomorphism of $(-\infty,\infty)$ onto $(0,\infty)$. We now define L to be the inverse function for E. Thus L is a homeomorphism of $(0,\infty)$ onto $(-\infty,\infty)$ and

$$L[E(x)]=x \qquad (-\infty<x<\infty); \qquad E[L(x)]=x \qquad (0<x<\infty). \tag{17}$$

Since $E(0)=1$ we have $L(1)=0$. If $L(x)=a$ and $L(y)=b$, then $x=E(a), y=E(b)$. Hence, by (16), $xy=E(a)E(b)=E(a+b)$. Thus $L(xy)=a+b=L(x)+L(y)$. We have shown

$$L(xy)=L(x)+L(y) \qquad [x,y\in(0,\infty)]. \tag{18}$$

For $y=x$ this shows $L(x^2)=2L(x)$. Similarly, $L(x^3)=L(x^2)+L(x)=3L(x)$. By induction we may show that

$$L(x^n)=nL(x) \qquad (0<x<\infty; n\in I). \tag{19}$$

If $y=1/x$, then from (18) we have $0=L(1)=L(x)+L(1/x)$. Thus

$$L\left(\frac{1}{x}\right)=-L(x) \qquad (0<x<\infty). \tag{20}$$

Hence,

$$L\left(\frac{x}{z}\right)=L\left(x\cdot\frac{1}{z}\right)=L(x)+L\left(\frac{1}{z}\right)=L(x)-L(z).$$

That is,

$$L\left(\frac{x}{z}\right)=L(x)-L(z) \qquad [x,z\in(0,\infty)]. \tag{21}$$

If $L(x)=y$, then $E(y)=x$. By 7.5E we have $L'(x)=1/E'(y)$. But from (15) we have $1/E'(y)=1/E(y)=1/x$. Thus

$$L'(x)=\frac{1}{x} \qquad (0<x<\infty). \tag{22}$$

Hence, by 7.8E,

$$L(x)-L(1)=\int_1^x L'(t)\,dt=\int_1^x \frac{1}{t}\,dt.$$

Since $L(1)=0$ this shows

$$L(x)=\int_1^x \frac{1}{t}\,dt \qquad (0<x<\infty). \tag{23}$$

From (22) we have $L'(1)=1$. That is,

$$1=\lim_{h\to 0}\frac{L(1+h)-L(1)}{h}=\lim_{h\to 0}\frac{L(1+h)}{h}.$$

This implies

$$1=\lim_{n\to\infty}\frac{L(1+1/n)}{1/n}=\lim_{n\to\infty} nL\left(1+\frac{1}{n}\right).$$

By (19) we then have $\lim_{n\to\infty} L[(1+1/n)^n]=1$. That is, the sequence

$$\left\{L\left[(1+1/n)^n\right]\right\}_{n=1}^{\infty}$$

converges to 1. Since E is continuous at 1 this shows that

$$\lim_{n\to\infty} E\left\{L\left[\left(1+\frac{1}{n}\right)^n\right]\right\}=E(1)$$

or, by (17),

$$\lim_{n\to\infty}\left(1+\frac{1}{n}\right)^n=E(1). \tag{24}$$

The left side of (24) is e as defined in (2) of 2.6C, while the right side is e^1 as defined in Section 8.2. This shows we have been consistent in our use of the e notation.

When we use the customary notation $L(x)=\log x$, (17) reads $\log e^x=x$ $(-\infty<x<\infty)$ and $e^{\log x}=x$ $(0<x<\infty)$; (18) reads $\log xy=\log x+\log y$; (20) reads $\log(1/x)=-\log x$, and so on.

Note that we have not proved that $\log x^a=a\log x$ except when a is a positive integer [equation (19)]. In fact we have not as yet defined x^a for irrational a. (As we mentioned in the introduction, we have always assumed that the definition of x^a for rational a is known from algebra.)

If x is any positive number and a is any real number, we now *define* x^a as

$$x^a=e^{a\log x}. \tag{25}$$

Thus $\log x^a=\log e^{a\log x}$ and so, by (17),

$$\log x^a=a\log x.$$

Also note that if $a=2$ we have

$$x^2=e^{2\log x}=e^{\log x+\log x}=e^{\log x}\cdot e^{\log x}=x\cdot x.$$

Thus even according to the definition (25), x^2 still means x times x. It may be shown similarly that if a is any rational number, then x^a as defined in (25) means the same as it did in high-school algebra.

Exercises 8.3

Use only results from Sections 8.1, 8.2, and 8.3 to work the following exercises.

1. (a) Show that L is strictly increasing on $(0,\infty)$.
   (b) Show that the graph of
   $$y=L(x) \qquad (0<x<\infty)$$
   is concave down.
   (c) Sketch the curve
   $$y=L(x) \qquad (0<x<\infty).$$
2. (a) If $a\neq 0$, prove that
   $$x^{-a}=\frac{1}{x^a} \qquad (0<x<\infty).$$
   (b) If $a\neq 0$, prove that
   $$\left(\frac{x}{y}\right)^a=\frac{x^a}{y^a}$$
   for all positive x and y.
3. (a) If $a\neq 0$ and
   $$f(x)=x^a \qquad (0<x<\infty),$$
   prove that
   $$f'(x)=ax^{a-1} \qquad (0<x<\infty).$$

(b) If $a>0$ and

$$g(x)=a^x \qquad (-\infty<x<\infty),$$

prove that

$$g'(x)=a^x\log a \qquad (-\infty<x<\infty).$$

4. Prove that $\lim_{x\to\infty}(\log x)/x=0$. [*Hint*: First show that

$$\frac{\log x}{x}\leqslant\frac{1}{\sqrt{x}}\int_1^x\frac{dt}{t^{3/2}} \qquad (x\geqslant 1.)]$$

5. If $a>0$, prove that

$$\lim_{x\to\infty}\frac{\log x}{x^a}=0.$$

6. Prove that for any $M>0$

$$\lim_{x\to\infty}\frac{x^M}{e^x}=0.$$

8.4 THE TRIGONOMETRIC FUNCTIONS

Before we begin our development of the trigonometric functions we would like to point out that, in most elementary calculus texts, the proof that the derivative of the sine function is the cosine function depends on the fact that $\lim_{x\to 0}(\sin x)/x=1$. This limit is usually established by a geometric argument that involves the formula for the area of a sector of a circle. This formula is derived from the formula for the total area of a circle. Hence, to find the derivative of the sine function by this approach we must know the formula for the area of a circle. However, one of the chief early applications of the trigonometric functions in a calculus course is the evaluation of

$$4\int_0^r\sqrt{r^2-x^2}\,dx$$

—the integral that represents the area of a circle. Thus in this overall approach, the "proof" that the area of a circle is πr^2 using trigonometric substitution is not valid, since it involves knowing that the derivative of the sine is the cosine, while this in turn involves knowing that the area of a circle is πr^2.

In our development of the trigonometric functions we assume no formulae from geometry. In fact, along with everything else, we define π itself (without geometry).

As we saw in 7.10B the improper integral

$$\int_0^1\frac{1}{\sqrt{1-t^2}}\,dt$$

converges. We define the real number π by

$$\frac{\pi}{2}=\int_0^1\frac{1}{\sqrt{1-t^2}}\,dt.$$

Thus if we define the function u by

$$u(x)=\int_0^x\frac{1}{\sqrt{1-t^2}}\,dt \qquad (-1\leqslant x\leqslant 1),$$

then $u(1)=\pi/2, u(-1)=-\pi/2$, and u is continuous on $[-1,1]$. Since $1/\sqrt{1-t^2}$ is continuous for $-1<t<1$, 7.8B implies

$$u'(x)=\frac{1}{\sqrt{1-x^2}} \qquad (-1<x<1). \tag{26}$$

Thus $u'(x)>0$ $(-1<x<1)$ and hence, by 7.7B, u is a $1-1$ function on $[-1,1]$. Thus by 6.7B, u is a homeomorphism of $[-1,1]$ onto $[-\pi/2,\pi/2]$. It is easy to show that $u(x)=-u(-x)$ for $-1\leqslant x\leqslant 1$. (Now draw a rough graph of u using information on u' and u''.)

Let s denote the inverse function for u. Then s is a homeomorphism of $[-\pi/2,\pi/2]$ onto $[-1,1]$ with $s(-\pi/2)=-1, s(\pi/2)=1$. Since $u(0)=0$ we also have $s(0)=0$. From the fact that $u(-y)=-u(y)$ for $-1\leqslant y\leqslant 1$ it follows that $s(-x)=-s(x)$ for $-\pi/2\leqslant x \leqslant \pi/2$. If $-\pi/2<x<\pi/2$ and $s(x)=y$, then $-1<y<1$ and $u(y)=x$. By 7.5E and (26) we have

$$s'(x)=\frac{1}{u'(y)}=\sqrt{1-y^2}=\sqrt{1-[s(x)]^2}\ .$$

That is,

$$s'(x)=\sqrt{1-[s(x)]^2} \qquad \left(-\frac{\pi}{2}<x<\frac{\pi}{2}\right). \tag{27}$$

This shows that $\lim_{x\to(\pi/2)-}s'(x)=\lim_{x\to(\pi/2)-}\sqrt{1-[s(x)]^2}=\sqrt{1-[s(\pi/2)]^2}=0$.Since

$$\frac{s(\pi/2)-s(x)}{\pi/2-x}=s'(c)$$

for some c such that $x<c<\pi/2$, it then follows that

$$s'(\pi/2)=\lim_{x\to(\pi/2)-}\frac{s(\pi/2)-s(x)}{\pi/2-x}=\lim_{x\to(\pi/2)-}s'(c)=0.$$

Thus $s'(\pi/2)=0$ [where $s'(\pi/2)$ denotes the left-hand derivative of s at $\pi/2$].* Similarly, it may be shown that $s'(-\pi/2)=0$, where this time the right-hand derivative is involved. This says that (27) holds also for $x=\pi/2$ and $x=-\pi/2$. That is,

$$s'(x)=\sqrt{1-[s(x)]^2} \qquad \left(-\frac{\pi}{2}\leqslant x\leqslant\frac{\pi}{2}\right). \tag{27a}$$

We now extend s by defining $s(y)$ for $\pi/2<y\leqslant 3\pi/2$ by the equation

$$s(x+\pi)=-s(x) \qquad \left(-\frac{\pi}{2}<x\leqslant\frac{\pi}{2}\right).$$

(Note that if $-\pi/2<x\leqslant\pi/2$, then $+\pi/2<x+\pi\leqslant 3\pi/2$.)
Then

$$s'(x+\pi)=-s'(x) \qquad \left(-\frac{\pi}{2}<x\leqslant\frac{\pi}{2}\right).$$

The right-hand derivative $s'(\pi/2)$ will thus be equal to negative of the right derivative $s'(-\pi/2)$, which we have shown to be equal to 0. Since the left-hand derivative of s at $\pi/2$ is also equal to 0, this shows that s, as we have extended it to $[-\pi/2,3\pi/2]$, has a (two-sided) derivative at $\pi/2$ and $s'(\pi/2)=0$.

We can extend s to all of $(-\infty,\infty)$ by simply requiring that

$$s(x+\pi)=-s(x) \qquad (-\infty<x<\infty). \tag{28}$$

* Here we could have used exercise 5 of Section 7.7.

The extended function s will then have a (two-sided) derivative at all points of $(-\infty,\infty)$. (This may easily be established by arguments similar to the one we have just given.) From (28) and the fact that $s(-x)=-s(x)$ for $-\pi/2\leqslant x\leqslant \pi/2$ it is easy to show that $s(-x)=-s(x)$ for all real x. (Draw a rough graph of s.)

We now define the function c on $[-\pi/2,\pi/2]$ by

$$c(x)=\sqrt{1-[s(x)]^2} \qquad \left(-\frac{\pi}{2}\leqslant x\leqslant \frac{\pi}{2}\right). \tag{29}$$

Then $c(0)=\sqrt{1-0^2}=1, c(\pi/2)=0=c(-\pi/2)$.

By (27a) we have $c(x)=s'(x)$ for $-\pi/2\leqslant x\leqslant\pi/2$. We then extend the function c to all of $(-\infty,\infty)$ by defining

$$c(x)=s'(x) \qquad (-\infty<x<\infty). \tag{30}$$

Since $s(-x)=-s(x)$ we have $-s'(-x)=-s'(x)$. That is, $c(-x)=c(x)$ for all real x. From (28) we have $s'(x+\pi)=-s'(x)$ for all x, and hence, $c(x+\pi)=-c(x)$ for all x.

From (27a) we have

$$s''(x)=\frac{-s(x)s'(x)}{\sqrt{1-[s(x)]^2}}=-s(x)$$

for $-\pi/2<x<\pi/2$. By methods already used it may then be easily shown that $s''(x)=-s(x)$ for all x—that is,

$$s''(x)=-s(x) \qquad (-\infty<x<\infty). \tag{31}$$

Hence, from (30),

$$c'(x)=-s(x) \qquad (-\infty<x<\infty),$$

and so

$$c''(x)=-c(x) \qquad (-\infty<x<\infty). \tag{32}$$

From (29) we have

$$[c(x)]^2+[s(x)]^2=1 \qquad \left(-\frac{\pi}{2}\leqslant x\leqslant\frac{\pi}{2}\right). \tag{33}$$

But then for any real y we have $y+k\pi\in[-\pi/2,\pi/2]$ for some integer k. Thus $s(y+k\pi)=-s[y+(k-1)\pi]=\cdots=(-1)^k s(y)$. Similarly, $c(y+k\pi)=(-1)^k c(y)$. Hence, $[s(y)]^2+[c(y)]^2=[s(y+k\pi)]^2+[c(y+k\pi)]^2=1$ [by (33)]. Thus

$$[c(y)]^2+[s(y)]^2=1 \qquad (-\infty<y<\infty). \tag{34}$$

We will now prove the identity

$$s(x+a)=s(x)c(a)+s(a)c(x) \qquad [a,x\in(-\infty,\infty)]. \tag{35}$$

Fix a. Let $F(x)=s(x+a)-s(x)c(a)-s(a)c(x)$. Then using (31) and (32) we have

$$F''(x)+F(x)=0 \qquad (-\infty<x<\infty).$$

Thus

$$2F'(x)F''(x)+2F(x)F'(x)=0 \qquad (-\infty<x<\infty).$$

Since $2F'F''+2FF'$ is the derivative of $(F')^2+F^2$, 7.8C implies

$$[F'(x)]^2+[F(x)]^2=\text{const.}=[F'(0)]^2+[F(0)]^2 \qquad (-\infty<x<\infty).$$

But since $s(0)=0, c(0)=1$, it is easy to show that $F(0)=0$. Similarly, $F'(0)=0$. Hence,

$$[F'(x)]^2+[F(x)]^2=0 \qquad (-\infty<x<\infty).$$

Thus both F and F' are identically 0. That F is identically 0 proves (35), while F' identically 0 proves

$$c(x+a)=c(x)c(a)-s(x)s(a) \qquad [a,x\in(-\infty,\infty)]. \tag{36}$$

For $a=x$ we then obtain the "double-angle" formulae

$$s(2x)=2s(x)c(x) \quad \text{and} \quad c(2x)=[c(x)]^2-[s(x)]^2. \tag{37}$$

Needless to say, s is what is usually called the sine function while c is the cosine function. Just about every trigonometric identity that exists can be derived from those we have already established [i.e., from (34)-(37)]. For example, from (37) we have

$$\cos 2x=\cos^2 x-\sin^2 x.$$

Using (34) we have

$$\cos 2x=2\cos^2 x-1,$$

$$\cos^2 x=\frac{1+\cos 2x}{2}.$$

Setting $2x=\theta$ we have

$$\cos\frac{\theta}{2}=\pm\left[\frac{(1+\cos\theta)}{2}\right]^{1/2},$$

the familiar "half-angle" formula.

Now to prove two more important identities.

From (36) we have

$$\cos(x+a)=\cos x\cos a-\sin x\sin a.$$

Replacing a by $-a$, and using the relations $\cos(-a)=\cos a, \sin(-a)=-\sin a$, we obtain

$$\cos(x-a)=\cos x\cos a+\sin x\sin a.$$

Thus

$$\cos(x+a)-\cos(x-a)=-2\sin x\sin a.$$

For any real θ and any $k\in I$, let $x=k\theta, a=\frac{1}{2}\theta$. Then

$$\cos(k+\tfrac{1}{2})\theta-\cos(k-\tfrac{1}{2})\theta=-2\sin k\theta\sin\tfrac{1}{2}\theta.$$

We now write the preceding equation for each value of k from $k=1$ to $k=n$:

$$\cos\tfrac{3}{2}\theta-\cos\tfrac{1}{2}\theta=-2\sin\theta\sin\tfrac{1}{2}\theta \qquad (k=1),$$

$$\cos\tfrac{5}{2}\theta-\cos\tfrac{3}{2}\theta=-2\sin 2\theta\sin\tfrac{1}{2}\theta \qquad (k=2),$$

$$\vdots$$

$$\cos(n+\tfrac{1}{2})\theta-\cos(n-\tfrac{1}{2})\theta=-2\sin n\theta\sin\tfrac{1}{2}\theta \qquad (k=n).$$

Adding, we then have

$$\cos(n+\tfrac{1}{2})\theta-\cos\tfrac{1}{2}\theta=-2\sin\tfrac{1}{2}\theta[\sin\theta+\sin 2\theta+\cdots+\sin n\theta].$$

Hence,

$$\sin\theta+\sin 2\theta+\cdots+\sin n\theta=\frac{\cos\frac{1}{2}\theta-\cos(n+\frac{1}{2})\theta}{2\sin\frac{1}{2}\theta} \tag{38}$$

(θ not a multiple of 2π).

[We used (38) in 3.8D.]

A similar identity is vital in the study of Fourier series. We begin with (35)

$$\sin(x+a)=\sin x\cos a+\sin a\cos x.$$

Changing a to $-a$ yields

$$\sin(x-a)=\sin x\cos a-\sin a\cos x.$$

Thus

$$\sin(x+a)-\sin(x-a)=2\sin a\cos x.$$

With $x=k\theta, a=\frac{1}{2}\theta$ we have

$$\sin(k+\tfrac{1}{2})\theta-\sin(k-\tfrac{1}{2})\theta=2\sin\tfrac{1}{2}\theta\cos k\theta.$$

Adding for $k=0,1,\ldots,n$ we obtain

$$\sin(n+\tfrac{1}{2})\theta-\sin(-\tfrac{1}{2}\theta)=2\sin\tfrac{1}{2}\theta\,(\cos 0\theta+\cos\theta+\cdots+\cos n\theta),$$

$$\sin(n+\tfrac{1}{2})\theta+\sin\tfrac{1}{2}\theta=2\sin\tfrac{1}{2}\theta\,(1+\cos\theta+\cdots+\cos n\theta),$$

$$\sin(n+\tfrac{1}{2})\theta=2\sin\tfrac{1}{2}\theta\,(\tfrac{1}{2}+\cos\theta+\cdots+\cos n\theta),$$

and finally

$$\tfrac{1}{2}+\cos\theta+\cdots+\cos n\theta=\frac{\sin(n+\frac{1}{2})\theta}{2\sin\frac{1}{2}\theta}\qquad(\theta\text{ not a multiple of }2\pi).\tag{39}$$

Exercises 8.4

Use only the results from Section 8.4 to work the following exercises.

1. Prove that for any real x
   (a) $\sin(\pi/2-x)=\cos x$.
   (b) $\cos(\pi/2-x)=\sin x$.
   [*Hint*: Use (35) and (36).]
2. Prove that for any real x
   (a) $\cos 2x=1-2\sin^2 x$.
   (b) $\sin 3x=3\sin x-4\sin^3 x$.
   (c) $\cos 3x=\cos x-4\sin^2 x\cos x$.
3. Show that
   (a) $\sin\dfrac{\pi}{4}=\cos\dfrac{\pi}{4}=\dfrac{\sqrt{2}}{2}$.
   (b) $\sin\dfrac{\pi}{6}=\cos\dfrac{\pi}{3}=\dfrac{1}{2}$.
   (c) $\sin\dfrac{\pi}{3}=\cos\dfrac{\pi}{6}=\dfrac{\sqrt{3}}{2}$.
   [*Hint*: To prove (b) first show that $\sin(\pi/6)=\cos(\pi/3)$. Then show that $\cos(\pi/3)=1-2\sin^2(\pi/6)$. Put these together.]
4. Show how the equation

   $$\sin(x+2\pi)=\sin x\qquad(-\infty<x<\infty)$$

   follows from (28). Then show that

   $$\cos(x+2\pi)=\cos x\qquad(-\infty<x<\infty).$$

5. If x is not an odd multiple of $\pi/2$, define

   $$\tan x=\frac{\sin x}{\cos x}$$

and

$$\sec x = \frac{1}{\cos x}.$$

Prove that

$$\sec^2 x = 1 + \tan^2 x.$$

6. Prove that
   (a) $\tan(-x) = -\tan x$
   (b) $\tan(x+a) = \dfrac{\tan x + \tan a}{1 - \tan x \tan a}$.
7. Show that if

$$t(x) = \tan x \qquad \left(-\frac{\pi}{2} < x < \frac{\pi}{2}\right),$$

then

$$t'(x) = \sec^2 x \qquad \left(-\frac{\pi}{2} < x < \frac{\pi}{2}\right).$$

If v is the inverse function of t, prove that

$$v : (-\infty, \infty) \Rightarrow \left(-\frac{\pi}{2}, \frac{\pi}{2}\right)$$

and that

$$v'(x) = \frac{1}{1+x^2} \qquad (-\infty < x < \infty).$$

8. If x is not a multiple of π, prove that

$$\cos x + \cos 3x + \cos 5x + \cdots + \cos(2n-1)x = \frac{\sin 2nx}{2\sin x},$$

and

$$\sin x + \sin 3x + \sin 5x + \cdots + \sin(2n-1)x = \frac{\sin^2 nx}{\sin x}.$$

8.5 TAYLOR'S THEOREM

8.5A. Suppose that for all x in some interval J the function f may be expressed as

$$f(x) = A_0 + A_1(x-a) + A_2(x-a)^2 + \cdots + A_n(x-a)^n + \cdots, \tag{1}$$

where $a \in J$. We say that (1) is an expansion of f in powers of $x-a$. Now we will *formally* show how to compute the coefficients $A_0, A_1, \ldots$.

If we set $x = a$ in (1), then all terms on the right vanish except the first one (A_0) and so

$$f(a) = A_0.$$

If we take the derivative on both sides of (1), we have

$$f'(x) = A_1 + 2A_2(x-a) + 3A_3(x-a)^2 + \cdots,$$

so that

$$f'(a) = A_1.$$

In the last step we made two unverified assumptions. First we assumed that $f'(x)$ exists, and second we assumed that the derivative of the right side of (1) may be computed by taking the derivative of each term separately. Since there are infinitely many terms on the right side of (1), this method is not justified by anything we have proved thus far.

Plunging on we have

$$f''(x)=2A_2+3\cdot 2A_3(x-a)+4\cdot 3(x-a)^2+\cdots$$

and so

$$f''(a)=2A_2.$$

In general, for any $n=0,1,2,\dots$ we have

$$f^{(n)}(x)=n!\,A_n+(n+1)(n)\cdots(2)A_{n+1}(x-a)$$
$$+(n+2)(n+1)\cdots(3)A_{n+2}(x-a)^2+\cdots$$

so that

$$f^{(n)}(a)=n!\,A_n.$$

(Here $f^{(0)}$ means f and $0!=1$, by definition.) We have thus formally shown that

$$A_n=\frac{f^{(n)}(a)}{n!} \qquad (n=0,1,2,\dots).$$

The right side of (1) thus becomes

$$f(a)+\frac{f'(a)}{1!}(x-a)+\frac{f''(a)}{2!}(x-a)^2+\cdots+\frac{f^{(n)}(a)}{n!}(x-a)^n+\cdots. \tag{2}$$

The series (2) is called the Taylor series (or Taylor expansion) about $x=a$ for the function f. The special case $a=0$—that is,

$$f(0)+\frac{f'(0)}{1!}x+\frac{f''(0)}{2!}x^2+\cdots+\frac{f^{(n)}(0)}{n!}x^n+\cdots$$

is sometimes called the Maclaurin series for f.

It is clear that $f^{(n)}(a)$ must exist for all $n=0,1,2,\dots$ before we can even write down the Taylor series for $f(x)$ about $x=a$. However, even if $f(x)$ *has* a Taylor series (2) about $x=a$, the series (2) may not converge to $f(x)$ for any x (except, of course, $x=a$). [In the next section we show that the Maclaurin series for $f(x)=e^{-1/x^2}$ has all terms equal to 0.]

Our investigation of Taylor series proceeds as follows: First we will establish the formula

$$f(x)=f(a)+\frac{f'(a)}{1!}(x-a)+\frac{f''(a)}{2!}(x-a)^2+\cdots+\frac{f^{(n)}(a)}{n!}(x-a)^n+R_{n+1}(x) \tag{3}$$

under suitable conditions on f. This formula is called Taylor's formula with remainder. The remainder term $R_{n+1}(x)$ can be expressed in various forms according to our needs. Thus for a given function f, showing that the Taylor series (2) converges to $f(x)$ can be accomplished by proving that

$$\lim_{n\to\infty} R_{n+1}(x)=0. \tag{4}$$

The proof of (4) may be easy, difficult, or impossible according to what function f is involved.

Let n denote any nonnegative integer and h any positive number.

8.5B. LEMMA. Let f be a real-valued function on the interval $[a,a+h]$ such that

$f^{(n+1)}(x)$ exists for every $x \in [a, a+h]$* and $f^{(n+1)}$ is continuous on $[a, a+h]$. Let

$$R_{k+1}(x) = \frac{1}{k!}\int_a^x (x-t)^k f^{(k+1)}(t)\,dt \qquad (x \in [a, a+h];\, k = 0, 1, \ldots, n).$$

Then

$$R_k(x) - R_{k+1}(x) = \frac{f^{(k)}(a)}{k!}(x-a)^k \qquad (x \in [a, a+h];\, k = 1, \ldots, n).$$

The same result holds if $h < 0$ and $[a, a+h]$ is replaced by $[a+h, a]$.

PROOF: Using integration by parts we have

$$\begin{aligned} R_{k+1}(x) &= \frac{1}{k!}\int_a^x (x-t)^k f^{(k+1)}(t)\,dt \\ &= \frac{(x-t)^k}{k!} f^{(k)}(t)\bigg|_{t=a}^{x} + \frac{k}{k!}\int_a^x (x-t)^{k-1} f^{(k)}(t)\,dt \\ &= \frac{-(x-a)^k}{k!} f^{(k)}(a) + R_k(x). \end{aligned}$$

The lemma follows.

We next establish Taylor's formula with the integral form of the remainder.

8.5C. THEOREM. Let f be a real-valued function on $[a, a+h]$ such that $f^{(n+1)}(x)$ exists for every $x \in [a, a+h]$ and $f^{(n+1)}$ is continuous on $[a, a+h]$. Then

$$\begin{aligned} f(x) = f(a) &+ \frac{f'(a)}{1!}(x-a) + \frac{f''(a)}{2!}(x-a)^2 + \cdots \\ &+ \frac{f^{(n)}(a)}{n!}(x-a)^n + R_{n+1}(x) \qquad (x \in [a, a+h]) \end{aligned}$$

where

$$R_{n+1}(x) = \frac{1}{n!}\int_a^x (x-t)^n f^{(n+1)}(t)\,dt.$$

The same result holds if $h < 0$ and $[a, a+h]$ is replaced by $[a+h, a]$.

PROOF: With R_k as in 8.5B we have

$$-R_1(x) = -\int_a^x f'(t)\,dt = f(a) - f(x).$$

Also, by 8.5B,

$$\begin{aligned} R_1(x) - R_2(x) &= \frac{f'(a)}{1!}(x-a), \\ R_2(x) - R_3(x) &= \frac{f''(a)}{2!}(x-a)^2, \\ &\vdots \\ R_n(x) - R_{n+1}(x) &= \frac{f^{(n)}(a)}{n!}(x-a)^n. \end{aligned}$$

* The existence of $f^{(n+1)}(x)$ for every $x \in [a, a+h]$ implies the existence of $f'(x), f''(x), \ldots, f^{(n)}(x)$ for every $x \in [a, a+h]$. By $f^{(k)}(a)$ and $f^{(k)}(a+h)$ $(k = 0, 1, \ldots, n)$ we mean, of course, one-sided derivatives.

If we add all these equations, we obtain

$$-R_{n+1}(x) = -f(x) + f(a) + \frac{f'(a)}{1!}(x-a) + \frac{f''(a)}{2!}(x-a)^2 + \cdots + \frac{f^{(n)}(a)}{n!}(x-a)^n.$$

The theorem follows.

Thus if f has derivatives of all orders on $[a, a+h]$, *and* if

$$\lim_{n\to\infty} R_{n+1}(x) = 0,$$

then

$$f(x) = f(a) + \frac{f'(a)}{1!}(x-a) + \frac{f''(a)}{2!}(x-a)^2 + \cdots + \frac{f^{(n)}(a)}{n!}(x-a)^n + \cdots.$$

That is, the Taylor series for f converges to $f(x)$. It is usually easier to handle the remainder term $R_{n+1}(x)$ when it is put in a different form. To accomplish this we need a result that is sometimes called the second mean-value theorem for integrals.

8.5D. THEOREM. Let φ be a continuous (real-valued) function on the closed bounded interval $[a,b]$, and let g be a continuous function on $[a,b]$ such that

$$g(t) \geqslant 0 \qquad (a \leqslant t \leqslant b).$$

Then there exists a number c with $a \leqslant c \leqslant b$ such that

$$\int_a^b \varphi(t) g(t)\,dt = \varphi(c) \int_a^b g(t)\,dt. \tag{1}$$

PROOF: By 6.6F the continuous function φ on the compact interval $[a,b]$ attains a maximum value M and a minimum value m. Then, since $g(t) \geqslant 0$ for all t,

$$m \int_a^b g(t)\,dt \leqslant \int_a^b \varphi(t) g(t)\,dt \leqslant M \int_a^b g(t)\,dt. \tag{2}$$

If g is identically zero, the theorem is obvious. We may therefore assume that $g(t) > 0$ for some t, so that

$$\int_a^b g(t)\,dt > 0$$

(why?). From (2) we then have

$$m \leqslant \theta \leqslant M$$

where

$$\theta = \frac{\int_a^b \varphi(t) g(t)\,dt}{\int_a^b g(t)\,dt}.$$

Since m and M are in the range of φ, theorem 6.2D implies that θ is in the range of φ. That is, $\varphi(c) = \theta$ for some $c \in [a,b]$. Equation (1) follows immediately.

We can now establish Taylor's formula with the Lagrange form of the remainder.

8.5E. THEOREM. Let f be a real-valued function on $[a, a+h]$ such that $f^{(n+1)}(x)$ exists for every $x \in [a, a+h]$ and $f^{(n+1)}$ is continuous on $[a, a+h]$. Then if $x \in [a, a+h]$, there exists a number c with $a \leqslant c \leqslant x$ such that

$$f(x)=f(a)+\frac{f'(a)}{1!}(x-a)+\frac{f''(a)}{2!}(x-a)^2$$

$$+\cdots+\frac{f^{(n)}(a)}{n!}(x-a)^n+\frac{f^{(n+1)}(c)}{(n+1)!}(x-a)^{n+1}.$$

The same result holds if $h<0$ and $[a,a+h]$ is replaced by $[a+h,a]$.

PROOF: By 8.5D [with $\varphi=f^{(n+1)}$ and $g(t)=(x-t)^n/n!$] we have

$$R_{n+1}(x)=\frac{1}{n!}\int_a^x f^{(n+1)}(t)(x-t)^n\,dt=\frac{f^{(n+1)}(c)}{n!}\int_a^x (x-t)^n\,dt$$

for some $c\in[a,x]$. Thus

$$R_{n+1}(x)=\frac{f^{(n+1)}(c)}{(n+1)!}(x-a)^{n+1}.$$

This and 8.5C complete the proof.

It is important to note that c depends on n (since it depends on $f^{(n+1)}$) as well as on x.

For an easy illustration consider $f(x)=e^x$ $(-\infty<x<\infty)$. Then $f^{(n)}(x)=e^x$ for all x and all $n=0,1,2,\ldots$. If we take $a=0$ in 8.5E, we have

$$e^x=1+\frac{x}{1!}+\frac{x^2}{2!}+\cdots+\frac{x^n}{n!}+\frac{e^c x^{n+1}}{(n+1)!}$$

where $0\leqslant c\leqslant x$ (or, if $x<0, x\leqslant c\leqslant 0$). Hence, no matter what n and x are, we have $0<e^c<1+e^x$. Since $\lim_{n\to\infty} x^{n+1}/n!=0$ (verify) we may let n approach infinity to obtain

$$e^x=1+\frac{x}{1!}+\frac{x^2}{2!}+\cdots+\frac{x^n}{n!}+\cdots=\sum_{n=0}^{\infty}\frac{x^n}{n!}$$

for every real x. Thus the Maclaurin series for e^x converges to e^x for all real x.

8.5F. If we take $x=a+h$ in 8.5E, then $c=a+\theta h$ where $0\leqslant\theta\leqslant 1$. The conclusion of 8.5E then reads: There exists θ with $0\leqslant\theta\leqslant 1$ such that

$$f(a+h)=f(a)+\frac{f'(a)}{1!}h+\frac{f''(a)}{2!}h^2+\cdots+\frac{f^{(n)}(a)}{n!}h^n+\frac{f^{(n+1)}(a+\theta h)}{(n+1)!}h^{n+1}. \tag{1}$$

Consider $f(x)=\log x$ $(x>0)$. Then

$$f'(x)=\frac{1}{x}, f''(x)=\frac{-1}{x^2},\ldots,f^{(n)}(x)=\frac{(-1)^{n-1}(n-1)!}{x^n}.$$

If we put $a=1$, then, for all $n, f^{(n)}(x)$ exists for every $x\in[1,1+h]$ if $h>0$. Also, if $-1<h<0$, then $f^{(n)}(x)$ exists for $x\in[1+h,1]$.

Now, for this f, (1) becomes

$$\log(1+h)=0+h-\frac{h^2}{2}+\frac{h^3}{3}+\cdots+\frac{(-1)^{n-1}h^n}{n}+\frac{(-1)^n h^{n+1}}{(n+1)(1+\theta h)^{n+1}}. \tag{2}$$

If $0 \leqslant h \leqslant 1$, then

$$\left|\frac{(-1)^n h^{n+1}}{(n+1)(1+\theta h)^{n+1}}\right| \leqslant \frac{1}{n},$$

and so the remainder term in (2) approaches 0 as n approaches infinity. Thus if $0 \leqslant h \leqslant 1$,

$$\log(1+h) = h - \frac{h^2}{2} + \frac{h^3}{3} - \cdots + \frac{(-1)^{n-1}h^n}{n} + \cdots = \sum_{n=1}^{\infty} \frac{(-1)^{n-1}h^n}{n}. \tag{3}$$

In particular, $1 - \frac{1}{2} + \frac{1}{3} - \frac{1}{4} + \cdots = \log 2$. (See the example after 3.3B.) If $-1 < h < 0$, it is still true that the remainder term in (2) approaches 0 as n approaches infinity, but it is not (as of now) easy to show. [Try it. The difficulty is that the denominator in $(-1)^n h^{n+1}/(n+1)(1+\theta h)^n$ may be very close to 0 if h is close to -1.] However, the proof that (3) holds for $-1 < h < 0$ will not be difficult after we introduce still another form of the remainder.

The next theorem establishes Taylor's formula with the Cauchy form of the remainder.

8.5G. THEOREM. Let f be a real-valued function on $[a, a+h]$ such that $f^{(n+1)}(x)$ exists for every $x \in [a, a+h]$ and $f^{(n+1)}$ is continuous on $[a, a+h]$. Then if $x \in [a, a+h]$, there exists a number* c with $a \leqslant c \leqslant x$ such that

$$f(x) = f(a) + \frac{f'(a)}{1!}(x-a) + \frac{f''(a)}{2!}(x-a)^2$$
$$+ \cdots + \frac{f^{(n)}(a)}{n!}(x-a)^n + \frac{f^{(n+1)}(c)}{n!}(x-c)^n(x-a).$$

The same result holds if $h < 0$ and $[a, a+h]$ is replaced by $[a+h, a]$.

PROOF: By 8.5D [with $\varphi(t) = f^{(n+1)}(t)(x-t)^n$ and $g(t) = 1$] we have

$$R_{n+1}(x) = \frac{1}{n!}\int_a^x f^{(n+1)}(t)(x-t)^n\,dt = \frac{f^{(n+1)}(c)(x-c)^n}{n!}\int_a^x 1\,dt$$

for some $c \in [a, x]$. Thus

$$R_{n+1}(x) = \frac{f^{(n+1)}(c)}{n!}(x-c)^n(x-a).$$

This and 8.5C complete the proof.

8.5H. If we take $x = a+h$ in 8.5G, then $c = a + \theta h$ where $0 \leqslant \theta \leqslant 1$. The conclusion of 8.5G then reads: There exists θ with $0 \leqslant \theta \leqslant 1$ such that

$$f(a+h) = f(a) + \frac{f'(a)}{1!}h + \frac{f''(a)}{2!}h^2 + \cdots + \frac{f^{(n)}(a)}{n!}h^n + \frac{f^{(n+1)}(a+\theta h)}{n!}(1-\theta)^n h^{n+1}.$$

Now consider, as in 8.5F, the special case where $f(x) = \log x$ $(x > 0)$ and $a = 1$. Then, if $|h| < 1$, we have, by 8.5G,

$$\log(1+h) = h - \frac{h^2}{2} + \cdots + \frac{(-1)^{n-1}h^n}{n} + \frac{(-1)^n(1-\theta)^n h^{n+1}}{(1+\theta h)^{n+1}}. \tag{1}$$

* The c in this theorem is not, in general, the same as the c in 8.5E.

Now suppose $-1<h<0$ (which was a case we could not handle with the Lagrange form of the remainder). Since $0 \leqslant \theta \leqslant 1$ we then have

$$1+\theta h \geqslant 1+h$$

and also

$$0 \leqslant \frac{1-\theta}{1+\theta h} \leqslant 1.$$

Using this we have

$$\left| \frac{(-1)^n (1-\theta)^n h^{n+1}}{(1+\theta h)^{n+1}} \right| = \left| \frac{1-\theta}{1+\theta h} \right|^n \cdot \frac{|h|^{n+1}}{|1+\theta h|} \leqslant \frac{|h|^{n+1}}{1+h}.$$

Thus the remainder term in (1) approaches zero as n approaches infinity. This shows that equation (3) of 8.5F also holds for $-1<h<0$.

We have deduced Taylor's theorem with the Cauchy (or the Lagrange) form of the remainder from 8.5C—Taylor's theorem with the integral form of the remainder. We can also prove Taylor's theorem with the Cauchy (or the Lagrange) form of the remainder directly from the generalized law of the mean 7.7C. Indeed, with this method of proof, we do not need to assume that $f^{(n+1)}$ is continuous but only that $f^{(n+1)}$ exists. We give details in the Cauchy case. The Lagrange case will be an exercise.

8.5I. THEOREM. Even if the hypothesis that $f^{(n+1)}$ is continuous on $[a, a+h]$ is dropped, theorem 8.5G remains true.

PROOF: Fix $x \in [a, a+h]$. Then if $t \in [a, x]$, let

$$F(t) = f(x) - f(t) - \frac{f'(t)}{1!}(x-t) - \cdots - \frac{f^{(n)}(t)}{n!}(x-t)^n, \tag{1}$$

$$G(t) = x - t.$$

Then easy computation shows that

$$F'(t) = \frac{-f^{(n+1)}(t)}{n!}(x-t)^n, \tag{2}$$

where F' means the derivative of F with respect to t.

Since $G'(t) = -1$ for all $t \in [a, x]$, the hypotheses of 7.7C are satisfied (with F, G instead of f, g). Hence, there exists $c \in (a, x)$ such that

$$\frac{F(x) - F(a)}{G(x) - G(a)} = \frac{F'(c)}{G'(c)}.$$

But $F(x) = G(x) = 0$, and $G'(c) = -1$. Then using (2) we obtain

$$\frac{F(a)}{G(a)} = -F'(c) = \frac{+f^{(n+1)}(c)}{n!}(x-c)^n,$$

and so

$$F(a) = \frac{+f^{(n+1)}(c)}{n!}(x-c)^n (x-a). \tag{3}$$

The theorem follows if we set $t = a$ in (1) and use (3).

A close examination of the proof shows that all that is really required is that $f^{(n)}$ be continuous on $[a, a+h]$ and that $f^{(n+1)}(x)$ exist for every x in $(a, a+h)$. For then F will be continuous on $[a, x]$ and $F'(t)$ will exist for every t in (a, x).

Exercises 8.5

1. Find the Taylor series about $x=2$ for

$$f(x)=x^3+2x+1 \qquad (-\infty<x<\infty).$$

Prove that the Taylor series converges to $f(x)$ for every real x.

2. Write Taylor's formula with the integral form of the remainder for

$$f(x)=\sin x \qquad (-\infty<x<\infty)$$

and $a=0$.
Show that

$$|R_{n+1}(x)|<\frac{|x|^{n+1}}{(n+1)!} \qquad (-\infty<x<\infty).$$

3. Show that the Taylor series about $x=0$ for $f(x)=\sin x$ converges to $\sin x$ for every real x.
4. Write Taylor's formula with the Lagrange form of the remainder in the following cases:
   (a) $f(x)=\log(1+x) \qquad (-1<x<\infty)$
   $a=2$
   $n=4$.
   (b) $f(x)=\tan^{-1}x \qquad \left(-\frac{\pi}{2}<x<\frac{\pi}{2}\right)$
   $a=0$
   $n=3$.
5. Write Taylor's formula with the Cauchy form of the remainder for

$$f(x)=(1-x)^{1/2} \qquad (-1<x<1)$$

and $a=0$.

6. If $a>0, h>0$, and $n\in I$, prove that there exist $\theta, 0\leqslant\theta\leqslant 1$ such that

$$\frac{1}{a+h}=\frac{1}{a}-\frac{h}{a^2}+\frac{h^2}{a^3}+\cdots+\frac{(-1)^{n-1}h^{n-1}}{a^n}+\frac{(-1)^n h^n}{(a+\theta h)^{n+1}}.$$

7. If f'' is continuous on $[a-\delta, a+\delta]$ for some $\delta>0$, prove that

$$\lim_{h\to 0}\frac{f(a+h)-2f(a)+f(a-h)}{h^2}=f''(a).$$

8.6 THE BINOMIAL THEOREM

8.6A. If $x\in R$, then the formulae

$$(1+x)^2=1+2x+x^2,$$
$$(1+x)^3=1+3x+3x^2+x^3,$$

are familiar from beginning algebra. In "college" algebra the more general formula

$$(1+x)^m=1+mx+\frac{m(m-1)}{2!}x^2+\cdots+x^m \tag{1}$$

is taught. Here m is any positive integer and the coefficient of x^n is

$$\frac{m!}{n!(m-n)!}=\frac{m(m-1)\cdots(m-n+1)}{1\cdot 2\cdots n}$$

for $n=1,\ldots,m$.

A proof of (1) may be easily given on the basis of Taylor's formula (with any form of the remainder).

If $m \in I$, let

$$f(x)=(1+x)^m \qquad (-\infty<x<\infty).$$

Then $f^{(n)}(x)$ exists for all x and all n. Indeed, if $n=1,\ldots,m$ then

$$f^{(n)}(x)=m(m-1)\cdots(m-n+1)(1+x)^{m-n},$$

while $f^{(n)}(x)=0$ if $n>m$. Hence,

$$f^{(n)}(0)=m(m-1)\cdots(m-n+1) \qquad (n=1,\ldots,m),$$
$$f^{(n)}(0)=0 \qquad (n>m).$$

If we now use Taylor's formula, we have

$$f(x)=f(0)+\frac{f'(0)}{1!}x+\frac{f''(0)}{2!}x^2+\cdots+\frac{f^m(0)}{m!}x^m+0.$$

That is,

$$(1+x)^m=1+mx+\frac{m(m-1)}{2!}x^2+\cdots+\frac{m(m-1)\cdots(m-m+1)}{m!}x^m,$$

which establishes (1).

8.6B. If m is not a nonnegative integer, there is still a formula for $(1+x)^m$ (provided $|x|<1$). This formula can also be derived from Taylor's formula, but much more difficulty is encountered.

THEOREM. If $m \in R$ is not a nonnegative integer, then

$$(1+x)^m=1+mx+\frac{m(m-1)}{2!}x^2+\cdots+\frac{m(m-1)\cdots(m-n+1)}{n!}x^n+\cdots \qquad (1)$$

provided that $|x|<1$.

FIRST PROOF: If $f(x)=(1+x)^m$ for $-1<x<1$, then

$$f^{(n)}(x)=m(m-1)\cdots(m-n+1)(1+x)^{m-n} \qquad (n=1,2,\ldots).$$

Thus for any n, Taylor's formula with the Cauchy form of the remainder (as in 8.5H with $a=0$, $-1<h<1$) yields

$$f(h)=1+mh+\frac{m(m-1)}{2!}h^2+\cdots+\frac{m(m-1)\cdots(m-n+1)}{n!}h^n+R_{n+1} \qquad (2)$$

where

$$R_{n+1}=\frac{m(m-1)\cdots(m-n)}{n!}\cdot(1+\theta h)^{m-n-1}(1-\theta)^n h^{n+1},$$

$$R_{n+1}=\frac{m(m-1)\cdots(m-n)}{n!}\left(\frac{1-\theta}{1+\theta h}\right)^n(1+\theta h)^{m-1}h^{n+1},$$

$$|R_{n+1}|\leqslant\left|\frac{m(m-1)\cdots(m-n)}{n!}\right|(1+\theta h)^{m-1}|h|^{n+1}. \qquad (3)$$

We emphasize that θ depends on n so that the behavior of $(1+\theta h)^{m-1}$ as n approaches infinity is not obvious. If $m>1$, then $m-1>0$, and so

$$0<(1+\theta h)^{m-1}\leqslant(1+|h|)^{m-1}.$$

If $m<1$, then

$$0<(1+\theta h)^{m-1}=\frac{1}{(1+\theta h)^{1-m}}\leqslant\frac{1}{(1-|h|)^{1-m}}$$
$$=(1-|h|)^{m-1}.$$

Hence, for any m,

$$(1+\theta h)^{m-1}\leqslant(1\pm|h|)^{m-1}.$$

From (3) we then have

$$|R_{n+1}|\leqslant(1\pm|h|)^{m-1}a_n$$

where

$$a_n=\frac{|m(m-1)\cdots(m-n)|\,|h|^{n+1}}{n!}.$$

We have thus removed the problem created by θ. Now the ratio test 3.6F shows that $\sum_{n=1}^{\infty}a_n<\infty$. Hence, $\lim_{n\to\infty}a_n=0$ and so $\lim_{n\to\infty}R_{n+1}=0$. This and (2) establish (1) (with h instead of x) and the theorem is proved.

[What we showed in 8.6A was simply that (1) of 8.6B holds for all x if m is a positive integer. For then all except the first $m+1$ terms on the right of (1) of 8.6B are zero.]

SECOND PROOF: If $f(x)=(1+x)^m$ for $-1<x<1$, then 8.5C yields

$$f(x)=1+mx+\frac{m(m-1)}{2!}x^2+\cdots$$
$$+\frac{m(m-1)\cdots(m-n+1)}{n!}x^n+R_{n+1}(x) \tag{4}$$

where

$$R_{n+1}(x)=\frac{m(m-1)\cdots(m-n)}{n!}\int_0^x(1+t)^{m-n-1}(x-t)^n\,dt$$
$$=\frac{m(m-1)\cdots(m-n)}{n!}\int_0^x(1+t)^{m-1}\left(\frac{x-t}{1+t}\right)^n dt.$$

If $0\leqslant x<1$, it is easy to see that the function

$$g(t)=\frac{x-t}{1+t}\qquad(0\leqslant t\leqslant x)$$

attains a maximum at $t=0$. If $-1<x<0$, then

$$G(t)=\frac{x-t}{1+t}\qquad(x\leqslant t\leqslant 0)$$

is nonpositive and nonincreasing on $[x,0]$. Hence, $|G|$ attains a maximum at $t=0$. Thus in either case,

$$\left|\frac{x-t}{1+t}\right|\leqslant|x|.$$

Hence,

$$|R_{n+1}(x)| = \left| \frac{m(m-1)\cdots(m-n)}{n!} \int_0^x (1+t)^{m-1}\left(\frac{x-t}{1+t}\right)^n dt \right|$$

$$\leqslant \left| \frac{m(m-1)\cdots(m-n)}{n!} \int_0^x (1+t)^{m-1}\left|\frac{x-t}{1+t}\right|^n dt \right|$$

$$\leqslant \left| \frac{m(m-1)\cdots(m-n)}{n!} \right| |x|^n \left| \int_0^x (1+t)^{m-1} dt \right|.$$

Since

$$\lim_{n\to\infty} \frac{m(m-1)\cdots(m-n)|x|^n}{n!} = 0$$

(as in the first proof) we have $\lim_{n\to\infty} R_{n+1}(x) = 0$. The theorem follows from (4).

Exercises 8.6

1. Let $\binom{m}{k}$ denote $\dfrac{m!}{k!(m-k)!}$ where k and m are nonnegative integers, $k \leqslant m$. Prove that

   (a) $2^m = \sum_{k=0}^{m} \binom{m}{k}$

   (b) $0 = \sum_{k=0}^{m} (-1)^k \binom{m}{k}$.

   [*Hint*: In (1) of 8.6A show that $\binom{m}{k}$ is the coefficient of x^k. Then take $x = \pm 1$.]
2. Give a proof of the binomial theorem 8.6B for $0 < x < 1$ using the Lagrange form of the remainder.

8.7 L'HOSPITAL'S RULE

In 4.1D we noted that

$$\lim_{x\to a} \frac{f(x)}{g(x)} = \frac{\lim_{x\to a} f(x)}{\lim_{x\to a} g(x)}$$

provided that both $\lim_{x\to a} f(x)$ and $\lim_{x\to a} g(x)$ exist *and* $\lim_{x\to a} g(x) \neq 0$. It sometimes happens that $\lim_{x\to a}[f(x)/g(x)]$ exists even if $\lim_{x\to a} g(x) = 0 = \lim_{x\to a} f(x)$.

L'Hospital's rule says, roughly, that if $\lim_{x\to a} f(x) = \lim_{x\to a} g(x) = 0$, then

$$\lim_{x\to a} \frac{f(x)}{g(x)} = \lim_{x\to a} \frac{f'(x)}{g'(x)}$$

provided the limit on the right exists.

We will confine our systematic investigation of this to the case of one-sided limits at $x=0$—that is, limits of the form $\lim_{x\to 0+} f(x)$. As we will see, all other cases can be easily handled on the basis of this single one.

8.7A. THEOREM. If $f'(x)$ and $g'(x)$ exist for every x in $(0,\delta]$, if

$$g'(x)\neq 0 \qquad (0<x\leqslant\delta),$$

if

$$\lim_{x\to 0+} f(x)=0=\lim_{x\to 0+} g(x), \tag{1}$$

and if

$$\lim_{x\to 0+}\frac{f'(x)}{g'(x)}=L, \tag{2}$$

then

$$\lim_{x\to 0+}\frac{f(x)}{g(x)}=L.$$

PROOF: By (1), both f and g will be continuous at 0 if we define $f(0)=0=g(0)$. By 7.7C, given $x\in(0,\delta]$ there exists $c\in(0,x)$ such that

$$\frac{f(x)-f(0)}{g(x)-g(0)}=\frac{f'(c)}{g'(c)},$$

where c, of course, depends on x. (How do we know $g(x)\neq g(0)$?)

Hence,

$$\frac{f(x)}{g(x)}=\frac{f'(c)}{g'(c)} \qquad (0<x<\delta). \tag{3}$$

Since c approaches 0 as $x\to 0+$ we have, by (2),

$$\lim_{x\to 0+}\frac{f'(c)}{g'(c)}=L.$$

The theorem then follows from (3).

For example, if $f(x)=\sin x$ and $g(x)=x$ for $0\leqslant x\leqslant 1$, the hypotheses of the theorem are satisfied. Hence,

$$\lim_{x\to 0+}\frac{\sin x}{x}=\lim_{x\to 0+}\frac{\cos x}{1}=1.$$

In some problems, two (or more) applications of the theorem are necessary. For example, if $f(x)=\sin x - x\cos x$ and $g(x)=x^2\sin x$, then $f'(x)=x\sin x, g'(x)=x^2\cos x+2x\sin x$, and so

$$\frac{f'(x)}{g'(x)}=\frac{\sin x}{x\cos x+2\sin x}=\frac{\varphi(x)}{\psi(x)}.$$

Here $\lim_{x\to 0+}\varphi(x)=0=\lim_{x\to 0+}\psi(x)$. But, by 8.7A

$$\lim_{x\to 0+}\frac{\varphi(x)}{\psi(x)}=\lim_{x\to 0+}\frac{\varphi'(x)}{\psi'(x)}=\lim_{x\to 0+}\frac{\cos x}{-x\sin x+3\cos x}=\frac{1}{3}$$

Hence,

$$\frac{1}{3}=\lim_{x\to 0+}\frac{\varphi'(x)}{\psi'(x)}=\lim_{x\to 0+}\frac{\varphi(x)}{\psi(x)}=\lim_{x\to 0+}\frac{f'(x)}{g'(x)}$$

$$=\lim_{x\to 0+}\frac{f(x)}{g(x)}=\lim_{x\to 0+}\frac{\sin x - c\cos x}{x^2\sin x},$$

where 8.7A was used at the second and fourth equality signs.

8.7B. If we wanted to prove the result for limits as $x \to a+$ corresponding to 8.7A, we merely need to use the fact that if $F(x) = f(x+a)$, then

$$\lim_{x\to 0+} F(x) = \lim_{x\to 0+} f(x+a) = \lim_{x\to a+} f(x),$$

and

$$\lim_{x\to 0+} F'(x) = \lim_{x\to a+} f'(x).$$

Thus 8.7A with $0+$ replaced by $a+$ may be proved by applying 8.7A (as it stands) to $F(x) = f(x+a)$ and $G(x) = g(x+a)$.

Similarly, if we wish to prove 8.7A with $0+$ replaced by ∞, we would consider

$$F(x) = f\left(\frac{1}{x}\right), \qquad G(x) = g\left(\frac{1}{x}\right).$$

For

$$\lim_{x\to 0+} F(x) = \lim_{x\to\infty} f(x).$$

Furthermore, $F'(x) = -(1/x^2)f'(1/x)$ and $G'(x) = -(1/x^2)g'(1/x)$. Hence,

$$\frac{F'(x)}{G'(x)} = \frac{f'(1/x)}{g'(1/x)},$$

and so

$$\lim_{x\to 0+} \frac{F'(x)}{G'(x)} = \lim_{x\to\infty} \frac{f'(x)}{g'(x)}.$$

The reader should now be able to formulate and prove the variants of 8.7A for $x \to a+$ and $x \to \infty$ *and* $x \to a-$.

8.7C. Under the hypotheses of 8.7A, the quotient $f(x)/g(x)$ is sometimes called an indeterminate form of type $0/0$, since both numerator and denominator approach 0 (as $x \to 0+$).

Another extremely important case is called indeterminate of type ∞/∞. This involves a quotient $f(x)/g(x)$ where both $f(x)$ and $g(x)$ approach infinity. in the following sense (compare with 2.4A).

DEFINITION. Let f be a real-valued function whose domain includes all points of some interval $(a-h, a+h)$ except possibly a itself. We say that $f(x)$ approaches infinity as x approaches a if given $M > 0$, there exists $\delta > 0$ such that

$$f(x) \geqslant M \qquad (0 < |x-a| < \delta).$$

In this case we write $f(x) \to \infty$ as $x \to a$.

Similar definitions apply to the statements

$$f(x) \to \infty \quad \text{as} \quad x \to a+,$$
$$f(x) \to \infty \quad \text{as} \quad x \to a-,$$
$$f(x) \to \infty \quad \text{as} \quad x \to \infty.$$

We leave these definitions to the reader.

Now we prove the second important case of L'Hospital's rule.

8.7D. THEOREM. If $f'(x)$ and $g'(x)$ exist for every x in $(0, \delta]$, if

$$g'(x) \neq 0 \qquad (0 < x \leqslant \delta),$$

if

$$f(x)\to\infty \quad \text{and} \quad g(x)\to\infty \quad \text{as} \quad x\to 0+,$$

and if

$$\lim_{x\to 0+}\frac{f'(x)}{g'(x)}=L, \tag{1}$$

then

$$\lim_{x\to 0+}\frac{f(x)}{g(x)}=L.$$

PROOF: Let $h(x)=f(x)-Lg(x)$ for $0<x\leqslant\delta$. Then $h'(x)=f'(x)-Lg'(x)$ and so, by (1),

$$\lim_{x\to 0+}\frac{h'(x)}{g'(x)}=0.$$

Given $\epsilon>0$, this and the hypothesis that $g(x)\to\infty$ as $x\to 0+$ imply the existence of $\delta_1>0$ such that

$$g(x)>0 \qquad (0<x\leqslant\delta_1) \tag{2}$$

and such that

$$\left|\frac{h'(c)}{g'(c)}\right|<\frac{\epsilon}{2}$$

for any $c\in(0,\delta_1)$. If $x\in(0,\delta_1)$, then

$$\frac{h(\delta_1)-h(x)}{g(\delta_1)-g(x)}=\frac{h'(c)}{g'(c)}$$

for some $c\in(x,\delta_1)$. Hence,

$$\left|\frac{h(x)-h(\delta_1)}{g(x)-g(\delta_1)}\right|<\frac{\epsilon}{2} \qquad (0<x<\delta_1). \tag{3}$$

Since $g(x)\to\infty$ as $x\to 0+$, there exists $\delta_2<\delta_1$ such that

$$g(x)>g(\delta_1) \qquad (0<x<\delta_2). \tag{4}$$

From (2) and (4) we thus have

$$0<g(x)-g(\delta_1)<g(x) \qquad (0<x<\delta_2). \tag{5}$$

From (3) and (5) we then conclude

$$\frac{|h(x)-h(\delta_1)|}{g(x)}<\frac{\epsilon}{2} \qquad (0<x<\delta_2). \tag{6}$$

Now choose $\delta_3<\delta_2$ such that

$$\frac{|h(\delta_1)|}{g(x)}<\frac{\epsilon}{2} \qquad (0<x<\delta_3). \tag{7}$$

If $0<x<\delta_3$, we then have

$$\frac{h(x)}{g(x)}=\frac{h(x)-h(\delta_1)}{g(x)}+\frac{h(\delta_1)}{g(x)},$$

$$\left|\frac{h(x)}{g(x)}\right|\leqslant\frac{|h(x)-h(\delta_1)|}{g(x)}+\frac{|h(\delta_1)|}{g(x)},$$

and so, by (6) and (7),

$$\left|\frac{h(x)}{g(x)}\right| < \epsilon \qquad (0 < x < \delta_3).$$

This proves

$$\lim_{x \to 0+} \frac{h(x)}{g(x)} = 0.$$

Since

$$\frac{f(x)}{g(x)} = \frac{h(x)}{g(x)} + L,$$

the theorem follows.

8.7E. In 8.7D, we may replace $x \to 0+$ by $x \to a+, x \to \infty$, or $x \to a-$. We omit the proof of this.

For an example, let $f(x) = \log x, g(x) = x$ for $x > 0$. Then $f(x) \to \infty$ and $g(x) \to \infty$ as $x \to \infty$. Since $f'(x) = 1/x, g'(x) = 1$ we have

$$\lim_{x \to \infty} \frac{f'(x)}{g'(x)} = 0.$$

Hence,

$$\lim_{x \to \infty} \frac{f(x)}{g(x)} = \lim_{x \to \infty} \frac{\log x}{x} = 0.$$

Now consider $\lim_{x \to \infty}(x^n/e^x)$ where $n \in I$. We have

$$\lim_{x \to \infty} \frac{x^n}{e^x} = \lim_{x \to \infty} \frac{nx^{n-1}}{e^x} = \cdots = \lim_{x \to \infty} \frac{n!}{e^x} = 0.$$

Thus

$$\lim_{x \to \infty} \frac{x^n}{e^x} = 0 \qquad (n \in I).$$

Using the fact that $\lim_{x \to \infty} f(x) = \lim_{x \to 0+} f(1/x)$ we deduce

$$\lim_{x \to 0+} \frac{1}{x^n e^{1/x}} = 0 \qquad (n \in I). \tag{1}$$

This enables us to give a very interesting example concerning Taylor series. Let

$$g(x) = e^{-1/x} \qquad (x > 0),$$
$$g(0) = 0.$$

Then from (1) we have

$$g'(0) = \lim_{x \to 0+} \frac{g(x) - g(0)}{x} = \lim_{x \to 0+} \frac{e^{-1/x}}{x} = 0.$$

Since

$$g'(x) = \frac{e^{-1/x}}{x^2} \qquad (x > 0),$$

(1) may also be used to show that $g''(0) = 0$. Indeed, since (for $x > 0$) $g^{(n)}(x)$ is a finite sum of terms of the form $e^{-1/x}/x^m$, an easy induction argument will show that

$$g(0) = g'(0) = \cdots = g^{(n)}(0) = \cdots = 0. \tag{2}$$

Hence, the Maclaurin series for g is identically zero, and therefore does *not* converge to $g(x)$ for any $x>0$.

This shows that the existence of all derivatives of a function f at a point a does not imply

$$f(x)=\sum_{k=0}^{\infty}\frac{f^{(k)}}{k!}(x-a)^k$$

for any $x\neq a$. That is why we must deal with the remainder term.

Note that all derivatives $g^{(n)}(0)$ in (2) are right-hand derivatives. The function h defined by

$$h(x)=e^{-1/x^2} \qquad (x\neq 0),$$
$$h(0)=0,$$

has $h(0)=h'(0)=\cdots=h^{(n)}(0)=\cdots$, where all these derivatives are two-sided. We leave the verification of this to the reader.

8.7F. Other kinds of indeterminate forms may often be handled by reducing them to one of the types $0/0$ or ∞/∞. For example, consider x^{-x} as $x\to 0+$. This may be called an indeterminate form of type 0^0. Now

$$\log x^{-x}=-x\log x=\frac{-\log x}{1/x}=\frac{\log(1/x)}{1/x},$$

and

$$\frac{\log(1/x)}{(1/x)}$$

is of type ∞/∞. Hence,

$$\lim_{x\to 0+}\log x^{-x}=\lim_{x\to 0+}\frac{\log(1/x)}{1/x}=\lim_{x\to 0+}\frac{-1/x}{-1/x^2}$$
$$=\lim_{x\to 0+}x=0,$$

where 8.7D was used at the second equality sign. But then, since the exponential function is continuous, we have

$$\lim_{x\to 0+}x^{-x}=\lim_{x\to 0+}e^{\log x^{-x}}=\exp\left(\lim_{x\to 0+}\log x^{-x}\right)=e^0=1.$$

More generally, an indeterminate form $[F(x)]^{G(x)}$ may be handled by considering

$$\log[F(x)]^{G(x)}=\frac{\log F(x)}{1/G(x)}.$$

As $x\to 0+$, the quantity $1/x^2-1/(x\tan x)$ is indeterminate of type $\infty-\infty$. However, with a little algebra we can write

$$\frac{1}{x^2}-\frac{1}{x\tan x}=\frac{\tan x-x}{x^2\tan x}=\frac{\sin x-x\cos x}{x^2\sin x}.$$

The quantity on the right is indeterminate of type $0/0$. We have already computed its limit (as $x\to 0+$) following theorem 8.7A.

Exercises 8.7

1. Evaluate the following limits.
   (a) $\lim_{x\to 0+}\dfrac{\tan x - x}{x - \sin x}$.
   (b) $\lim_{x\to 0+}\dfrac{10^x - 5^x}{x}$.
   (c) $\lim_{x\to 0}\dfrac{\log[(1+x)/(1-x)]}{x}$.
2. Do the same for
   (a) $\lim_{x\to\infty}\dfrac{\log(1+e^{3x})}{x}$.
   (b) $\lim_{x\to\infty}\dfrac{\log x}{x^{0.0001}}$.
   (c) $\lim_{x\to\infty} x(\sqrt{x^2+4} - x)$.

   (*Hint*: Rationalize the "numerator.")
3. Do the same for
   (a) $\lim_{x\to 1+}\dfrac{x - 5x^5 + 4x^6}{(1-x)^2}$.
   (b) $\lim_{x\to 1}\dfrac{1 - 4\sin^2(\pi x/6)}{1 - x^2}$.
4. Same for
   (a) $\lim_{x\to 0}\left(\dfrac{1}{x} - \dfrac{1}{\sin x}\right)$.
   (b) $\lim_{x\to\infty} x^{1/x}$.
   (c) $\lim_{x\to\infty}\left(1 + \dfrac{1}{x}\right)^x$.
   (d) $\lim_{x\to 3}\dfrac{1}{x-3}\displaystyle\int_3^x e^{\sqrt{1+t^2}}\,dt$.

9

SEQUENCES AND SERIES OF FUNCTIONS

9.1 POINTWISE CONVERGENCE OF SEQUENCES OF FUNCTIONS

In Chapters 2 and 3 we discussed convergence of sequences and series of real numbers. In this chapter we discuss the convergence of sequences and series of functions. We deal almost exclusively with real-valued functions.

9.1A. DEFINITION. Let $\{f_n\}_{n=1}^{\infty}$ be a sequence of real-valued functions on a set E. We say that $\{f_n\}_{n=1}^{\infty}$ converges to the function f on E if

$$\lim_{n\to\infty} f_n(x)=f(x) \qquad (x\in E). \tag{1}$$

If (1) holds, we sometimes say that $\{f_n\}_{n=1}^{\infty}$ converges *pointwise* to f on E. For if (1) holds, then, for every *point* x of E, the sequence $\{f_n(x)\}_{n=1}^{\infty}$ of real numbers converges to $f(x)$. Here are several examples.

If

$$f_n(x)=x^n \qquad (0\leqslant x\leqslant 1),$$

then $\{f_n\}_{n=1}^{\infty}$ converges to f on $[0,1]$ where

$$f(x)=0 \qquad (0\leqslant x<1),$$
$$f(1)=1.$$

See Figure 25.

For a second example let

$$g_n(x)=\frac{x}{1+nx} \qquad (0\leqslant x<\infty).$$

If $x>0$, then $0<g_n(x)\leqslant x/nx=1/n$. Hence,

$$\lim_{n\to\infty} g_n(x)=0 \qquad (x>0).$$

Also, since $g_n(0)=0$ for each $n\in I$, it is clear that $\{g_n\}_{n=1}^{\infty}$ converges to 0 (the function identically 0) on $[0,\infty)$.

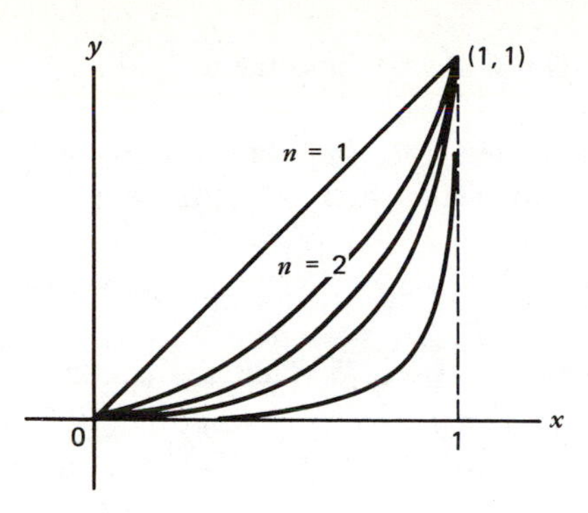

FIGURE 25. Graph of $y=f_n(x)$, where $f_n(x) = x^n$ $(0 \leqslant x \leqslant 1)$

Next, let

$$h_n(x)=\frac{nx}{1+n^2x^2} \qquad (0 \leqslant x < \infty).$$

Then if $x>0$, we have

$$h_n(x)=\frac{1/nx}{(1/n^2x^2)+1}$$

and hence, $\lim_{n\to\infty} h_n(x)=0$. Since $h_n(0)=0$ for each $n\in I$ we see that $\{h_n\}_{n=1}^{\infty}$ converges to 0 on $[0, \infty)$. See Figure 26.

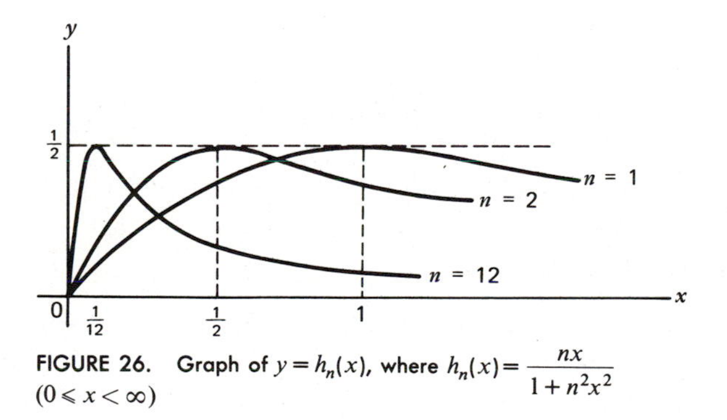

FIGURE 26. Graph of $y=h_n(x)$, where $h_n(x)=\dfrac{nx}{1+n^2x^2}$ $(0 \leqslant x < \infty)$

For a fourth example let χ_n denote the characteristic function of $[-n,n]$. For any $x\in R^1$ we have $\chi_n(x)=\chi_{n+1}(x)=\chi_{n+2}(x)=\cdots=1$ provided $n\geqslant |x|$. (For then $x\in [-n,n]$). Hence,

$$\lim_{n\to\infty} \chi_n(x)=1 \qquad (x\in R^1),$$

and so $\{\chi_n\}_{n=1}^{\infty}$ converges to 1 on $(-\infty,\infty)$.

9.1B. According to definition 9.1A, the sequence of functions $\{f_n\}_{n=1}^{\infty}$ converges to f on the set E if, for each $x\in E$, given $\epsilon>0$ there exists $N\in I$ such that

$$|f_n(x)-f(x)|<\epsilon \qquad (n\geqslant N). \tag{1}$$

In general, the number N depends on both ϵ *and* x. It is not always possible to find an N such that (1) holds for *all* x in E simultaneously.

For example, if $f_n(x)=x^n$ $(0\leqslant x\leqslant 1)$, then, as we have seen, $\{f_n\}_{n=1}^{\infty}$ converges to f on $[0,1]$ where $f(x)=0$ $(0\leqslant x<1)$ and $f(1)=1$. With $\epsilon=\frac{1}{2}$, then, for each $x\in E$, there exists $N\in I$ such that

$$|f_n(x)-f(x)|<\tfrac{1}{2} \qquad (n\geqslant N). \tag{2}$$

If $x=0$ or $x=1$, then (2) is true for $N=1$. However, if $x=\frac{3}{4}=0.75$, then the smallest value of N for which (2) is true is $N=3$. For, if $x=\frac{3}{4}$, then $f_n(x)=(\frac{3}{4})^n$ while $f(x)=0$. Thus $|f_n(x)-f(x)|=(\frac{3}{4})^n$, and $(\frac{3}{4})^n<\frac{1}{2}$ if and only if $n\geqslant 3$. Similarly, if $x=0.9$, then the smallest value of N for which (2) is true is $N=7$.

Indeed, there is no $N\in I$ such that (2) holds for *all* $x\in[0,1]$. For, if such an N existed, we would have

$$x^n<\tfrac{1}{2} \qquad (n\geqslant N)$$

for all x in $[0,1)$. This implies $x^N<\frac{1}{2}$ $(0\leqslant x<1)$. Letting $x\to 1^-$ we obtain the contradiction $1\leqslant\frac{1}{2}$.

For the second example in 9.1A the story is different. For, if

$$g_n(x)=\frac{x}{1+nx} \qquad (0\leqslant x<\infty),$$

then $0\leqslant g_n(x)\leqslant 1/n$ $(0\leqslant x<\infty)$. Hence, for any $\epsilon>0$ the statement

$$|g_n(x)-0|<\epsilon \qquad (n\geqslant N) \tag{3}$$

is true for *all* x in $[0,\infty)$, provided only that $N>1/\epsilon$. (For in this case $|g_n(x)-0|\leqslant 1/n \leqslant 1/N<\epsilon$ for all x in $[0,\infty)$.) Thus for this sequence $\{g_n\}_{n=1}^{\infty}$ an $N\in I$ can be found such that (3) holds for all $x\in I$. This N depends only on ϵ and not on x.

Now consider

$$h_n(x)=\frac{nx}{1+n^2x^2} \qquad (0\leqslant x<\infty).$$

We have seen that $\{h_n\}_{n=1}^{\infty}$ converges to 0 on $[0,\infty)$. Given $\epsilon>0$, we know therefore that for each $x\in[0,\infty)$ there exists $N\in I$ such that

$$|h_n(x)-0|<\epsilon \qquad (n\geqslant N). \tag{4}$$

However, note that $h_n(1/n)=\frac{1}{2}$. Hence, if $\epsilon=\frac{1}{2}$, there is no single $N\in I$ such that (4) holds for all $x\in[0,\infty)$. For if such an N existed, we would have

$$h_N(x)<\tfrac{1}{2} \qquad (0\leqslant x<\infty).$$

But if $x=1/N$, this is a contradiction.

We leave it to the reader to show that if $\epsilon<1$, then there is no $N\in I$ such that the statement

$$|\chi_n(x)-1|<\epsilon \qquad (n\geqslant N)$$

holds for all real x, where χ_n is as in the fourth example following 9.1A.

Exercises 9.1

1. Let

$$f_n(x)=\frac{\sin nx}{n} \qquad (0\leqslant x\leqslant 1).$$

Does there exist $N \in I$ such that

$$|f_n(x) - 0| < \tfrac{1}{10} \qquad (n \geqslant N)$$

for all $x \in [0, 1]$ simultaneously?

2. Let

$$f_n(x) = \frac{x^n}{1 + x^n} \qquad (0 \leqslant x \leqslant 1).$$

Show that $\{f_n\}_{n=1}^{\infty}$ converges pointwise on $[0, 1]$. If

$$f(x) = \lim_{n \to \infty} f_n(x) \qquad (0 \leqslant x \leqslant 1),$$

is there an $N \in I$ such that

$$|f_n(x) - f(x)| < \tfrac{1}{4} \qquad (n \geqslant N)$$

for all $x \in [0, 1]$ simultaneously?

3. Let χ_n be the characteristic function of the open interval $(0, 1/n)$, and let

$$f_n(x) = n\chi_n(x) \qquad (0 \leqslant x \leqslant 1).$$

(a) Show that $\{f_n\}_{n=1}^{\infty}$ converges to 0 on $[0, 1]$.
(b) Does there exist $N \in I$ such that

$$|f_n(x) - 0| < \tfrac{1}{2} \qquad (n \geqslant N)$$

for all $x \in [0, 1]$ simultaneously?
(c) Calculate $\lim_{n \to \infty} \int_0^1 f_n$.
(d) Compare (a) and (c).

4. For $n \in I$ let

$$f_n(x) = nx(1 - x^2)^n \qquad (0 \leqslant x \leqslant 1).$$

(a) Show that $\{f_n\}_{n=1}^{\infty}$ converges to 0 on $[0, 1]$.
(b) Show that $\left\{ \int_0^1 f_n \right\}_{n=1}^{\infty}$ converges to $\tfrac{1}{2}$.
(c) Compare (a) and (b).

5. Let

$$f_n(x) = \frac{x}{n} e^{-x/n} \qquad (0 \leqslant x < \infty).$$

(a) Prove that $\{f_n\}_{n=1}^{\infty}$ converges to 0 on $[0, \infty)$.
(b) Does there exist $N \in I$ such that

$$|f_n(x) - 0| < \tfrac{1}{10} \qquad (n \geqslant N) \tag{*}$$

for all $x \in [0, \infty)$ simultaneously? (*Answer*; No.)
(c) If $A > 0$, does there exist N such that (*) holds for all x in $[0, A]$ simultaneously? (*Answer*: Yes.)

9.2 UNIFORM CONVERGENCE OF SEQUENCES OF FUNCTIONS

We have agreed to say the $\{f_n\}_{n=1}^{\infty}$ converges (pointwise) to f on E if, for each $x \in E$, given $\epsilon > 0$ there exists $N \in I$ such that

$$|f_n(x) - f(x)| < \epsilon \qquad (n \geqslant N). \tag{1}$$

We have seen several examples in which it is impossible to find an N such that (1) holds for all $x \in E$ simultaneously.

If for each $\epsilon > 0$ it *is* possible to find an N such that (1) holds for all $x \in E$, then we say that $\{f_n\}_{n=1}^{\infty}$ converges uniformly to f on E.

9.2A. DEFINITION. Let $\{f_n\}_{n=1}^{\infty}$ be a sequence of real-valued functions on a set E. We say that $\{f_n\}_{n=1}^{\infty}$ converges uniformly to the function f on E if given $\epsilon > 0$, there exists $N \in I$ such that

$$|f_n(x) - f(x)| < \epsilon \qquad (n \geqslant N; x \in E). \tag{2}$$

The wording of this definition implies that N depends on ϵ but not on x. It is clear that if $\{f_n\}_{n=1}^{\infty}$ converges uniformly to f on E, then $\{f_n\}_{n=1}^{\infty}$ converges pointwise to f on E.

Thus if $g_n(x) = x/(1+nx)$ $(0 \leqslant x < \infty)$, then our work in Section 9.1 shows that $\{g_n\}_{n=1}^{\infty}$ converges uniformly to 0 on $[0, \infty)$. For we have already shown that given $\epsilon > 0$ there exists $N \in I$ such that

$$|g_n(x) - 0| < \epsilon \qquad (n \geqslant N; 0 \leqslant x < \infty).$$

(Any N such that $N > 1/\epsilon$ will do.)

It is not too easy to state what it means for the sequence $\{f_n\}_{n=1}^{\infty}$ *not* to converge uniformly to f on E. We shall now do this.

9.2B. COROLLARY. The sequence $\{f_n\}_{n=1}^{\infty}$ does not converge uniformly to f on E if and only if there exists some $\epsilon > 0$ such that there is no $N \in I$ for which the statement

$$|f_n(x) - f(x)| < \epsilon \qquad (n \geqslant N; x \in E)$$

holds.

The reader should not proceed until he is convinced that 9.2B is equivalent to 9.2A.

If $f_n(x) = x^n$ $(0 \leqslant x \leqslant 1)$ and $f(x) = 0$ $(0 \leqslant x \leqslant 1), f(1) = 1$, then we have seen that $\{f_n\}_{n=1}^{\infty}$ converges pointwise to f on $[0,1]$. However, $\{f_n\}_{n=1}^{\infty}$ does *not* converge uniformly to f on $[0,1]$. For, as we saw in 9.1B, if $\epsilon = \frac{1}{2}$, then there is no $N \in I$ such that

$$|f_n(x) - f(x)| < \epsilon \qquad (n \geqslant N; 0 \leqslant x \leqslant 1).$$

Similarly, the sequences $\{h_n\}_{n=1}^{\infty}$ and $\{\chi_n\}_{n=1}^{\infty}$ of Section 2.8 do *not* converge uniformly on their domains to 0 and 1, respectively (even though $\{h_n\}_{n=1}^{\infty}$ and $\{\chi_n\}_{n=1}^{\infty}$ do converge pointwise to 0 and 1). (Verify!)

9.2C. Note that (2) can be expressed as

$$f(x) - \epsilon < f_n(x) < f(x) + \epsilon \qquad (n \geqslant N; x \in E).$$

We can describe uniform convergence (of a sequence of real-valued functions whose domain is a set E of real numbers) in the following geometric terms: In order for $\{f_n\}_{n=1}^{\infty}$ to converge uniformly to f on E, given $\epsilon > 0$ there must exist $N \in I$ such that if $n \geqslant N$, then the entire graph of $y = f_n(x)$ must lie between the graphs of $y = f(x) - \epsilon$ and $y = f(x) + \epsilon$. See Figure 27.

This geometric criterion, with $\epsilon < \frac{1}{2}$, gives us an alternate demonstration that the sequences $\{f_n\}_{n=1}^{\infty}$ and $\{h_n\}_{n=1}^{\infty}$ in Figures 25 and 26 do not converge uniformly. For $\{f_n\}_{n=1}^{\infty}$ in Figure 25 converges pointwise to f where $f(1) = 1$ and $f(x) = 0$ for $0 \leqslant x < 1$. However, no matter what n is, there will be points on the graph of $y = f_n(x)$ for x sufficiently near (but not equal to) 1 that are not between the graphs $y = f(x) - \epsilon$ and $y = f(x) + \epsilon$. Similarly, $\{h_n\}_{n=1}^{\infty}$ in Figure 26 converges pointwise to 0. However, for each n, the point $\langle \frac{1}{n}, \frac{1}{2} \rangle$ is on the graph of $y = h_n(x)$ but this point does not lie between the graphs of $y = -\epsilon$ and $y = \epsilon$.

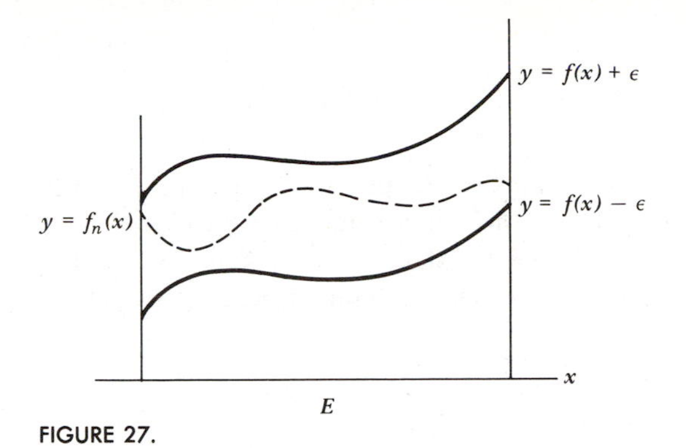

FIGURE 27.

9.2D. Here is still another way to view uniform convergence. If $\{f_n\}_{n=1}^{\infty}$ converges uniformly to 0 on E, then given $\epsilon>0$ there exists N such that

$$|f_n(x)|<\epsilon \qquad (n\geqslant N; x\in E).$$

This implies

$$\underset{x\in E}{\text{l.u.b.}}\,|f_n(x)|\leqslant\epsilon \qquad (n\geqslant N).$$

Hence, if $\{f_n\}_{n=1}^{\infty}$ converges uniformly to zero on E, then

$$\lim_{n\to\infty}\underset{x\in E}{\text{l.u.b.}}\,|f_n(x)|=0. \tag{1}$$

Conversely, it is not difficult to show that if (1) holds, then $\{f_n\}_{n=1}^{\infty}$ converges uniformly to 0 on E.

This readily proves that the sequence $\{h_n\}_{n=1}^{\infty}$ of Section 9.1 does not converge uniformly to zero on $(-\infty,\infty)$. For

$$\underset{-\infty<x<\infty}{\text{l.u.b.}}\,|h_n(x)|\geqslant\left|h_n\left(\frac{1}{n}\right)\right|=\frac{1}{2} \qquad (n=1,2,\ldots),$$

and hence, $\text{l.u.b.}_{-\infty<x<\infty}|h_n(x)|$ cannot approach zero as $n\to\infty$.

9.2E. From 9.2A it follows immediately that $\{f_n\}_{n=1}^{\infty}$ converges uniformly to f on E if and only if $\{f_n-f\}_{n=1}^{\infty}$ converges uniformly to 0 on E. From 9.1D we then have

THEOREM. The sequence of functions $\{f_n\}_{n=1}^{\infty}$ converges uniformly to f on E if and only if

$$\underset{x\in E}{\text{l.u.b.}}\,|f_n(x)-f(x)|\to 0 \quad \text{as} \quad n\to\infty.$$

The next result is called the Cauchy criterion for uniform convergence. It is analogous to the result that a sequence of real numbers is convergent if and only if it is Cauchy.

9.2F. THEOREM. Let $\{f_n\}_{n=1}^{\infty}$ be a sequence of real-valued functions on a set E. Then $\{f_n\}_{n=1}^{\infty}$ is uniformly convergent on E (to some function f) if and only if given $\epsilon>0$, there exists $N\in I$ such that

$$|f_m(x)-f_n(x)|<\epsilon \qquad (m,n\geqslant N; x\in E). \tag{1}$$

PROOF: Suppose first that $\{f_n\}_{n=1}^{\infty}$ is a uniformly convergent sequence of functions on E, converging to f on E. Then, given $\epsilon>0$, there exists $N\in I$ such that

$$|f_n(x)-f(x)|<\frac{\epsilon}{2} \qquad (n\geqslant N; x\in E).$$

Thus if $m,n \geqslant N$, we have, for any $x \in E$,

$$|f_m(x)-f_n(x)| \leqslant |f_m(x)-f(x)|+|f(x)-f_n(x)| < \frac{\epsilon}{2}+\frac{\epsilon}{2}$$

and hence, (1) holds for this N.

Conversely let $\{f_n\}_{n=1}^{\infty}$ be any sequence of functions on E such that, given $\epsilon > 0$, there exists $N \in I$ such that (1) holds. We must show that there is a function f on E such that $\{f_n\}_{n=1}^{\infty}$ converges uniformly to f on E. From (1) we see that, for each fixed $x \in E$, the sequence of real numbers $\{f_n(x)\}_{n=1}^{\infty}$ is a Cauchy sequence. Hence, $\lim_{n\to\infty} f_n(x)$ exists for each $x \in E$. Define f by

$$f(x) = \lim_{n\to\infty} f_n(x) \qquad (x \in E).$$

Keeping m fixed in (1) and letting $n \to \infty$ we obtain

$$|f_m(x)-f(x)| \leqslant \epsilon \qquad (m \geqslant N; x \in E).$$

Since ϵ was arbitrary, this shows that $\{f_m\}_{m=1}^{\infty}$ converges uniformly to f on E, and the proof is complete.

The next result, called Dini's theorem, shows that under a very special set of circumstances a sequence of *continuous* functions must converge uniformly.

9.2G. THEOREM. Let $\{f_n\}_{n=1}^{\infty}$ be a sequence of continuous real-valued functions on the *compact* metric space $\langle M, \rho \rangle$ such that

$$f_1(x) \leqslant f_2(x) \leqslant \cdots \leqslant f_n(x) \leqslant \cdots \qquad (x \in M). \tag{1}$$

If $\{f_n\}_{n=1}^{\infty}$ converges (pointwise) on M to the continuous function f, then $\{f_n\}_{n=1}^{\infty}$ converges *uniformly* to f on M.

PROOF: For each $n \in I$ let $g_n = f - f_n$. Then from (1) we have

$$g_1(x) \geqslant g_2(x) \geqslant \cdots \geqslant g_n(x) \geqslant \cdots \geqslant 0 \qquad (x \in M). \tag{2}$$

Also, since $\{f_n\}_{n=1}^{\infty}$ converges to f on M we have

$$\lim_{n\to\infty} g_n(x) = 0 \qquad (x \in M). \tag{3}$$

We must show that $\{g_n\}_{n=1}^{\infty}$ converges uniformly to 0 on M.

Fix $\epsilon > 0$. If $x \in M$, then (3) assures us of the existence of $N(x) \in I$ such that

$$g_{N(x)}(x) < \frac{\epsilon}{2}.$$

Since $g_{N(x)}$ is continuous at x, there is an open ball B_x about x such that

$$g_{N(x)}(y) < \epsilon \qquad (y \in B_x).$$

The B_x for all $x \in M$ form an open covering of M. By 6.5G a finite number of the B_x—say

$$B_{x_1}, B_{x_2}, \ldots B_{x_k}$$

—also cover M. Let $N = \max[N(x_1), \ldots, N(x_k)]$. Now if y is any point in M, then $y \in B_{x_j}$ for some $j = 1, \ldots, k$. Hence,

$$g_{N(x_j)}(y) < \epsilon.$$

But since $N(x_j) \leqslant N$, (2) implies

$$g_N(y) \leqslant g_{N(x_j)}(y).$$

Hence,

$$0 \leqslant g_N(y) < \epsilon$$

for all $y \in M$. But then (2) shows that

$$0 \leqslant g_n(y) < \epsilon \qquad (n \geqslant N; y \in M),$$

and so $\{g_n\}_{n=1}^{\infty}$ converges uniformly to 0 on M. This completes the proof.

It is clear that 9.2G remains true if the inequality signs in (1) are all reversed. For then we could set $g_n = f_n - f$ and proceed as above.

Exercises 9.2

1. If $\{f_n\}_{n=1}^{\infty}$ and $\{g_n\}_{n=1}^{\infty}$ converge uniformly on E, prove that $\{f_n + g_n\}_{n=1}^{\infty}$ converges uniformly on E.
2. Let

$$g_n(x) = \frac{1}{n} e^{-nx} \qquad (0 \leqslant x < \infty).$$

Prove that $\{g_n\}_{n=1}^{\infty}$ converges uniformly to 0 on $[0, \infty)$.
3. Let f be a uniformly continuous real-valued function on $(-\infty, \infty)$, and for each $n \in I$ let

$$f_n(x) = f\left(x + \frac{1}{n}\right) \qquad (-\infty < x < \infty).$$

Prove that $\{f_n\}_{n=1}^{\infty}$ converges uniformly on $(-\infty, \infty)$ to f.
4. Let

$$f_n(x) = \frac{x^n}{1 + x^n} \qquad (0 \leqslant x \leqslant 1).$$

(a) Show that $\{f_n\}_{n=1}^{\infty}$ converges uniformly on $[0, \frac{1}{2}]$.
(b) Does $\{f_n\}_{n=1}^{\infty}$ converge uniformly on $[0, 1]$?
5. Let

$$f_n(x) = \frac{x}{n} e^{-x/n} \qquad (0 \leqslant x < \infty).$$

(a) Does $\{f_n\}_{n=1}^{\infty}$ converge uniformly to 0 on $[0, \infty)$?
(b) Does $\{f_n\}_{n=1}^{\infty}$ converge uniformly to 0 on $[0, 500]$?
6. Let $\{f_n\}_{n=1}^{\infty}$ be a sequence of continuous real-valued functions that converges uniformly on the closed bounded interval $[a, b]$. For each $n \in I$ let

$$F_n(x) = \int_a^x f_n(t)\,dt \qquad (a \leqslant x \leqslant b).$$

Show that $\{F_n\}_{n=1}^{\infty}$ converges uniformly on $[a, b]$. (*Hint*: Use 9.2F.)
7. Let

$$f_n(x) = \frac{x^n}{n} \qquad (0 \leqslant x \leqslant 1).$$

(a) Show that $\{f_n\}_{n=1}^{\infty}$ converges uniformly to 0 on $[0, 1]$.
(b) Does $\{f_n'\}_{n=1}^{\infty}$ converge to 0 on $[0, 1]$?
8. Let $\{f_n\}_{n=1}^{\infty}$ be a sequence of continuous functions $[0, 1]$ that converges uniformly.
(a) Show that there exists $M > 0$ such that

$$|f_n(x)| \leqslant M \qquad (n \in I; 0 \leqslant x \leqslant 1).$$

(b) Does the result in part (a) hold if uniform convergence is replaced by pointwise convergence?

9. Show by example that Dini's theorem is no longer true if we omit the hypothesis that M is compact.
10. If $\{f_n\}_{n=1}^{\infty}$ is a sequence of functions that converges uniformly to the continuous function f on $(-\infty, \infty)$, prove that

$$\lim_{n\to\infty} f_n\left(x+\frac{1}{n}\right)=f(x) \qquad (-\infty<x<\infty).$$

11. Let A be a dense subset of the metric space M. If $\{f_n\}_{n=1}^{\infty}$ is a sequence of continuous functions on M, and if $\{f_n\}_{n=1}^{\infty}$ converges uniformly on A, prove that $\{f_n\}_{n=1}^{\infty}$ converges uniformly on M.

9.3 CONSEQUENCES OF UNIFORM CONVERGENCE

9.3A. If $f_n(x)=x^n$ $(0\leqslant x\leqslant 1)$ for $n\in I$, then f_n is continuous on $[0,1]$. However, $\{f_n\}_{n=1}^{\infty}$ converges (pointwise) to f on $[0,1]$ where $f(x)=0$ $(0\leqslant x<1), f(1)=1$. The function f is not continuous on $[0,1]$. This shows that a sequence of continuous functions may converge pointwise to a discontinuous function.

However, if $\{f_n\}_{n=1}^{\infty}$ is a sequence of continuous functions that converges *uniformly* to f, then f must be continuous. This result will be a corollary to the following theorem.

9.3B. THEOREM. Let $\{f_n\}_{n=1}^{\infty}$ be a sequence of real-valued functions on a metric space M that converges uniformly to the function f on M. If each f_n $(n\in I)$ is continuous at $a\in M$, then f is also continuous at a.

PROOF: Given $\epsilon>0$ we may choose $N\in I$ such that

$$|f_n(x)-f(x)|<\frac{\epsilon}{3} \qquad (n\geqslant N; x\in M).$$

Since f_N is continuous at a there exists $\delta>0$ such that

$$|f_N(x)-f_N(a)|<\frac{\epsilon}{3} \qquad [\rho(x,a)<\delta].$$

where ρ is the metric for M.

Hence, if $\rho(x,a)<\delta$, we have

$$\begin{aligned}|f(x)-f(a)| &\leqslant |f(x)-f_N(x)|+|f_N(x)-f_N(a)|+|f_N(a)-f(a)|\\ &<\frac{\epsilon}{3}+\frac{\epsilon}{3}+\frac{\epsilon}{3}=\epsilon.\end{aligned}$$

Thus

$$|f(x)-f(a)|<\epsilon \qquad [\rho(x,a)<\delta],$$

which proves the theorem.

9.3C. COROLLARY. If $\{f_n\}_{n=1}^{\infty}$ is a sequence of continuous real-valued functions on the metric space M that converges uniformly to f on M, then f is also continuous on M.

9.3D. Just as a sequence of continuous functions may converge pointwise to a discontinuous function, a sequence of Riemann integrable functions may converge pointwise to a function that is not Riemann integrable.

For example, let $A=\{r_1,r_2,\ldots\}$ be the set of all rational numbers in $[0,1]$ and let χ_n be the characteristic function of the finite subset $\{r_1,\ldots,r_n\}$. Then χ_n is bounded on $[0,1]$

and χ_n is continuous at all points of $[0,1]$ except at $r_1,\ldots,r_n$. Hence, for each $n\in I$, we have $\chi_n \in \mathcal{R}[0,1]$. But $\{\chi_n\}_{n=1}^{\infty}$ converges pointwise to χ_A, the characteristic function of A. Since χ_A is discontinuous at each point of $[0,1], \chi_A \notin \mathcal{R}[0,1]$.

On the other hand, with uniform convergence we have the following positive result.

9.3E. THEOREM. If $\{f_n\}_{n=1}^{\infty}$ is a sequence of functions in $\mathcal{R}[a,b]$, and if $\{f_n\}_{n=1}^{\infty}$ converges uniformly to f on $[a,b]$, then f is also in $\mathcal{R}[a,b]$.

PROOF: For $\epsilon=1$ there exists $N\in I$ such that

$$|f_n(x)-f(x)|<1 \qquad (n\geqslant N; x\in[a,b]).$$

In particular

$$|f_N(x)-f(x)|<1 \qquad (x\in[a,b]),$$

and so

$$|f(x)|\leqslant|f_N(x)|+|f(x)-f_N(x)|<|f_N(x)|+1$$

for all $x\in[a,b]$. Now f_N is bounded, since $f_N\in\mathcal{R}[a,b]$. Clearly, then, f is also bounded on $[a,b]$.

For each $n\in I$, let E_n be the set of points of $[a,b|)$ at which f_n is not continuous, and let $E=\cup_n^{\infty}=1E_n$. By 7.3A, each set E_n is of measure zero. Hence, by 7.1B, E is also of measure zero. But if $x\in[a,b]-E$, then x is in no E_n, and so every f_n is continuous at x. By 9.3B, the function f is also continuous at x. Hence, f is continuous at almost every point of $[a,b]$. Since f is bounded, 7.3A implies $f\in\mathcal{R}[a,b]$, and the proof is complete.

We have just seen that uniform convergence is a sufficient condition for a sequence of Riemann integrable functions to converge to a Riemann integrable function. It must not be supposed, however, that uniform convergence is a *necessary* condition. For example, if $f_n(x)=x^n$ $(0\leqslant x\leqslant 1)$, then $\{f_n\}_{n=1}^{\infty}$ converges on $[0,1]$ to a Riemann integrable function, even though the convergence is *not* uniform.

Similarly, uniform convergence is a sufficient but *not* a necessary condition that a sequence of continuous functions converge to a continuous function. (Verify.)

9.3F. Suppose now that $\{f_n\}_{n=1}^{\infty}$ is a sequence of functions in $\mathcal{R}[a,b]$ that converges (pointwise) to a function f on $[a,b]$. We now ask: If we assume that $f\in\mathcal{R}[a,b]$, must it be true that $\left\{\int_a^b f_n\right\}_{n=1}^{\infty}$ converges to $\int_a^b f$? In other "words," if

$$\lim_{n\to\infty} f_n(x)=f(x),$$

is

$$\lim_{n\to\infty}\int_a^b f_n(x)\,dx=\int_a^b f(x)\,dx?$$

This is equivalent to asking if

$$\lim_{n\to\infty}\int_a^b f_n(x)\,dx=\int_a^b \lim_{n\to\infty} f_n(x)\,dx. \tag{1}$$

We sometimes express this by asking: "Is it permissible to interchange $\lim_{n\to\infty}$ and $\int_a^b$?" Again, if only pointwise convergence is assumed, undesirable phenomena may occur.

For example, let

$$f_n(x) = 2n \qquad \left(\frac{1}{n} \leqslant x \leqslant \frac{2}{n}\right),$$

$$f_n(x) = 0 \qquad \text{for all other } x \in [0,1].$$

Then, for each $n \in I$, we have

$$\int_0^1 f_n(x)\,dx = \int_{1/n}^{2/n} 2n\,dx = 2n\left(\frac{2}{n} - \frac{1}{n}\right) = 2.$$

Hence,

$$\lim_{n\to\infty} \int_0^1 f_n(x)\,dx = 2.$$

On the other hand we have $\lim_{n\to\infty} f_n(x) = 0$ $(0 \leqslant x \leqslant 1)$. For $f_n(0) = 0$ $(n \in I)$, while, if $x > 0$, then $f_N(x) = f_{N+1}(x) = \cdots = 0$ if $2/N < x$. Hence,

$$\int_0^1 \lim_{n\to\infty} f_n(x)\,dx = 0.$$

Thus (1) does not hold for this sequence $\{f_n\}_{n=1}^{\infty}$.

Once more, uniform convergence makes things work out in a desirable fashion. That is, (1) will hold if $\{f_n\}_{n=1}^{\infty}$ converges *uniformly* on $[a,b]$.

9.3G. THEOREM. Let $\{f_n\}_{n=1}^{\infty}$ be a sequence of functions in $\mathcal{R}[a,b]$ that converges uniformly to the function f on $[a,b]$. Then $f \in \mathcal{R}[a,b]$ and

$$\lim_{n\to\infty} \int_a^b f_n(x)\,dx = \int_a^b f(x)\,dx.$$

PROOF: That $f \in \mathcal{R}[a,b]$ follows from 9.3E. Given $\epsilon > 0$ there exists $N \in I$ such that

$$|f_n(x) - f(x)| < \frac{\epsilon}{b-a} \qquad (n \geqslant N; a \leqslant x \leqslant b). \tag{1}$$

Using 7.4C and 7.4F we have

$$\left|\int_a^b f_n(x)\,dx - \int_a^b f(x)\,dx\right| = \left|\int_a^b [f_n(x) - f(x)]\,dx\right| \leqslant \int_a^b |f_n(x) - f(x)|\,dx.$$

Hence, by (1), if $n \geqslant N$ we obtain

$$\left|\int_a^b f_n(x)\,dx - \int_a^b f(x)\,dx\right| \leqslant \int_a^b \frac{\epsilon}{b-a}\,dx = \epsilon.$$

This shows that $\left\{\int_a^b f_n(x)\,dx\right\}_{n=1}^{\infty}$ converges to $\int_a^b f(x)\,dx$, which is what we wished to prove.

9.3H. Finally we take up the following question: Suppose $\{f_n\}_{n=1}^{\infty}$ is a sequence of functions on $[a,b]$ such that, for each $n \in I, f_n'(x)$ exists for all $x \in [a,b]$. Suppose also that $\{f_n\}_n^{\infty} = 1$ converges to f on $[a,b]$. We ask (1) does $f'(x)$ exist for all x? and (2) if so, does $\{f_n'\}_{n=1}^{\infty}$ converge to f'?

First of all, it is possible that a sequence of differentiable functions can converge *uniformly* to a function that does *not* have a derivative at any point. For, in Section 9.7, we will show that there exists a function F on $[0,1]$ which is continuous on $[0,1]$ but does not have a derivative at any point of $[0,1]$. But from 10.2A it will follow that there is a sequence of polynomial functions $\{P_n\}_{n=1}^{\infty}$ that converges uniformly to F. Here $P_n'(x)$ exists for every $n \in I$ and $x \in [0,1]$, and $\{P_n\}_{n=1}^{\infty}$ converges uniformly to F on $[0,1]$, but $F'(x)$ exists for no x in $[0,1]$. This shows that the answer to the first question is "no."

Even if $\{f_n\}_{n=1}^{\infty}$ converges uniformly to f, and f_n' and f' exist for all $x \in [a,b]$, it may happen that $\{f_n'\}_{n=1}^{\infty}$ does not converge to f' at some x. For example, if

$$f_n(x) = \frac{x^n}{n} \qquad (0 \leqslant x \leqslant 1),$$

then $\{f_n\}_{n=1}^{\infty}$ converges uniformly to $f=0$, but $\{f_n'(1)\}_{n=1}^{\infty}$ does not converge to $f'(1)$. Thus in this example, the equation

$$\lim_{n\to\infty} f_n'(x) = \left(\lim_{n\to\infty} f_n\right)'(x)$$

does not hold for $x=1$.

What *can* be said is the following: If $\{f_n\}_{n=1}^{\infty}$ converges on $[a,b]$, if each f_n' is continuous, and if $\{f_n'\}_{n=1}^{\infty}$ converges *uniformly* on $[a,b]$, then

$$\lim_{n\to\infty} f_n' = \left(\lim_{n\to\infty} f_n\right)'$$

This is a consequence of the following theorem.

9.3I. THEOREM. If (for each $n \in I$) $f_n'(x)$ exists for each $x \in [a,b]$, if f_n' is continuous on $[a,b]$, if $\{f_n\}_{n=1}^{\infty}$ converges on $[a,b]$ to f, and if $\{f_n'\}_{n=1}^{\infty}$ converges uniformly on $[a,b]$ to g, then

$$g(x) = f'(x) \qquad (a \leqslant x \leqslant b).$$

That is,

$$\lim_{n\to\infty} f_n'(x) = f'(x) \qquad (a \leqslant x \leqslant b).$$

PROOF: Since $\{f_n'\}_{n=1}^{\infty}$ converges uniformly to g on $[a,b]$, then g is continuous on $[a,b]$, by 9.3C. Moreover, $\{f_n'\}_{n=1}^{\infty}$ converges uniformly to g on $[a,y]$ where y is any point in $[a,b]$. By 9.3G,

$$\lim_{n\to\infty} \int_a^y f_n'(x)\,dx = \int_a^y g(x)\,dx.$$

Thus by 7.8E,

$$\lim_{n\to\infty} \left[f_n(y) - f_n(a) \right] = \int_a^y g(x)\,dx.$$

But, by hypothesis, $\lim_{n\to\infty} f_n(y) = f(y)$ and $\lim_{n\to\infty} f_n(a) = f(a)$. Thus

$$f(y) - f(a) = \int_a^y g(x)\,dx \qquad (a \leqslant y \leqslant b).$$

By 7.8A we then have

$$f'(y) = g(y) \qquad (a \leqslant y \leqslant b),$$

and the theorem is proved.

Exercises 9.3

1. By examining $f(x) = \lim_{n\to\infty} f_n(x)$ for $0 \leqslant x < \infty$, where

$$f_n(x) = \frac{1}{1+x^n},$$

prove that $\{f_n\}_{n=1}^{\infty}$ does not converge uniformly on $[0,\infty)$.

2. Let g be a continuous function on the closed bounded interval $[a,b]$. Let $\{f_n\}_{n=1}^{\infty}$ be a sequence of continuous functions that converges uniformly on $[a,b]$ to f. Prove that

$$\lim_{n\to\infty}\int_a^b f_n g=\int_a^b fg.$$

3. Let

$$f_n(x)=nx(1-x^2)^n \qquad (0\leqslant x\leqslant 1).$$

Use the result of exercise 4 of Section 9.1 to show that $\{f_n\}_{n=1}^{\infty}$ does *not* converge uniformly on $[0,1]$, even though $\{f_n\}_{n=1}^{\infty}$ converges pointwise.

4. Let

$$f_n(x)=\frac{\sin nx}{n} \qquad (0\leqslant x\leqslant 1).$$

Show that $\{f_n\}_{n=1}^{\infty}$ converges uniformly to 0 on $[0,1]$, but that $\{f_n'\}_{n=1}^{\infty}$ does not converge (even) pointwise to 0 on $[0,1]$.

5. Let $\{f_n\}_{n=1}^{\infty}$ be a sequence of functions on $[a,b]$ such that $f_n'(x)$ exists for every $x\in[a,b]$ $(n\in I)$ and
(1) $\{f_n(x_0)\}_{n=1}^{\infty}$ converges for some $x_0\in[a,b]$.
(2) $\{f_n'\}_{n=1}^{\infty}$ converges uniformly on $[a,b]$.
Prove that $\{f_n\}_{n=1}^{\infty}$ converges uniformly on $[a,b]$. Show how this result may be used to weaken the hypothesis of 9.3I. [*Hint*: For $x\in[a,b]$ write

$$f_n(x)-f_m(x)=\{[f_n(x)-f_m(x)]-[f_n(x_0)-f_m(x_0)]\}+[f_n(x_0)-f_m(x_0)].$$

Apply 7.7A to obtain

$$f_n(x)-f_m(x)=[f_n'(c)-f_m'(c)](x-x_0)+[f_n(x_0)-f_m(x_0)].$$

Now use (1) and (2).]

9.4 CONVERGENCE AND UNIFORM CONVERGENCE OF SERIES OF FUNCTIONS

Just as the convergence of a series of real numbers is defined to mean the convergence of the sequence of partial sums, the convergence of a series of functions is also defined in terms of the sequence of partial sums.

9.4A. DEFINITION. Let $u_1,u_2,\ldots$ be real-valued functions on a set E. We say that $\sum_{n=1}^{\infty}u_n$ converges to the function f on E if the sequence of functions $\{s_n\}_{n=1}^{\infty}$ converges to f on E, where $s_n=u_1+u_2+\cdots+u_n$. In this case we write

$$\sum_{n=1}^{\infty}u_n=f$$

or

$$\sum_{n=1}^{\infty}u_n(x)=f(x) \qquad (x\in E).$$

For example, if $u_n(x)=x^n$ $(-1<x<1)$, then $\sum_{n=1}^{\infty}u_n$ converges to f on $(-1,1)$ where $f(x)=x/(1-x)$ $(-1<x<1)$. For

$$\sum_{n=1}^{\infty}u_n(x)=\sum_{n=1}^{\infty}x^n=\frac{x}{1-x}=f(x) \qquad (-1<x<1).$$

We next define uniform convergence of series of functions.

9.4B. DEFINITION. If $u_1, u_2, \ldots$ are real-valued functions on a set E, we say that $\Sigma_{n=1}^{\infty} u_n$ converges uniformly to f on E if $\{s_n\}_{n=1}^{\infty}$ converges uniformly to f on E where $s_n = u_1 + u_2 + \cdots + u_n$. In this case we write

$$\sum_{n=1}^{\infty} u_n = f \quad \text{uniformly}$$

or

$$\sum_{n=1}^{\infty} u_n(x) = f(x) \quad \text{uniformly} \qquad (x \in E).$$

From 9.3B we deduce the next result.

9.4C. THEOREM. Let $u_1, u_2, \ldots$ be real-valued functions on the metric space M. If $\Sigma_{n=1}^{\infty} u_n$ converges uniformly to f on M, and if each u_n is continuous at the point $a \in M$, then f is also continuous at a.

PROOF: The sequence of functions $\{s_n\}_{n=1}^{\infty}$ converges uniformly to f on M, where $s_n = u_1 + u_2 + \cdots + u_n$. Now each u_k is continuous at a. By 5.3E, then, each s_n is continuous at a. Thus by 9.3B, f is continuous at a, which is what we wished to show.

9.4D. COROLLARY. If $u_1, u_2, \ldots$ are continuous real-valued functions on the metric space M, and if $\Sigma_{n=1}^{\infty} u_n$ converges uniformly to f on M, then f is continuous on M.

The series

$$\sum_{n=0}^{\infty} x(1-x)^n$$

converges on $[0,1]$ to the function f where $f(0)=0$ and $f(x)=1$ $(0<x \leqslant 1)$. (For, if $0<x<1$, then

$$\left.\sum_{n=0}^{\infty} x(1-x)^n = x \sum_{n=0}^{\infty} (1-x)^n = x\left[\frac{1}{1-(1-x)}\right] = 1.\right)$$

Now, if

$$u_n(x) = x(1-x)^n \qquad (0 \leqslant x \leqslant 1),$$

then u_n is continuous on $[0,1]$ and $\Sigma_{n=1}^{\infty} u_n = f$. Since f is *not* continuous on $[0,1]$, corollary 9.4D assures us that $\Sigma_{n=1}^{\infty} u_n$ does *not* converge uniformly on $[0,1]$.

Here is a famous test for uniform convergence called the Weierstrass M test.

9.4E. THEOREM. Let $\Sigma_{k=1}^{\infty} u_k$ be a series of real-valued functions on a set E. If there exist positive numbers $M_1, M_2, \ldots,$ with $\Sigma_{k=1}^{\infty} M_k < \infty$ such that

$$\sum_{k=1}^{\infty} u_k(x) \ll \sum_{k=1}^{\infty} M_k \qquad (x \in E),^*$$

then $\Sigma_{k=1}^{\infty} u_k$ converges uniformly on E.

* That is, there exists $N_1 \in I$ such that for each $k \geqslant N_1$ we have $|u_k(x)| \leqslant M_k$ for all $x \in E$. See 3.6A.

PROOF: Let $s_n = \sum_{k=1}^{n} u_k$, $t_n = \sum_{k=1}^{n} M_k$. Then, for $m > n \geqslant N_1$,

$$|s_m(x) - s_n(x)| = \left| \sum_{k=n+1}^{m} u_k(x) \right| \leqslant \sum_{k=n+1}^{m} |u_k(x)|$$

$$\leqslant \sum_{k=n+1}^{m} M_k = t_m - t_n \qquad (x \in E). \tag{1}$$

Since $\sum_{k=1}^{\infty} M_k < \infty, \{t_n\}_{n=1}^{\infty}$ is a convergent sequence and hence, a Cauchy sequence. Thus given $\epsilon > 0$, there exists $N \geqslant N_1$ such that

$$|t_m - t_n| < \epsilon \qquad (m, n \geqslant N).$$

But then (1) implies

$$|s_m(x) - s_n(x)| < \epsilon \qquad (m, n \geqslant N; x \in E).$$

By 9.2F, $\{s_n\}_{n=1}^{\infty}$ converges uniformly on E. This means that $\sum_{k=1}^{\infty} u_k$ converges uniformly on E, and the proof is complete.

For example, for all real x the series $\sum_{n=1}^{\infty} \sin nx / n^2$ is dominated by $\sum_{n=1}^{\infty} 1/n^2$ which converges. Hence, $\sum_{n=1}^{\infty} \sin nx / n^2$ converges uniformly $(-\infty < x < \infty)$. From 9.4D we then know that the sum of $\sum_{n=1}^{\infty} \sin nx / n^2$ is continuous on $(-\infty, \infty)$.

We emphasize that in 9.4E the numbers M_k must be independent of x.

The M test enables us to prove an important result on power series.

9.4F. THEOREM. If the power series

$$\sum_{k=0}^{\infty} a_k x^k \tag{1}$$

converges for $x = x_0$ (where $x_0 \neq 0$), then (1) converges uniformly on $[-x_1, x_1]$ where x_1 is any number such that $0 < x_1 < |x_0|$.

PROOF: By 3.6I, if (1) converges for $x = x_0$, then (1) converges absolutely for any x with $|x| < x_0|$. In particular, if $0 < x_1 < |x_0|$, then (1) converges absolutely for $x = x_1$. That is,

$$\sum_{k=0}^{\infty} |a_k| x_1^k < \infty.$$

But

$$\sum_{k=1}^{\infty} a_k x^k \ll \sum_{k=0}^{\infty} |a_k| x_1^k \qquad (|x| \leqslant x_1).$$

By 9.4E (with $M_k = |a_k| x_1^k$), the series (1) converges uniformly for $|x| \leqslant x_1$, which is what we wished to show.

For example, the series $\sum_{k=1}^{\infty} x^k / k$ converges for $-1 \leqslant x < 1$. From 9.4F it follows that $\sum_{k=1}^{\infty} x^k / k$ converges uniformly for $-a \leqslant x \leqslant a$ where a is any number such that $0 < a < 1$.

The series $\sum_{k=0}^{\infty} x^k / k!$ converges (to e^x) for all real x. This series does not converge uniformly on $(-\infty, \infty)$. (Verify.) By 9.4F, however, for any $R > 0$ the series $\sum_{k=0}^{\infty} x^k / k!$ *does* converge uniformly on $[-R, R]$.

Dini's theorem 9.2G yields the following result on series of nonnegative continuous functions. This is called Dini's theorem for series.

9.4G. THEOREM. Let $\Sigma_{n=1}^{\infty} u_n$ be a series of continuous nonnegative-valued functions on the compact metric space M. If $\Sigma_{n=1}^{\infty} u_n$ converges on M to the continuous function f, then $\Sigma_{n=1}^{\infty} u_n$ converges to f uniformly on M.

PROOF: For $n \in I$ let $s_n = u_1 + \cdots + u_n$. Since each $u_k(x) \geqslant 0$ for all $x \in M$, we have

$$s_1(x) \leqslant s_2(x) \leqslant \cdots \leqslant s_n(x) \leqslant \cdots \qquad (x \in M).$$

Also, $\lim_{n\to\infty} s_n(x) = f(x)$ $(x \in M)$. By 9.2G, then, the sequence $\{s_n\}_{n=1}^{\infty}$ converges uniformly to f on M. Hence, $\Sigma_{n=1}^{\infty} u_n$ converges uniformly to f on M. This completes the proof.

Exercises 9.4

1. Show that each of the following series converges uniformly on the interval indicated.
   (a) $\Sigma_{n=1}^{\infty} \dfrac{1}{n^2 + x^2}$ $\qquad (0 \leqslant x < \infty)$,
   (b) $\Sigma_{n=1}^{\infty} e^{-nx} x^n$ $\qquad (0 \leqslant x \leqslant 10)$.
   (*Hint*: Find the maximum of xe^{-x} on the interval.)
2. Does the series

$$\sum_{n=0}^{\infty} \frac{x^2}{(1+x^2)^n}$$

   converge uniformly on $(-\infty, \infty)$? (*Hint*: Find the sum of the series for all x.)
3. If $\Sigma_{n=0}^{\infty} |a_n| < \infty$, prove that $\Sigma_{n=0}^{\infty} a_n x^n$ converges uniformly for $0 \leqslant x \leqslant 1$.
4. If the series $\Sigma_{n=0}^{\infty} a_n$ converges and

$$f(x) = \sum_{n=0}^{\infty} a_n x^n \qquad (-1 < x < 1),$$

   prove that f is continuous on $(-1, 1)$.
5. Show that the series

$$\sum_{n=1}^{\infty} \frac{nx^2}{n^3 + x^3}$$

   is uniformly convergent on $[0, A]$ for any $A > 0$.
   Prove that

$$\lim_{x\to 1} \left[\sum_{n=1}^{\infty} \frac{nx^2}{n^3 + x^3} \right] = \sum_{n=1}^{\infty} \frac{n}{n^3 + 1}.$$

6. Let A be a dense subset of the metric space M. If $u_1, u_2, \ldots$ are continuous functions on M, and if $\Sigma_{n=1}^{\infty} u_n$ converges uniformly on A, prove that $\Sigma_{n=1}^{\infty} u_n$ converges uniformly on M.
7. Translate 9.2F into a criterion for the uniform convergence of a *series* of functions.
8. Let $\{u_n\}_{n=1}^{\infty}$ be a sequence of functions on E such that

$$|s_n(x)| \leqslant M \qquad (n \in I; x \in E),$$

   where $s_n = u_1 + \cdots + u_n$. Let $\{b_n\}_{n=1}^{\infty}$ be a nonincreasing sequence of nonnegative numbers that converges to 0.
   Prove that $\Sigma_{n=1}^{\infty} b_n u_n$ converges uniformly on E.
   (*Hint*: See 3.8C.)

9. Use the preceding exercise to show that

$$\frac{\sin x}{1}+\frac{\sin 2x}{2}+\frac{\sin 3x}{3}+\cdots$$

converges uniformly on $[\delta,\pi/2]$ for any $\delta>0$.

9.5 INTEGRATION AND DIFFERENTIATION OF SERIES OF FUNCTIONS

Using results from Section 9.3 on integration and differentiation of sequences of functions, we now investigate similar problems for series of functions.

9.5A. THEOREM. Let $\sum_{k=1}^{\infty}u_k$ be a series of functions in $\mathfrak{R}[a,b]$ that converges uniformly to f on $[a,b]$. Then $f\in\mathfrak{R}[a,b]$ and

$$\int_a^b f(x)\,dx=\sum_{k=1}^{\infty}\int_a^b u_k(x)\,dx.$$

That is,

$$\int_a^b \sum_{k=1}^{\infty} u_k(x)\,dx=\sum_{k=1}^{\infty}\int_a^b u_k(x)\,dx.$$

PROOF: Let $s_n=u_1+\cdots+u_n$. Then $s_n\in\mathfrak{R}[a,b]$ and $\{s_n\}_{n=1}^{\infty}$ converges uniformly to f on $[a,b]$. By 9.3G, $f\in\mathfrak{R}[a,b]$ and

$$\lim_{n\to\infty}\int_a^b s_n(x)\,dx=\int_a^b f(x)\,dx. \tag{1}$$

But, by 7.4C,

$$\begin{aligned}\int_a^b s_n(x)\,dx&=\int_a^b [u_1(x)+\cdots+u_n(x)]\,dx\\&=\sum_{k=1}^{n}\int_a^b u_k(x)\,dx.\end{aligned}$$

Hence,

$$\begin{aligned}\lim_{n\to\infty}\int_a^b s_n(x)\,dx&=\lim_{n\to\infty}\sum_{k=1}^{n}\int_a^b u_k(x)\,dx\\&=\sum_{k=1}^{\infty}\int_a^b u_k(x)\,dx. \qquad (2)\end{aligned}$$

The theorem follows from (1) and (2).

Theorem 9.5A says that a uniformly convergent series of functions may be integrated term by term. That is, if

$$u_1+u_2+\cdots+u_n+\cdots \tag{3}$$

converges uniformly on $[a,b]$, then the integral over $[a,b]$ of (3) is equal to

$$\int_a^b u_1+\int_a^b u_2+\cdots+\int_a^b u_n+\cdots.$$

For example, we have

$$1-x+x^2-x^3+\cdots=\frac{1}{1+x},$$

the series converging for $-1<x<1$. If $|y|<1$, then, by 9.4F, the series converges uniformly on $[0,y]$ (or on $[y,0]$ if $y<0$). By 9.5A we may integrate from 0 to y term by term to obtain

$$\int_0^y 1\,dx-\int_0^y x\,dx+\int_0^y x^2\,dx-\cdots=\int_0^y \frac{1}{1+x}\,dx.$$

Thus

$$y-\frac{y^2}{2}+\frac{y^3}{3}-\cdots=\log(1+y) \qquad (|y|<1).$$

This result was previously obtained in 8.5F and 8.5G.

We next prove a theorem on term-by-term differentiation of series.

9.5B. THEOREM. If $u_1,u_2,\ldots$ are functions each of which has a derivative at every point of $[a,b]$, if u_k' is continuous on $[a,b]$ for $k=1,2,\ldots$, if $\Sigma_{k=1}^{\infty}u_k$ converges to f on $[a,b]$, and if $\Sigma_{k=1}^{\infty}u_k'$ converges uniformly on $[a,b]$, then

$$f'(x)=\sum_{k=1}^{\infty}u_k'(x) \qquad (a\leqslant x\leqslant b).$$

PROOF: Let $s_n=u_1+\cdots+u_n$. Then $\{s_n\}_{n=1}^{\infty}$ converges to f on $[a,b]$. Since $s_n'=u_1'+\cdots+u_n'$, the sequence $\{s_n'\}_{n=1}^{\infty}$ converges uniformly to g on $[a,b]$ where $g=\Sigma_{k=1}^{\infty}u_k'$. Thus by 9.3I,

$$f'(x)=g(x) \qquad (a\leqslant x\leqslant b),$$

which is what we wished to show.

Thus under the conditions of 9.5B, the derivative of

$$u_1+u_2+\cdots+u_n+\cdots$$

is

$$u_1'+u_i'+\cdots+u_n'+\cdots.$$

For example, we have

$$1+x+x^2+\cdots+x^n+\cdots=\frac{1}{1-x}, \tag{1}$$

the series converging for $|x|<1$. By 9.5B, if $0<a<1$, we may differentiate term by term to obtain

$$1+2x+3x^2+\cdots nx^{n-1}+\cdots=\frac{1}{(1-x)^2} \qquad (-a\leqslant x\leqslant a) \tag{2}$$

provided that the series in (2) converges uniformly on $[-a,a]$. But the ratio test may be applied to show that the series in (2) converges for $-1<x<1$. Hence, by 9.4F, the series in (2) *does* converge uniformly on $[-a,a]$ so that our term-by-term differentiation of (1) is justified. Note that, since a was any number between 0 and 1, it follows that

$$1+2x+3x^2+\cdots+nx^{n-1}+\cdots=\frac{1}{(1-x)^2}$$

for all x with $|x|<1$.

This last example illustrates the following general theorem on power series.

9.5C. THEOREM. If $\sum_{n=0}^{\infty} a_n x^n$ converges on $(-S, S)$ for some $S>0$, and if

$$f(x)=\sum_{n=0}^{\infty} a_n x^n \qquad (-S<x<S), \tag{1}$$

then $f'(x)$ exists for $-S<x<S$ and

$$f'(x)=\sum_{n=1}^{\infty} n a_n x^{n-1} \qquad (-S<x<S). \tag{2}$$

PROOF: By 3.6H the series (1) will diverge if $|x|>1/L$, where

$$L=\limsup_{n\to\infty} \sqrt[n]{|a_n|}$$

(or will converge for all real x if $L=0$). Hence, if $L>0$,

$$S \leqslant \frac{1}{L}.$$

But the root test 3.6H applied to

$$\sum_{n=1}^{\infty} n a_n x^{n-1} \tag{3}$$

shows that (3) will converge for $|x|<1/L$ if $L>0$, and for all x if $L=0$. Hence, (3) converges for $|x|<S$. By 9.4F, (3) will converge uniformly on $[-T, T]$ where T is any positive number less than S. Thus by 9.5B term-by-term differentiation of (1) is justified for $|x| \leqslant T$. We obtain

$$f'(x)=\sum_{n=1}^{\infty} n a_n x^{n-1} \qquad (-T \leqslant x \leqslant T). \tag{4}$$

The conclusion (2) follows since (4) holds for any T less than S.

By applying 9.5C to f' (instead of f) we obtain

$$f''(x)=\sum_{n=2}^{\infty} n(n-1) a_n x^{n-2} \qquad (-S<x<S).$$

Proceeding in this fashion we may show that, under the hypotheses of 9.5C, the function f has derivatives of all orders and

$$f^{(k)}(x)=\sum_{n=k}^{\infty} n(n-1)\cdots(n-k+1) a_n x^{n-k}. \tag{5}$$

If we set $x=0$, then all terms of the series vanish except the term for which $n=k$. We obtain

$$f^{(k)}(0)=k!\, a_k.$$

We thus have the following corollary of 9.5C.

9.5D. COROLLARY. If

$$f(x)=\sum_{n=0}^{\infty} a_n x^n \qquad (-S<x<S)$$

for some $S>0$, then, for any $k\in I, f^{(k)}(x)$ exists for $-S<x<S$ and

$$f^{(k)}(x)=\sum_{n=k}^{\infty} n(n-1)\cdots(n-k+1)a_n x^{n-k}.$$

Moreover,

$$a_k=\frac{f^{(k)}(0)}{k!} \qquad (k\in I).$$

That is, if $f(x)=\sum_{n=0}^{\infty}a_n x^n$ $(-S<x<S)$, then $\sum_{n=0}^{\infty}a_n x^n$ must be the Maclaurin series for f.

Exercises 9.5

1. If $\sum_{n=0}^{\infty}|a_n|<\infty$, prove that

$$\int_0^1\left(\sum_{n=0}^{\infty}a_n x^n\right)dx=\sum_{n=0}^{\infty}\frac{a_n}{n+1}.$$

2. Use theorem 9.5C to deduce the equation

$$\cos x=1-\frac{x^2}{2!}+\frac{x^4}{4!}-\cdots \qquad (-\infty<x<\infty)$$

from the equation

$$\sin x=x-\frac{x^3}{3!}+\frac{x^5}{5!}-\cdots \qquad (-\infty<x<\infty).$$

3. Without finding the sum $f(x)$ of the series

$$1+\frac{x^2}{1!}+\frac{x^4}{2!}+\cdots+\frac{x^{2n}}{n!}+\cdots \qquad (-\infty<x<\infty),$$

show that $f'(x)=2xf(x)$ $(-\infty<x<\infty)$.

9.6 ABEL SUMMABILITY

We now take up another method of summability of series called Abel summability. The proof that Abel summability is regular (see 3.9B) involves uniform convergence, which accounts for the presence of this section in the current chapter.

9.6A. The Abel method of summability is illustrated by the following example. The series

$$1-1+1-1+1-\cdots \tag{1}$$

is divergent. We form a power series by multiplying the nth term $(n=0,1,2,\ldots)$ of (1) by x^n to obtain

$$1-x+x^2-x^3+\cdots. \tag{2}$$

For $0\leqslant x<1$ the series (2) converges, and its sum is $1/(1+x)$. Now

$$\lim_{x\to 1^-}\frac{1}{1+x}=\frac{1}{2}.$$

Thus although taking the limit as $x\to 1^-$ in (2) term by term would give the divergent series (1), taking the limit as $x \to 1^-$ of the *sum* of the series (2) yields the limit $\frac{1}{2}$. We will say, therefore, that (1) is Abel summable to $\frac{1}{2}$. [Note that (1) is also $(C,1)$ summable to $\frac{1}{2}$.]

Here is the general definition of Abel summability.

9.6B. DEFINITION. We say that the series $\Sigma_{n=0}^{\infty} a_n$ is Abel summable to L if $\lim_{x\to 1^-} f(x) = L$, where

$$f(x) = \sum_{n=0}^{\infty} a_n x^n \qquad (0 \leqslant x < 1).^*$$

In this case we write

$$\sum_{n=0}^{\infty} a_n = L \qquad (A).$$

In the example 9.6A we thus have $a_n = (-1)^n$ $(n = 0,1,2,\ldots), f(x) = 1/(1+x)$. From 9.6A it follows that

$$1 - 1 + 1 - 1 + 1 - \cdots = \tfrac{1}{2} \qquad (A).$$

Consider now the series

$$1 - 2 + 3 - 4 + \cdots \tag{1}$$

which, of course, diverges. In this case we have

$$f(x) = 1 - 2x + 3x^2 - 4x^3 + \cdots. \tag{2}$$

(The ratio test shows that this power series converges for $-1 < x < 1$.) But since

$$\frac{-1}{1+x} = -1 + x - x^2 + x^3 - x^4 + \cdots \qquad (-1 < x < 1),$$

theorem 9.5C implies

$$\frac{1}{(1+x)^2} = 1 - 2x + 3x^2 - 4x^3 + \cdots.$$

Comparing this with (2) we see that $f(x) = 1/(1+x)^2$. Hence, $\lim_{x\to 1^-} f(x) = \frac{1}{4}$. By 9.6B we conclude

$$1 - 2 + 3 - 4 + \cdots = \tfrac{1}{4} \qquad (A).$$

[From Sections 2.11 and 3.9 we know that (1) is not $(C,1)$ summable but *is* $(C,2)$ summable to $\frac{1}{4}$.]

The next theorem (called Abel's theorem) yields the regularity of the Abel summability method as a corollary.

9.6C. THEOREM. If $\Sigma_{k=0}^{\infty} a_n$ converges, then $\Sigma_{k=0}^{\infty} a_k x^k$ converges uniformly for $0 \leqslant x \leqslant 1$.

PROOF: Given $\epsilon > 0$, we may choose $N \in I$ such that

$$\left| \sum_{k=m+1}^{n} a_k \right| < \epsilon \qquad (m, n \geqslant N).$$

* The Abel summability method thus makes sense only for series $\Sigma_{n=0}^{\infty} a_n$ such that $\Sigma_{n=0}^{\infty} a_n x^n$ converges for $0 \leqslant x < 1$.

That is,

$$-\epsilon < a_{m+1} + a_{m+2} + \cdots + a_n < \epsilon \qquad (m,n \geqslant N).$$

If $0 \leqslant x \leqslant 1$, then Abel's lemma 3.8B (applied to $\{a_k\}_{k=m+1}^{\infty}$ and $\{x^k\}_{k=m+1}^{\infty}$) implies

$$-\epsilon x^{m+1} < a_{m+1}x^{m+1} + a_{m+2}x^{m+2} + \cdots + a_n x^n < \epsilon x^{m+1}.$$

Hence,

$$\left| \sum_{k=m+1}^{n} a_k x^k \right| < \epsilon x^{m+1} \leqslant \epsilon \qquad (m,n \geqslant N; 0 \leqslant x \leqslant 1). \tag{1}$$

If $f_n(x) = \sum_{k=0}^{n} a_k x^k$ $(0 \leqslant x \leqslant 1)$, then (1) implies that $|f_n(x) - f_m(x)| < \epsilon$ for $m,n \geqslant N$ and $0 \leqslant x \leqslant 1$. By 9.2F, $\{f_n\}_{n=0}^{\infty}$ converges uniformly on $[0,1]$. Since $f_n(x)$ is the nth partial sum of $\sum_{k=0}^{\infty} a_k x^k$, the theorem follows.

9.6D. COROLLARY. Abel summability is regular.

PROOF: Suppose $\sum_{k=0}^{\infty} a_n$ converges to L. We must show that $\sum_{n=0}^{\infty} a_n$ is Abel summable to L. That is, we must show that $\lim_{x\to 1^-} f(x) = L$ where

$$f(x) = \sum_{n=0}^{\infty} a_n x^n.$$

But, by 9.6C, $\sum_{n=0}^{\infty} a_n x^n$ converges uniformly for $0 \leqslant x \leqslant 1$. Hence, by 9.4D, f is continuous on $[0,1]$. In particular, f is continuous at 1. Thus $\lim_{x\to 1^-} f(x) = f(1)$. Since $f(1) = \sum_{n=0}^{\infty} a_n = L$, we have $\lim_{x\to 1^-} f(x) = L$, which is what we wished to show.

Abel summability is a stronger method than the (C,k) method for any k. That is, for any $k \in I$, if $\sum_{n=0}^{\infty} a_n$ is (C,k) summable to L, then $\sum_{n=0}^{\infty} a_n$ is also Abel summable to L.* We will content ourselves with the proof of this for $k=1$.

9.6E. THEOREM. If

$$\sum_{n=0}^{\infty} a_n = L \qquad (C,1), \tag{1}$$

then

$$\sum_{n=0}^{\infty} a_n = L \qquad (A).$$

PROOF: Let $s_n = a_0 + a_1 + \cdots + a_n$ $(n = 0,1,2,\ldots)$. Then (1) means that the sequence $s_0, s_1, s_2, \ldots$ is $(C,1)$ summable to L. That is,

$$\lim_{n\to\infty} \sigma_n = L$$

where†

$$\sigma_n = \frac{s_0 + s_1 + \cdots + s_n}{n+1} \qquad (n = 0,1,2,\ldots).$$

Since $(n+1)\sigma_n - n\sigma_{n-1} = (s_0 + \cdots + s_n) - (s_0 + \cdots + s_{n-1}) = s_n$, we have

$$\lim_{n\to\infty} \frac{s_n}{n} = \lim_{n\to\infty} \left(\frac{n+1}{n} \sigma_n - \sigma_{n-1} \right) = L - L = 0.$$

Hence,

$$\lim_{n\to\infty} \frac{a_n}{n} = \lim_{n\to\infty} \frac{s_n - s_{n-1}}{n} = \lim_{n\to\infty} \frac{s_n}{n} - \lim_{n\to\infty} \left(\frac{n-1}{n} \right) \left(\frac{s_{n-1}}{n-1} \right) = 0 - 0 = 0.$$

* There exist, however, series that are Abel summable but that are not (C,k) summable for any k.

† Whether we denote $(s_0 + s_1 + \cdots + s_n)/(n+1)$ by σ_n or σ_{n+1} is immaterial.

By 2.5B, the sequence $\{a_n/n\}_{n=1}^{\infty}$ is bounded. This shows that, for $0 \leqslant x < 1$, the $\sum_{n=0}^{\infty} a_n x^n$ is dominated by $\sum_{n=0}^{\infty} Mnx^n$ for some $M > 0$. Hence, by 3.6B, $\sum_{n=0}^{\infty} a_n x^n$ converges absolutely for $0 \leqslant x < 1$. Let

$$f(x) = a_0 + a_1 x + a_2 x^2 + \cdots \qquad (0 \leqslant x < 1). \tag{2}$$

Then, since

$$\frac{1}{1-x} = 1 + x + x^2 + \cdots \tag{3}$$

is also absolutely convergent for $0 \leqslant x < 1$, we have by 3.5H,

$$\frac{f(x)}{1-x} = \sum_{n=0}^{\infty} c_n x^n \qquad (0 \leqslant x < 1)$$

where $c_n = a_0 \cdot 1 + a_1 \cdot 1 + \cdots + a_n \cdot 1$. That is, $c_n = s_n$, so that

$$\frac{f(x)}{1-x} = \sum_{n=0}^{\infty} s_n x^n \qquad (0 \leqslant x < 1). \tag{4}$$

Using 3.5H once more, we multiply (4) by (3) to obtain

$$\frac{f(x)}{(1-x)^2} = \sum_{n=0}^{\infty} (n+1)\sigma_n x^n \qquad (0 \leqslant x < 1).$$

Hence,

$$f(x) = (1-x)^2 \sum_{n=0}^{\infty} (n+1)\sigma_n x^n \qquad (0 \leqslant x < 1). \tag{5}$$

But

$$\sum_{n=0}^{\infty} (n+1)x^n = 1 + 2x + 3x^2 + \cdots = \frac{1}{(1-x)^2} \qquad (-1 < x < 1),$$

so that

$$(1-x)^2 \sum_{n=0}^{\infty} (n+1)x^n = 1 \qquad (-1 < x < 1),$$

$$L = (1-x)^2 \sum_{n=0}^{\infty} (n+1)Lx^n \qquad (-1 < x < 1). \tag{6}$$

From (5) and (6) we have

$$f(x) - L = (1-x)^2 \sum_{n=0}^{\infty} (n+1)(\sigma_n - L)x^n \qquad (0 \leqslant x < 1). \tag{7}$$

Given $\epsilon > 0$ choose $N \in I$ such that

$$|\sigma_n - L| < \frac{\epsilon}{2} \qquad (n \geqslant N).$$

(We can do this since $\lim_{n\to\infty} \sigma_n = L$.) Then, from (7)

$$\begin{aligned} |f(x) - L| &\leqslant (1-x)^2 \sum_{n=1}^{N} (n+1)|\sigma_n - L|x^n + \frac{\epsilon}{2}(1-x)^2 \sum_{n=N+1}^{\infty} (n+1)x^n \\ &\leqslant (1-x)^2 \sum_{n=1}^{N} (n+1)|\sigma_n - L| + \frac{\epsilon}{2}(1-x)^2 \sum_{n=0}^{\infty} (n+1)x^n \\ &\leqslant (1-x)^2 \cdot A + \frac{\epsilon}{2} \end{aligned}$$

where $A=\Sigma_{n=1}^{N}(n+1)|\sigma_n-L|$. Since $(1-x)^2<\epsilon/2A$ if $1-\delta<x<1$, where $\delta=\sqrt{\epsilon/2A}$, we have

$$|f(x)-L|<\frac{\epsilon}{2A}\cdot A+\frac{\epsilon}{2}=\epsilon \qquad (1-\delta<x<1).$$

This proves $\lim_{x\to 1^-}f(x)=L$, and the proof is complete.

Theorem 3.9C is an immediate consequence of 9.6E and 9.6D.

A series may, of course, be Abel summable even though it is divergent. The following theorem, called Tauber's theorem, gives a condition on a series which, together with the Abel summability of the series, will ensure that the series converges. This theorem is the ancestor of a large class of theorems called Tauberian theorems.

9.6F. THEOREM. If

$$\sum_{n=0}^{\infty} a_n=L \qquad (A), \tag{1}$$

and if

$$\lim_{n\to\infty} na_n=0, \tag{2}$$

then $\Sigma_{n=0}^{\infty}a_n$ converges to L.

PROOF: By (2) and 2.11B, the sequence $\{|na_n|\}_{n=1}^{\infty}$ is $(C,1)$ summable to 0. That is,

$$\lim_{n\to\infty}\frac{1}{n}\sum_{k=1}^{n}|ka_k|=0. \tag{3}$$

Given $\epsilon>0$, it follows from (2), (3), and (1) that there exists $N\in I$ such that

$$|na_n|<\frac{\epsilon}{3} \qquad (n\geqslant N), \tag{4}$$

$$\frac{1}{n}\sum_{k=1}^{n}|ka_k|<\frac{\epsilon}{3} \qquad (n\geqslant N), \tag{5}$$

and such that

$$\left|L-\sum_{k=0}^{\infty}a_kx^k\right|<\frac{\epsilon}{3} \qquad \left(1-\frac{1}{N}<x<1\right). \tag{6}$$

For any $n\in I$ and $x\in(0,1)$ we have

$$\begin{aligned}L-\sum_{k=0}^{n}a_k&=L-\sum_{k=0}^{\infty}a_kx^k+\sum_{k=0}^{\infty}a_kx^k-\sum_{k=0}^{n}a_k\\&=L-\sum_{k=0}^{\infty}a_kx^k+\sum_{k=1}^{n}a_k(x^k-1)+\sum_{k=n+1}^{\infty}a_kx^k.\end{aligned}$$

Hence,

$$\begin{aligned}\left|L-\sum_{k=0}^{n}a_k\right|&\leqslant\left|L-\sum_{k=0}^{\infty}a_kx^k\right|+\sum_{k=1}^{n}|a_k|\cdot(1-x^k)+\sum_{k=n+1}^{\infty}|a_k|x^k.\\&=I_1+I_2+I_3, \quad \text{say.}\end{aligned} \tag{7}$$

For any $n\geqslant N$, choose x such that $1-1/n<x<1-1/(n+1)$. Then $1-1/N\leqslant 1-1/n<x<1$, and so, by (6),

$$I_1=\left|L-\sum_{k=0}^{\infty}a_kx^k\right|<\frac{\epsilon}{3}.$$

Now $1-x^k=(1-x)(1+x+x^2+\cdots+x^{k-1})\leqslant k(1-x)$, for any $k\in I$. Hence, since $1-x<1/n$, we have

$$1-x^k\leqslant k(1-x)<\frac{k}{n}. \tag{8}$$

By (8) and (5) we then have (since $n\geqslant N$)

$$I_2=\sum_{k=1}^{n}|a_k|(1-x^k)<\frac{1}{n}\sum_{k=1}^{n}|ka_k|<\frac{\epsilon}{3}.$$

To estimate I_3 we have, using (4),

$$I_3=\sum_{k=n+1}^{\infty}|ka_k|\frac{x^k}{k}<\frac{\epsilon}{3}\sum_{k=n+1}^{\infty}\frac{x^k}{k}\leqslant\frac{\epsilon}{3(n+1)}\sum_{k=n+1}^{\infty}x^k$$

$$\leqslant\frac{\epsilon}{3(n+1)}\sum_{k=0}^{\infty}x^k=\frac{\epsilon}{3(n+1)(1-x)}.$$

But $x<1-1/(n+1)$ and so $1-x>1/(n+1)$. Thus $(n+1)(1-x)>1$ and so

$$I_3\leqslant\frac{\epsilon}{3(n+1)(1-x)}<\frac{\epsilon}{3}.$$

From (7) we then have

$$\left|\sum_{k=0}^{n}a_k-L\right|\leqslant I_1+I_2+I_3<\epsilon \qquad (n\geqslant N),$$

which proves that $\sum_{k=0}^{\infty}a_k$ converges to L.

In view of 9.6E, we see that the theorem in 3.9D is a consequence of 9.6F.

Exercises 9.6

1. Show that the following series are Abel summable:
   (a) $1-\frac{1}{3}+\frac{1}{5}-\frac{1}{7}+\cdots$.
   (b) $1-3+6-10+15-\cdots$.
2. If

$$\sum_{n=0}^{\infty}a_n=L \qquad (\text{A})$$

and

$$\sum_{n=0}^{\infty}b_n=M \qquad (\text{A}),$$

prove that

$$\sum_{n=0}^{\infty}(a_n+b_n)=L+M \qquad (\text{A}).$$

3. If

$$a_0+a_1+a_2+\cdots=L \qquad (\text{A}),$$

prove that

$$0+a_0+0+a_1+0+a_2+\cdots=L \qquad (\text{A}).$$

4. If $\sum_{n=0}^{\infty}a_nL^n$ converges, where $L>0$, show that $\sum_{n=0}^{\infty}a_nz^n$ converges uniformly $(0\leqslant z\leqslant L)$.

5. Prove that if

$$f(x)=\sum_{n=1}^{\infty}\frac{x^n}{n} \qquad (0\leqslant x<1), \qquad (*)$$

then

$$\int_0^1 f(x)\,dx=\sum_{n=1}^{\infty}\frac{1}{n(n+1)}.$$

Is the integral proper or improper? [*Hint*: Integrate (*) from 0 to y where $0<y<1$. Then use 9.6C to evaluate the limit of the right side as $y\to 1^-$.]

6. If $\Sigma_{n=0}^{\infty}a_n$ and $\Sigma_{n=0}^{\infty}b_n$ are convergent series of real numbers, if

$$c_n=\sum_{k=0}^{n}a_k b_{n-k} \qquad (n=0,1,2,\ldots),$$

and if $\Sigma_{n=1}^{\infty}c_n$ converges, show that

$$\left(\sum_{n=0}^{\infty}a_n\right)\left(\sum_{n=0}^{\infty}b_n\right)=\sum_{n=0}^{\infty}c_n.$$

[*Hint*: First show that

$$\left(\sum_{n=0}^{\infty}a_n x^n\right)\left(\sum_{n=0}^{\infty}b_n x^n\right)=\sum_{n=0}^{\infty}c_n x^n$$

for $0\leqslant x<1$.] This result becomes false if the hypothesis that $\Sigma_{n=0}^{\infty}c_n$ converges is deleted.

9.7 A CONTINUOUS, NOWHERE-DIFFERENTIABLE FUNCTION

As another application of uniform convergence we construct a real-valued function F that is continuous on R^1 but does not have a derivative at any point of R^1. We do this by defining F as the sum $u_1+u_2+\cdots+u_n+\cdots$ of a uniformly convergent series of continuous functions, where the graph of u_n has roughly $2\cdot 10^n$ sharp corners in every interval $[x,x+1]$ and hence, wiggles quite a bit. The cumulative effect of all this wiggling yields the desired result.

9.7A. THEOREM. There exists a real-valued function F such that F is continuous on R^1, but such that $F'(a)$ exists for no $a\in R^1$.

PROOF: In the proof we use only decimal expansions that do not end in a string of 9's.

We begin by defining the function f_0 as follows: For any real x let $f_0(x)$ be the distance from x to the nearest integer. For example, if $x=7.3$, then the integer nearest to x is 7. Hence, $f_0(x)=$ distance from 7.3 to $7=0.3$. Similarly, $f_0(-6)=0, f_0(1.83)=0.17$. [The graph of f_0 consists of straight line segments joining $\langle m,0\rangle$ to $\langle m+\frac{1}{2},\frac{1}{2}\rangle$ and $\langle m+\frac{1}{2},\frac{1}{2}\rangle$ to $\langle m+1,0\rangle$ for $m=0,\pm 1,\pm 2,\ldots$.] It is clear that

$$f_0(x+1)=f_0(x) \qquad (x\in R^1).$$

Now define $f_1(x)$ as the distance from $10x$ to the nearest integer. For example, $f_1(7.64)=$ distance from 76.4 to $76=0.4$. We note that

$$f_1(x)=f_0(10x) \qquad (x\in R^1).$$

Continuing, we define $f_2(x)$ as the distance from $100x$ to the nearest integer. For example, $f_2(0.678)=0.2$. In general for $k=0,1,2,\ldots$ we define

$$f_k(x)=f_0(10^k x) \qquad (x\in R^1).$$

Now define F as

$$F(x)=\sum_{k=0}^{\infty}\frac{f_k(x)}{10^k} \qquad (x\in R^1). \tag{1}$$

Since

$$\sum_{k=0}^{\infty}\frac{f_k(x)}{10^k}\ll\sum_{k=0}^{\infty}\frac{\frac{1}{2}}{10^k} \qquad (x\in R^1),$$

the M test 9.4E shows that the series in (1) converges uniformly on R^1. Since each f_k is continuous on R^1, it follows from 9.4D that F is continuous on R^1.

Now we will show that if $a\in R^1$, then $F'(a)$ does not exist. To do this it is enough to show the existence of a sequence $\{x_n\}_{n=1}^{\infty}$ such that $\lim_{n\to\infty}x_n=a$ but such that

$$\lim_{n\to\infty}\frac{F(x_n)-F(a)}{x_n-a}$$

does not exist.

Suppose $a=a_0\cdot a_1a_2a_3\cdots a_n\cdots$. For $n\in I$ let

$$x_n=a_0\cdot a_1a_2\cdots a_{n-1}b_na_{n+1}\cdots$$

where $b_n=a_n+1$ if $a_n\neq 4$ or 9 and $b_n=a_n-1$ if $a_n=4$ or $a_n=9$. Thus $x_n-a=\pm 10^{-n}$, and so $\lim_{n\to\infty}x_n=a$. For example, if

$$a=0.27451,$$

then

$$\begin{aligned}
x_1&=0.37451\\
x_2&=0.28451\\
x_3&=0.27351\\
x_4&=0.27461.
\end{aligned}$$

For this example we have

$$\begin{aligned}
f_0(x_3)-f_0(a)&=-0.001,\\
f_1(x_3)-f_1(a)&=+0.01,\\
f_2(x_3)-f_2(a)&=-0.1,\\
f_3(x_3)-f_3(a)&=0,\\
f_k(x_3)-f_k(a)&=0 \qquad (k\geqslant 3).
\end{aligned}$$

[To see why we want $x_3=0.27351$ and not 0.27551, calculate $f_2(0.27551)-f_2(a)$.]

This is a numerical example of the following fact. For any $n\in I$,

$$\begin{aligned}
f_k(x_n)-f_k(a)&=\pm 10^{k-n} \qquad (k=0,1,\ldots,n-1)\\
f_k(x_n)-f_k(a)&=0 \qquad (k\geqslant n).
\end{aligned} \tag{2}$$

Thus

$$\begin{aligned}
\frac{F(x_n)-F(a)}{x_n-a}&=\sum_{k=0}^{\infty}\frac{f_k(x_n)-f_k(a)}{10^k(x_n-a)}=\sum_{k=0}^{n-1}\frac{\pm 10^{k-n}}{10^k(\pm 10^{-n})}\\
&=\sum_{k=0}^{n-1}\pm 1.
\end{aligned}$$

That is, for any $n \in I$, $[F(x_n) - F(a)]/(x_n - a)$ is the sum of n terms each of which is $+1$ or -1. It follows that $[F(x_n) - F(a)]/(x_n - a)$ is an odd integer if n is odd while $[F(x_n) - F(a)]/(x_n - a)$ is an even integer if n is even. This proves that $\lim_{n\to\infty}[F(x_n) - F(a)]/(x_n - a)$ does not exist, which is what we wished to show.

9.7B. The above example of a continuous, nowhere-differentiable function is due to van der Waerden. An earlier example, due to Weierstrass, is given by the function

$$F(x) = \sum_{n=0}^{\infty} \frac{\cos 3^n x}{2^n} \qquad (x \in R^1).$$

Since the series is uniformly convergent on R^1, it follows from 9.4D that F is continuous on R^1. We omit the proof that $F'(x)$ exists for no x. However, note that if we differentiate the series term by term we obtain $-\sum_{n=0}^{\infty}(\frac{3}{2})^n \sin 3^n x$, which diverges when x is not a multiple of π. This gives reason to believe that F is nowhere differentiable.

Again it is the fact that the graph of the function $\cos 3^n x$ wiggles rapidly (for large n) that makes this example work.

Exercises 9.7

1. Let $a = 0.39261$
   (a) Calculate x_1, x_2, x_3 for this a.
   (b) Calculate $f_0(x_2) - f_0(a)$,
   $f_1(x_2) - f_1(a)$,
   $f_2(x_2) - f_2(a)$,
   $f_3(x_2) - f_3(a)$.
   (c) Do the same for x_3.
   (d) Use (b) and (c) to show that $[F(x_2) - F(a)]/(x_2 - a)$ is even and that $[F(x_3) - F(a)]/(x_3 - a)$ is odd.

10

THREE FAMOUS THEOREMS

10.1 THE METRIC SPACE $C[a,b]$

In each of the next three sections we will prove a well-known and important theorem in analysis. Each of these theorems may be viewed as a theorem about a particular metric space—namely, the space of continuous functions on a closed bounded interval endowed with a metric, which we now proceed to define. We begin by defining the norm of a continuous real-valued function f on a closed bounded interval $[a,b]$. (Note that the absolute value $|f|$ of such a function is also continuous and hence, by 6.6G, attains a maximum at some point of $[a,b]$.)

10.1A. DEFINITION. Let f be a continuous real-valued function on the closed bounded interval $[a,b]$. We define $\|f\|$, called the norm of f, as

$$\|f\| = \max_{a \leqslant x \leqslant b} |f(x)|.$$

This norm has properties similar to the norm for ℓ^2. Specifically (compare with 3.10E),

10.1B. THEOREM. The norm for continuous real-valued functions f,g on the closed bounded interval $[a,b]$ has the following properties:

$$\|f\| \geqslant 0, \tag{1}$$

$$\|f\| = 0 \quad \text{if and only if } f(x) = 0 \quad (a \leqslant x \leqslant b), \tag{2}$$

$$\|cf\| = |c| \cdot \|f\| \qquad (c \text{ a real number}), \tag{3}$$

$$\|f+g\| \leqslant \|f\| + \|g\| \tag{4}$$

PROOF: All but property (4) are obvious. To establish (4) we have, for any $x \in [a,b]$,

$$|f(x)+g(x)| \leqslant |f(x)| + |g(x)| \leqslant \|f\| + \|g\|,$$

and hence,

$$\max_{a \leqslant x \leqslant b} |f(x)+g(x)| \leqslant \|f\| + \|g\|.$$

The left side of this inequality is precisely $\|f+g\|$, and (4) follows.

10.1C. From 10.1B it follows easily that if we define ρ by

$$\rho(\varphi,\psi)=\|\varphi-\psi\|,$$

then ρ is a metric for the space of continuous real-valued functions on $[a,b]$. For example, if φ,ψ,η are such functions, then

$$\rho(\varphi,\eta)=\|\varphi-\eta\|=\|(\varphi-\psi)+(\psi-\eta)\|.$$

By (4) of 10.1B we then have

$$\rho(\varphi,\eta)\leqslant\|\varphi-\psi\|+\|\psi-\eta\|=\rho(\varphi,\psi)+\rho(\psi,\eta).$$

Hence, ρ satisfies the triangle inequality. The other requirements for a metric are easily verified. This leads us to the following definition.

DEFINITION. By $C[a,b]$ we mean the metric space of all continuous, real-valued functions on the closed bounded interval $[a,b]$ with metric ρ defined by

$$\rho(f,g)=\|f-g\| \qquad (f,g\in C[a,b]).$$

The "points" (or elements) of $C[a,b]$ are thus functions on $[a,b]$.

It turns out that convergence of a sequence $\{f_n\}_{n=1}^{\infty}$ in $C[a,b]$ with respect to the metric of $C[a,b]$ (see 4.3C) is precisely the same as uniform convergence on $[a,b]$ of the sequence of functions $\{f_n\}_{n=1}^{\infty}$.

10.1D. THEOREM. The sequence $\{f_n\}_{n=1}^{\infty}$ in $C[a,b]$ converges to $f\in C[a,b]$ (with respect to the metric ρ) if and only if $\{f_n\}_{n=1}^{\infty}$ converges uniformly to f on $[a,b]$.

PROOF: Suppose that $\{f_n\}_{n=1}^{\infty}$ converges to f with respect to the metric ρ. This means that given $\epsilon>0$ there exists $N\in I$ such that

$$\rho(f_n,f)=\|f_n-f\|<\epsilon \qquad (n\geqslant N). \tag{1}$$

or

$$\max_{a\leqslant x\leqslant b}|f_n(x)-f(x)|<\epsilon \qquad (n\geqslant N). \tag{2}$$

But (2) is equivalent to

$$|f_n(x)-f(x)|<\epsilon \qquad (n\geqslant N;a\leqslant x\leqslant b). \tag{3}$$

(Why?) From (3) we see that $\{f_n\}_{n=1}^{\infty}$ converges uniformly to f on $[a,b]$.

Conversely, the fact that (3) implies (1) shows that if $\{f_n\}_{n=1}^{\infty}$ converges uniformly to f on $[a,b]$, then $\{f_n\}_{n=1}^{\infty}$ converges to f with respect to the metric ρ of $C[a,b]$. This completes the proof.

From 10.1D we deduce the following important result.

10.1E. THEOREM. The metric space $C[a,b]$ is complete.

PROOF: Let $\{f_n\}_{n=1}^{\infty}$ be a Cauchy sequence in $C[a,b]$. (See 4.3D.) Then given $\epsilon>0$ there exists $N\in I$ such that

$$\rho(f_m,f_n)<\epsilon \qquad (m,n\geqslant N).$$

That is,

$$\|f_m-f_n\|<\epsilon \qquad (m,n>N)$$

or

$$\max_{a\leqslant x\leqslant b}|f_m(x)-f_n(x)|<\epsilon \qquad (m,n\geqslant N).$$

This implies

$$|f_m(x)-f_n(x)|<\epsilon \qquad (m,n\geqslant N; a\leqslant x\leqslant b).$$

By 9.2F, $\{f_n\}_{n=1}^{\infty}$ converges uniformly on $[a,b]$ to some function f. Moreover, $f\in C[a,b]$ by 9.3C. But then, by 10.1D, $\{f_n\}_{n=1}^{\infty}$ is convergent to f with respect to ρ. Hence, every Cauchy sequence in $C[a,b]$ converges to a point in $C[a,b]$, which proves that $C[a,b]$ is complete.

We will need the following corollary.

10.1F. COROLLARY. Let l and m be any real numbers with $l<m$. Let C^* be the subset of $C[a,b]$ consisting of all $f\in C[a,b]$ such that

$$l\leqslant f(x)\leqslant m \qquad (a\leqslant x\leqslant b).$$

Then C^* (with the metric of $C[a,b]$) is a complete metric space.

PROOF: Since $C^*\subset C[a,b]$ and $C[a,b]$ is complete, it is sufficient, by 6.4C, to prove that C^* is a closed subset of $C[a,b]$. Accordingly, suppose $f\in C[a,b]$ is a limit point of C^*. Then there exists a sequence $\{f_n\}_{n=1}^{\infty}$ in C^* that converges (with respect to the metric of $C[a,b]$) to f. Hence, by 10.1D, $\{f_n\}_{n=1}^{\infty}$ converges uniformly to f on $[a,b]$, and so $\{f_n\}_{n=1}^{\infty}$ converges pointwise to f on $[a,b]$. That is,

$$\lim_{n\to\infty} f_n(x)=f(x) \qquad (a\leqslant x\leqslant b). \tag{1}$$

But each f_n is in C^* and so

$$l\leqslant f_n(x)\leqslant m \qquad (a\leqslant x\leqslant b; n\in I). \tag{2}$$

From (1), (2), and 2.7E, it follows that $l\leqslant f(x)\leqslant m$ for all x in $[a,b]$. Thus $f\in C^*$. We have shown that C^* contains all its limit points. Hence, C^* is closed, which is what we wished to show.

Exercises 10.1

1. If L is the real-valued function on $C[a,b]$ defined by

$$L(f)=\int_a^b f \qquad (f\in C[a,b]),$$

prove that L is continuous on $C[a,b]$.

2. Let T be a contraction (6.4E) on $C[a,b]$. Show that T is then a uniformly continuous function on $C[a,b]$.
3. Let $C^1[0,1]$ be the subspace of $C[0,1]$ consisting of all $f\in C[0,1]$ such that f is differentiable and $f'\in C[0,1]$. Give $C^1[0,1]$ the metric of $C[0,1]$. Define $T: C^1[0,1]\to C[0,1]$ by

$$Tf=f' \qquad (f\in C^1[0,1]).$$

Show that T is not continuous.

4. Let $m,n\in R$ with $m<n$. Let

$$A=\{f\in C[a,b]\,\big|\,m<f(x)<n\} \qquad (a\leqslant x\leqslant b).$$

Prove that A is an open subset of $C[a,b]$.

5. Let $B[a,b]$ be the set of all real-valued bounded functions on $[a,b]$. For $f\in B[a,b]$ let

$$\|f\|_B=\operatorname*{l.u.b.}_{a\leqslant x\leqslant b}|f(x)|,$$

and define a metric ρ for $B[a,b]$ by

$$\rho(f,g)=\|f-g\|_B \qquad (f,g\in B[a,b]).$$

Prove that $B[a,b]$ is complete.

6. Prove that $C[a,b]$ is not compact.

10.2 THE WEIERSTRASS APPROXIMATION THEOREM

By a polynomial function (or, more simply, a polynomial) we mean a function P that can be expressed as

$$P(x)=a_0x^n+a_1x^{n-1}+\cdots+a_{n-1}x+a_n \qquad (x\in R^1)$$

where n is some nonnegative integer and $a_0,a_1,\ldots,a_n$ are real numbers. The restriction of a polynomial function to the closed bounded interval $[a,b]$ is clearly continuous on $[a,b]$. On the other hand, there exist functions in $C[a,b]$ that are not polynomials.* (For example, a function in $C[a,b]$ that fails to have a derivative at some point of $[a,b]$ cannot be a polynomial.) Weierstrass showed, however, that every function in $C[a,b]$ is uniformly approximable by polynomials. This result is called the Weierstrass approximation theorem, which we now state.

10.2A. THEOREM. Let f be any function in $C[a,b]$. Then, given $\epsilon>0$, there exists a polynomial P such that

$$|P(x)-f(x)|<\epsilon \qquad (a\leqslant x\leqslant b). \tag{1}$$

10.2B. We will give a proof of 10.2A presently. However, we will first give some reformulations. Note that (1) of 10.2A may be written

$$\|P-f\|<\epsilon$$

or, in terms of the metric ρ for $C[a,b]$,

$$\rho(P,f)<\epsilon.$$

Thus 10.2A says that any open ball $B[f;\epsilon]$ about f in $C[a,b]$ contains a polynomial. By 5.5D, this means that every $f\in C[a,b]$ is the limit point of the set $\mathcal{P}$ of all polynomials. Hence, 10.2A may be rephrased as

I. The set $\mathcal{P}$ of all polynomials is dense in the metric space $C[a,b]$.

Another reformulation of 10.2A is obtained as follows. The statement I is equivalent to saying that for each $f\in C[a,b]$ there is a sequence $\{P_n\}_{n=1}^{\infty}$ of polynomials that converges in $C[a,b]$ to f. By 10.1D this happens if and only if

II. For each $f\in C[a,b]$ there is a sequence of polynomials $\{P_n\}_{n=1}^{\infty}$ such that $\{P_n\}_{n=1}^{\infty}$ converges uniformly to f on $[a,b]$.

Hence, 10.2A, I, and II are all equivalent.

10.2C. Before we give a proof of 10.2A we wish to observe the following lemma.

LEMMA. It is sufficient to prove 10.2A for the special case in which $[a,b]=[0,1]$.

* More precisely, there exist functions in $C[a,b]$ that are not the restriction to $[a,b]$ of polynomials.

PROOF: Suppose 10.2A were true for $C[0,1]$. We will show that 10.2A would then be true for $C[a,b]$, where $[a,b]$ is any closed bounded interval. Thus if $f\in C[a,b]$ and $\epsilon>0$, we must find a polynomial P such that

$$|P(x)-f(x)|<\epsilon \qquad (a\leqslant x\leqslant b). \tag{1}$$

Define g by

$$g(x)=f(a+[b-a]x) \qquad (0\leqslant x\leqslant 1).$$

Then $g(0)=f(a), g(1)=f(b)$. Indeed, by 5.3D, g is continuous on $[0,1]$. Thus by our assumption that 10.2A holds for $C[0,1]$, there is a polynomial Q such that

$$|g(y)-Q(y)|<\epsilon \qquad (0\leqslant y\leqslant 1).$$

If we set $y=(x-a)/(b-a)$, then

$$g(y)=g\left(\frac{x-a}{b-a}\right)=f\left(a+(b-a)\frac{x-a}{b-a}\right)=f(x).$$

We then have

$$\left|f(x)-Q\left(\frac{x-a}{b-a}\right)\right|<\epsilon \qquad (a\leqslant x\leqslant b). \tag{2}$$

If we define P by

$$P(x)=Q\left(\frac{x-a}{b-a}\right),$$

then, by the binomial theorem, P is a polynomial (because Q is). Inequality (1) then follows from (2), and the proof is complete.

10.2D. We now prove 10.2A for $C[0,1]$. This proof is due to Bernstein and makes use of the so-called Bernstein polynomials.

For any $f\in C[0,1]$ we define a sequence of polynomials $\{B_n\}_{n=1}^{\infty}$ as follows:

$$B_n(x)=\sum_{k=0}^{n}\binom{n}{k}x^k(1-x)^{n-k}f\left(\frac{k}{n}\right) \qquad (0\leqslant x\leqslant 1; n\in I). \tag{1}$$

Here

$$\binom{n}{k}=\frac{n!}{k!(n-k)!}.$$

The polynomial B_n is called the nth Bernstein polynomial for f.

Given $\epsilon>0$ we will show that there exists $N\in I$ such that

$$\|f-B_n\|<\epsilon \qquad (n\geqslant N). \tag{2}$$

This will show that $\{B_n\}_{n=1}^{\infty}$ converges uniformly to f on $[0,1]$ and, in particular, will prove 10.2A.

We need a considerable amount of preliminary computation. For any $p,q\in R$ we have, by the binomial theorem

$$\sum_{k=0}^{n}\binom{n}{k}p^kq^{n-k}=(p+q)^n \qquad (n\in I). \tag{3}$$

Differentiating with respect to p we obtain

$$\sum_{k=0}^{n}\binom{n}{k}kp^{k-1}q^{n-k}=n(p+q)^{n-1},$$

which implies

$$\sum_{k=0}^{n} \frac{k}{n}\binom{n}{k} p^k q^{n-k} = p(p+q)^{n-1} \qquad (n \in I). \tag{4}$$

Differentiating once more we have

$$\sum_{k=0}^{n} \frac{k^2}{n}\binom{n}{k} p^{k-1} q^{n-k} = p(n-1)(p+q)^{n-2} + (p+q)^{n-1},$$

and so

$$\sum_{k=0}^{n} \frac{k^2}{n^2}\binom{n}{k} p^k q^{n-k} = p^2\left(1-\frac{1}{n}\right)(p+q)^{n-2} + \frac{p}{n}(p+q)^{n-1}. \tag{5}$$

Now, if $x \in [0,1]$, set $p = x, q = 1-x$. Then (3), (4), and (5) yield

$$\sum_{k=0}^{n} \binom{n}{k} x^k (1-x)^{n-k} = 1. \tag{6}$$

$$\sum_{k=0}^{n} \frac{k}{n}\binom{n}{k} x^k (1-x)^{n-k} = x,$$

$$\sum_{k=0}^{n} \frac{k^2}{n^2}\binom{n}{k} x^k (1-x)^{n-k} = x^2\left(1-\frac{1}{n}\right) + \frac{x}{n}.$$

From these equations it follows easily [on expanding $(k/n - x)^2$] that

$$\sum_{k=0}^{n} \left(\frac{k}{n} - x\right)^2 \binom{n}{k} x^k (1-x)^{n-k} = \frac{x(1-x)}{n} \qquad (0 \leqslant x \leqslant 1). \tag{7}$$

By 6.8C, f is uniformly continuous on $[0,1]$. Hence, given $\epsilon > 0$ there exists $\delta > 0$ such that

$$|f(x) - f(y)| < \frac{\epsilon}{2} \qquad (|x-y| < \delta; x,y \in [0,1]).$$

Now choose $N \in I$ such that

$$\frac{1}{\sqrt[4]{N}} < \delta \tag{8}$$

and such that

$$\frac{1}{\sqrt{N}} < \frac{\epsilon}{4\|f\|}. \tag{9}$$

(We may assume, of course, that $\|f\| > 0$.)

Fix $x \in [0,1]$. Multiplying (6) by $f(x)$ and subtracting (1) we obtain, for any $n \in I$

$$f(x) - B_n(x) = \sum_{k=0}^{n} \left[f(x) - f\left(\frac{k}{n}\right)\right]\binom{n}{k} x^k (1-x)^{n-k} = \Sigma' + \Sigma'' \tag{10}$$

where Σ' is the sum over those values of k such that

$$\left|\frac{k}{n} - x\right| < \frac{1}{\sqrt[4]{n}}, \tag{11}$$

while Σ'' is the sum over the other values of k. If k does not satisfy (11), that is, if $|k/n-x| \geqslant 1/\sqrt[4]{n}$, then $(k-nx)^2 = n^2|k/n-x|^2 \geqslant \sqrt{n^3}$. Hence,

$$\begin{aligned}|\Sigma''| &= \left|\Sigma''\left[f(x)-f\left(\frac{k}{n}\right)\right]\binom{n}{k}x^k(1-x)^{n-k}\right| \\ &\leqslant \Sigma''\left[|f(x)|+\left|f\left(\frac{k}{n}\right)\right|\right]\binom{n}{k}x^k(1-x)^{n-k} \\ &\leqslant 2\|f\|\Sigma''\binom{n}{k}x^k(1-x)^{n-k} \leqslant \frac{2\|f\|}{\sqrt{n^3}}\Sigma''(k-nx)^2\binom{n}{k}x^k(1-x)^{n-k} \\ &\leqslant \frac{2\|f\|}{\sqrt{n^3}}\sum_{k=0}^{n}(k-nx)^2\binom{n}{k}x^k(1-x)^{n-k}.\end{aligned}$$

Hence, by (7)

$$|\Sigma''| \leqslant \frac{2\|f\|}{\sqrt{n^3}}\cdot nx(1-x) \leqslant \frac{2\|f\|}{\sqrt{n}}.$$

If $n \geqslant N$, it follows from (9) that $1/\sqrt{n} < \epsilon/4\|f\|$ and so

$$|\Sigma''| < \frac{\epsilon}{2}.$$

Moreover, if $n \geqslant N$ and if k satisfies (11), then, by (8) and (11), $|k/n-x| < \delta$ and so

$$\left|f(x)-f\left(\frac{k}{n}\right)\right| < \frac{\epsilon}{2}.$$

Thus

$$|\Sigma'| = \left|\sum{}'\left[f(x)-f\left(\frac{k}{n}\right)\right]\binom{n}{k}x^k(1-x)^{n-k}\right| < \frac{\epsilon}{2}\sum{}'\binom{n}{k}x^k(u-x)^{n-k},$$

and so by (6)

$$|\Sigma'| < \frac{\epsilon}{2}.$$

Thus from (10),

$$|f(x)-B_n(x)| \leqslant |\Sigma'| + |\Sigma''| < \frac{\epsilon}{2} + \frac{\epsilon}{2} = \epsilon.$$

Since x was any point in $[0,1]$, and n any integer with $n \geqslant N$, this shows that

$$|f(x)-B_n(x)| < \epsilon \qquad (0 \leqslant x \leqslant 1; n \geqslant N).$$

This establishes (2), and the proof is complete.

10.2E. The theory of probability throws some light on the preceding proof. Suppose a coin is tossed and that the probability of heads is x while the probability of tails is, accordingly, $1-x$. If the coin is tossed n times, then the probability of exactly k heads in the n tosses is $\binom{n}{k}x^k(1-x)^{n-k}$. [This expression occurs in the definition of $B_n(x)$.]

The expected number of heads in n tosses is nx. (This is a technical probabilistic fact that is surely believable even to those who do not know the precise definition of "expected number.") Indeed, one feels sure that one is more likely to obtain precisely k heads in n tosses for the value (or values) of k close to nx than for those k that are far from nx. Thus Σ' refers to the k for which precisely k heads in n tosses is "more probable," while Σ'' refers to the k for which precisely k heads in n tosses is "less probable."

Indeed, the proof of 10.2A that we have given is essentially the same as one of the more familiar proofs of the "weak law of large numbers."

10.2F. We have proved that the set of *all* polynomials is dense in $C[0,1]$. It is natural to ask whether we actually need *all* polynomials.

Specifically, let $N=\{n_i\}_{i=1}^{\infty}$ be a strictly increasing sequence of positive integers, and let $\mathcal{P}_N$ be the set of all polynomials P of the form

$$P(x)=a_0+a_1x^{n_1}+a_2x^{n_2}+\cdots+a_kx^{n_k}.$$

That is, $P\in\mathcal{P}_N$ if P is a constant plus a polynomial whose exponents all belong to N. Here is a striking result (whose proof is beyond the scope of this book).

THE MUNTZ-SZASZ THEOREM. The set $\mathcal{P}_N$ is dense in $C[0,1]$ if and only if

$$\sum_{i=1}^{\infty}\frac{1}{n_i}=\infty.$$

Thus for example, the set of all polynomials of the form

$$a_0+a_1x^2+a_2x^4+a_3x^8+\cdots+a_kx^{2^k}$$

is not dense in $C[0,1]$ since here, $n_i=2^i$ so that $\Sigma_{i=1}^{\infty}\frac{1}{n_i}=\Sigma_{i=1}^{\infty}\frac{1}{2^i}<\infty$. On the other hand, the set of all polynomials with even exponents *is* dense in $C[0,1]$.

Exercises 10.2

1. Calculate B_1, B_2, and B_3 for f where

$$f(x)=x^2 \qquad (0\leqslant x\leqslant 1).$$

Then graph these functions.

10.3 PICARD EXISTENCE THEOREM FOR DIFFERENTIAL EQUATIONS

Many problems in a course in elementary differential equations involve the solution of equations of the form

$$\frac{dy}{dx}=f(x,y) \tag{1}$$

with initial condition

$$y(x_0)=y_0. \tag{2}$$

Here f is, of course, some real-valued function defined on all or part of R^2. By a solution we mean a function φ with domain containing some interval $[x_0-\delta,x_0+\delta]$ such that $\varphi(x_0)=y_0$ and

$$\varphi'(x)=f[x,\varphi(x)] \qquad (|x-x_0|\leqslant\delta). \tag{3}$$

This is equivalent (via integration) to the equation

$$\varphi(x)=y_0+\int_{x_0}^{x}f[t,\varphi(t)]\,dt \qquad (|x-x_0|\leqslant\delta). \tag{4}$$

Thus the question of the existence of a solution to the problem posed by (1) and (2) is equivalent to the existence of a function φ satisfying (4) for some δ. We now prove a theorem, due to Picard, which gives conditions on f sufficient to ensure both the existence and the uniqueness of a function φ satisfying (4).

THEOREM. If f is continuous on some rectangle $D \subset R^2$ whose interior contains $\langle x_0, y_0 \rangle$,* and if there exists $M > 0$ such that †

$$|f(x,y_1) - f(x,y_2)| \leqslant M|y_1 - y_2| \qquad (\langle x,y_1 \rangle, \langle x,y_2 \rangle \in D), \tag{5}$$

then there exists $\delta > 0$ and a unique function φ such that (4) holds.

PROOF: Since f is continuous on the compact set D, we know (by 6.6B) that there exists $k > 0$ such that

$$|f(x,y)| \leqslant k \qquad (\langle x,y \rangle \in D).$$

Choose $\delta > 0$ such that

$$\langle x,y \rangle \in D \quad \text{if} \quad |x - x_0| \leqslant \delta, |y - y_0| \leqslant k\delta \tag{6}$$

and such that

$$M\delta < 1 \tag{7}$$

where M is as in (5).

Let C^* be the subset of $C[x_0 - \delta, x_0 + \delta]$ consisting of all functions φ that are continuous on $[x_0 - \delta, x_0 + \delta]$ and such that

$$|\varphi(x) - y_0| \leqslant k\delta. \qquad (|x - x_0| \leqslant \delta).$$

Then, by 10.1F, C^* is a complete metric space. Note that by (6) we have $\langle t, \varphi(t) \rangle \in D$ if $|t - x_0| \leqslant \delta$ and $\varphi \in C^*$.

We now define a function T on C^* as follows: For $\varphi \in C^*$ define $T\varphi = \psi$ as

$$\psi(x) = y_0 + \int_{x_0}^{x} f[t, \varphi(t)]\,dt \qquad (|x - x_0| \leqslant \delta). \tag{8}$$

Then φ satisfies (4) if and only if $T\varphi = \varphi$.

We will show that T is a contraction (6.4E) on the complete metric space C^*. First we show that T maps C^* into C^*. Indeed, if $\varphi \in C^*$ and $\psi = T\varphi$, it is easy to show that ψ is continuous on $[x_0 - \delta, x_0 + \delta]$. Moreover, if $|x - x_0| \leqslant \delta$, then

$$|\psi(x) - y_0| = \left|\int_{x_0}^{x} f[t, \varphi(t)]\,dt\right| \leqslant k|x - x_0| \leqslant k\delta.$$

Hence, $|\psi(x) - y_0| \leqslant k\delta$ if $|x - x_0| \leqslant \delta$, and so $\psi \in C^*$. Hence, $T: C^* \to C^*$.

To show that T is a contraction suppose $\varphi_1, \varphi_2 \in C^*$ and let $\psi_1 = T\varphi_1, \psi_2 = T\varphi_2$. Then, from (8), if $|x - x_0| \leqslant \delta$, we have

$$\psi_1(x) - \psi_2(x) = \int_{x_0}^{x} \{f[t, \varphi_1(t)] - f[t, \varphi_2(t)]\}\,dt,$$

and so

$$|\psi_1(x) - \psi_2(x)| \leqslant \left|\int_{x_0}^{x} |f[t, \varphi_1(t)] - f[t, \varphi_2(t)]|\,dt\right|.$$

* To be precise, say $D = \{\langle x,y \rangle \mid |x - x_0| \leqslant a, |y - y_0| \leqslant b\}$ for some $a > 0, b > 0$.
† The condition (5) is called a Lipschitz condition.

Using (5) we then obtain

$$|\psi_1(x)-\psi_2(x)|\leqslant\left|M\int_{x_0}^{x}|\varphi_1(t)-\varphi_2(t)|\,dt\right|$$

$$\leqslant\left|M\int_{x_0}^{x}\|\varphi_1-\varphi_2\|\,dt\right|\leqslant M\delta\,\|\varphi_1-\varphi_2\|.$$

Thus

$$\|\psi_1-\psi_2\|\leqslant M\delta\,\|\varphi_1-\varphi_2\|,$$

or

$$\|T\varphi_1-T\varphi_2\|\leqslant M\delta\,\|\varphi_1-\varphi_2\|.$$

With respect to the metric ρ for C^* this reads

$$\rho(T\varphi_1,T\varphi_2)\leqslant M\delta\rho(\varphi_1,\varphi_2).$$

In view of (7) this proves that T is a contraction on C^*. Hence, by 6.4F, there is precisely one $\varphi\in C^*$ such that $T\varphi=\varphi$. But the definition (8) of T shows that $T\varphi=\varphi$ means that (4) holds. This completes the proof.

Our proof uses both the continuity of f and the Lipschitz condition (5) to show that a solution to

$$\frac{dy}{dx}=f(x,y),\qquad y(x_0)=y_0$$

exists and is unique.

It is possible, using a different method of proof, to show that a solution *exists* assuming only the continuity of f but not the Lipschitz condition. See Section 10.6.

However, the Lipschitz condition *is* necessary in order to prove that the solution is unique. For example, both $\varphi_1(x)=0$ and $\varphi_2(x)=x^3/27$ are solutions to

$$\frac{dy}{dx}=y^{2/3},\qquad y(0)=0. \tag{9}$$

There is thus no unique solution to (9). We leave it to the reader to show that $f(x,y)=y^{2/3}$ does not satisfy a Lipschitz condition in any rectangle D about $\langle 0,0\rangle$. (Show that

$$|f(0,y)-f(0,0)|\leqslant My\qquad(\langle 0,y\rangle\in D)$$

holds for no M.)

Exercises 10.3

1. Show that φ is a solution to

$$\frac{dy}{dx}=x+y,\qquad y(0)=0$$

if and only if

$$\varphi(x)=\int_0^x[t+\varphi(t)]\,dt.$$

Define T as follows: For any φ let $T\varphi=\psi$ where

$$\psi(x)=\int_0^x[t+\varphi(t)]\,dt.$$

Let $\varphi_0=0$. If $\varphi_1=T\varphi_0$, show that $\varphi_1(x)=x^2/2!$. If $\varphi_2=T\varphi_1$, show that $\varphi_2(x)=x^2/2!+x^3/3!$. In general, if $\varphi_n=T\varphi_{n-1}$, show that $\varphi_n(x)=x^2/2!+x^3/3!+\cdots+x^n/n!$. Then show that

$$\lim_{n\to\infty}\varphi_n(x)=\lim_{n\to\infty}T^n\varphi_0(x)=e^x-x-1.$$

Verify that $\varphi(x)=e^x-x-1$ is a solution to the original problem.

Compare this method of solution with the proof of 6.4F.

2. Use the same method to solve

$$y'=y, \qquad y(1)=1.$$

3. Use the same method to solve

$$y'=x-y, \qquad y(0)=1.$$

10.4 THE ARZELA THEOREM ON EQUICONTINUOUS FAMILIES

10.4A. In higher analysis it is often useful to know when a sequence of continuous functions on $[a,b]$ will have a uniformly convergent subsequence.

For an example where this does not happen consider $\{f_n\}_{n=1}^{\infty}$ where

$$f_n(x)=x^n \qquad (0\leqslant x\leqslant 1).$$

As we have seen, $\{f_n\}_{n=1}^{\infty}$ converges pointwise on $[0,1]$ to the discontinuous function f where

$$f(x)=0 \qquad (0\leqslant x\leqslant 1),$$
$$f(1)=1.$$

Any subsequence of $\{f_n\}_{n=1}^{\infty}$ must, therefore, also converge pointwise to f. Hence, by 9.3C, no subsequence of $\{f_n\}_{n=1}^{\infty}$ can converge uniformly on $[0,1]$, since f is not continuous.

The condition we are seeking involves the concept of equicontinuity.

10.4B. DEFINITION. Let $[a,b]$ be a closed bounded interval. The subset $\mathscr{F}$ of $C[a,b]$ is said to be equicontinuous if given $\epsilon>0$, there exists $\delta>0$ such that

$$|f(x)-f(y)|<\epsilon \qquad (|x-y|<\delta; f\in\mathscr{F}).$$

That is, $\mathscr{F}$ is an equicontinuous subset of $C[a,b]$ if, given $\epsilon>0$, there exists δ independent of f such that $|f(x)-f(y)|<\epsilon$ if $|x-y|<\delta$. The same δ must work for all $f\in\mathscr{F}$ as well as for all x,y.

A condition sufficient to ensure that a sequence of continuous functions on $[a,b]$ has a uniformly convergent subsequence will come out of the next result (which is known as Arzela's theorem or as Ascoli's theorem).

10.4C. THEOREM. Let $\mathscr{F}$ be a bounded equicontinuous subset of the metric space $C[a,b]$. Then $\mathscr{F}$ is totally bounded.

PROOF: Given $\epsilon>0$ we must show that there are a finite number of subsets A_i of $C[a,b]$ such that $\operatorname{diam} A_i<\epsilon$ and $\mathscr{F}\subset\bigcup A_i$.

Since $\mathscr{F}$ is bounded there exists $M>0$ such that

$$\rho(f,0)=\|f\|\leqslant M \qquad (f\in\mathscr{F}).$$

Since $\mathcal{F}$ is equicontinuous there exists $\delta > 0$ such that

$$|f(x) - f(x')| < \frac{\epsilon}{15} \qquad (|x - x'| < \delta; f \in \mathcal{F}). \tag{1}$$

Subdivide $[a,b]$ by points $a = x_0 < x_1 < \cdots < x_n = b$ where $x_{j+1} - x_j < \delta$ $(j = 0, 1, \ldots, n-1)$. Subdivide $[-M, M]$ (thought of as an interval on the y-axis) by points $-M = y_0 < y_1 < \cdots < y_m = M$ such that $y_{k+1} - y_k < \epsilon/15$ $(k = 0, 1, \ldots, m-1)$. Thus the rectangle $a \leqslant x \leqslant b$, $-M \leqslant y \leqslant M$ is subdivided into subrectangles of base less than δ and height less than $\epsilon/15$. For any $f \in \mathcal{F}$ define $g \in C[a,b]$ as follows: For each x_j $(j = 0, 1, \ldots, n)$ let $g(x_j) = y_{k(j)}$ where $k(j)$ is chosen so that

$$|g(x_j) - f(x_j)| = |y_{k(j)} - f(x_j)| < \frac{\epsilon}{15}. \tag{2}$$

This is possible because of the way we chose the y's. Then define g so that the graph of g consists of straight-line segments joining successively the points $\langle x_0, g(x_0)\rangle, \langle x_1, g(x_1)\rangle, \ldots, \langle x_n, g(x_n)\rangle$. Then, for $j = 0, 1, \ldots, n-1$ we have

$$|g(x_{j+1}) - g(x_j)| \leqslant |g(x_{j+1}) - f(x_{j+1})| + |f(x_{j+1}) - f(x_j)| + |f(x_j) - g(x_j)|,$$

and so, by (1) and (2)

$$|g(x_{j+1}) - g(x_j)| < \frac{\epsilon}{15} + \frac{\epsilon}{15} + \frac{\epsilon}{15} = \frac{\epsilon}{5}.$$

Since the restriction of g to $[x_j, x_{j+1}]$ is linear, it follows immediately that

$$|g(x) - g(x_j)| < \frac{\epsilon}{5} \qquad (x_j \leqslant x \leqslant x_{j+1}; j = 0, 1, \ldots, n-1). \tag{3}$$

Now for any $x \in [a,b]$ choose j so that $x_j \leqslant x \leqslant x_{j+1}$. Then

$$|g(x) - f(x)| \leqslant |g(x) - g(x_j)| + |g(x_j) - f(x_j)| + |f(x_j) - f(x)|.$$

Thus by (3), (2), and (1),

$$|g(x) - f(x)| < \frac{\epsilon}{5} + \frac{\epsilon}{15} + \frac{\epsilon}{15} = \frac{\epsilon}{3}.$$

Thus for each $f \in \mathcal{F}$, we have shown there is a g such that

$$\|g - f\| < \frac{\epsilon}{3}.$$

The open balls $B[g; \epsilon/3]$ have diameter $\leqslant 2\epsilon/3$ and $\mathcal{F}$ is contained in their union. But there can be only a finite number of distinct g! For each g is determined by its values at the $n+1$ points $x_0, x_1, \ldots, x_n$. Moreover, for each $j, g(x_j)$ must be one of the $m+1$ numbers $y_0, y_1, \ldots, y_m$. Hence, there are at most $(m+1)^{n+1}$ functions g. Thus there are a *finite* number of the open balls $B[g; \epsilon/3]$, each with diameter less than ϵ, and their union contains $\mathcal{F}$. This completes the proof.

Here is the result we have been looking for.

10.4D. COROLLARY. If $\{f_n\}_{n=1}^{\infty}$ is a sequence in $C[a,b]$, and if the functions in $\{f_n\}_{n=1}^{\infty}$ form a bounded equicontinuous subset of $C[a,b]$, then $\{f_n\}_{n=1}^{\infty}$ has a subsequence that converges uniformly to some function in $C[a,b]$.

PROOF: Let $\mathcal{F}$ be the set of functions in $\{f_n\}_{n=1}^{\infty}$. That is, $\mathcal{F} = \{f_1, f_2, \ldots\}$. By 10.4C, $\mathcal{F}$ is totally bounded. Hence, by 6.3H, $\{f_n\}_{n=1}^{\infty}$ has a Cauchy subsequence $\{f_{n_k}\}_{k=1}^{\infty}$ (with respect to the metric for $C[a,b]$). Since $C[a,b]$ is complete (10.1E), the sequence $\{f_{n_k}\}_{k=1}^{\infty}$ is convergent to some $f \in C[a,b]$. By 10.1D, this implies that $\{f_{n_k}\}_{k=1}^{\infty}$ converges uniformly to f on $[a,b]$. This proves the corollary.

Exercises 10.4

1. Prove that the family $\{\sin nx\}_{n=1}^{\infty}$ is not an equicontinuous subset of $C[0,\pi]$.
2. Let $\{\varphi_n\}_{n=1}^{\infty}$ be any sequence of functions on $[0,1]$ such that
$$|\varphi_n(x)| \leqslant M \qquad (0 \leqslant x \leqslant 1; n \in I)$$
and
$$|\varphi_n'(x)| \leqslant M \qquad (0 \leqslant x \leqslant 1; n \in I)$$
for some $M > 0$. Prove that $\{\varphi_n\}_{n=1}^{\infty}$ has uniformly convergent subsequence.
3. Let $\{f_n\}_{n=1}^{\infty}$ be a sequence of continuous functions on $[0,1]$ that converges uniformly on $[0,1]$. Prove that the family of functions f_n is equicontinuous.
4. Give an example of a sequence of functions $f_n \in C[a,b]$ that forms an equicontinuous family but such that $\{f_n\}_{n=1}^{\infty}$ has no uniformly convergent subsequence.

10.5 NOTES AND ADDITIONAL EXERCISES FOR CHAPTERS 9 AND 10.

I. The Stone-Weierstrass theorem and an application.

10.5A Weierstrass proved his theorem 10.2A in 1885. The proof we gave was published by Bernstein in 1912. In 1937 M. H. Stone presented a truly remarkable generalization of the Weierstrass theorem. This result applies to an arbitrary compact space in place of $[a,b]$. Moreover, in place of the set of polynomials it treats an arbitrary algebra of continuous functions that contains the constant functions and separates points. We will now define "algebra" and "separates points."

Let E be any set and let F be a family of real-valued functions on E. We say that F is an algebra if it is closed under the operations of addition, multiplication, and multiplication by constants. That is, F is an algebra if

$$f,g \in F \quad \text{imply} \quad f+g \in F \quad \text{and} \quad fg \in F.$$
$$f \in F, c \in R \quad \text{imply} \quad cf \in F.$$

We say that a family G of real-valued functions on E separates points of E if, whenever x,y are distinct points of E, there exists $g \in G$ such that $g(x) \neq g(y)$.

10.5B If M is a compact metric space, we denote by $C(M)$ the set of all continuous real-valued functions on M. If we define

$$\|f\| = \max_{x \in M} |f(x)| \qquad (f \in C(M))$$

and

$$\rho(f,g) = \|f-g\| \qquad (f,g \in C(M)),$$

then ρ is a metric for $C(M)$. This may be shown in precisely the same way as the special case $M=[a,b]$ is handled in 10.1. Moreover, as with $C[a,b]$, convergence with respect to the metric for $C(M)$ is precisely uniform convergence on M.

EXERCISE. Let M be a compact metric space and suppose $A \subset C(M)$. If A is an algebra, prove that $\bar{A}$ (the closure of A in M) is also an algebra. (This is important for the Stone-Weierstrass theorem.)

10.5C Here is one version of the Stone-Weierstrass theorem.

THEOREM. Let M be a compact metric space. Let A be a subset of $C(M)$ such that

$$A \quad \text{is an algebra,} \tag{1}$$

$$A \quad \text{separates points of } M, \tag{2}$$

$$A \quad \text{contains the constant functions.} \tag{3}$$

Then $\overline{A}=C(M)$. That is, A is dense in $C(M)$.

We outline the proof of the Stone-Weierstrass theorem in a sequence of lemmas, some of which are supplied with a sketch of proof.

EXERCISE. Give all details in the proofs of the following lemmas.

We assume that M is compact and that A is a subset of $C(M)$ such that (1), (2), and (3) hold.

LEMMA 1. If $f\in\overline{A}$, then $|f|\in\overline{A}$.

SKETCH OF PROOF: Given $\epsilon>0$ there exists a polynomial P such that

$$\big||x|-P(x)\big|<\epsilon \qquad (-\|f\|\leqslant x\leqslant\|f\|).$$

Hence,

$$\big||f(t)|-P[f(t)]\big|<\epsilon \qquad (t\in M).$$

But $P\circ f\in\overline{A}$. Hence, $|f|\in\overline{A}$.

LEMMA 2. Suppose $f,g\in\overline{A}$. Then $\max(f,g)\in\overline{A}$ and $\min(f,g)\in\overline{A}$.

LEMMA 3. Suppose $x_1,x_2\in M$ and $x_1\neq x_2$. Then there exists $f\in\overline{A}$ such that

$$0\leqslant f(x)\leqslant 1 \qquad (x\in M),$$

$$f(x_1)=1,$$

$$f(x)=0 \text{ for all } x \text{ in some open ball about } x_2.$$

SKETCH OF PROOF: By assumption (2) there exists $P\in A$ such that $P(x_1)\neq P(x_2)$. Let

$$\Phi(x)=\frac{P(x)-P(x_2)}{P(x_1)-P(x_2)} \qquad (x\in M).$$

Then $\Phi(x_1)=1$ and $\Phi(x)<\frac{1}{2}$ for all x in some open ball about x_2. Let

$$\Psi(x)=2\max\left[\Phi(x)-\tfrac{1}{2},0\right] \qquad (x\in M),$$

and

$$f(x)=\min\left[\Psi(x),1\right] \qquad (x\in M).$$

Then $f\in\overline{A}$, and f has all the required properties.

LEMMA 4. Suppose K is a proper compact subset of M, and $x_1\in M-K$. Then there exists $f\in\overline{A}$ such that

$$0\leqslant f(x)\leqslant 1 \qquad (x\in M),$$

$$f(x_1)=1,$$

$$f(x)=0 \qquad (x\in K).$$

SKETCH OF PROOF: Lemma 3, theorem 6.5G, and lemma 2.

LEMMA 5. Suppose that K_1 and K_2 are disjoint compact subsets of M. Then there exists $f \in \bar{A}$ such that

$$\begin{aligned} 0 \leqslant f(x) \leqslant 1 \qquad & (x \in M), \\ f(x) = 1 \qquad & (x \in K_1), \\ f(x) = 0 \qquad & (x \in K_2). \end{aligned}$$

SKETCH OF PROOF: For each $x_1 \in K_1$ there exists $F \in \bar{A}$ such that

$$\begin{aligned} 0 \leqslant F(x) \leqslant 1 \qquad & (x \in M), \\ F(x) = 0 \qquad & (x \in K_2), \end{aligned}$$

$F(x) > \frac{1}{2}$ for all x in some open ball B about x_1.

Let

$$\Phi(x) = 2 \min\left[F(x), \tfrac{1}{2} \right] \qquad (x \in M).$$

Then

$$\begin{aligned} 0 \leqslant \Phi(x) \leqslant 1 \qquad & (x \in M), \\ \Phi(x) = 0 \qquad & (x \in K_2), \\ \Phi(x) = 1 \qquad & (x \in B). \end{aligned}$$

Now use Heine-Borel and lemma 2.

Here is the completion of the proof of the theorem.

LEMMA 6. If $f \in C(M)$, then $f \in \bar{A}$.

SKETCH OF PROOF: We may assume f is not constant. Then $g = f + \|f\|$ is nonnegative valued and not identically zero. Let $h = \dfrac{g}{\|g\|}$. Then $0 \leqslant h(x) \leqslant 1$ for all x. It is sufficient to show that $h \in \bar{A}$. Given $\epsilon > 0$ choose $n \in I$ such that $\dfrac{1}{n} < \epsilon$. For $k = 0, 1, \ldots, n-1$ let

$$E_k = \left\{ x \in M \,|\, h(x) \leqslant \frac{k}{n} \right\}, \qquad F_k = \left\{ x \in M \,|\, h(x) \geqslant \frac{k+1}{n} \right\}.$$

Then there exists $f_k \in \bar{A}$ such that

$$\begin{aligned} 0 \leqslant f_k(x) \leqslant 1 \qquad & (x \in M), \\ f_k(x) = 0 \qquad & (x \in E_k), \\ f_k(x) = 1 \qquad & (x \in F_k). \end{aligned}$$

Let

$$P(x) = \frac{1}{n} \sum_{k=0}^{n-1} f_k(x) \qquad (x \in M).$$

Then $P \in \bar{A}$ and

$$|h(x) - P(x)| < \frac{1}{n} \qquad (x \in M).$$

Hence, $\|h - P\| < \epsilon$ so that $h \in \bar{A}$.

10.5D We mentioned in 10.3 that the Lipschitz condition (5) is not necessary to prove the existence (as opposed to the uniqueness) of a solution to (1) and (2) of that section.

We now state this as a theorem and outline the proof (to which the Stone-Weierstrass theorem and the Arzela theorem are relevant). First we state a

LEMMA. Let

$$D=\{\langle x,y\rangle \mid |x-x_0|\leqslant a, |y-y_0|\leqslant b\}.$$

Under the hypotheses of the theorem in 10.3, there exists a solution φ to

$$\varphi(x)=y_0+\int_{x_0}^{x} f[t,\varphi(t)]\,dt$$

which is defined on the entire interval $[x_0-h, x_0+h]$ where

$$h=\min\left(a,\frac{b}{k}\right)$$

and k is any upper bound for $\{|f(x,y)| \mid \langle x,y\rangle\in D\}$.

The importance of the lemma is that h depends only on D and the bound for f but not on the constant M in the Lipschitz condition.

EXERCISE. Prove the lemma and fill in the details of the proof of the following.

THEOREM. Let D be as above. If f is continuous on D, then there exists $\delta>0$ and a function φ such that

$$\varphi(x)=y_0+\int_{x_0}^{x} f[t,\varphi(t)]\,dt \qquad (|x-x_0|\leqslant\delta). \tag{*}$$

(Hence, $y=\varphi(x)$ is a solution to $dy/dx=f(x,y), y(x_0)=y_0$.)

SKETCH OF PROOF: There exists a sequence $\{P_n\}_{n=1}^{\infty}$ of polynomials in x and y such that $\{P_n\}_{n=1}^{\infty}$ converges uniformly to f on D.

There exists $k>0$ such that

$$|f(x,y)|\leqslant k \qquad (\langle x,y\rangle\in D)$$

and

$$|P_n(x,y)|\leqslant k \qquad (\langle x,y\rangle\in D; n=1,2,\ldots).$$

Each P_n satisfies a Lipschitz condition on D. Hence, there exists φ_n such that

$$\varphi_n(x)=y_0+\int_{x_0}^{x} P_n[t,\varphi_n(t)]\,dt \qquad (|x-x_0|\leqslant\delta)$$

where $\delta=\min(a,b/k)$. These φ_n form an equicontinuous family on $[x_0-\delta, x_0+\delta]$ so that a subsequence of $\{\varphi_n\}$ converges uniformly on $[x_0-\delta, x_0+\delta]$ to a function φ. This φ is a solution to (*).

II. The Tietze extension theorem

This is a theorem about the extendability of continuous functions.

10.5E THEOREM. Let F be a closed subset of the metric space M. Let f be a continuous, bounded, real-valued function on F. Then there exists a continuous real-valued function φ on all of M such that

$$\varphi(x)=f(x) \qquad (x\in F),$$

and

$$\underset{x \in M}{\text{l.u.b.}} |\varphi(x)| = \underset{x \in F}{\text{l.u.b.}} |f(x)|.$$

We first need a

LEMMA. Let F be a closed subset of the metric space M. Let g be a continuous, bounded, real-valued function on F. If

$$\theta = \underset{x \in F}{\text{l.u.b.}} |g(x)|,$$

then there exists a continuous real-valued function h on M such that

$$|h(x)| \leqslant \frac{\theta}{3} \qquad (x \in M)$$

and

$$|g(x) - h(x)| \leqslant \frac{2\theta}{3} \qquad (x \in F).$$

BEGINNING OF PROOF: Let

$$A = \left\{ x \in F \,|\, -\theta \leqslant g(x) \leqslant -\frac{\theta}{3} \right\},$$

$$B = \left\{ x \in F \,|\, -\frac{\theta}{3} < g(x) < \frac{\theta}{3} \right\},$$

$$C = \left\{ x \in F \,|\, \frac{\theta}{3} \leqslant g(x) \leqslant \theta \right\},$$

and let

$$h(x) = \frac{\theta}{3} \cdot \frac{\rho(x,A) - \rho(x,C)}{\rho(x,A) + \rho(x,C)} \cdot \qquad (x \in M),$$

where $\rho(x,A)$ is the distance from the point x to the closed set A.

EXERCISE. Finish the proof.

SKETCH OF PROOF OF THEOREM: It is sufficient to consider an f such that

$$\underset{x \in F}{\text{l.u.b.}} f(x) = 1, \qquad \underset{x \in F}{\text{g.l.b.}} f(x) = -1.$$

Define a sequence $\{ f_n \}_{n=1}^{\infty}$ of continuous functions on M as follows:

Let f_1 be the h of the lemma corresponding to $g = f$.

If $f_1, \ldots, f_{n-1}$ have been defined, let f_n be the h of the lemma corresponding to $g = f - (f_1 + \cdots + f_{n-1})$. Then for every $n = 1, 2, \ldots$, we have

$$|f_n(x)| \leqslant \frac{1}{3}\left(\frac{2}{3}\right)^{n-1} \qquad (x \in M)$$

and

$$|f(x) - [f_1(x) + \cdots + f_n(x)]| \leqslant \left(\frac{2}{3}\right)^n \qquad (x \in F).$$

Then $\sum_{n=1}^{\infty} f_n$ converges uniformly on M to a function φ.

EXERCISE. Fill in the details and finish the proof.

10.5F The following corollary is known as Urysohn's lemma.

EXERCISE. Let A, B be disjoint closed subsets of a metric space M. Prove that there exists a continuous function f on M such that

$$\begin{aligned} 0 \leqslant f(x) \leqslant 1 &\qquad (x \in M), \\ f(x) = 0 &\qquad (x \in A), \\ f(x) = 1 &\qquad (x \in B). \end{aligned}$$

III. MISCELLANEOUS EXERCISES

1. Let

$$\begin{aligned} f_n(x) = x^n(1-x) &\qquad (0 \leqslant x \leqslant 1; n \in I) \\ g_n(x) = x^n(1-x^n) &\qquad (0 \leqslant x \leqslant 1; n \in I). \end{aligned}$$

Does $\{f_n\}_{n=1}^{\infty}$ converge uniformly on $[0,1]$? Does $\{g_n\}_{n=1}^{\infty}$?

2. Let $\{f_n\}_{n=1}^{\infty}$ be a sequence in $C[0,1]$ such that $\lim_{n\to\infty} f_n(0)$ exists. Suppose that for every $g \in C[0,1]$ such that $g(0)=0$, the sequence $\{f_n g\}_{n=1}^{\infty}$ converges uniformly on $[0,1]$.

Must $\{f_n\}_{n=1}^{\infty}$ converge uniformly on $[0,1]$?

3. Let

$$f_n(x) = (1+x^n)^{1/n} \qquad (0 \leqslant x \leqslant 2; n \in I)$$

so that each f_n is differentiable. Show that $\{f_n\}_{n=1}^{\infty}$ converges uniformly to a function that fails to be differentiable at some point on $(0,2)$.

4. Let $r_1, r_2, \ldots$ be an enumeration of the rationals in $[0,1]$. For $n = 1, 2, \ldots$ let

$$\begin{aligned} t_n(x) = 0 &\qquad (0 \leqslant x < r_n), \\ t_n(x) = \frac{1}{n^2} &\qquad (r_n \leqslant x \leqslant 1) \end{aligned}$$

and define

$$f(x) = \sum_{n=1}^{\infty} t_n(x) \qquad (0 \leqslant x \leqslant 1).$$

Prove that f is continuous at every irrational. (See 6.9G.)

5. Suppose $\{f_n\}_{n=1}^{\infty}$ is a uniformly convergent sequence of real-valued functions on a subset of E of R^1 such that

$$f_n(x) \leqslant A \qquad (x \in E; n \in I)$$

for some $A > 0$. If $f(x) = \lim_{n\to\infty} f_n(x)$ for $x \in E$, prove that

$$\lim_{n\to\infty} \left[\operatorname*{l.u.b.}_{x \in E} f_n(x) \right] = \operatorname*{l.u.b.}_{x \in E} f(x).$$

6. Prove that

$$1 - \frac{1}{5} + \frac{1}{9} - \frac{1}{13} + \cdots = \frac{1}{4\sqrt{2}} \left[\pi + 2\log(1+\sqrt{2}) \right].$$

7. Prove that

$$\log(x+\sqrt{x^2+1}\,)=x-\frac{1}{2}\cdot\frac{x^3}{3}+\frac{1.3}{2.4}\frac{x^5}{5}-\frac{1.3.5}{2.4.6}\cdot\frac{x^7}{7}+\cdots$$

for $0\leqslant x\leqslant 1$.

8. Let

$$f_n(x)=\left(1+\frac{x}{n}\right)^n \qquad (x\in R^1;n\in I).$$

Prove that

$$\lim_{n\to\infty} f_n(x)=e^x$$

uniformly on every closed bounded interval of R^1.

9. Let $\{a_n\}_{n=1}^{\infty}$ be a nondecreasing sequence of positive numbers. Suppose

$$\sum_{k=1}^{\infty} a_k \sin kx$$

converges uniformly on $[-1,1]$. Prove that $\lim_{n\to\infty} na_n=0$.
(Begin this way: For $-1\leqslant x\leqslant 1$ and $n\in I$ let

$$s_n(x)=\sum_{k=1}^{n} a_k \sin kx$$

and let $t_n=s_{2n}-s_n$. Show that

$$t_n\left(\frac{\pi}{4n}\right)\geqslant na_n \sin\frac{\pi}{4}.$$

10. Show that

$$\sum_{n=1}^{\infty}\frac{x^n \sin nx}{n}$$

converges uniformly on $[-1,1]$.

11. Let $f\in C[a,b]$. If

$$\int_0^1 x^n f(x)\,dx=0 \qquad (n=0,1,2,\ldots),$$

show that $f(x)=0$ for all $x\in[a,b]$.

12. Prove that $C[0,1]$ has a countable dense subset.
13. Prove that $C[a,b]$ is connected.
14. Suppose $\{f_n\}_{n=1}^{\infty}$ is a convergent sequence in $C[0,1]$. For each $x\in[0,1]$ let

$$h(x)=\text{l.u.b.}\{f_1(x),f_2(x),\ldots\}.$$

Prove that $h\in C[0,1]$. Show that this need not be true if only pointwise convergence for $\{f_n\}$ is assumed.

15. Define $T:C[0,1]\to C[0,1]$ as follows: For $\varphi\in C[0,1]$ let $T\varphi=\psi$ where

$$\psi(x)=\int_0^x \varphi(t)\,dt \qquad [0\leqslant x\leqslant 1].$$

Show that T is not a contraction but that $T^2=T\circ T$ is. Note that T has a fixed point.

16. Suppose that $T:M\to M$ where M is a complete metric space. Assume also that T^n is a contraction for some $n\in I$. Prove that T has a fixed point.

11

THE LEBESGUE INTEGRAL

There are many ways to develop the Lebesgue integral. The development we will give is by no means the most elegant one, but it is probably the easiest to follow.

It is the utilization of the concept of measure that makes the Lebesgue integral different from the Riemann integral. Where the definition of the Riemann integral involves subdivision of $[a,b]$ into closed intervals, the definition of the Lebesgue integral involves subdivisions of $[a,b]$ into much more general kinds of sets called measurable sets.

We begin by leading up to the definition of "measurable set."

11.1 LENGTH OF OPEN SETS AND CLOSED SETS

11.1A. Let $[a,b]$ be a closed bounded interval in R^1. If G is a nonempty open subset of (the metric space) $[a,b]$, then, by 6.1B, G is the intersection of $[a,b]$ with an open subset of R^1. From 5.4F it then follows that G is the union of finitely or countable many pairwise disjoint* intervals I_n where each I_n is open in $[a,b]$. (Remember that intervals of the form $[a,c)$ or $(c,b]$ as well as (c,d) are open in $[a,b]$.) In any event, if G is a nonempty open subset of $[a,b]$, then $G = \cup_n I_n$ where each I_n is an interval and no two of the I_n have a point in common. This enables us to define the length $|G|$ of an open subset G of $[a,b]$.

DEFINITION. If G is an open subset of $[a,b], G = \cup I_n$, then the length $|G|$ of G is defined as

$$|G| = \sum_n |I_n|$$

where $|I_n|$ denotes the length of the interval I_n.

Thus $|G|$ is the sum of the lengths of the intervals comprising G.

* Saying that the I_n are pairwise disjoint means that no two of the I_n have a point in common.

We leave it to the reader as one of the exercises to show that if G_1 and G_2 are open subsets of $[a,b]$ with $G_1 \subset G_2$, then $|G_1| \leqslant |G_2|$. It then follows that $|G| \leqslant b-a$ for any set G open in $[a,b]$.

It is also easy to verify that

$$|I_1 \cup I_2 \cup \cdots \cup I_n| \leqslant |I_1| + |I_2| + \cdots + |I_n|$$

where the I_j are *intervals*. (See exercise 2 of this section.) We now generalize this result.

11.1B. THEOREM. If $G_1, G_2, \ldots$ are open subsets of $[a,b]$, then*

$$\left| \bigcup_{n=1}^{\infty} G_n \right| \leqslant \sum_{n=1}^{\infty} |G_n|. \tag{1}$$

PROOF: For each $n \in I$ we have $G_n = \cup_{k=1}^{\infty} I_k^n$. If $G = \cup_{n=1}^{\infty} G_n$, then G is open (by 5.4C) and so $G = \cup_{n=1}^{\infty} J_n$ where the J_n are pairwise disjoint intervals. Given $\epsilon > 0$ choose $N \in I$ so that $\Sigma_{n=N+1}^{\infty} |J_n| < \epsilon$. Then $|G| < \Sigma_{n=1}^{N} |J_n| + \epsilon$. Now for each $n = 1, \ldots, N$ let K_n be an open interval such that $\overline{K}_n \subset J_n$ and $|J_n| < |K_n| + \epsilon/N$. Then

$$|G| < \sum_{n=1}^{N} |K_n| + 2\epsilon. \tag{2}$$

Moreover, the union $\cup_{n=1}^{N} \overline{K}_n$ is closed (5.5G) and bounded, and is thus compact. Since $\cup_{n=1}^{N} \overline{K}_n$ is contained in $\cup_{n=1}^{\infty} G_n$ there are a finite number of the I_k^n—say

$$I_{k_1}^{n_1}, I_{k_2}^{n_2}, \ldots, I_{k_r}^{n_r}$$

—such that

$$\bigcup_{n=1}^{N} \overline{K}_n \subset I_{k_1}^{n_1} \cup \cdots \cup I_{k_r}^{n_r}.$$

Hence,

$$\bigcup_{n=1}^{N} K_n \subset I_{k_1}^{n_1} \cup \cdots \cup I_{k_r}^{n_r}.$$

Since the K_n are disjoint (why?) we have

$$\sum_{n=1}^{N} |K_n| \leqslant |I_{k_1}^{n_1} \cup \cdots \cup I_{k_r}^{n_r}| \leqslant |I_{k_1}^{n_1}| + \cdots + |I_{k_r}^{n_r}|$$

and hence,

$$\sum_{n=1}^{N} |K_n| \leqslant \sum_{n=1}^{\infty} |G_n|. \tag{3}$$

From (2) and (3) it follows that $|G| \leqslant \Sigma_{n=1}^{\infty} |G_n| + 2\epsilon$. Since ϵ was arbitrary we must have $|G| \leqslant \Sigma_{n=1}^{\infty} |G_n|$ which is precisely the desired inequality (1).

The next result will enable us to define the length of a closed set.

11.1C. THEOREM. If G_1 and G_2 are open subsets of $[a,b]$, then

$$|G_1| + |G_2| = |G_1 \cup G_2| + |G_1 \cap G_2|. \tag{1}$$

* It is possible, of course, that $\Sigma_{n=1}^{\infty} |G_n| = \infty$.

PROOF: Consider first the case where both G_1 and G_2 are unions of a *finite* number of intervals. Then $G_1 \cup G_2$ and $G_1 \cap G_2$ are also unions of a finite number of intervals. It is then easy to see that the characteristic functions

$$\chi_{G_1}, \chi_{G_2}, \chi_{G_1 \cup G_2}, \chi_{G_1 \cap G_2}$$

of these sets are Riemann integrable on $[a,b]$. Moreover, for any $x \in [a,b]$,

$$\chi_{G_1}(x) + \chi_{G_2}(x) = \chi_{G_1 \cup G_2}(x) + \chi_{G_1 \cap G_2}(x). \tag{2}$$

Indeed, if x is in both G_1 and G_2, then both sides of (2) are equal to 2. If x is in neither G_1 nor G_2, then both sides of (2) are equal to 0. Finally, if x is in one but not both of G_1 and G_2, then both sides of (2) are equal to 1. If we now integrate (2) from a to b, we obtain (1). Hence, (1) holds when G_1 and G_2 are unions of a finite number of intervals.

Now for the general case. We have $G_1 = \cup_{n=1}^{\infty} I_n, G_2 = \cup_{n=1}^{\infty} J_n$. Given $\epsilon > 0$ choose $N \in I$ such that

$$\sum_{n=N+1}^{\infty} |I_n| < \epsilon, \qquad \sum_{n=N+1}^{\infty} |J_n| < \epsilon.$$

Let

$$G_1^* = \bigcup_{n=1}^{N} I_n, \quad G_1^{**} = \bigcup_{n=N+1}^{\infty} I_n, \quad G_2^* = \bigcup_{n=1}^{N} J_n, \quad G_2^{**} = \bigcup_{n=N+1}^{\infty} J_n.$$

Then $|G_i| = |G_i^*| + |G_i^{**}|$ for $i = 1,2$, and $|G_i^{**}| < \epsilon$. Hence,

$$\begin{aligned} |G_1| + |G_2| &= |G_1^*| + |G_1^{**}| + |G_2^*| + |G_2^{**}| \\ &< |G_1^*| + |G_2^*| + 2\epsilon. \end{aligned}$$

Since G_1^*, G_2^* are *finite* unions of intervals, we can use the first part of the proof to obtain

$$|G_1| + |G_2| < |G_1^* \cup G_2^*| + |G_1^* \cap G^*| + 2\epsilon.$$

But $G_1^* \cup G_2^* \subset G_1 \cup G_2$, and $G_1^* \cap G_2^* \subset G_1 \cap G_2$. Hence,

$$|G_1| + |G_2| < |G_1 \cup G_2| + |G_1 \cap G_2| + 2\epsilon.$$

Since ϵ was arbitrary this implies

$$|G_1| + |G_2| \leqslant |G_1 \cup G_2| + |G_1 \cap G_2|. \tag{3}$$

On the other hand, we have $G_1 \cup G_2 = (G_1^* \cup G_2^*) \cup G_1^{**} \cup G_2^{**}$ and so, by 11.1B,

$$|G_1 \cup G_2| \leqslant |G_1^* \cup G_2^*| + 2\epsilon. \tag{4}$$

Similarly, $G_1 \cap G_2 = (G_1^* \cup G_1^{**}) \cap (G_2^* \cup G_2^{**}) \subset (G_1^* \cap G_2^*) \cup G_1^{**} \cup G_2^{**}$, and so

$$|G_1 \cap G_2| \leqslant |(G_1^* \cap G_2^*) \cup G_1^{**} \cup G_2^{**}| \leqslant |G_1^* \cap G_2^*| + 2\epsilon. \tag{5}$$

Hence, again using the first part of the proof together with (4) and (5), we obtain

$$\begin{aligned} |G_1| + |G_2| &\geqslant |G_1^*| + |G_2^*| = |G_1^* \cup G_2^*| + |G_1^* \cap G_2^*| \\ &\geqslant (|G_1 \cup G_2| - 2\epsilon) + (|G_1 \cap G_2| - 2\epsilon). \end{aligned}$$

Since ϵ was arbitrary, this implies

$$|G_1| + |G_2| \geqslant |G_1 \cup G_2| + |G_u \cap G_2|. \tag{6}$$

The conclusion (1) now follows from (3) and (6).

11.1D. If F is a closed subset of $[a,b]$, then F' (the complement of F relative to $[a,b]$) is an open subset of $[a,b]$ by 5.5I. Consequently, if G is any open subset of $[a,b]$, then $G - F = G \cap F'$ is open by 5.4E, and thus $|G - F|$ is already defined. We then define $|F|$ as follows.

DEFINITION. Let F be any closed subset of $[a,b]$. Then the length $|F|$ of F is defined as

$$|F| = |G| - |G - F| \tag{1}$$

where G is any open subset of $[a,b]$ such that $G \supset F$.

It is necessary to show that this definition of $|F|$ does not depend on which open set G is used. That is, we must show that if G_1 and G_2 are both open sets containing F, then

$$|G_1| - |G_1 - F| = |G_2| - |G_2 - F| \tag{2}$$

so that $|F|$ can be calculated using either side of (2). Now by 11.1C (with $G_2 - F$ instead of G_2) we have, since $F \subset G_1, F \subset G_2$,

$$|G_1| + |G_2 - F| = |G_1 \cup G_2| + |(G_1 \cap G_2) - F|.$$

Also, by 11.1C,

$$|G_1 - F| + |G_2| = |G_1 \cup G_2| + |(G_1 \cap G_2) - F|.$$

Equating the left sides of the last two equations will establish (2). Hence, $|F|$ in (1) does not depend on G.

In particular, it follows that

$$|F| = b - a - |F'|$$

for any closed subset F of $[a,b]$.

Exercises 11.1

1. If G_1 and G_2 are open subsets of $[a,b]$, and if $G_1 \subset G_2$, prove that $|G_1| \leqslant |G_2|$. (*Hint*: Write $G_1 = \bigcup_k I_k$, $G_2 = \bigcup_n J_n$. First show that every I_k is contained in some J_n. Then group an arbitrary finite number of the I_k according to the J_n in which they are contained.)
2. If $I_1, \ldots, I_k$ are open subintervals of $[a,b]$, prove that

$$|I_1 \cup \cdots \cup I_k| \leqslant |I_1| + \cdots + |I_k|. \tag{*}$$

Do not use 11.1B as (*) was used in the proof of 11.1B. (*Hint*: For $j = 1, \ldots, k$ let χ_j be the characteristic function of I_j, and let χ be the characteristic function of $I_1 \cup \cdots \cup I_k$. Show that $\chi, \chi_1, \ldots, \chi_n$ are in $\mathscr{R}[a,b]$ and that

$$\chi(x) \leqslant \chi_1(x) + \cdots + \chi_k(x) \qquad (a \leqslant x \leqslant b).$$

Then integrate.)
3. True or false? If G is an open subset of $[a,b]$ and $|G| = 0$, then $G = \varnothing$.
4. True or false? If F is a closed subset of $[a,b]$ and $|F| = 0$, then $F = \varnothing$.
5. Show that the Cantor set of 1.6D has length 0.

11.2 INNER AND OUTER MEASURE. MEASURABLE SETS

We now define the inner measure and outer measure of an arbitrary subset E of $[a,b]$.

11.2A. DEFINITION. If $E \subset [a,b]$, then $\overline{m}E$, called the outer measure of E, is defined as

$$\overline{m}E = \text{g.l.b.} |G|$$

where the g.l.b. is taken over all open sets G that contain E.

The inner measure $\underline{m}E$ of E is defined as

$$\underline{m}E = \text{l.u.b.} |F|$$

where the l.u.b. is taken over all closed set F contained in E.

Thus $\overline{m}E$ is computed by open sets "closing down" on E. From the definition it follows that $\overline{m}E \leqslant |G|$ whenever G is open and $G \supset E$. Similarly, $\underline{m}E \geqslant |F|$ if F is closed and $F \subset E$.

11.2B. DEFINITION. The set $E \subset [a,b]$ is said to be measurable if $\overline{m}E = \underline{m}E$. In this case we define mE, the measure of E, as

$$mE = \overline{m}E = \underline{m}E.$$

Thus $\underline{m}E$ and $\overline{m}E$ are defined for all subsets E of $[a,b]$. However, mE is defined only for those E whose inner measure and outer measure are equal. We will presently show that there exists a set E such that $\underline{m}E \neq \overline{m}E$—that is, there exists a nonmeasurable set.

We now begin to develop properties of inner and outer measure.

11.2C. THEOREM. If $E \subset [a,b]$, then $\underline{m}E \leqslant \overline{m}E$.

PROOF: Let F be any closed set contained in E and let G be any open set containing E. Then $F \subset E \subset G$, and so $|F| = |G| - |G - F|$. It follows that

$$|F| \leqslant |G|.$$

Take the g.l.b. over all such G. We obtain $|F| \leqslant \overline{m}E$. Now, we take the l.u.b. over all closed $F \subset E$. We obtain $\underline{m}E \leqslant \overline{m}E$, which is what we wished to show.

11.2D. THEOREM. If $E \subset [a,b]$, then

$$\overline{m}E + \underline{m}E' = b - a$$

(where $E' = [a,b] - E$).

PROOF: Let G be any open set containing E. Then G' is closed, and $G' \subset E'$. We have

$$|G| + \underline{m}E' \geqslant |G| + |G'|.$$

Hence, since $|G| + |G'| = b - a$, we have

$$|G| + \underline{m}E' \geqslant b - a.$$

Taking the g.l.b. over all open $G \supset E$ we obtain

$$\overline{m}E + \underline{m}E' \geqslant b - a. \tag{1}$$

Now if F is any closed set with $F \subset E'$, we have $F' \supset E$ and so

$$\overline{m}E + |F| \leqslant |F'| + |F|,$$

$$\overline{m}E + |F| \leqslant b - a,$$

$$\overline{m}E + \underline{m}E' \leqslant b - a. \tag{2}$$

The theorem follows from (1) and (2).

As a consequence of the preceding two theorems we obtain the following corollary.

11.2E. COROLLARY. Let $E \subset [a,b]$. Then E is measurable if and only if

$$\overline{m}E + \overline{m}E' \leqslant b - a. \tag{1}$$

PROOF: For any set E we have (by 11.2D with E and E' interchanged)

$$\overline{m}E' + \underline{m}E = b - a. \tag{2}$$

Now, if E is measurable, then $\underline{m}E = \overline{m}E$. Hence,

$$\overline{m}E' + \overline{m}E = b - a,$$

which implies (1).

Conversely, suppose (1) holds for some $E \subset \neq a,b]$. Since (2) also holds for E we may subtract (2) from (1) to obtain $\overline{m}E - \underline{m}E \leqslant 0$. This, together with 11.2C, implies $\overline{m}E = \underline{m}E$. Hence, E is measurable, and the proof is complete.

The following criterion for measurability is sometimes used as the definition of measurability.

11.2F. THEOREM. The subset E of $[a,b]$ is measurable if and only if given $\epsilon > 0$ there exist open sets G_1 and G_2 such that $G_1 \supset E, G_2 \supset E'$, and $|G_1 \cap G_2| < \epsilon$.

PROOF: First suppose that E is measurable. Given $\epsilon > 0$ there exist open sets G_1 and G_2 such that $G_1 \supset E, G_2 \supset E'$ and such that $|G_1| < \overline{m}E + \epsilon/2, |G_2| < \overline{m}E' < \epsilon/2$. (This is just a consequence of the definition of outer measure.) Then, by 11.2C,

$$|G_1 \cap G_2| = |G_1| + |G_2| - |G_1 \cup G_2|$$

and so

$$|G_1 \cap G_2| < \overline{m}E + \overline{m}E' + \epsilon - |G_1 \cup G_2|.$$

But since $G_1 \supset E, G_2 \supset E'$, we have $G_1 \cup G_2 = [a,b]$. Hence,

$$|G_1 \cap G_2| < \overline{m}E + \overline{m}E' - (b-a) + \epsilon.$$

Since we are supposing that E is measurable, 11.2E then implies

$$|G_1 \cap G_2| < \epsilon.$$

This proves half the theorem.

Conversely, suppose $E \subset [a,b]$ and that for any $\epsilon > 0$ there exists open sets G_1, G_2 with $G_1 \supset E, G_2 \supset E', |G_1 \cap G_2| < \epsilon$. We have

$$\overline{m}E + \overline{m}E' \leqslant |G_1| + |G_2| = |G_1 \cup G_2| + |G_1 \cap G_2|,$$

and so

$$\overline{m}E + \overline{m}E' \leqslant b - a + \epsilon.$$

Since ϵ was arbitrary, this implies $\overline{m}E + \overline{m}E' \leqslant b - a$. By 11.2E, the set E is measurable. This completes the proof.

11.2G. COROLLARY. If E is a measurable subset of $[a,b]$, then E' is also measurable and

$$mE' = (b-a) - mE. \tag{1}$$

PROOF: Since the criterion for measurability in 11.2F is symmetric with respect to E and its complement E', it is clear that the measurability of E implies the measurability of E'. But then $\overline{m}E = mE$ and $\underline{m}E' = mE'$, and so (1) follows from 11.2D.

We next show that open and closed sets are measurable.

11.2H. THEOREM. If G is an open subset of $[a,b]$, then G is measurable and $mG = |G|$. Also, if F is a closed subset of $[a,b]$, then F is measurable and $mF = |F|$.

PROOF: If G is open, it is obvious that $\overline{m}G = |G|$. Now $G = \cup_{n=1}^{\infty} J_n$ where the J_n are intervals open in $[a,b]$. Given $\epsilon > 0$ we can define $K_1, \ldots, K_N$ as in the proof of 11.1B.

Then $\cup_{n=1}^{N}\overline{K}_n$ is closed and is contained in G. Also, by (2) of 11.1B,

$$\overline{m}G < \left|\bigcup_{n=1}^{N} K_n\right| + 2\epsilon.$$

Hence, $\overline{m}G < \underline{m}G + 2\epsilon$. Since ϵ was arbitrary, it follows that $\overline{m}G \leqslant \underline{m}G$. This and 11.2C prove that G is measurable. Moreover, $mG = \overline{m}G = |G|$.

Now let F be a closed subset of $[a,b]$. Then the complement G of F is open and hence, measurable. By 11.2G, F is measurable and $mF = (b-a) - mG = (b-a) - |G|$. But by definition, $|F| = (b-a) - |G|$. Hence, $mF = |F|$, and the proof is complete.

11.2I. We leave it for the reader to show that a subset E of $[a,b]$ is measurable and $mE = 0$ if and only if E is of measure zero according to 7.1A.

Consider, for example, the Cantor set K of 1.6D. The complement K' of K (relative to $[0,1]$) is the union of open intervals whose lengths add up to 1. Hence, K' is open and $mK' = |K'| = 1$. Therefore, by 11.2G, K is measurable and $mK = (1-0) - mK' = 0$. This shows that the Cantor set is measurable and has measure zero (even though K is not countable).

It is obvious that if $mE = 0$, then any subset of E is measurable and has measure zero.

Exercises 11.2

1. Show that $E \subset [a,b]$ is measurable if and only if given $\epsilon > 0$ there exist a closed set $F \subset E$ and an open set $G \supset E$ such that $|G| - |F| < \epsilon$.
2. If c and d are in (a,b) and $c < d$, prove that $[c,d)$ is measurable.
3. If $E \subset [a,b]$ and $\overline{m}E = 0$, prove that E is measurable and $mE = 0$.
4. Let E be a measurable subset of $[a,b]$. Prove that
$$\overline{m}A = \overline{m}(A \cap E) + \overline{m}(A \cap E')$$
for every subset A of $[a,b]$. (*Hint*: Use 11.2F.)
5. If $E \subset [a,b]$, if $x \in E'$, and if $E \cup \{x\}$ is measurable, prove that E is measurable.
6. If E_1 is a measurable subset of $[a,b]$ and if $mE_2 = 0$, prove that $E_1 \cup E_2$ is measurable.
7. If $E_1, E_2 \subset [a,b]$, if $mE_2 = 0$, and if $E_1 \cup E_2$ is measurable, prove that E_1 is measurable.
8. If $a < c < b$ and E is a measurable subset of $[a,b]$, show that $E \cap [a,c]$ is a measurable subset of $[a,c]$.
9. Prove that the characteristic function of the Cantor set is in $\mathcal{R}[0,1]$.
10. If $E \subset [a,b]$, show that there exists a subset H of E such that H is of type F_σ and $\underline{m}H = \underline{m}E$.

11.3 PROPERTIES OF MEASURABLE SETS

In this section we will show (among other things) that both the union and the intersection of a countable number of measurable sets are again measurable sets. The first theorem is needed to show that the union and the intersection of two measurable sets are measurable.

11.3A. THEOREM. If E_1 and E_2 are subsets of $[a,b]$, then

$$\overline{m}E_1 + \overline{m}E_2 \geqslant \overline{m}(E_1 \cup E_2) + \overline{m}(E_1 \cap E_2), \tag{1}$$

and

$$\underline{m}E_1 + \underline{m}E_2 \leqslant \underline{m}(E_1 \cup E_2) + \underline{m}(E_1 \cap E_2). \tag{2}$$

PROOF: Given $\epsilon > 0$, choose open sets G_1 and G_2 such that $G_1 \supset E_1, G_2 \supset E_2$, and

$$mG_1 < \overline{m}E_1 + \frac{\epsilon}{2}, \qquad mG_2 < \overline{m}E_2 + \frac{\epsilon}{2}.$$

Then

$$\overline{m}E_1 + \overline{m}E_2 + \epsilon > mG_1 + mG_2 + |G_1| + |G_i|.$$

By 11.1C, this implies

$$\overline{m}E_1 + \overline{m}E_2 + \epsilon > |G_1 \cup G_2| + |G_1 \cap G_2|.$$

But $G_1 \cup G_2$ and $G_1 \cap G_2$ are open sets containing $E_1 \cup E_2$ and $E_1 \cap E_2$, respectively. Hence,

$$\overline{m}E_1 + \overline{m}E_2 + \epsilon > \overline{m}(E_1 \cup E_2) + \overline{m}(E_1 \cap E_2).$$

Since ϵ was arbitrary, (1) follows.

Applying (1) to E_1' and E_2' we obtain

$$\overline{m}E_1' + \overline{m}E_2' \geqslant \overline{m}(E_1' \cup E_2') + \overline{m}(E_1' \cap E_2').$$

In view of 1.2H this implies

$$\overline{m}E_1' + \overline{m}E_2' \geqslant \overline{m}(E_1 \cap E_2)' + \overline{m}(E_1 \cup E_2)'.$$

From 11.2D we then have

$$(b-a-\underline{m}E_1) + (b-a-\underline{m}E_2) \geqslant \left[b-a-\underline{m}(E_1 \cap E_2)\right] + \left[b-a-\underline{m}(E_1 \cup E_2)\right]$$

which, on simplification, yields (2). This completes the proof.

11.3B. COROLLARY. If E_1 and E_2 are measurable subsets of $[a,b]$, then both $E_1 \cup E_2$ and $E_1 \cap E_2$ are also measurable. Moreover,

$$mE_1 + mE_2 = m(E_1 \cup E_2) + m(E_1 \cap E_2). \tag{1}$$

PROOF: By hypothesis $\overline{m}E_i = \underline{m}E_i = mE_i$ for $i = 1, 2$. From 11.3A and 11.2C we then have

$$mE_1 + mE_2 \geqslant \overline{m}(E_1 \cup E_2) + \overline{m}(E_1 \cap E_2) \geqslant \underline{m}(E_1 \cup E_2) + \underline{m}(E_1 \cap E_2) \geqslant mE_1 + mE_2. \tag{2}$$

Since the extreme left and right sides of (2) are equal, we may put equals signs in place of the $\geqslant$ signs in (2). Hence,

$$\overline{m}(E_1 \cup E_2) + \overline{m}(E_1 \cap E_2) = \underline{m}(E_1 \cup E_2) + \underline{m}(E_1 \cap E_2).$$

By 11.2C this implies $\overline{m}(E_1 \cup E_2) = \underline{m}(E_1 \cup E_2)$ and $\overline{m}(E_1 \cap E_2) = \underline{m}(E_1 \cap E_2)$. Hence, $E_1 \cup E_2$ and $E_1 \cap E_2$ are measurable. We may then substitute m for $\overline{m}$ and $\underline{m}$ in (2) to obtain (1).

11.3C. COROLLARY. If E_1 and E_2 are measurable subsets of $[a,b]$, then $E_1 - E_2$ is also measurable. In addition, if $E_2 \subset E_1$, then $m(E_1 - E_2) = mE_1 - mE_2$.

PROOF: By 11.2G, E_2' is measurable. But $E_1 - E_2 = E_1 \cap E_2'$. Apply 11.3B to show that $E_1 - E_2$ is measurable.

Now if $E_2 \subset E_1$, then $E_1 = E_2 \cup (E_1 - E_2)$ and $m[E_2 \cap (E_1 - E_2)] = m\varnothing = 0$. Hence,

$$mE_1 = m\left[E_2 \cup (E_1 - E_2)\right] = m\left[E_2 \cup (E_1 - E_2)\right] + m\left[E_2 \cap (E_1 - E_2)\right].$$

Applying 11.3B to the right-hand side we then have

$$mE_1 = mE_2 + m(E_1 - E_2),$$

which is what we wished to show.

11.3D. THEOREM.

(a) If $E_1, E_2, \ldots$ are any subsets of $[a,b]$, then

$$\overline{m}\left(\bigcup_{n=1}^{\infty} E_n\right) \leqslant \sum_{n=1}^{\infty} \overline{m}E_n.$$

(b) If $E_1, E_2, \ldots$ are *pairwise disjoint* subsets of $[a,b]$, then

$$\underline{m}\left(\bigcup_{n=1}^{\infty} E_n\right) \geqslant \sum_{n=1}^{\infty} \underline{m}E_n.$$

PROOF: (a) Given $\epsilon > 0$, choose the open set G_n so that $G_n \supset E_n$ and $|G_n| < \overline{m}E_n + \epsilon/2^n$. Then $\bigcup_{n=1}^{\infty} G_n$ is an open set (5.4C) that contains $\bigcup_{n=1}^{\infty} E_n$. Hence, $\overline{m}(\bigcup_{n=1}^{\infty} E_n) \leqslant |\bigcup_{n=1}^{\infty} G_n|$. By 11.1B we then have

$$\overline{m}\left(\bigcup_{n=1}^{\infty} E_n\right) \leqslant \sum_{n=1}^{\infty} |G_n| < \sum_{n=1}^{\infty} \left(\overline{m}E_n + \frac{\epsilon}{2^n}\right)$$

and so

$$\overline{m}\left(\bigcup_{n=1}^{\infty} E_n\right) < \sum_{n=1}^{\infty} \overline{m}E_n + \epsilon.$$

Since ϵ was arbitrary, this proves (a).

(b) Now suppose $E_1, E_2, \ldots$ are pairwise disjoint. Then, by (2) of 11.3A,

$$\underline{m}(E_1 \cup E_2) \geqslant \underline{m}E_1 + \underline{m}E_2.$$

By induction, it is easy to show that

$$\underline{m}(E_1 \cup E_2 \cup \cdots \cup E_N) \geqslant \underline{m}E_1 + \underline{m}E_2 + \cdots + \underline{m}E_N \qquad (N \in I). \tag{1}$$

For any $N \in I$ we then have $\bigcup_{n=1}^{\infty} E_n \supset \bigcup_{n=1}^{N} E_n$, and so $\underline{m}(\bigcup_{n=1}^{\infty} E_n) \geqslant \underline{m}(\bigcup_{n=1}^{N} E_n)$. This and (1) prove that

$$\underline{m}\left(\bigcup_{n=1}^{\infty} E_n\right) \geqslant \sum_{n=1}^{N} \underline{m}E_n.$$

Letting N approach infinity establishes (b).

The crucial results on measurable sets now follow. The first deals with the union of countably many *disjoint* measurable sets. The next two deal with "increasing unions" and "decreasing intersections" of measurable sets. The last states the measurability of the union and intersection of countably many measurable sets.

11.3E. COROLLARY. If $E_1, E_2, \ldots$ are pairwise disjoint measurable subsets of $[a,b]$, then $\bigcup_{n=1}^{\infty} E_n$ is measurable and

$$m\left(\bigcup_{n=1}^{\infty} E_n\right) = \sum_{n=1}^{\infty} mE_n.$$

PROOF: By 11.3D and 11.2C we have (since $\underline{m}E_n = \overline{m}E_n = mE_n$)

$$\sum_{n=1}^{\infty} mE_n \leqslant \underline{m}\left(\bigcup_{n=1}^{\infty} E_n\right) \leqslant \overline{m}\left(\bigcup_{n=1}^{\infty} E_n\right) \leqslant \sum_{n=1}^{\infty} mE_n.$$

Since the extreme left and right sides are equal in (1), all four quantities in (1) must be equal. The corollary follows.

11.3F. COROLLARY. If $E_1, E_2, \ldots$ are measurable subsets of $[a,b]$, and if $E_1 \subset E_2 \subset E_3 \subset \cdots$, then $\bigcup_{n=1}^{\infty} E_n$ is measurable and

$$m\left(\bigcup_{n=1}^{\infty} E_n\right) = \lim_{n\to\infty} mE_n. \tag{1}$$

PROOF: The sets $E_1, E_2 - E_1, E_3 - E_2, \ldots, E_n - E_{n-1}, \ldots$ are measurable by 11.3C, and are pairwise disjoint. Hence, by 11.3E,

$$E_1 \cup (E_2 - E_1) \cup \cdots \cup (E_n - E_{n-1}) \cup \cdots$$

is measurable and

$$\begin{aligned} m[&E_1 \cup (E_2 - E_1) \cup \cdots \cup (E_n - E_{n-1}) \cup \cdots] \\ &= mE_1 + \sum_{k=2}^{\infty} m(E_k - E_{k-1}) = mE_1 + \lim_{n\to\infty} \sum_{k=2}^{n} m(E_k - E_{k-1}). \end{aligned} \tag{2}$$

But $E_1 \cup (E_2 - E_1) \cup \cdots \cup (E_n - E_{n-1}) \cup \cdots$ is precisely $\bigcup_{n=1}^{\infty} E_n$ (verify). Moreover, by 11.3C,

$$\begin{aligned} \sum_{k=2}^{n} m(E_k - E_{k-1}) &= \sum_{k=2}^{n} (mE_k - mE_{k-1}) \\ &= (mE_2 - mE_1) + (mE_3 - mE_2) + \cdots + (mE_n - mE_{n-1}) \\ &= mE_n - mE_1. \end{aligned}$$

Thus $\bigcup_{n=1}^{\infty} E_n$ is measurable and, from (2),

$$m\left(\bigcup_{n=1}^{\infty} E_n\right) = mE_1 + \lim_{n\to\infty} [mE_n - mE_1].$$

This implies (1), and the proof is complete.

11.3G. COROLLARY. If $E_1, E_2, \ldots$ are measurable subsets of $[a,b]$ and if $E_1 \supset E_2 \supset E_3 \supset \cdots$, then $\bigcap_{n=1}^{\infty} E_n$ is measurable and

$$m\left(\bigcap_{n=1}^{\infty} E_n\right) = \lim_{n\to\infty} mE_n.$$

PROOF: For each $n \in I$ the set E_n' is measurable by 11.2G, and $mE_n' = b - a - mE_n$. Moreover, $E_1' \subset E_2' \subset E_3' \subset \cdots$. Thus by 11.3F, $\bigcup_{n=1}^{\infty} E_n'$ is measurable and

$$m\left(\bigcup_{n=1}^{\infty} E_n'\right) = \lim_{n\to\infty} mE_n' = b - a - \lim_{n\to\infty} mE_n.$$

But $\bigcup_{n=1}^{\infty} E_n'$ is the complement of $\bigcap_{n=1}^{\infty} E_n$. Hence, $\bigcap_{n=1}^{\infty} E_n$ is measurable and

$$b - a - m\left(\bigcap_{n=1}^{\infty} E_n\right) = b - a - \lim_{n\to\infty} mE_n.$$

The corollary follows.

Finally we show that $\cup_{n=1}^{\infty} E_n$ and $\cap_{n=1}^{\infty} E_n$ are measurable for any measurable sets $E_1, E_2, \ldots$.

11.3H. COROLLARY. If $E_1, E_2, \ldots$ are any measurable subsets of $[a,b]$, then $\cup_{n=1}^{\infty} E_n$ is measurable and

$$m\left(\bigcup_{n=1}^{\infty} E_n\right) \leqslant \sum_{n=1}^{\infty} mE_n. \tag{1}$$

Moreover, $\cap_{n=1}^{\infty} E_n$ is measurable.

PROOF: We have

$$\bigcup_{n=1}^{\infty} E_n = E_1 \cup [E_2 - E_1] \cup [E_3 - (E_1 \cup E_2)] \cup \cdots$$
$$\cup [E_n - (E_1 \cup E_2 \cup \cdots \cup E_{n-1})] \cup \cdots. \tag{2}$$

The sets on the right of (2) are measurable and are pairwise disjoint. Thus by 11.3E, the right side of (2) is a measurable set. Hence, $\cup_{n=1}^{\infty} E_n$ is measurable. The inequality (1) then follows from part (a) of 11.3D, since we can replace $\overline{m}$ by m.

That $\cap_{n=1}^{\infty} E_n$ is measurable then follows from the fact that $\cap_{n=1}^{\infty} E_n$ is the complement of $\cup_{n=1}^{\infty} E_n'$.

11.3I. By the symmetric difference of the sets A and B we mean the set of points that are in one of the sets but not in both. That is, the symmetric difference of A and B is the union of the sets $A - B$ and $B - A$. The following theorem shows that sets of measure zero have no influence on measurability.

THEOREM. If E_1 and E_2 are subsets of $[a,b]$, if the symmetric difference of E_1 and E_2 has measure zero, and if E_1 is measurable, then E_2 is measurable.

Moreover, $mE_2 = mE_1$.

PROOF: We have

$$E_2 = [E_1 \cup (E_2 - E_1)] - (E_1 - E_2). \tag{1}$$

By hypothesis, both $E_2 - E_1$ and $E_1 - E_2$ are measurable and have measure zero. Since E_1 and $E_2 - E_1$ are disjoint, 11.3E implies that $E_1 \cup (E_2 - E_1)$ is measurable and $m[E_1 \cup (E_2 - E_1)] = mE_1 + 0 = mE_1$. But, since

$$E_1 - E_2 \subset [E_1 \cup (E_2 - E_1)],$$

it follows from (1) and 11.3C that E_2 is measurable and

$$mE_2 = m[E_1 \cup (E_2 - E_1)] - m(E_1 - E_2) = mE_1 - 0 = mE_1.$$

This completes the proof.

11.3J. The preceding theorems show that any operations involving countable unions and countable intersections of measurable sets will yield measurable sets. For this reason, it is far from elementary to show that there is such a thing as a set which is not measurable. We now indicate how to show the existence of a nonmeasurable set. It is convenient to construct this set on a circle C of circumference 1. This circle can be identified with the interval $[0, 1)$ in an obvious way.

If $x, y \in C$, we say that $x \sim y$ if the arc length from x to y is a rational number. It is then clear that

$$x \sim x; \tag{1}$$

$$\text{if} \quad x \sim y, \quad \text{then} \quad y \sim x; \tag{2}$$

and

$$\text{if} \quad x \sim y \quad \text{and} \quad y \sim z, \quad \text{then} \quad x \sim z. \tag{3}$$

We then divide C into subsets E_α such that x and y are in the same E_α if and only if $y \sim x$. Each E_α thus contains a countable number of points (since the rationals are countable). Moreover, on the basis of (1), (2), (3) it is easy to show that the E_α are pairwise disjoint. Hence, there must be an uncountable number of E_α. Now let V be any subset of C such that V contains precisely one element x_α from each E_α. Thus no two distinct elements x_α, x_β in V can satisfy $x_\alpha \sim x_\beta$. (We must use the so-called axiom of choice in order to be assured of the existence of such a set V.) We will show that V is not measurable.

Let $r_1, r_2, \ldots$ be the rationals in $[0,1)$ with $r_1 = 0$. For each $n \in I$ let V_n be the subset of C obtained by rotating V (counterclockwise) through an arc length of r_n. Then $V_1 = V$. Moreover, the V_n are congruent to one another. We will now show that the V_n are pairwise disjoint. If not, then, for some m and n with $m \neq n$, $V_m \cap V_n$ contains a point y. But since $y \in V_m$ there exists $x_\alpha \in V$ such that the arc length from x_α to y is r_m. Similarly, since $y \in V_n$ there exists $x_\beta \in V$ such that the arc length from x_β to y is r_n. The arc length from x_α to x_β is thus rational. It follows that $x_\alpha = x_\beta$. But then we have $r_m = r_n$, which is a contradiction. This shows that the V_n are pairwise disjoint.

Finally, every $x \in C$ lies in some V_n. For $x \in E_\alpha$ for some α, and thus the arc length from x_α to x is equal to some r_n. Hence, $x \in V_n$. We thus have the following situation.

(a) $C = V_1 \cup V_2 \cup V_3 \cup \cdots$,
(b) the V_n are pairwise disjoint,
(c) the V_n are congruent to one another.

If $V = V_1$ were measurable, then, by (c), all the V_n would be measurable and $mV_n = mV$, for all n. But then, by (a), (b), and 11.3E,

$$mC = mV_1 + mV_2 + mV_3 + \cdots,$$

$$mC = mV + mV + mV + \cdots. \tag{4}$$

The left side of (4) is finite. Hence, $mV + mV + mV + \cdots$ must converge. This is possible only if $mV = 0$. But then (4) would imply $mC = 0$, which is a contradiction, since $mC = 1$. This contradiction proves that V is not measurable.

Exercises 11.3

1. Prove that every subset of $[a,b]$ that is of type F_σ is measurable.
2. True or false? The union of uncountably many measurable subsets of $[a,b]$ must be measurable.
3. If E_1 and E_2 are measurable subsets of $[a,b]$, prove that the symmetric difference of E_1 and E_2 is also measurable.
4. If E_1 and E_2 are measurable subsets of $[0,1]$, and if $mE_1 = 1$, prove that

$$m(E_1 \cap E_2) = mE_2.$$

5. If $E_1, E_2, \ldots$ are measurable subsets of $[a,b]$, prove that

$$\limsup_{n\to\infty} E_n \quad \text{and} \quad \liminf_{n\to\infty} E_n$$

are measurable.

6. Show that there exists a closed nowhere-dense subset E of $[a,b]$ such that $mE > 0$.

11.4 MEASURABLE FUNCTIONS

We will see that a function may be Lebesgue integrable on $[a,b]$ even if the function is not continuous at any point of $[a,b]$. Indeed, for a *bounded* function to be Lebesgue integrable we will see that the function need only satisfy a condition much less restrictive than continuity—namely, measurability. The definition of "measurable function" applies to unbounded as well as bounded functions, and we ultimately extend the definition of Lebesgue integral to a wide class of unbounded but measurable functions.

We will discuss only real-valued functions.

11.4A. DEFINITION. Let f be a function on $[a,b]$. We say that f is a measurable function if, for every $s \in R$, the set

$$\{x \mid f(x) > s\}$$

is a measurable set.

That is, f is a measurable function if, for every real s, the inverse image under f of (s,∞) is a measurable set. It follows immediately that every continuous function g on $[a,b]$ is measurable! For (s,∞) is an open set. If g is continuous, then, by 5.4G, the inverse image under g of (s,∞) is open. But, by 11.2H, open sets are measurable. Hence, $\{x \mid g(x) > s\}$ is a measurable set, and so g is a measurable function.

On the other hand, some functions that are discontinuous at every point are still measurable. For example, if χ is the characteristic function of the rational numbers in $[0,1]$, then $\{x \mid \chi(x) > s\}$ is empty if $s \geqslant 1$, $\{x \mid \chi(x) > s\}$ is the set of rationals in $[0,1]$ if $0 \leqslant s < 1$, while $\{x \mid \chi(x) > s\} = [0,1]$ if $s < 0$. In any case $\{x \mid \chi(x) > s\}$ is a measurable set.

Here are other criteria for measurability equivalent to 11.4A.

11.4B. THEOREM. The function f on $[a,b]$ is measurable if and only if any one (and hence all) of the following statements hold.

(a) For every $s \in R$ the set $\{s \mid f(x) \geqslant s\}$ is a measurable set.
(b) For every $s \in R$ the set $\{x \mid f(x) < s\}$ is a measurable set.
(c) For every $s \in R$ the set $\{x \mid f(x) \leqslant s\}$ is a measurable set.

PROOF: Suppose f is measurable. Then, by 11.2A, $\{x \mid f(x) > s\}$ is a measurable set. But $\{x \mid f(x) \leqslant s\}$ is the complement of $\{x \mid f(x) > s\}$. Hence, by 11.2G, $\{x \mid f(x) \leqslant s\}$ is a measurable set. Hence, if f is measurable, then (c) holds.

We now show that (a) holds if f is measurable. For if f is measurable and $s \in R$, then, by 11.4A, each of the sets $\{x \mid f(x) > s - 1/n\}$ is measurable $(n = 1, 2, \ldots)$. But then, by 11.3H, $\cap_{n=1}^{\infty}\{x \mid f(x) > s - 1/n\}$ is measurable. However, it is easy to verify that

$$\{x \mid f(x) \geqslant s\} = \bigcap_{n=1}^{\infty} \left\{x \mid f(x) > s - \frac{1}{n}\right\}.$$

Hence, the set on the left is measurable, and so (a) holds.

The remainder of the proof is left to the reader.

11.4C. COROLLARY. If f is a measurable function on $[a,b]$, then the inverse image under f of any interval (bounded, unbounded, closed, open, half-open, etc.) is a measurable set.

PROOF: Let $[\lambda,\mu)$ be a bounded half-open interval. Then

$$[\lambda,\mu)=[\lambda,\infty)\cap(-\infty,\mu).$$

By (a) of 11.4B, the set $f^{-1}([\lambda,\infty))$ is measurable. By (b) of 11.4B, the set $f^{-1}((-\infty,\mu))$ is measurable. But, by 1.3F,

$$f^{-1}([\lambda,\mu))=f^{-1}([\lambda,\infty))\cap f^{-1}((-\infty,\mu)).$$

Since each of the two sets on the right is measurable, it follows from 11.3B that $f^{-1}([\lambda,\mu))$ is measurable. The theorem is thus proved for the case of intervals of the form $[\lambda,\mu)$. All other cases may be handled in identical fashion.

Sets of measure zero do not affect measurability for functions.

11.4D. THEOREM. If f and g are functions on $[a,b]$, if

$$f(x)=g(x) \quad \text{almost everywhere*} \qquad (a\leqslant x\leqslant b), \tag{1}$$

and if f is measurable, then g is also measurable.

PROOF: To show that g is measurable we must show that, if $s\in R$, then the set

$$E_1=\{x|g(x)>s\}$$

is measurable. Since f is measurable, we know that the set

$$E_2=\{x|f(x)>s\}$$

is measurable. But, by (1), the symmetric difference of E_1 and E_2 has measure zero. Hence, by the theorem in 11.3I, the measurability of E_1 follows from that of E_2.

We now set about showing that sums, products and limits of sequences of measurable functions are again measurable functions. Indeed, just about anything you can do with measurable functions will yield measurable functions. This fact leads to better theorems than can be obtained for Riemann-integrable functions.

11.4E. THEOREM. If f is a measurable function on $[a,b]$, and if $c\in R$, then the functions $f+c$ and cf are measurable.

PROOF: If $s\in R$, then

$$\{x|f(x)+c>s\}=\{x|f(x)>s-c\}.$$

The set on the right is measurable since f is a measurable function. Hence, the set on the left is measurable, which shows that the function $f+c$ is measurable.

If $c<0$, then $cf(x)>s$ if and only if $f(x)<s/c$. Hence,

$$\{x|cf(x)>s\}=\left\{x|f(x)<\frac{s}{c}\right\}.$$

The set on the right is measurable by (b) of 11.4B. Hence, the set on the left is measurable. This shows that cf is a measurable function if $c<0$. The cases $c>0$ and $c=0$ may be handled in similar fashion to complete the proof.

* This means that the set of x in $[a,b]$ for which the statement $f(x)=g(x)$ does not hold has measure zero. See 7.1D.

From 11.4E it follows that $-f$ is a measurable function whenever f is. Also, if f is measurable, then $c-f$ is a measurable function for any $c \in R$. We next treat sums, products, and so forth, of measurable functions.

11.4F. THEOREM. If f and g are measurable functions on $[a,b]$, then so are $f+g$, $f-g$, and fg. Furthermore, if $g(x) \neq 0$ $(a \leqslant x \leqslant b)$, then f/g is also measurable.

PROOF: Let $r_1, r_2, r_3, \ldots$ be an enumeration of the set of all rational numbers. If $x \in [a,b]$ and $s \in R$, it is clear that $f(x) > s - g(x)$ if and only if there is a rational number r_n such that $f(x) > r_n$ *and* $r_n > s - g(x)$. Hence,

$$\{x \mid f(x)+g(x)>s\} = \bigcup_{n=1}^{\infty} [\{x \mid f(x)>r_n\} \cap \{x \mid s-g(x)<r_n\}]. \tag{1}$$

For any $n \in I$, the set $\{x \mid f(x) > r_n\}$ is measurable, since f is a measurable function. The set $\{x \mid s-g(x)<r_n\}$ is also measurable, since, by 11.4E, $s-g$ is a measurable function. Thus by 11.3B and 11.3H, the set on the right of (1) is measurable. This proves that $f+g$ is a measurable function.

It then follows from 11.4E that $f-g=f+(-g)$ is measurable.

To show that fg is measurable we imitate a trick used in the first proof of 2.7G. We will prove first that the square of a measurable function is measurable. Indeed if h is a measurable function on $[a,b]$ and $s<0$, then the set $\{x \mid [h(x)]^2 > s\}$ is equal to $[a,b]$ and is thus measurable. If $s \geqslant 0$, then

$$\{x \mid [h(x)]^2 > s\} = \{x \mid h(x) > \sqrt{s}\} \cup \{x \mid h(x) < -\sqrt{s}\}.$$

Since h is a measurable function, each of the sets on the right is measurable. Thus $\{x \mid [h(x)]^2 > s\}$ is measurable for any $s \in R$, which proves that h^2 is a measurable function.

Now if f and g are measurable functions, then, by what we have already proved, $(f+g)^2$ and $(f-g)^2$ are also measurable. Since

$$fg = \tfrac{1}{4}\left[(f+g)^2 - (f-g)^2\right],$$

it follows that fg is measurable.

We leave the proof of the assertion concerning f/g to the reader. Note that since $f/g = f(1/g)$, it suffices to show that $1/g$ is measurable whenever g is a measurable function such that $g(x) \neq 0$ $(a \leqslant x \leqslant b)$.

The next results deal with sequences of measurable functions.

11.4G. THEOREM. Suppose $\{f_n\}_{n=1}^{\infty}$ is a sequence of measurable functions on $[a,b]$ such that the sequence $\{f_n(x)\}_{n=1}^{\infty}$ is bounded for every $x \in [a,b]$. Let

$$M(x) = \text{l.u.b.}\{f_1(x), f_2(x), f_3(x), \ldots\} \qquad (a \leqslant x \leqslant b)$$

and

$$m(x) = \text{g.l.b.}\{f_1(x), f_2(x), f_3(x), \ldots\} \qquad (a \leqslant x \leqslant b).$$

Then the functions M and m are both measurable.

PROOF: If $s \in R$ and $x \in [a,b]$, then $m(x) < s$ if and only if $f_n(x) < s$ for some n. Hence,

$$\{x \mid m(x) < s\} = \bigcup_{n=1}^{\infty} \{x \mid f_n(x) < s\}.$$

By (b) of 11.4B, each of the sets $\{x|\, f_n(x)<s\}$ is measurable. Hence, by 11.3H, $\{x|m(x)<s\}$ is measurable. This, again by (b) of 11.4B, shows that the function m is measurable.

That M is measurable follows from the equation

$$\{x|M(x)>s\} = \bigcup_{n=1}^{\infty} \{x|\, f_n(x)>s\}.$$

The special case in which $f_2=f_3=f_4=\cdots$ shows that $\max(f_1,f_2)$ and $\min(f_1,f_2)$ are measurable functions if f_1 and f_2 are measurable.

11.4H. THEOREM. If $\{f_n\}_{n=1}^{\infty}$ is a sequence of measurable functions on $[a,b]$ such that the sequence $\{f_n(x)\}_{n=1}^{\infty}$ is bounded for every $x\in[a,b]$, and if

$$f^*(x) = \limsup_{n\to\infty} f_n(x) \qquad (a\leqslant x\leqslant b),$$

$$f_*(x) = \liminf_{n\to\infty} f_n(x) \qquad (a\leqslant x\leqslant b),$$

then the functions f^* and f_* are both measurable. In particular, if $\{f_n\}_{n=1}^{\infty}$ converges pointwise to f on $[a,b]$, then f is measurable.

PROOF: For $n\in I$ let

$$g_n(x) = \text{l.u.b.}\{f_n(x), f_{n+1}(x), f_{n+2}(x), \dots\} \qquad (a\leqslant x\leqslant b).$$

Then, by 11.4G, each g_n is a measurable function. Moreover, by 2.9A,

$$f^*(x) = \lim_{n\to\infty} g_n(x) \qquad (a\leqslant x\leqslant b).$$

Also, for any $x\in[a,b]$,

$$g_1(x)\geqslant g_2(x)\geqslant g_3(x)\geqslant\cdots.$$

Hence, if $s\in R$,

$$\{x|\, f^*(x)<s\} = \bigcup_{n=1}^{\infty} \{x|\, g_n(x)<s\}.$$

From 11.4B and 11.3H it follows that f^* is measurable.

That f_* is measurable may be proved similarly. Finally, if $\{f_n\}_{n=1}^{\infty}$ converges pointwise to f, then, by 2.9C and 2.9F, $f=f^*=f_*$ and so f is measurable. This completes the proof.

We conclude this section by showing that in 11.4H pointwise convergence may be replaced by pointwise convergence almost everywhere.

11.4I. THEOREM. If $\{f_n\}_{n=1}^{\infty}$ is a sequence of measurable functions on $[a,b]$, and if

$$\lim_{n\to\infty} f_n(x) = f(x) \quad \text{almost everywhere} \qquad (a\leqslant x\leqslant b),$$

then f is measurable.

PROOF: Let E be the set of x in $[a,b]$ at which the statement

$$\lim_{n\to\infty} f_n(x) = f(x)$$

does not hold. Then, by hypothesis, E has measure zero. Define the functions $g_n (n\in I)$

and g as follows:

$$g_n(x)=f_n(x) \quad (x\notin E); \qquad g(x)=f(x) \quad (x\notin E)$$
$$g_n(x)=0 \quad (x\in E); \qquad g(x)=0 \quad (x\in E).$$

Then each g_n is measurable, by 11.4D. Now, if $x\in E$, then

$$\lim_{n\to\infty} g_n(x)=0=g(x).$$

Also, if $x\notin E$, then

$$\lim_{n\to\infty} g_n(x)=\lim_{n\to\infty} f_n(x)=f(x)=g(x).$$

Hence, $\{g_n\}_{n=1}^{\infty}$ converges pointwise (everywhere) to g on $[a,b]$. Since each g_n is measurable, it follows from 11.4H that g is measurable. Another application of 11.4D shows that f is measurable, and the proof is complete.

Exercises 11.4

1. If

$$f(x)=\frac{1}{x} \qquad (0<x<1),$$
$$f(0)=5,$$
$$f(1)=7,$$

prove that f is measurable on $[0,1]$.
2. Show that the subset E of $[a,b]$ is measurable if and only if its characteristic function χ_E is measurable.
3. Does there exist a nonmeasurable function on $[a,b]$?
4. If J_1 and J_2 are intervals of real numbers, and if f is a measurable function on $[a,b]$, show that $f^{-1}(J_1\cup J_2)$ is a measurable subset of $[a,b]$.
5. If $F'(x)$ exists for every x in $[a,b]$ and

$$f(x)=F'(x) \qquad (a\leqslant x\leqslant b),$$

prove that f is a measurable function. (*Hint*: Define $F(x)=F(b)$ for $x>b$. Then let

$$f_n(x)=\frac{F(x+1/n)-F(x)}{1/n} \qquad (a\leqslant x\leqslant b; n\in I).$$

Show that each f_n is measurable and note that

$$f(x)=\lim_{n\to\infty} f_n(x) \qquad (a\leqslant x<b).)$$

6. If G is an open subset of R^1 and if f is a measurable function on $[a,b]$, prove that $f^{-1}(G)$ is a measurable subset of $[a,b]$.

11.5 DEFINITION AND EXISTENCE OF THE LEBESGUE INTEGRAL FOR BOUNDED FUNCTIONS

Our definition of the Lebesgue integral parallels that of the Riemann integral. We begin by defining $M[f;E]$ and $m[f;E]$ for a bounded function f and a subset E of the closed bounded interval $[a,b]$. This will generalize 7.2A.

11.5A. DEFINITION. Let f be a bounded function on $[a,b]$, and let E be a subset of $[a,b]$. Then we define $M[f;E]$ and $m[f;E]$ as

$$M[f;E] = \underset{x \in E}{\text{l.u.b.}}\, f(x),$$

$$m[f;E] = \underset{x \in E}{\text{g.l.b.}}\, f(x).$$

Instead of dividing $[a,b]$ into intervals (as in 7.2B) we will partition $[a,b]$ into measurable subsets.

11.5B. DEFINITION. By a measurable partition P of $[a,b]$ we mean a finite collection $\{E_1, E_2, \ldots, E_n\}$ of measurable subsets of $[a,b]$ such that

$$\bigcup_{k=1}^{n} E_k = [a,b]$$

and such that

$$m(E_j \cap E_k) = 0 \qquad (j,k, = 1, \ldots, n; j \neq k).$$

The sets $E_1, E_2, \ldots, E_n$ are called the components of P.

If P and Q are measurable partitions, then Q is called a refinement of P if every component of Q is wholly contained in some component of P. (That is, if the components of Q are obtained by breaking up the components of P.)

Thus a measurable partition P is a finite collection of subsets whose union is all of $[a,b]$ and whose intersections with one another have measure zero.

It is then clear that if $\sigma = \{x_0, x_1, \ldots, x_n\}$ is a subdivision of $[a,b]$ (as in 7.2B) with component intervals $I_1, I_2, \ldots, I_n$, then $\{I_1, I_2, \ldots, I_n\}$ is a measurable partition of $[a,b]$. However, there are many measurable partitions of $[a,b]$ whose components are not intervals. For example, if E_1 is the set of rationals in $[a,b]$, and E_2 is the set of irrationals in $[a,b]$, then $\{E_1, E_2\}$ is a measurable partition of $[a,b]$.

We next generalize 7.2C.

11.5C. DEFINITION. Let f be a bounded function on $[a,b]$ and let $P = \{E_1, \ldots, E_n\}$ be any measurable partition of $[a,b]$. We define the upper sum $U[f;P]$ as

$$U[f;P] = \sum_{k=1}^{n} M[f;E_k] \cdot mE_k.$$

Similarly, we define the lower sum $L[f;P]$ as*

$$L[f;P] = \sum_{k=1}^{n} m[f;E_k] \cdot mE_k.$$

Note that if $E_1, \ldots, E_n$ are the component intervals of a subdivision σ, then $U[f;P]$ as defined here is precisely the same as $U[f;\sigma]$ as defined in 7.2C. Hence, the set of numbers $U[f;\sigma]$ for all subdivisions σ is a subset of the set of numbers $U[f;P]$ for all measurable partitions P.

Corresponding to 7.2D we have the following result.

11.5D. LEMMA. Let f be a bounded function on $[a,b]$. Then every upper sum for f is greater than or equal to every lower sum for f. That is, if P and Q are any two measurable partitions of $[a,b]$, then $U[f;P] \geqslant L[f;Q]$.

* Do not confuse the two uses of m.

PROOF: We will not give many details of this proof since it follows that of 7.2D very closely. First one should show that if P^* is any refinement of P, then

$$U[f;P] \geqslant U[f;P^*]. \tag{1}$$

The case where $P = \{E_1, \ldots, E_k, \ldots, E_n\}$ and $P^* = \{E_1, \ldots, E_k^*, E_k^{**}, \ldots, E_n\}$ is proved as in 7.2D, and the general case of (1) follows by induction.

Similarly, if Q^* is a refinement of Q, then

$$L[f;Q] \leqslant L[f;Q^*].$$

Now, if the components of P are $E_1, \ldots, E_n$, and the components of Q are $F_1, \ldots, F_m$, let T be the measurable partition whose components are the $n \cdot m$ subsets $E_i \cap F_j (i = 1, \ldots, n; j = 1, \ldots, m)$. Then T is a refinement of both P and Q. Hence, by (1) and (2) we have

$$U[f;P] \geqslant U[f;T] \geqslant L[f;T] \geqslant L[f;Q],$$

and the lemma is proved.

11.5E. Exactly as in 7.2E we may now show that

$$\underset{P}{\text{g.l.b.}}\, U[f;P] \geqslant \underset{P}{\text{l.u.b.}}\, L[f;P], \tag{1}$$

where the g.l.b. and l.u.b. are taken over all measurable partitions P of $[a,b]$. (Verify.) This puts us in a position to define the Lebesgue upper and lower integrals of a bounded function f on $[a,b]$. To avoid ambiguity, we will denote by

$$\mathfrak{R}\overline{\int_a^b} f \quad \text{and} \quad \mathfrak{R}\underline{\int_a^b} f$$

the Riemann upper and lower integrals of f as defined in 7.2E, while the Lebesgue upper and lower integrals of f, which we are about to define, will be denoted by

$$\mathfrak{L}\overline{\int_a^b} f \quad \text{and} \quad \underline{\int_a^b} f.$$

DEFINITION. Let f be a bounded function on $[a,b]$. We define

$$\mathfrak{L}\overline{\int_a^b} f(x)\,dx,$$

called the Lebesgue upper integral of f over $[a,b]$, as

$$\mathfrak{L}\overline{\int_a^b} f(x)\,dx = \underset{P}{\text{g.l.b.}}\, U[f;P]$$

where the g.l.b. is taken over all measurable partitions P of $[a,b]$. Similarly, we define

$$\mathfrak{L}\underline{\int_a^b} f(x)\,dx,$$

called the Lebesgue lower integral of f over $[a,b]$, as

$$\mathfrak{L}\underline{\int_a^b} f(x)\,dx = \underset{P}{\text{l.u.b.}}\, L[f;P].$$

For simplicity we sometimes denote the upper and lower integrals of f by

$$\mathfrak{L}\overline{\int_a^b} f \quad \text{and} \quad \mathfrak{L}\underline{\int_a^b} f.$$

From the inequality (1) it follows that

$$L\underline{\int_a^b} f \leqslant L\overline{\int_a^b} f. \tag{2}$$

From the remark following definition 11.5C (namely, that every $U[f;\sigma]$ is a $U[f;P]$) it follows that

$$L\overline{\int_a^b} f = \underset{P}{\text{g.l.b.}}\, U[f;P] \leqslant \underset{\sigma}{\text{g.l.b.}}\, U[f;\sigma] = \mathfrak{R}\overline{\int_a^b} f. \tag{3}$$

(That is, roughly, the bigger the set, the smaller the g.l.b. of the set.) Similarly,

$$\mathfrak{L}\underline{\int_a^b} f = \underset{P}{\text{l.u.b.}}\, L[f;P] \geqslant \underset{\sigma}{\text{l.u.b.}}\, L[f;\sigma] = \mathfrak{R}\underline{\int_a^b} f. \tag{4}$$

Thus from (4), (2), and (3) we conclude that for any bounded function f on $[a,b]$,

$$\mathfrak{R}\underline{\int_a^b} f \leqslant \mathfrak{L}\underline{\int_a^b} f \leqslant \mathfrak{L}\overline{\int_a^b} f \leqslant \mathfrak{R}\overline{\int_a^b} f. \tag{5}$$

We will now denote the (Riemann) integral as defined in 7.2F by $\mathfrak{R}\int_a^b f$, and the Lebesgue integral, which we will now define, by $\mathfrak{L}\int_a^b f$.

11.5F. DEFINITION. If f is a bounded function on $[a,b]$, we say that f is Lebesgue integrable on $[a,b]$ if

$$\mathfrak{L}\underline{\int_a^b} f = \mathfrak{L}\overline{\int_a^b} f.$$

In this case, we define $\mathfrak{L}\int_a^b f(x)\,dx \left(\text{or } \mathfrak{L}\int_a^b f\right)$ as

$$\mathfrak{L}\int_a^b f = \mathfrak{L}\underline{\int_a^b} f = \mathfrak{L}\overline{\int_a^b} f.$$

If f is Lebesgue integrable on $[a,b]$, we write $f \in \mathfrak{L}[a,b]$.

In Section 11.7 we define the Lebesgue integral for a wide class of unbounded functions. Thus ultimately, the statement $f \in \mathfrak{L}[a,b]$ will not imply that f is bounded. Hence, in Section 11.6, we include the boundedness in the hypotheses of many theorems on (bounded) functions in $\mathfrak{L}[a,b]$.

We next prove the extremely important result which states that if a bounded function f is Riemann integrable, then f must be Lebesgue integrable and the two integrals of f are equal!

11.5G. THEOREM. Let f be a bounded function on $[a,b]$. If $f \in \mathfrak{R}[a,b]$, then $f \in \mathfrak{L}[a,b]$ and

$$\mathfrak{R}\int_a^b f = \mathfrak{L}\int_a^b f. \tag{1}$$

PROOF: From (5) of 11.5D we have

$$\mathfrak{R}\underline{\int_a^b} f \leqslant \mathfrak{L}\underline{\int_a^b} f \leqslant \mathfrak{L}\overline{\int_a^b} f \leqslant \mathfrak{R}\overline{\int_a^b} f. \tag{2}$$

If $f \in \mathfrak{R}[a,b]$, then, by definition, the extreme left- and right-hand terms in (2) must be equal. It follows that all four terms in (2) are equal—that is,

$$\mathfrak{R}\underline{\int_a^b} f = \mathfrak{L}\underline{\int_a^b} f = \mathfrak{L}\overline{\int_a^b} f = \mathfrak{R}\overline{\int_a^b} f.$$

Thus $f \in \mathfrak{L}[a,b]$ and the equation (1) follows immediately.

Thus there is no need to distinguish between

$$\mathfrak{R}\int_a^b f \quad \text{and} \quad \mathfrak{L}\int_a^b f,$$

for when both integrals exist they must be equal. Henceforth we will write $\int_a^b f \left(\text{or } \int_a^b f(x)\,dx \right)$ for the Riemann or the Lebesgue integral of f and, by 11.5G, no ambiguity will arise.

Theorem 11.5G states that any bounded function that is Riemann integrable must also be Lebesgue integrable. After we duscuss the existence of the Lebesgue integral it will be clear that many bounded functions that are Lebesgue integrable are not Riemann integrable.

The proof of the next theorem is almost identical to that of 7.2G (with subdivisions replaced by partitions), and we therefore omit it.

11.5H. THEOREM. Let f be a bounded function on $[a,b]$. Then $f \in \mathfrak{L}[a,b]$ if and only if for each $\epsilon > 0$ there exists a measurable partition P of $[a,b]$ such that

$$U[f;P] < L[f;P] + \epsilon. \tag{1}$$

To illustrate 11.5H let χ be the characteristic function of the *irrational* numbers in $[0,1]$. Let E_1 be the set of irrational numbers in $[0,1]$, and let E_2 be the set of rational numbers in $[0,1]$. Then $P = \{E_1, E_2\}$ is a measurable partition of $[0,1]$. Moreover, χ is identically 1 on E_1 and χ is identically 0 on E_2. Hence, $M[\chi; E_1] = m[\chi; E_1] = 1$, while $M[\chi; E_2] = m[\chi; E_2] = 0$. Hence, $U[\chi; P] = 1 \cdot mE_1 + 0 \cdot mE_2 = 1$. Similarly, $L[\chi; P] = 1$. Since $U[\chi; P] = L[\chi; P]$, it follows from 11.5H that $\chi \in \mathfrak{L}[0,1]$. Moreover, since

$$L[\chi;P] \leqslant \int_0^1 \chi \leqslant U[\chi;P],$$

we have

$$\int_0^1 \chi = 1.$$

Note that $\chi \notin \mathfrak{R}[0,1]$.

For most functions f there is no one partition P for which $U[f;P] = L[f;P]$. (The function χ of the preceding paragraph is an exception.) If $f \in \mathfrak{L}[a,b]$, the partition P such that (1) holds usually depends on ϵ.

Next we show the important role played by measurable functions. We will prove that every bounded measurable function f is Lebesgue integrable. Note that the proof involves a subdivision of an interval containing the range of f. That is, we subdivide on the y axis instead of on the x axis as with the Riemann integral.

11.5I. THEOREM. If f is a bounded measurable function on $[a,b]$, then $f \in \mathfrak{L}[a,b]$.

PROOF: Since f is bounded there exists $M>0$ such that the range of f is contained in the half-open interval $[-M,M)$. Given $\epsilon>0$ there exist a finite number of points $y_0,y_1,\ldots,y_n$ such that $-M=y_0<y_1<\cdots<y_n=M$ and such that $y_k-y_{k-1}<\epsilon/(b-a)$ $(k=1,\ldots,n)$. (That is, $\{y_0,y_1,\ldots,y_n\}$ is a subdivision of $[-M,M]$ such that the distance between any two successive points of subdivision is less than $\epsilon/(b-a)$.) For each $k=1,2,\ldots,n$, let E_k be the inverse image of $[y_{k-1},y_k)$ under f. (That is, $x\in E_k$ if and only if $y_{k-1}\leqslant f(x)<y_k$.) Then each E_k is measurable, by 11.4C. It is then easy to verify that $P=\{E_1,E_2,\ldots,E_n\}$ is a measurable partition of $[a,b]$. Since $M[f;E_k]\leqslant y_k$ we have

$$U[f;P]=\sum_{k=1}^{n} M[f;E_k]\cdot mE_k\leqslant \sum_{k=1}^{\infty} y_k\cdot mE_k.$$

Also, since $m[f;E_k]\geqslant y_{k-1}$ we have

$$L[f;P]=\sum_{k=1}^{n} m[f;E_k]\cdot mE_k\geqslant \sum_{k=1}^{n} y_{k-1}\cdot mE_k.$$

Thus

$$U[f;P]-L[f;P]\leqslant \sum_{k=1}^{\infty}(y_k-y_{k-1})\cdot mE_k<\frac{\epsilon}{b-a}\sum_{k=1}^{\infty} mE_k. \tag{1}$$

Since the E_k are pairwise disjoint and $\cup_{k=1}^{n}E_k=[a,b]$, we have, by 11.3E,

$$\sum_{k=1}^{n} mE_k=m\left(\bigcup_{k=1}^{n} E_k\right)=b-a. \tag{2}$$

From (1) and (2) we then obtain

$$U[f;P]-L[f;P]<\epsilon.$$

By 11.5H we then have $f\in\mathcal{L}[a,b]$, and the proof is complete.

For emphasis, we repeat the result of 11.5I. *Every bounded measurable function on* $[a,b]$ *is Lebesgue integrable.*

Thus if f is bounded on $[a,b]$, the measurability of f is a *sufficient* condition that $f\in\mathcal{L}[a,b]$.

We will show in the next section (theorem 11.6N) that for a bounded function f, measurability is also a *necessary* condition that $f\in\mathcal{L}[a,b]$. That is, if f is bounded and $f\in\mathcal{L}[a,b]$, then f must be measurable.

Exercises 11.5

1. Write out detailed proofs of 11.5D and 11.5H.
2. What can you say about the function f on $[a,b]$ if there exists a measurable partition P of $[a,b]$ such that

$$U[f;P]=L[f;P]?$$

3. Let

$$\begin{aligned} f(x)&=2 \qquad (0\leqslant x<1),\\ f(x)&=4 \qquad (1\leqslant x<2),\\ f(x)&=3 \qquad (2\leqslant x<3),\\ f(x)&=2 \qquad (3\leqslant x\leqslant 4). \end{aligned}$$

(a) If σ is the subdivision $\{0,1,2,3,4\}$ of $[0,4]$, calculate $U[f;\sigma]$.

(b) For $k=2,3,4$, let E_k be the inverse image under f of $[k,k+1)$. Show that $P=\{E_2,E_3,E_4\}$ is a measurable partition of $[0,4]$.
(c) Calculate $U[f;P]$ and $L[f;P]$.

11.6 PROPERTIES OF THE LEBESGUE INTEGRAL FOR BOUNDED MEASURABLE FUNCTIONS

11.6A. THEOREM. If f is a bounded measurable function (and hence in $\mathcal{L}[a,b]$), and if $a<c<b$, then $f\in\mathcal{L}[a,c], f\in\mathcal{L}[c,b]$, and

$$\int_a^b f=\int_a^c f+\int_c^b f.$$

PROOF: We must first show that f is Lebesgue integrable on $[a,c]$ and $[c,b]$, or, more precisely, that the restriction of f to these intervals is Lebesgue integrable. To show that the restriction of f to $[a,c]$ is a measurable function on $[a,c]$ we must show that for any $s\in R$, the set $E=\{x\in[a,c] \| f(x)>s\}$ is measurable. But $E=[a,c]\cap E^*$ where $E^*=\{x\in[a,b] \| f(x)>s\}$. Since f is measurable on $[a,b]$, the set E^* is measurable. Hence, E is measurable (see exercise 8 of Section 11.2). Thus f is bounded and measurable on $[a,c]$, and is therefore Lebesgue integrable on $[a,c]$. Similarly, f is Lebesgue on $[c,b]$. The proof that

$$\int_a^b f=\int_a^c f+\int_c^b f$$

is then an imitation of the corresponding part of the proof of 7.4A with subdivisions replaced by partitions.

The next two results may be proved in exactly the same manner as 7.4B and 7.4C. Note that if f is measurable on $[a,b]$ and $\lambda\in R$, then λf is measurable by 11.4E. Also, if f and g are measurable on $[a,b]$, then $f+g$ is measurable by 11.4E. Hence, λf and $f+g$ will be bounded and measurable if both f and g are bounded and measurable.

11.6B. THEOREM. If f is a bounded measurable function on $[a,b]$, and if $\lambda\in R$, then $\lambda f\in\mathcal{L}[a,b]$ and

$$\int_a^b \lambda f=\lambda\int_a^b f.$$

11.6C. THEOREM. If f and g are bounded measurable functions on $[a,b]$, then $f+g$ is in $\mathcal{L}[a,b]$ and

$$\int_a^b (f+g)=\int_a^b f+\int_a^b g.$$

The next result shows one of the great advantages of the Lebesgue integral over the Riemann integral. If f is a bounded measurable function on $[a,b]$ (and hence, Lebesgue integrable on $[a,b]$), then changing the values of f on a set of measure zero has no effect either on the (Lebesgue) integrability of f or on the value of $\int_a^b f$. (On the other hand, changing the values of a Riemann integrable function on a set of measure zero may destroy the Riemann integrability of the function. For example, if $f(x)=1$ $(0\leqslant x\leqslant 1)$ and if χ is the characteristic function of the irrationals in $[0,1]$, then χ may be obtained from f by changing the values of f on a set of measure zero, namely, on the set of rationals in $[0,1]$. But f is Riemann integrable and χ is not.)

11.6D. THEOREM. If f is a bounded measurable function on $[a,b]$, and if g is a bounded function on $[a,b]$ such that

$$f(x)=g(x) \quad \text{almost everywhere} \qquad (a\leqslant x\leqslant b),$$

then

$$g\in\mathfrak{L}[a,b] \quad \text{and} \quad \int_a^b g=\int_a^b f.$$

PROOF: By 11.4D, g is measurable. Since g is bounded it follows from 11.5I that g is integrable. If E is the set of $x\in[a,b]$ for which $f(x)\neq g(x)$, then $m(E)=0$ by hypothesis. Then, if $E'=[a,b]-E$, we have $f(x)=g(x)$ for $x\in E'$. Let $P=\{E,E'\}$. Then

$$U[g-f;P]=M[g-f;E]\cdot mE+M[g-f;E']\cdot mE'=M[g-f;E]\cdot 0+0\cdot mE'=0.$$

Similarly, $L[g-f;P]=0$. Since

$$0=L[g-f;P]\leqslant\int_a^b (g-f)\leqslant U[g-f;P]=0,$$

we have

$$\int_a^b (g-f)=0.$$

Hence,

$$\int_a^b g=\int_a^b (g-f)+\int_a^b f=0+\int_a^b f,$$

and the theorem is proved.

From 11.6D it follows immediately that a bounded function that is zero almost everywhere must be Lebesgue integrable and have integral zero.

11.6E. THEOREM. If f is a bounded measurable function on $[a,b]$, and if

$$f(x)\geqslant 0 \quad \text{almost everywhere} \qquad (a\leqslant x\leqslant b),$$

then

$$\int_a^b f\geqslant 0.$$

PROOF: By 11.6D we may assume that $f(x)\geqslant 0$ for all x in $[a,b]$. (For this requires changing the values of f only on a set of measure zero and hence, cannot affect $\int_a^b f$.) But it is then obvious that $U[f;P]\geqslant 0$ for any measurable partition P, and hence,

$$\mathfrak{L}\overline{\int_a^b} f=\underset{P}{\text{g.l.b.}}\, U[f;P]\geqslant 0.$$

Since f is integrable we have

$$\int_a^b f=\mathfrak{L}\overline{\int_a^b} f\geqslant 0,$$

and the theorem is proved.

We may then deduce the next corollary exactly as 7.4E was deduced from 7.4D.

11.6F. COROLLARY. If f and g are bounded measurable functions on $[a,b]$, and if

$$f(x) \leqslant g(x) \quad \text{almost everywhere} \qquad (a \leqslant x \leqslant b),$$

then

$$\int_a^b f \leqslant \int_a^b g.$$

From 11.6F we deduce the following result.

11.6G. COROLLARY. If f is a bounded measurable function on $[a,b]$, then $|f| \in \mathcal{L}[a,b]$ and

$$\left|\int_a^b f\right| \leqslant \int_a^b |f|.$$

PROOF: Since $|f| = \max(f,0) - \min(f,0)$, it follows from 11.4G and 11.4F that $|f|$ is measurable. Since $|f|$ is clearly bounded,, $|f|$ must be integrable. The remainder of the proof follows that of 7.4F.

11.6H. DEFINITION. If $b < a$, we define

$$\int_a^b f \quad \text{to be} \quad -\int_b^a f,$$

provided that f is integrable on $[b,a]$.

11.6I. We leave it to the reader to prove that

$$\int_a^c f + \int_c^b f = \int_a^b,$$

regardless of the order of a,b,c.

11.6J. We will now give the definition of $\int_E f$ where f is a bounded measurable function on $[a,b]$ and E is a measurable subset of $[a,b]$. Note that in this situation the function $f\chi_E$ will be bounded and measurable on $[a,b]$ and hence, integrable on $[a,b]$. (Here χE is, of course, the characteristic function of E.)

DEFINITION. Let E be a measurable subset of $[a,b]$ and let f be a bounded measurable function on $[a,b]$. Then $\int_E f$ is defined as

$$\int_E f = \int_a^b f\chi_E.$$

The integral $\int_E f$ then has the same elementary properties as those we have just proved for $\int_a^b f$. We now list these properties, after which we indicate how they may be demonstrated.

11.6K. THEOREM.

1. If E_1 and E_2 are disjoint measurable subsets of $[a,b]$, and if f is a bounded measurable function on $[a,b]$, then

$$\int_{E_1 \cup E_2} f = \int_{E_1} f + \int_{E_2} f.$$

2. If E is a measurable subset of $[a,b]$, if f is a bounded measurable function on $[a,b]$, and if $\lambda \in R$, then

$$\int_E \lambda f = \lambda \int_E f.$$

3. If E is a measurable subset of $[a,b]$, and if f and g are bounded measurable functions on $[a,b]$, then

$$\int_E (f+g) = \int_E f + \int_E g.$$

4. If E is a measurable subset of $[a,b]$, and if f and g are bounded measurable functions on $[a,b]$ such that

$$f(x) = g(x) \quad \text{almost everywhere} \qquad (x \in E),$$

then

$$\int_E f = \int_E g.$$

5. If E is a measurable subset of $[a,b]$, and if f is a bounded measurable function on $[a,b]$ such that

$$f(x) \geqslant 0 \quad \text{almost everywhere} \qquad (x \in E),$$

then

$$\int_E f \geqslant 0.$$

6. If E is a measurable subset of $[a,b]$, if f and g are bounded measurable functions on $[a,b]$, and if

$$f(x) \leqslant g(x) \quad \text{almost everywhere} \qquad (x \in E),$$

then

$$\int_E f \leqslant \int_E g.$$

PROOF: To prove (1), for example, note that $\chi_{E_1 \cup E_2} = \chi_{E_1} + \chi_{E_2}$ since E_1 and E_2 are disjoint. Hence, $f\chi_{E_1 \cup E_2} = f\chi_{E_1} + f\chi_{E_2}$, and so we obtain, using 11.6J and 11.6C,

$$\begin{aligned}\int_{E_1 \cup E_2} f &= \int_a^b f\chi_{E_1 \cup E_2} = \int_a^b (f\chi_{E_1} + f\chi_{E_2}) \\ &= \int_a^b f\chi_{E_1} + \int_a^b f\chi_{E_2} = \int_{E_1} f + \int_{E_2} f.\end{aligned}$$

This proves (1).

To prove (4) note that if $f(x)=g(x)$ for almost every x in E, then

$$f(x)\chi_E(x)=g(x)\chi_E(x) \quad \text{almost everywhere} \qquad (a\leqslant x\leqslant b).$$

But then, by 11.6D,

$$\int_a^b f\chi_E=\int_a^b g\chi_E.$$

Thus by 11.6J,

$$\int_E f=\int_E g.$$

This proves (4). The other assertions follow just as easily.

The following assertion is elementary but important.

11.6L. THEOREM. If E is a measurable subset of $[a,b]$, then*

$$\int_E 1=mE.$$

PROOF: We have

$$\int_E 1=\int_a^b \chi_E. \tag{1}$$

If $E'=[a,b]-E$ and if $P=\{E,E'\}$, it is easy to verify that $U[\chi_E;P]=mE=L[\chi_E;P]$. Hence,

$$mE=L[\chi;P]\leqslant\int_a^b \chi_E\leqslant U[\chi;P]=mE,$$

and so

$$\int_a^b \chi_E=mE. \tag{2}$$

The theorem follows from (1) and (2).
The following result is quite useful.

11.6M. THEOREM. If f is a bounded measurable function on $[a,b]$ such that

$$f(x)\geqslant 0 \quad \text{almost everywhere} \qquad (a\leqslant x\leqslant b),$$

and if

$$\int_a^b f=0, \tag{1}$$

then

$$f(x)=0 \quad \text{almost everywhere} \quad (a\leqslant x\leqslant b).$$

PROOF: Suppose the theorem were false. Then the set $E=\{x|f(x)>0\}$ would be measurable and $mE>0$. Now $E=\cup_{n=1}^{\infty}E_n$ where $E_n=\{x|f(x)>1/n\}$. Since E_n is measurable and since $E_1\subset E_2\subset\cdots$, it follows from 11.3F that $\lim_{n\to\infty}mE_n=mE$ and

* The integral $\int_E 1$ means, of course, $\int_E f$ where $f(x)=1$ $(a\leqslant x\leqslant b)$.

hence, $\lim_{n\to\infty} mE_n > 0$. Thus $mE_N > 0$ for some $N \in I$. But then

$$\int_a^b f \geq \int_a^b f\chi_{E_N} = \int_{E_N} f \geq \int_{E_N} \frac{1}{N} = \frac{1}{N}\cdot mE_N > 0.$$

This contradicts (1), and the contradiction shows that the theorem must be true.

Finally, we show that a bounded Lebesgue integrable function must be measurable.

11.6N THEOREM. If f is bounded and $f \in \mathcal{L}[a,b]$, then f is measurable.

PROOF: For each $n = 1,2,\ldots$ there exists, by 11.5H, a measurable partition P_n of $[a,b]$ such that

$$U[f;P_n] - L[f;P_n] < \frac{1}{n}. \tag{1}$$

We may assume that P_{n+1} is a refinement of P_n. (For otherwise we could take the intersections of the components of P_n with those of P_{n+1} to obtain a partition P'_{n+1} which is a refinement of both P_n and P_{n+1}. We would then have

$$L[f;P_{n+1}] \leq L[f;P'_{n+1}] \leq U[f;P'_{n+1}] \leq U[f;P_{n+1}],$$

so that

$$U[f;P'_{n+1}] - L[f;P'_{n+1}] \leq U[f;P_{n+1}] - L[f;P_{n+1}] < \frac{1}{n+1}.$$

Hence, P'_{n+1} is a refinement of P_n that satisfies (1).)

Fix n. Suppose $P_n = \{E_n^1, E_n^2, \ldots, E_n^k\}$. We may assume that the E_n^j are pairwise disjoint. Define functions g_n, h_n on $[a,b]$ by

$$h_n(x) = M[f;E_n^j] \qquad (x \in E_n^j; j = 1,\ldots,k.)$$

$$g_n(x) = m[f;E_n^j] \qquad (x \in E_n^j; j = 1,\ldots,k.)$$

so that

$$g_n(x) \leq f(x) \leq h_n(x) \qquad (a \leq x \leq b).$$

Both g_n and h_n are measurable since they are constant on each E_n^j. Moreover, since P_{n+1} is a refinement of P_n, we have

$$g_n(x) \leq g_{n+1}(x), \qquad h_n(x) \geq h_{n+1}(x)$$

for all $n = 1,2,\ldots$, and all x in $[a,b]$. By 2.6B and 2.6E the sequences of functions $\{g_n\}_{n=1}^{\infty}$ converge pointwise on $[a,b]$. Let

$$g(x) = \lim_{n\to\infty} g_n(x) \qquad (a \leq x \leq b)$$

$$h(x) = \lim_{n\to\infty} h_n(x) \qquad (a \leq x \leq b).$$

Then g and h are measurable, by 11.4I, and

$$g_n(x) \leq g(x) \leq f(x) \leq h(x) \leq h_n(x) \qquad (n = 1,2,\ldots; a \leq x \leq b). \tag{2}$$

Hence,

$$\int_a^b g_n \leq \int_a^b g \leq \int_a^b f \leq \int_a^b h \leq \int_a^b h_n \qquad (n = 1,2,\ldots). \tag{3}$$

But $h_n = M[f; E_n^j]$ on E_n^j so that, by 11.6K,

$$\int_a^b h_n = \int_{E_n^1} h_n + \cdots + \int_{E_n^k} h_n = M[f; E_n^1]\cdot mE_n^1 + \cdots + M[f; E_n^k]\cdot mE_n^k.$$

Hence,

$$\int_a^b h_n = U[f; P_n]. \tag{4}$$

Similarly,

$$\int_a^b g_n = L[f; P_n]. \tag{5}$$

From (1), (4), and (5) we have

$$\int_a^b h_n - \int_a^b g_n < \frac{1}{n} \qquad (n = 1, 2, \ldots).$$

From (3) we then have

$$\int_a^b h - \int_a^b g < \frac{1}{n} \qquad (n = 1, 2, \ldots).$$

Hence, $\int_n^b (h-g) = 0$. Since $g(x) \leqslant h(x)$ for all x, it follows from 11.6M that $g(x) = h(x)$ a.e. Thus from (2), $f(x) = h(x)$ a.e. But then, since h is measurable, 11.4D implies that f is measurable and the proof is complete.

From 11.5I and 11.6N we see that if f is a bounded function on $[a,b]$, then $f \in \mathfrak{L}\{a,b]$ if and only if f is measurable.

Exercises 11.6

1. If $f \in \mathfrak{L}[a,b]$, if $E \subset [a,b]$ and $mE = 0$, show that

$$\int_E f = 0.$$

2. If E_1 and E_2 are measurable subsets of $[a,b]$, and if $f \subset \mathfrak{L}[a,b]$, prove that

$$\int_{E_1} f + \int_{E_2} f = \int_{E_1 \cup E_2} f + \int_{E_1 \cap E_2} f.$$

3. If f is a bounded measurable function on $[a,b]$, and $\int_a^b [f(x)]^2\, dx = 0$, prove that $f(x) = 0$ for almost all x in $[a,b]$.
4. Write out the details of the proofs of 11.6D and 11.6G.
5. Let f be a bounded measurable function on $[a,b]$ such that

$$f(x) \geqslant 0 \quad \text{almost everywhere} \qquad (a \leqslant x \leqslant b).$$

If E and F are measurable subsets of $[a,b]$ such that $E \subset F$, prove that

$$\int_E f \leqslant \int_F f.$$

6. Let $E_1, E_2, \ldots, E_n$ be measurable subsets of $[0,1]$. If each point of $[0,]$ belongs to at least three of these sets, show that at least one of the sets has measure $\geqslant 3/n$. [*Hint*: Let $\chi_1, \ldots, \chi_n$ be the characteristic functions of $E_1, \ldots, E_n$. First show that

$$\chi_1(x) + \cdots + \chi_n(x) \geqslant 3 \qquad (0 \leqslant x \leqslant 1).$$

11.7 THE LEBESGUE INTEGRAL FOR UNBOUNDED FUNCTIONS

We now extend the definition of the Lebesgue integral to a large class of unbounded measurable functions. We begin by considering nonnegative-valued functions.

11.7A. DEFINITION. If f is a nonnegative-valued function on $[a,b]$, and if $n \in I$, we define the function ${}^n f$ on $[a,b]$ as follows. For each x in $[a,b]$ let

$$
\begin{aligned}
{}^n f(x) &= f(x) \quad \text{if} \quad 0 \leqslant f(x) \leqslant n, \\
{}^n f(x) &= n \qquad \text{if} \quad f(x) > n.
\end{aligned}
$$

That is,

$$ {}^n f(x) = \min\left[f(x), n \right] \qquad (a \leqslant x \leqslant b). $$

Thus the graph of ${}^n f$ is obtained by truncating the graph of f. For example, if

$$
\begin{aligned}
f(x) &= \frac{1}{\sqrt[3]{x}} \qquad (0 < x \leqslant 1), \\
f(0) &= 0,
\end{aligned}
$$

then

$$
\begin{aligned}
{}^4 f(x) &= \frac{1}{\sqrt[3]{x}} \qquad \left(\frac{1}{64} \leqslant x \leqslant 1 \right), \\
{}^4 f(x) &= 4 \qquad \left(0 < x < \frac{1}{4} \right), \\
{}^4 f(0) &= 0.
\end{aligned}
$$

11.7B. Now suppose that f is a nonnegative-valued unbounded *measurable* function. Then, for each $n \in I$, the function ${}^n f$ is a bounded function and, by 11.4G, ${}^n f$ is measurable. Hence, by 11.5I, ${}^n f$ is Lebesgue integrable. It is then clear that

$$ \left\{ \int_a^b {}^n f \right\}_{n=1}^{\infty} $$

is a nondecreasing sequence of real numbers, and hence, either converges or diverges to infinity.

DEFINITION. Let f be a nonnegative-valued unbounded measurable function on $[a,b]$. If

$$ \lim_{n \to \infty} \int_a^b {}^n f $$

exists, then we say that f is (Lebesgue) integrable on $[a,b]$ and define $\int_a^b f$ as

$$ \int_a^b f = \lim_{n \to \infty} \int_a^b {}^n f. $$

If f is Lebesgue integrable, we write $f \in \mathcal{L}[a,b]$.

If f is the function in the example after 11.7A, then*

$$ \int_0^1 {}^n f = \int_{1/n^3}^1 \frac{1}{\sqrt[3]{x}}\, dx + \int_0^{1/n^3} n\, dx = \left(\frac{3}{2} - \frac{3}{2n^2} \right) + \frac{1}{n^2}. $$

* The value of ${}^n f$ at 0 does not affect our computations (11.6D).

Hence,

$$\lim_{n\to\infty} \int_0^1 {}^n f = \frac{3}{2}.$$

According to the definition we will thus say that f is integrable on $[0,1]$ and

$$\int_0^1 f = \int_0^1 \frac{1}{\sqrt[3]{x}}\, dx = \frac{3}{2}.$$

Thus, although

$$\int_0^1 \frac{1}{\sqrt[3]{x}}\, dx$$

is an improper Riemann integral, as a Lebesgue integral it is perfectly "proper" even though the integrand is unbounded. Note that, when viewed as an improper Riemann integral,

$$\int_0^1 \frac{1}{\sqrt[3]{x}}\, dx$$

also has the value $\frac{3}{2}$.

We leave it to the reader to show that if $f(x)=1/x$ $(0<x\leqslant 1)$, $f(0)=19$, then f is not integrable on $[0,1]$.

From 11.7B it is easy to show that if f is a nonnegative-valued measurable function on $[a,b]$, and if

$$f(x)\leqslant g(x) \qquad (a\leqslant x\leqslant b),$$

where $g\in \mathfrak{L}[a,b]$, the f is also in $\mathfrak{L}[a,b]$. (See exercise 3.)

If f is a nonnegative-valued *bounded* measurable function on $[a,b]$, then ${}^n f=f$ for all sufficiently large n. Hence, the equation

$$\int_a^b f = \lim_{n\to\infty} \int_a^b {}^n f$$

also holds for bounded f.

In order to define the Lebesgue integral for measurable functions that take both positive and negative values, we show that such functions can be written as the difference of two nonnegative-valued measurable functions.

11.7C. DEFINITION. Let f be any real-valued function on $[a,b]$. We define the functions f^+ and f^-, called respectively the positive and negative parts of f, as

$$f^+ = \max(f,0).$$

$$f^- = \max(-f,0).$$

Fix $x\in[a,b]$. If $f(x)>0$, then $f^+(x)=f(x)$ and $f^-(x)=0$. If $f(x)<0$, then $f^+(x)=0$ and $f^-(x)=-f(x)$. If $f(x)=0$, then $f^+(x)=f^-(x)=0$. From these considerations it is clear that the following corollary is true.

11.7D. COROLLARY. If f is any real-valued function on $[a,b]$, then

$$f=f^+-f^- \quad \text{and} \quad |f|=f^++f^-.$$

From 11.7C it is clear that f^+ and f^- are nonnegative-valued (even though f^- is called the negative part of f). The graph of f^+ consists of those portions of the graph of f that lie above the x-axis, together with portions of the x-axis. The graph of f^- is obtained in a similar fashion from the graph of $-f$.

For an example, if

$$f(x)=x^2-1 \qquad (-2 \leqslant x \leqslant 2),$$

then

$$\begin{aligned} f^+(x)&=f(x) && (-2 \leqslant x \leqslant -1),\\ f^+(x)&=0 && (-1<x<1),\\ f^+(x)&=f(x) && (1 \leqslant x \leqslant 2). \end{aligned}$$

Similarly,

$$f^-(x)=-f(x) \quad (-1<x<1),$$

and so on.

11.7E. Now if the real-valued function f on $[a,b]$ is measurable, it follows from 11.4G that both f^+ and f^- are also measurable. Also, both f^+ and f^- are nonnegative valued. Hence, whether or not f^+ and f^- are integrable can be determined from previous definitions. This leads us to the following definition of the Lebesgue integral for arbitrary measurable functions.

DEFINITION. Let f be a measurable function on $[a,b]$. If *both* f^+ and f^- are Lebesgue integrable on $[a,b]$, then we say that f is Lebesgue intregrable on $[a,b]$. In this case we write $f \in \mathcal{L}[a,b]$ and define $\int_a^b f$ as

$$\int_a^b f=\int_a^b f^+-\int_a^b f^-. \tag{$*$}$$

We leave it to the reader to show that if f is bounded then $(*)$ is consistent with previous results.

Thus the class $\mathcal{L}[a,b]$ contains all bounded measurable functions and, in addition, all unbounded measurable functions f such that f^+ and f^- are both integrable according to definition 11.7B or 11.5F.*

Note that the statement $f \in \mathcal{L}[a,b]$ implies that f is measurable !!

Most of the elementary properties of the Lebesgue integral for arbitrary measurable functions may be easily established by use of corresponding results for bounded measurable functions. (Certain results for the case of unbounded functions, however, take quite a bit of work to establish.) We demonstrate most of these properties for integrals on a measurable subset E of $[a,b]$.

If f is a nonnegative-valued measurable function on $[a,b]$, and if E is a measurable subset of $[a,b]$, then, for any $n \in I$, it is easy to verify that

$$^n f \cdot \chi_E = {}^n(f\chi_E).$$

Integrating both sides we have

$$\int_E {}^n f=\int_a^b {}^n(f\chi_E). \tag{1}$$

*If f is unbounded, one (but not both) of the functions f^+ and f^- *may* be bounded.

If $f \in \mathcal{L}[a,b]$, then both sides of (1) are not greater than $\int_a^b f$. Hence, as $n \to \infty$, the limit of each side of (1) exists. We then have

$$\lim_{n\to\infty} \int_E {}^n f = \lim_{n\to\infty} \int_a^b {}^n(f\chi_E).$$

But by definition 11.7B, the quantity on the right is equal to $\int_a^b f\chi_E$. Hence,

$$\lim_{n\to\infty} \int_E {}^n f = \int_a^b f\chi_E. \tag{2}$$

Either side of (2), therefore, may be used to define $\int_E f$ where f is a nonnegative, measurable, but not necessarily bounded function in $\mathcal{L}[a,b]$.

11.7F. DEFINITION. Let E be any measurable subset of $[a,b]$. If f is a nonnegative-valued function in $\mathcal{L}[a,b]$, we define $\int_E f$ as

$$\int_E f = \int_a^b f\chi_E = \lim_{n\to\infty} \int_E {}^n f.$$

Furthermore, if f is an arbitrary measurable function in $\mathcal{L}[a,b]$, we define $\int_E f$ as

$$\int_E f = \int_E f^+ - \int_E f^-.$$

We now come to the properties of the integral.

11.7G. THEOREM. Let E be any measurable subset of $[a,b]$. If $f \in \mathcal{L}[a,b]$, $g \in \mathcal{L}[a,b]$, and if

$$f(x) = g(x) \quad \text{almost everywhere} \qquad (x \in E), \tag{1}$$

then

$$\int_E g = \int_E f.$$

PROOF: Suppose first that f and g are nonnegative-valued. From (1) it follows that, for any $n \in I$,

$${}^n f(x) = {}^n g(x) \quad \text{almost everywhere} \qquad (x \in E).$$

By (4) of 11.6K we then have

$$\int_E {}^n f = \int_E {}^n g.$$

Letting $n \to \infty$ and using 11.7F we then have

$$\int_E f = \int_E g.$$

Thus the theorem is true for nonnegative-valued f and g.

Now suppose f and g are arbitrary functions in $\mathcal{L}[a,b]$ such that (1) holds. Then

$$f^+(x) = g^+(x) \quad \text{almost everywhere} \qquad (x \in E).$$

By the first part of the proof we then have

$$\int_E f^+ = \int_E g^+.$$

Similarly,

$$\int_E f^- = \int_E g^-.$$

Hence,

$$\int_E f = \int_E f^+ - \int_E f^- = \int_E g^+ - \int_E g^- = \int_E g,$$

and the proof is complete.

The proof of 11.7G illustrates the pattern of proof common to many theorems about properties of the Lebesgue integral for arbitrary measurable functions. The pattern is as follows: A property P is known for the integral of bounded measurable functions. By a limit process P is then shown to hold for the integral of nonnegative-valued, measurable (but possibly unbounded) functions. Finally, via the equation

$$\int_E f = \int_E f^+ - \int_E f^-,$$

P is shown to be true for the integral of arbitrary measurable functions in $\mathcal{L}[a,b]$.

This pattern may be used to prove the assertions in the next two theorems.

11.7H. THEOREM. Let E_1 and E_2 be disjoint measurable subsets of $[a,b]$. If $f \in \mathcal{L}[a,b]$, then

$$\int_{E_1 \cup E_2} f = \int_{E_1} f + \int_{E_2} f.$$

11.7I. THEOREM. Let E be a measurable subset of $[a,b]$, and let λ be any real number. If $f \in \mathcal{L}[a,b]$, the $\lambda f \in \mathcal{L}[a,b]$ and

$$\int_E \lambda f = \lambda \int_E f.$$

11.7J. The extension of (3) of 11.6K to arbitrary functions in $\mathcal{L}[a,b]$ is *not* easy. We first need a lemma.

LEMMA. Let f and g be nonnegative-valued functions on $[a,b]$. If f, $g \in \mathcal{L}[a,b]$, then $f+g \in \mathcal{L}[a,b]$ and

$$\int_a^b (f+g) = \int_a^b f + \int_a^b g. \tag{1}$$

Also, $f-g \in \mathcal{L}[a,b]$ and

$$\int_a^b (f-g) = \int_a^b f - \int_a^b g. \tag{2}$$

PROOF: Let $h = f+g$, so that h is nonnegative-valued and measurable. It is then easy to verify that, for any $n \in I$,

$${}^n h(x) \leqslant {}^n f(x) + {}^n g(x) \leqslant {}^{2n} h(x) \qquad (a \leqslant x \leqslant b).$$

Using (6) and (3) of 11.6K we then have

$$\int_a^b {}^n h \leqslant \int_a^b {}^n f + \int_a^b {}^n g \leqslant \int_a^b {}^{2n} h. \tag{3}$$

Since

$$\lim_{n\to\infty} \int_a^b {}^n f = \int_a^b f, \qquad \lim_{n\to\infty} \int_z^b {}^n g = \int_a^b g,$$

it follows that

$$\lim_{n\to\infty} \int_a^b {}^n h$$

exists. Hence, $h = f + g \in \mathfrak{L}[a,b]$. Letting $n\to\infty$ in (3) we then have

$$\int_a^b h \leqslant \int_a^b f + \int_a^b g \leqslant \int_a^b h.$$

Hence,

$$\int_a^b h = \int_a^b f + \int_a^b g,$$

which is precisely (1).

To prove the second part of the lemma, let $k = f - g$ and let

$$E_1 = \{x \in [a,b] \,|\, k(x) \geqslant 0\}.$$

Then, for $x \in E_1$, the values $f(x)$, $g(x)$, and $k(x)$ are all nonnegative. Hence, $f\chi_{E_1}$, $g\chi_{E_1}$, $k\chi_{E_1}$ are nonnegative-valued on $[a,b]$. Also

$$f\chi_{E_1} = g\chi_{E_1} + k\chi_{E_1}. \tag{4}$$

This shows that $k\chi_{E_1}(x) \leqslant f\chi_{E_1}(x)$ for all $x \in [a,b]$, which implies that $k\chi_{E_1} \in \mathfrak{L}[a,b]$. By the first part of the theorem we may integrate from a to b in (4) to obtain

$$\int_a^b f\chi_{E_1} = \int_a^b g\chi_{E_1} + \int_a^b k\chi_{E_1}.$$

But (by definition of E_1) $k\chi_{E_1} = k^+$. Hence, $k^+ \in \mathfrak{L}[a,b]$ and we have

$$\int_a^b f\chi_{E_1} = \int_a^b g\chi_{E_1} + \int_a^b k^+. \tag{5}$$

Now let $E_2 = \{x \in [a,b] \,|\, k(x) < 0\}$. Then, for $x \in E_2$, the values $f(x), g(x)$, and $-k(x)$ are all nonnegative. Also

$$g\chi_{E_2} = f\chi_{E_2} + (-k\chi_{E_2}).$$

Since $-k\chi_{E_2} = k^-$ it follows that $k^- \in \mathfrak{L}[a,b]$ and

$$\int_a^b g\chi_{E_2} = \int_a^b f\chi_{E_2} + \int_a^b k^-. \tag{6}$$

But, by (1),

$$\int_a^b f\chi_{E_1} + \int_a^b f\chi_{E_2} = \int_a^b f(\chi_{E_1} + \chi_{E_2}) = \int_a^b f.$$

Similarly,

$$\int_a^b g\chi_{E_1} + \int_z^b g\chi_{E_2} = \int_a^b g.$$

Also, since $k^+, k^- \in \mathfrak{L}[a,b]$, it follows that $k \in \mathfrak{L}[a,b]$ and

$$\int_a^b k = \int_a^b k^+ - \int_a^b k^-.$$

Subtracting (6) from (5) we thus obtain

$$\int_a^b (f-g) = \int_a^b k = \int_a^b f - \int_a^b g.$$

This completes the proof.

We now prove that

$$\int_E (f+g) = \int_E f + \int_E g$$

for arbitrary f,g. More precisely,

11.7K. THEOREM. Let E be a measurable subset of $[a,b]$. If $f,g \in \mathfrak{L}[a,b]$, then $f+g \in \mathfrak{L}[a,b]$ and

$$\int_E (f+g) = \int_E f + \int_E g. \tag{1}$$

PROOF: By definition 11.7E, the functions f^+, f^-, g^+, g^- are all in $\mathfrak{L}[a,b]$. If $h = f+g$, then $h = (f^+ - f^-) + (g^+ - g^-)$, and hence, $h = (f^+ + g^+) - (f^- + g^-)$. Now, by the first part of the lemma 11.7J, both $(f^+ + g^+)$ and $(f^- + g^-)$ are in $\mathfrak{L}[a,b]$. Hence, by the second part of the lemma, $h \in \mathfrak{L}[a,b]$. We then have

$$\begin{aligned}\int_a^b h &= \int_a^b [(f^+ + g^+) - (f^- + g^-)] = \int_a^b (f^+ + g^+) - \int_a^b (f^- + g^-)\\ &= \int_a^b f^+ + \int_a^b g^+ - \int_a^b f^- - \int_a^b g^-\\ &= \left(\int_a^b f^+ - \int_a^b f^-\right) + \left(\int_a^b g^+ - \int_a^b g^-\right).\end{aligned}$$

That is,

$$\int_a^b (f+g) = \int_a^b f + \int_a^b g. \tag{2}$$

If we now replace f,g in (2) by $f\chi_E, g\chi_E$, we obtain (1).

From 11.7K and 11.7I it then follows that

$$\int_E (f-g) = \int_E f - \int_E g$$

(under the hypotheses of 11.7K).

We next prove two more extensions of theorems previously established for bounded measurable functions.

11.7L. THEOREM. Let E be a measurable subset of $[a,b]$. If $f,g \in \mathfrak{L}[a,b]$ and if

$$f(x) \leqslant g(x) \quad \text{almost everywhere} \qquad (x \in E),$$

then

$$\int_E f \leqslant \int_E g.$$

PROOF: By hypothesis we have, for any $n \in I$,

$$^{n}f(x) \leqslant {}^{n}g(x) \quad \text{almost everywhere} \qquad (x \in E).$$

By (6) of 11.6K we then have

$$\int_E {}^{n}f \leqslant \int_E {}^{n}g.$$

The theorem follows on letting $n \to \infty$.

The proof of the next theorem is equally easy and is omitted.

11.7M. THEOREM. If $f \in \mathcal{L}[a,b]$, if

$$f(x) \geqslant 0 \quad \text{almost everywhere} \qquad (a \leqslant x \leqslant b),$$

and if

$$\int_a^b f = 0,$$

then

$$f(x) = 0 \quad \text{almost everywhere} \qquad (a \leqslant x \leqslant b).$$

The final result of this section is particularly important.

11.7N. THEOREM. Let f be a measurable function on $[a,b]$. Then $f \in \mathcal{L}[a,b]$ if and only if $|f| \in \mathcal{L}[a,b]$. Moreover, if $f \in \mathcal{L}[a,b]$, then

$$\left| \int_a^b f \right| \leqslant \int_a^b |f|,$$

PROOF: If $f \in \mathcal{L}[a,b]$, then, by 11.7E, both f^+ and f^- are in $\mathcal{L}[a,b]$. But, by 11.7D, $|f| = f^+ + f^-$. Hence, $|f| \in \mathcal{L}[a,b]$, by 11.7K. The inequality

$$\left| \int_a^b f \right| \leqslant \int_a^b |f|.$$

then follows as in the proof of 7.4F.

Conversely, suppose f is measurable and suppose $|f| \in \mathcal{L}[a,b]$. Since

$$0 \leqslant f^+(x) \leqslant |f(x)| \qquad (a \leqslant x \leqslant b),$$

it follows that $f^+ \in \mathcal{L}[a,b]$. Similarly, $f^- \in \mathcal{L}[a,b]$ and hence, $f \in \mathcal{L}[a,b]$. This completes the proof.

Exercises 11.7

1. If

$$f(x) = \log \frac{1}{x} \qquad (0 < x \leqslant 1),$$

find ^{2}f.

2. If

$$f(x) = \frac{1}{x^p} \qquad (0 < x \leqslant 1),$$

prove that $f \in \mathcal{L}[0,1]$ if $p < 1$ and that

$$\int_0^1 f = \frac{1}{1+p}.$$

3. If f is a nonnegative-valued measurable function on $[a,b]$ and if
$$f(x) \leqslant g(x) \qquad (a \leqslant x \leqslant b)$$
where $g \in \mathcal{L}[a,b]$, prove that $f \in \mathcal{L}[a,b]$.
4. If
$$f(x) = \tfrac{1}{2} + \sin x \qquad (0 \leqslant x < 2\pi),$$
find f^+ and f^-.
5. If $g \in \mathcal{L}[a,b]$ and if f is a measurable function on $[a,b]$ such that
$$|f(x)| \leqslant |g(x)| \qquad (a \leqslant x \leqslant b),$$
prove that $f \in \mathcal{L}[a,b]$.
6. True or false? If f and g are in $\mathcal{L}[a,b]$, the $fg \in \mathcal{L}[a,b]$.
7. Let f be a nonnegative-valued function in $\mathcal{L}[a,b]$. For each $n=0,1,2,\ldots$, let $E_n = \{x \mid n \leqslant f(x) < n+1\}$. Prove that $\sum_{n=0}^{\infty} n \cdot mE_n < \infty$.
8. If $f(x)=0$ for every x in the Cantor set K, and $f(x)=k$ for x in each of the intervals of length $1/3^k$ in K', prove that f is Lebesgue integrable on $[0,1]$ and that
$$\int_0^1 f = 3.$$
9. Let $f \in \mathcal{L}[a,b]$. Given $\epsilon > 0$, show that there exists a bounded measurable function g on $[a,b]$ such that
$$\int_a^b |f-g| < \epsilon.$$
(*Hint*: Do this first for nonnegative-valued f.)
10. If $f \in \mathcal{L}[a,b]$ and if, for some $c \in R$,
$$g(t) = f(t-c) \qquad (a+c \leqslant t \leqslant b+c),$$
show that $g \in \mathcal{L}[a+c, b+c]$ and
$$\int_{a+c}^{b+c} g(t)\,dt = \int_{a+c}^{b+c} f(t-c)\,dt = \int_a^b f(t)\,dt.$$
11. Show that
$$\int_a^b f(t)\,dt = \int_{-b}^{-a} f(-t)\,dt$$
if $f \in \mathcal{L}[a,b]$.

11.8 SOME FUNDAMENTAL THEOREMS

Lebesgue integration gives us a very general set of conditions under which a sequence may be integrated term by term. We first prove a lemma.

11.8A. LEMMA. Let $f \in \mathcal{L}[a,b]$. Then given $\epsilon > 0$ there exists $\delta > 0$ such that
$$\left| \int_E f \right| < \epsilon$$
whenever E is a measurable subset of $[a,b]$ with $mE < \delta$.

PROOF: Consider first the case in which f is nonnegative-valued. Then, by 11.7B,
$$\lim_{n\to\infty} \int_a^b {}_n f = \int_a^b f.$$

Thus given $\epsilon>0$, there exists $N\in I$ such that

$$\int_a^b f-\int_a^b {}^N f<\frac{\epsilon}{2}.$$

That is,

$$\int_a^b (f-{}^N f)<\frac{\epsilon}{2}. \tag{1}$$

Now choose any $\delta>0$ with $\delta<\epsilon/2N$. If E is a measurable subset of $[a,b]$ and $mE<\delta$, then we have

$$\int_E {}^N f\leqslant \int_E N=N(mE)<N\delta<\frac{\epsilon}{2}. \tag{2}$$

Hence, using (1) and (2) we have

$$\int_E f=\int_E (f-{}^N f)+\int_E {}^N f\leqslant \int_a^b (f-{}^N f)+\int_E {}^N f<\frac{\epsilon}{2}+\frac{\epsilon}{2}=\epsilon,$$

which proves the lemma for nonnegative-valued f.

For an arbitrary measurable function f in $\mathfrak{L}[a,b]$ we have $f=f^+-f^-$. By the first part of the proof, given $\epsilon>0$, there exists $\delta_1>0$ such that

$$\int_E f^+<\frac{\epsilon}{2}$$

when $mE<\delta_1$. Similarly, there exists $\delta_2>0$ such that

$$\int_E f^-<\frac{\epsilon}{2}$$

when $mE<\delta_2$. Thus if $mE<\delta=\min(\delta_1,\delta_2)$, we have

$$\left|\int_E f\right|\leqslant \int_E |f|=\int_E f^+ +\int_E f^- <\frac{\epsilon}{2}+\frac{\epsilon}{2}=\epsilon.$$

This completes the proof.

The following theorem, called the Lebesgue dominated convergence theorem, shows that a sequence of integrable functions may be integrated term by term under much less restrictive conditions than the uniform convergence required in 9.3G.

11.8B. LEBESGUE DOMINATED CONVERGENCE THEOREM. Let $\{f_n\}_{n=1}^{\infty}$ be a sequence of functions in $\mathfrak{L}[a,b]$ such that

$$\lim_{n\to\infty} f_n(x)=f(x) \quad \text{almost everywhere} \qquad (a\leqslant x\leqslant b). \tag{1}$$

Suppose there exists $g\in\mathfrak{L}[a,b]$ such that

$$|f_n(x)|\leqslant g(x) \quad \text{almost everywhere} \qquad (a\leqslant x\leqslant b; n\in I). \tag{2}$$

Then $f\in\mathfrak{L}[a,b]$ and

$$\lim_{n\to\infty}\int_a^b f_n=\int_a^b f. \tag{3}$$

PROOF: Since each f_n belongs to $\mathfrak{L}[a,b]$ it follows that each f_n must be measurable (by 11.6N or 11.7E). Hence, by (1) and 11.4I, f is measurable. By (1) and (2) we have $|f(x)|\leqslant g(x)$ for almost all x. It follows that $|f|\in\mathfrak{L}[a,b]$. Hence, by 11.7N, $f\in\mathfrak{L}[a,b]$.

Fix $\epsilon>0$. For each $N\in I$ let E_N be the set of all $x\in[a,b]$ such that

$$|f_n(x)-f(x)|<\frac{\epsilon}{2(b-a)} \qquad (n\geqslant N). \tag{4}$$

Then $E_1 \subset E_2 \subset E_3 \subset \cdots$ and, by (1), almost every point of $[a,b]$ lies in one (and hence, infinitely many) of the E_N. That is, $\cup_{N=1}^{\infty} E_N$ has measure $b-a$. By 11.3F we have

$$\lim_{N\to\infty} mE_N = b-a. \tag{5}$$

By lemma 11.8A there exists $\delta > 0$ such that

$$\int_E g < \frac{\epsilon}{4}$$

if $mE < \delta$. It then follows from (2) and (1) that

$$\int_E |f_n| < \frac{\epsilon}{4} \quad \text{and} \quad \int_E |f| < \frac{\epsilon}{4} \qquad (mE < \delta). \tag{6}$$

By (5) there exists $M \in I$ such that $b-a-mE_M < \delta$. That is, $mE_M' < \delta$, where $E_M' = [a,b] - E_M$. Now if $n \geqslant M$, we have, by (4) and (6),

$$\int_a^b |f_n - f| = \int_{E_M} |f_n - f| + \int_{E_M'} |f_n - f|$$
$$\leqslant \int_{E_M} \frac{\epsilon}{2(b-a)} + \int_{E_M'} |f_n| + \int_{E_M'} |f| < \frac{\epsilon}{2} + \frac{\epsilon}{4} + \frac{\epsilon}{4}.$$

That is,

$$\int_a^b |f_n - f| < \epsilon \qquad (n \geqslant M). \tag{7}$$

This implies, by 11.7N,

$$\left| \int_a^b f_n - \int_a^b f \right| < \epsilon \qquad (n \geqslant N),$$

which proves (3).

In view of (7) we have actually proved the following result, which is a little stronger than 11.8B.

11.8C. THEOREM. Under the hypotheses of 11.8B we have

$$\lim_{n\to\infty} \int_a^b |f_n(x) - f(x)|\, dx = 0.$$

Theorem 11.8B has many advantages over theorem 9.3G. For example, in 11.8B the functions f_n need not be bounded. Moreover, uniform convergence is not required.

Indeed, it is easy to give an example of a sequence $\{f_n\}_{n=1}^{\infty}$ that satisfies the hypotheses of 11.8B but does not converge uniformly. For $n \in I$ let

$$f_n(x) = \sqrt{n} \qquad \left(\frac{1}{n} \leqslant x \leqslant \frac{2}{n}\right),$$
$$f_n(x) = 0 \qquad \left(x \in \left[0, \frac{1}{n}\right) \cup \left(\frac{2}{n}, 2\right]\right).$$

Then $\lim_{n\to\infty} f_n(x) = 0$ for all $x \in [0,2]$, so that (1) of 11.8B is satisfied with $f = 0$. Moreover, for every $n \in I$, we have $|f_n(x)| \leqslant g(x) (0 \leqslant x \leqslant 2)$ where

$$g(x) = \frac{\sqrt{2}}{\sqrt{x}} \qquad (0 < x \leqslant 2),$$
$$g(0) = 0.$$

Since $g \in \mathcal{L}[0,2]$, the hypothesis (2) of 11.8B is also satisfied. It is clear, however, that $\{f_n\}_{n=1}^{\infty}$ does not converge uniformly on $[0,2]$.

The next result on integration of sequences, due to Fatou, is also well known. For historical reasons it is called a lemma.

11.8D. THEOREM (FATOU'S LEMMA). Let $\{f_n\}_{n=1}^{\infty}$ be a sequence of nonnegative-valued measurable functions in $\mathcal{L}[a,b]$. Suppose

$$\lim_{n\to\infty} f_n(x) = f(x) \quad \text{almost everywhere} \qquad (a \leqslant x \leqslant b). \tag{1}$$

Then

$$\liminf_{n\to\infty} \int_a^b f_n \geqslant \int_a^b f$$

if $f \in \mathcal{L}[a,b]$, while

$$\liminf_{n\to\infty} \int_a^b f_n = \infty$$

if $f \notin \mathcal{L}[a,b]$.

PROOF: For any $m \in I$ we have from (1)

$$\lim_{n\to\infty} {}^m f_n(x) = {}^m f(x) \quad \text{almost everywhere} \qquad (a \leqslant x \leqslant b).$$

For fixed m the sequence $\{{}^m f_n\}_{n=1}^{\infty}$ thus obeys the hypotheses of 11.8B (take $g = m$). Hence,

$$\lim_{n\to\infty} \int_a^b {}^m f_n = \int_a^b {}^m f. \tag{2}$$

But, for every $x \in [a,b]$, we have ${}^m f_n(x) \leqslant f_n(x)$. Hence,

$$\lim_{n\to\infty} \int_a^b {}^m f_n = \liminf_{n\to\infty} \int_a^b {}^m f_n \leqslant \liminf_{n\to\infty} \int_a^b f_n. \tag{3}$$

Hence, from (2) and (3) we have

$$\liminf_{n\to\infty} \int_a^b f_n \geqslant \int_a^b {}^m f.$$

The conclusion of the theorem follows on letting $m \to \infty$.

In 7.8B we proved that if $f \in \mathcal{R}[a,b]$ and

$$F(x) = \int_a^x f(t)\,dt \qquad (a \leqslant x \leqslant b),$$

then $F'(x_0) = f(x_0)$ if f is continuous at x_0. Since Riemann integrable functions are continuous almost everywhere it follows that

$$F'(x) = f(x) \quad \text{almost everywhere} \qquad (a \leqslant x \leqslant b). \tag{$*$}$$

Now if $f \in \mathcal{L}[a,b]$ and

$$F(x) = \int_a^x f(t)\,dt,$$

then it may still be proved that $(*)$ holds even though f may not be continuous at any point. To establish this result, however, requires a tremendous amount of work, and we therefore omit the proof. It is given in detail in the references at the end of the chapter.

11.8E. THEOREM. If $f \in \mathcal{L}[a,b]$ and if

$$F(x)=\int_a^x f(t)\,dt \qquad (a \leqslant x \leqslant b),$$

then

$$F'(x)=f(x) \quad \text{almost everywhere} \qquad (a \leqslant x \leqslant b).$$

Exercises 11.8

1. If $f \in \mathcal{L}[a,b]$ and if

$$F(x)=\int_a^x f(t)\,dt \qquad (a \leqslant x \leqslant b),$$

prove that F is continuous on $[a,b]$.

2. If $\{f_n\}_{n=1}^{\infty}$ is a sequence of functions in $\mathcal{L}[a,b]$, if

$$|f_n(x)| \leqslant g(x) \quad \text{almost everywhere} \qquad (a \leqslant x \leqslant b; n \in I)$$

where $g \in \mathcal{L}[a,b]$, if

$$\lim_{n\to\infty} f_n(x)=f(x) \quad \text{almost everywhere} \qquad (a \leqslant x \leqslant b),$$

and if h is any bounded measurable function on $[a,b]$, prove that

$$\lim_{n\to\infty} \int_a^b f_n h = \int_a^b fh.$$

3. Suppose $\{f_n\}_{n=1}^{\infty}$ is a sequence of functions in $\mathcal{L}[a,b]$ such that

$$0 \leqslant f_1(x) \leqslant f_2(x) \leqslant \cdots \leqslant f_n(x) \leqslant \cdots \qquad (a \leqslant x \leqslant b).$$

Suppose also that

$$\lim_{n\to\infty} f_n(x)=f(x) \qquad (a \leqslant x \leqslant b).$$

(a) If $f \in \mathcal{L}[a,b]$, show that

$$\lim_{n\to\infty} \int_a^b f_n = \int_a^b f.$$

(b) If $f \not\in \mathcal{L}[a,b]$, show that

$$\left\{ \int_a^b f_n \right\}_{n=1}^{\infty}$$

diverges to ∞.

This result is known as the Monotone convergence theorem.

4. If $f \in \mathcal{L}[a,b]$ and if, for each $n=0,1,2,\ldots,$

$$E_n=\{x \in [a,b] \,|\, |f(x)| \geqslant n\},$$

prove that

$$\sum_{n=0}^{\infty} mE_n < \infty. \qquad (*)$$

[*Hint*: Show that $mE_n = mH_n + mH_{n+1} + \cdots$ where $H_k=\{x \,|\, k \leqslant |f(x)| < k+1\}$. Then $\Delta mE_n = mE_{n+1} - mE_n = -mH_n$. Apply formula (2) of 3.8A with $s_k=k, b_k=mE_k$.]

5. For $n \in I$ let

$$f_n(x) = 2n \left(\frac{1}{2n} \leqslant x \leqslant \frac{1}{n}\right),$$

$$f_n(x) = 0 \left[x \in \left(0, \frac{1}{2n}\right) \cup \left(\frac{1}{n}, 1\right)\right].$$

Calculate

$$\int_0^1 \left[\lim_{n\to\infty} f_n(x)\right] dx \quad \text{and} \quad \lim_{n\to\infty} \int_0^1 f_n(x)\, dx.$$

Show that Fatou's lemma applies but that the Lebesgue dominated convergence theorem does not.

11.9 THE METRIC SPACE $\mathcal{L}^2[a,b]$

In this section we introduce a class of functions on $[a,b]$ that has many properties in common with the class l^2 of sequences.

11.9A. DEFINITION. Let f be a measurable function on the closed bounded interval $[a,b]$. We say that f is square integrable on $[a,b]$ if $f^2 \in \mathcal{L}[a,b]$.

Next we prove the Schwarz and Minkowski inequalities for square integrable functions. (Compare with 2.10B and 2.10C.)

11.9B. THEOREM (THE SCHWARZ INEQUALITY). Let f and g be measurable functions on $[a,b]$. If f and g are square integrable, the $fg \in \mathcal{L}[a,b]$ and

$$\left|\int_a^b fg\right| \leqslant \left(\int_a^b f^2\right)^{1/2} \left(\int_a^b g^2\right)^{1/2}. \tag{1}$$

PROOF: For any $x \in [a,b]$ we have $[|f(x)| - |g(x)|]^2 \geqslant 0$. Hence,

$$|f(x)\,g(x)| \leqslant \tfrac{1}{2}([f(x)]^2 + [g(x)]^2) \qquad (a \leqslant x \leqslant b).$$

Since, by hypothesis, f^2 and g^2 are in $\mathcal{L}[a,b]$, it follows that $|fg| \in \mathcal{L}[a,b]$, Hence, by 11.7N, $fg \in \mathcal{L}[a,b]$.

For any $\lambda \in R$ we then have

$$\int_a^b (\lambda f + g)^2 \geqslant 0$$

or, equivalently,

$$\lambda^2 \int_a^b f^2 + 2\lambda \int_a^b fg + \int_a^b g^2 \geqslant 0.$$

This can be written $A\lambda^2 + B\lambda + C \geqslant 0$ where

$$A = \int_a^b f^2, \qquad B = 2\int_a^b fg, \qquad C = \int_a^b g^2.$$

If $A = 0$, then, by 11.7M, we would have $[f(x)]^2 = 0$ for almost all x in $[a,b]$. That is, $f(x) = 0$ for almost all x in $[a,b]$. In this case both sides of (1) are equal to zero.

Otherwise $A \neq 0$. We may then set $\lambda = -B/2A$ to obtain $B^2 \leqslant 4AC$. But this says

$$\left(\int_a^b fg\right)^2 \leqslant \left(\int_a^b f^2\right)\left(\int_a^b g^2\right),$$

from which (1) follows on taking square roots.

11.9C. THEOREM (THE MINKOWSKI INEQUALITY). Let f and g be measurable functions on $[a,b]$. If f and g are square integrable, the $f+g$ is also square integrable and

$$\left[\int_a^b (f+g)^2\right]^{1/2} \leqslant \left(\int_a^b f^2\right)^{1/2} + \left(\int_a^b g^2\right)^{1/2}. \tag{1}$$

PROOF: We have

$$(f+g)^2 = f^2 + 2fg + g^2. \tag{2}$$

By hypothesis, both f^2 and g^2 are in $\mathcal{L}[a,b]$. Also, $fg \in \mathcal{L}[a,b]$ by 11.9B. Hence, by (2) and 11.7K, the function $(f+g)^2$ is in $\mathcal{L}[a,b]$ [that is, $(f+g)$ is square integrable] and

$$\int_a^b (f+g)^2 = \int_a^b f^2 + 2\int_a^b fg + \int_a^b g^2.$$

Using 11.9B we then obtain

$$\int_a^b (f+g)^2 \leqslant \int_a^b f^2 + 2\left(\int_a^b f^2\right)^{1/2} \cdot \left(\int_a^b g^2\right)^{1/2} + \int_a^b g^2 = \left[\left(\int_a^b f^2\right)^{1/2} + \left(\int_a^b g^2\right)^{1/2}\right]^2,$$

and (1) follows.

We next define the norm of a square integrable function.

11.9D DEFINITION. If f is a square integrable function on $[a,b]$, we define $\|f\|_2$ as

$$\|f\|_2 = \left(\int_a^b f^2\right)^{1/2}.$$

We are using the same notation for the norm of a square integrable function as we did for a sequence in l^2. However, no confusion should arise.

11.9E. THEOREM. The norm for square integrable functions has the following properties:

$$\|f\|_2 \geqslant 0 \quad \text{for any square integrable function } f \text{ on } [a,b]. \tag{1}$$

$$\|f\|_2 = 0 \quad \text{if and only if} \quad f(x) = 0 \quad \text{almost everywhere} \qquad (a \leqslant x \leqslant b). \tag{2}$$

$$\|cf\|_2 = |c| \cdot \|f\|_2 \quad \text{if} \quad f \text{ is square integrable and } c \in R. \tag{3}$$

$$\|f+g\|_2 \leqslant \|f\|_2 + \|g\|_2 \quad \text{if} \quad f, g \text{ are square integrable.} \tag{4}$$

PROOF: Properties (1) and (3) are immediate consequences of the definition of norm. Property (2) is a consequence of 11.7M. Finally, property (4) is a restatement of the Minkowski inequality 11.9C.

11.9F. We now wish to make the collection of all square integrable functions into a metric space. If we were to proceed parallel to our treatment of l^2, we would define

$\rho(f,g)=\|f-g\|_2$ for f and g square integrable. However, unless we make one more assumption, with this definition ρ would not be a metric! For according to (2) of 11.9E, we would have $\rho(f,g)=0$ for two *distinct* square integrable functions f and g, provided that $f(x)=g(x)$ for almost all (but not all) values of x in $[a,b]$. In order to manufacture our metric space we must therefore regard any two functions whose values are equal almost everywhere as representing the same point in our space. That is, the points in the space—which we denote by $\mathcal{L}^2[a,b]$—are, by definition, *classes* of square integrable functions, the functions in any one class differing from one another only on sets of measure zero.

It is, however, a time-honored custom to refer to the individual elements of $\mathcal{L}^2[a,b]$ as functions instead of classes of functions. Thus we speak of a square integrable function f as being "in $\mathcal{L}^2[a,b]$" instead of being "a representative of a class of functions in $\mathcal{L}^2[a,b]$," and we write $f\in\mathcal{L}^2[a,b]$.

If the square integrable functions f and g are (that is, represent) distinct elements of $\mathcal{L}^2[a,b]$, then $\|f-g\|_2>0$ by (2) of 11.9E. For if f and g are (that is, represent) distinct elements of $\mathcal{L}^2[a,b]$, then the values of $f(x)$ and $g(x)$ differ for all x in some set of positive measure. Hence, if we now define ρ by

$$\rho(f,g)=\|f-g\|_2 \qquad (f,g\in\mathcal{L}^2[a,b]), \tag{*}$$

then $\rho(f,g)>0$ if $f\neq g$. The other requirements for a metric are easily verified with the aid of 11.9E, and we have

THEOREM. If ρ is defined by (*), then ρ is a metric for $\mathcal{L}^2[a,b]$.
We denote the metric space $\langle\mathcal{L}^2[a,b],\rho\rangle$ simply by $\mathcal{L}^2[a,b]$.

Next we prove that $\mathcal{L}^2[a,b]$ is complete. That is, we show that if $\{f_n\}_{n=1}^{\infty}$ is a Cauchy sequence in $\mathcal{L}^2[a,b]$, then there exists $f\in\mathcal{L}^2[a,b]$ such that $\{f_n\}_{n=1}^{\infty}$ converges to f with respect to the metric of $\mathcal{L}^2[a,b]$. This result is sometimes called the Riesz-Fischer theorem.

11.9G. THEOREM. The metric space $\mathcal{L}^2[a,b]$ is complete.

PROOF: Let $\{f_n\}_{n=1}^{\infty}$ be a Cauchy sequence in $\mathcal{L}^2[a,b]$. Then, given $\epsilon>0$, there exists $N\in I$ such that

$$\|f_m-f_n\|_2=\left[\int_a^b[f_m(x)-f_n(x)]^2\,dx\right]^{1/2}<\epsilon \qquad (n\geqslant N).$$

For each $\nu\in I$ let n_ν be the smallest positive integer such that

$$\int_a^b[f_m(x)-f_n(x)]^2\,dx<\frac{1}{3^\nu} \qquad (m,n\geqslant n_\nu). \tag{1}$$

Then $n_1\leqslant n_2\leqslant\cdots\leqslant n_\nu\leqslant\cdots$. In particular,

$$\int_a^b[f_{n_{\nu+1}}(x)-f_{n_\nu}(x)]^2\,dx<\frac{1}{3^\nu} \qquad (\nu\in I).$$

Let E_ν be the set of all $x\in[a,b]$ such that

$$|f_{n_{\nu+1}}(x)-f_{n_\nu}(x)|>2^{-\nu/2}.$$

Then

$$mE_\nu = \int_{E_\nu} 1 \leqslant 2^\nu \int_{E_\nu} |f_{n_{\nu+1}}(x) - f_{n_\nu}(x)|^2 \, dx$$

$$\leqslant 2^\nu \int_a^b \left[f_{n_{\nu+1}}(x) - f_{n_\nu}(x) \right]^2 dx < \left(\tfrac{2}{3}\right)^\nu.$$

For each $N \in I$ let F_N be the complement of $E_N \cup E_{N+1} \cup \cdots$. Then $F_1 \subset F_2 \subset \cdots \subset F_N \subset \cdots$. Also, since $m(E_N \cup E_{N+1} \cup \cdots) \leqslant (\tfrac{2}{3})^N + (\tfrac{2}{3})^{N+1} + \cdots$, it follows that

$$\lim_{N \to \infty} m(E_N \cup E_{N+1} \cup \cdots) = 0.$$

Consequently,

$$m\left(\bigcup_{N=1}^{\infty} F_N \right) = \lim_{N \to \infty} mF_N = b - a.$$

That is, almost every x in $[a,b]$ is in $\cup_{N=1}^{\infty} F_N$. But if $x \in \cup_{N=1}^{\infty} F_N$, then, for some N, x is not in any of the sets $E_N, E_{N+1}, \ldots$. Hence, if $x \in \cup_{N=1}^{\infty} F_N$, then

$$\sum_{\nu=1}^{\infty} \left[f_{n_{\nu+1}}(x) - f_{n_\nu}(x) \right] \ll \sum_{\nu=1}^{\infty} 2^{-\nu/2}$$

(see 3.6A). From 3.6B it follows that

$$\sum_{\nu=1}^{\infty} \left[f_{n_{\nu+1}}(x) - f_{n_\nu}(x) \right] \quad \text{converges almost everywhere} \qquad (a \leqslant x \leqslant b).$$

That is,

$$\lim_{\nu \to \infty} \left\{ \left[f_{n_2}(x) - f_{n_1}(x) \right] + \left[f_{n_3}(x) - f_{n_2}(x) \right] + \cdots + \left[f_{n_\nu}(x) - f_{n_{\nu-1}}(x) \right] \right\}$$

exists for almost all x in $[a,b]$. Thus

$$\lim_{\nu \to \infty} \left[f_{n_\nu}(x) - f_{n_1}(x) \right]$$

exists for almost all x, and hence, $\lim_{\nu \to \infty} f_{n_\nu}(x)$ exists for almost all x. For every such x let

$$f(x) = \lim_{\nu \to \infty} f_{n_\nu}(x). \tag{2}$$

[Let $f(x) = 0$ if the limit in (2) does not exist.] Then, for fixed ν,

$$\lim_{\mu \to \infty} \left[f_{n_\nu}(x) - f_{n_\mu}(x) \right]^2 = \left[f_{n_\nu}(x) - f(x) \right]^2 \quad \text{almost everywhere} \qquad (a \leqslant x \leqslant b),$$

and so, by Fatou's lemma 11.8D and (1), we have $(f_{n_\nu} - f) \in \mathfrak{L}^2[a,b]$ and

$$\int_a^b \left[f_{n_\nu}(x) - f(x) \right]^2 dx \leqslant \liminf_{\mu \to \infty} \int_a^b \left[f_{n_\nu}(x) - f_{n_\mu}(x) \right]^2 dx \leqslant \frac{1}{3^\nu}. \tag{3}$$

Hence, $f = (f - f_{n_\nu}) + f_{n_\nu}$ is in $\mathfrak{L}^2[a,b]$, by 11.9C. From (3) it follows that

$$\rho(f_{n_\nu}, f) = \| f_{n_\nu} - f \|_2 < \frac{1}{\sqrt{3^\nu}}$$

while from (1) we have

$$\rho(f_n, f_{n_\nu}) = \| f_n - f_{n_\nu} \|_2 < \frac{1}{\sqrt{3^\nu}} \qquad (n \geqslant n_\nu).$$

Hence,

$$\rho(f_n,f)<\frac{2}{\sqrt{3^\nu}} \qquad (n\geqslant n_\nu),$$

and it follows easily that $\{f_n\}_{n=1}^\infty$ is convergent to f. This completes the proof.

The completeness of $\mathfrak{L}^2[a,b]$ has far-reaching consequences in higher analysis. If we had only Riemann integration at our disposal, we could not prove such a result (see exercise 7).

As the final main result in this section we prove that the set of continuous functions on $[a,b]$ is dense in $\mathfrak{L}^2[a,b]$. This will be an immediate consequence of the following two lemmas.

11.9H. LEMMA. If f is a bounded measurable function on $[a,b]$, then, given $\epsilon>0$, there is a continuous function g on $[a,b]$ such that $\|f-g\|_2<\epsilon$.

PROOF: For $x\in(b,b+1]$ define $f(x)=0$, so that f is defined on $[a,b+1]$. Let

$$F(x)=\int_a^x f(t)\,dt \qquad (a\leqslant x\leqslant b+1).$$

Then F is continuous on $[a,b+1]$. [For

$$|F(x+h)-F(h)|=\left|\int_x^{x+h} f(t)\,dt\right|\leqslant M|h|,$$

where $M=\text{l.u.b.}_{x\in[a,b]}|f(x)|$.] For each $n\in I$ let

$$G_n(x)=n\int_x^{x+1/n} f(t)\,dt \qquad (a\leqslant x\leqslant b). \tag{1}$$

Then

$$G_n(x)=\frac{F(x+1/n)-F(x)}{1/n} \qquad (a\leqslant x\leqslant b).$$

It follows that G_n is continuous on $[a,b]$. By 11.8E we have

$$\lim_{n\to\infty} G_n(x)=F'(x)=f(x) \quad \text{almost everywhere} \qquad (a\leqslant x\leqslant b),$$

and so,

$$\lim_{n\to\infty}\left[G_n(x)-f(x)\right]^2=0 \quad \text{almost everywhere} \qquad (a\leqslant x\leqslant b). \tag{2}$$

From (1) we have

$$\begin{aligned}|G_n(x)| &\leqslant n\int_x^{x+1/n}|f(t)|\,dt\\ &\leqslant n\int_x^{x+1/n} M\,dt=M,\end{aligned}$$

and hence,

$$\left[G_n(x)-f(x)\right]^2\leqslant(M+M)^2=4M^2 \qquad (a\leqslant x\leqslant b). \tag{3}$$

From (2), (3), and 11.8B we have

$$\lim_{n\to\infty}\int_a^b (G_n-f)^2=\lim_{n\to\infty}\|G_n-f\|_2^2=0.$$

Hence, given $\epsilon > 0$, we have $\|G_n - f\|_2 < \epsilon$ for n sufficiently large, and the lemma is proved (with $g = G_n$).

11.9I. LEMMA. If $h \in \mathcal{L}^2[a,b]$, then, given $\epsilon > 0$, there is a bounded measurable function f on $[a,b]$ such that $\|h - f\|_2 < \epsilon$.

PROOF: Consider first the case in which h is a nonnegative-valued function in $\mathcal{L}^2[a,b]$. We have

$$\lim_{m \to \infty} \left[h(x) - {}^m h(x) \right]^2 = 0 \qquad (a \leqslant x \leqslant b).$$

Since $0 \leqslant {}^m h(x) \leqslant h(x)$ $(a \leqslant x \leqslant b)$, we also have

$$\left[h(x) - {}^m h(x) \right]^2 \leqslant \left[h(x) \right]^2 \qquad (a \leqslant x \leqslant b).$$

Since $h^2 \in \mathcal{L}[a,b]$, 11.8B then implies

$$\lim_{m \to \infty} \int_a^b (h - {}^m h)^2 = 0.$$

Hence, $\|h - {}^m h\|_2 < \epsilon$ for sufficiently large m. Since ${}^m h$ is bounded, the lemma is proved for the case of nonnegative-valued h.

Now if h is an arbitrary function in $\mathcal{L}^2[a,b]$, we have $h = h^+ - h^-$. Since $[h^+(x)]^2 \leqslant [h(x)]^2$ $(a \leqslant x \leqslant b)$, it is clear that $h^+ \in \mathcal{L}^2[a,b]$. By the first part of the lemma, given $\epsilon > 0$, there exists a bounded measurable function f_1 such that $\|h^+ - f_1\|_2 < \frac{\epsilon}{2}$. Similarly, there exists a bounded measurable function f_2 such that $\|h^- - f_2\|_2 < \epsilon/2$. If $f = f_1 - f_2$, then f is bounded and

$$\begin{aligned} \|h - f\|_2 &= \|(h^+ - h^-) - (f_1 - f_2)\|_2 = \|(h^+ - f_1) - (h^- - f_2)\|_2 \\ &\leqslant \|h^+ - f_1\|_2 + \|h^- - f_2\|_2 < \epsilon. \end{aligned}$$

This completes the proof.

From 11.9H and 11.9I we immediately deduce the following theorem.

11.6J. THEOREM. If $h \in \mathcal{L}^2[a,b]$, then, given $\epsilon > 0$, there exists a continuous function g on $[a,b]$ such that $\|h - g\|_2 < \epsilon$.

Theorem 11.9J states that every open ball $B[h; \epsilon]$ about h in $\mathcal{L}^2[a,b]$ contains a continuous function. That is, every $h \in \mathcal{L}^2[a,b]$ is a limit point of the set of continuous functions. In other words, the set of continuous functions is dense in $\mathcal{L}^2[a,b]$.

Our proof of 11.9J depends on 11.8E (why?). However, with more work we could have established 11.9J without the use of 11.8E.

The next result may be easily deduced from 11.9J. We leave the proof to the reader.

11.9K. THEOREM. If $h \in \mathcal{L}^2[a,b]$ and $\epsilon > 0$, there exists a continuous function g on $[a,b]$ such that $g(a) = g(b)$ and such that $\|h - g\|_2 < \epsilon$.

Exercises 11.9

1. (a) Show that $C[a,b] \subset \mathcal{L}^2[a,b]$.
   (b) If $\{f_n\}_{n=1}^{\infty}$ is a sequence in $C[a,b]$ that converges uniformly to f on $[a,b]$, show that $\lim_{n \to \infty} \|f_n - f\|_2 = 0$.
2. Let $\{f_n\}_{n=1}^{\infty}$ be a sequence of functions in $\mathcal{L}^2[a,b]$ that converges in the metric of $\mathcal{L}^2[a,b]$ to f.

If $g \in \mathcal{L}^2[a,b]$, show that

$$\lim_{n\to\infty} \int_a^b f_n g = \int_a^b fg.$$

(*Hint*: Apply the Schwarz inequality to $\int_a^b (f_n - f)g$.)

3. If f is square integrable on $[a,b]$, prove that $f \in \mathcal{L}[a,b]$.
4. If $f \in \mathcal{L}^2[a,b]$ and if $f(t)=0$ for $t \notin [a,b]$, prove that

$$\lim_{x\to 0} \int_a^b [f(x+t) - f(t)]^2 dt = 0.$$

(*Hint*: Prove this first for continuous f. Then use 11.9J.)

5. If $f \in \mathcal{L}^2[a,b]$, if $f(x)=0$ for $x \notin [a,b]$, and if

$$F(x) = \int_a^b f(x+t)\cdot f(t)dt \qquad (-\infty < x < \infty),$$

prove that F is continuous at 0. (*Hint*: Apply the Schwarz inequality to

$$F(x) - F(0) = \int_a^b [f(x+t) - f(t)]\cdot f(t)dt.$$

Then use the preceding exercise.)

6. If E is a subset of $[a,b]$, then the difference set $D(E)$ is defined as the set of all numbers of the form $x-y$ for $x,y \in E$. [That is, $D(E)$ is the set of all differences of points of E.] Fill in the details of this proof of a famous THEOREM. If E is a measurable subset of $[a,b]$ and if $mE > 0$, then $D(E)$ contains an open interval about the point 0.

PROOF: Let χ be the characteristic function of E, and let

$$F(x) = \int_a^b \chi(t)\chi(x+t)dt \qquad (-\infty < x < \infty).$$

(a) Then F is continuous at 0 (why?).
(b) Since $F(0) > 0$ (why?) it follows that there exists δ such that

$$F(x) = \int_a^b \chi(t)\chi(x+t)dt > 0 \qquad (-\delta < x < \delta).$$

(c) If $x \in (-\delta, \delta)$, then there exists t_0 (depending on x) such that $\chi(t_0)\chi(x+t_0) > 0$ (why?).
(d) This implies that $t_0 \in E$ and $x + t_0 \in E$ (why?).
(e) Hence, since $x = (x+t_0) - t_0$, we have $x \in D(E)$. This proves $D(E) \supset (-\delta, \delta)$. (Proof due to A. Calderon.)

7. Let $r_1, r_2, \ldots$ be an enumeration of the rational numbers in $[0,1]$. For each $n \in I$ let J_n be an open subinterval of $[0,1]$ such that $r_n \in J_n$ and $|J_n| < 1/2^n$.
   (a) Let χ be the characteristic function of $\cup_{n=1}^{\infty} J_n$. Show that $\chi = \chi^2$ is Lebesgue integrable.
   (b) By showing that χ is discontinuous at any point in the complement of $\cup_{n=1}^{\infty} J_n$, show that $\chi = \chi^2$ is not Riemann integrable, and that no square integrable function that is equal almost everywhere to χ can be Riemann integrable.
   (c) If χ_n is the characteristic function of $J_1 \cup \cdots \cup J_n$, show that

$$\chi_n = \chi_n^2 \in \mathcal{R}[a,b].$$

(d) Show that $\{\chi_n\}_{n=1}^{\infty}$ converges in the metric for $\mathfrak{L}^2[a,b]$ to χ.

(e) From all this deduce that 11.9G would not be true if we dealt only with the Riemann integral.

11.10 THE INTEGRAL ON $(-\infty, \infty)$ AND IN THE PLANE

11.10A. We now define $\mathfrak{L}[-\infty,\infty]$. Suppose f is a *nonnegative*-valued function on $(-\infty,\infty)$. If $f\in\mathfrak{L}[-N,N]$ for every $N\in I$, and if

$$\lim_{N\to\infty}\int_{-N}^{N} f$$

exists, then we define $\int_{-\infty}^{\infty} f$ as

$$\int_{-\infty}^{\infty} f = \lim_{N\to\infty}\int_{-N}^{N} f.$$

In this case we write $f\in\mathfrak{L}[-\infty,\infty]$.

For a function f that takes both positive and negative values, we say $f\in\mathfrak{L}[-\infty,\infty]$ if and only if both f^+ and f^- are in $\mathfrak{L}[-\infty,\infty]$, and we define $\int_{-\infty}^{\infty} f$ as

$$\int_{-\infty}^{\infty} f = \int_{-\infty}^{\infty} f^+ - \int_{-\infty}^{\infty} f^-.$$

11.10B. If f is measurable on every bounded interval, it is then easy to show that $f\in\mathfrak{L}[-\infty,\infty]$ if and only if $|f|\in\mathfrak{L}[-\infty,\infty]$. Indeed, if $f\in\mathfrak{L}[-\infty,\infty]$, then both f^+ and f^- are in $\mathfrak{L}[-\infty,\infty]$. Since $|f|=f^+ + f^-$, and

$$\int_{-N}^{N}|f| = \int_{-N}^{N} f^+ + \int_{-N}^{N} f^-,$$

it follows easily that

$$\lim_{N\to\infty}\int_{-N}^{N}|f|$$

exists and hence, that $|f|\in\mathfrak{L}[-\infty,\infty]$. Conversely, if f is measurable on every bounded interval and $|f|\in\mathfrak{L}[-\infty,\infty]$, then, for any $n\in I$,

$$\int_{-N}^{N} f^+ \leqslant \int_{-N}^{N}|f| \leqslant \lim_{N\to\infty}\int_{-N}^{N}|f| = \int_{-\infty}^{\infty}|f| = A.$$

Hence,

$$\lim_{N\to\infty}\int_{-N}^{N} f^+$$

exists (and is $\leqslant A$). Thus $f^+\in\mathfrak{L}[-\infty,\infty]$. Similarly, $f^-\in\mathfrak{L}[-\infty,\infty]$ and hence, $f\in\mathfrak{L}[-\infty,\infty]$.

The fact that $f\in\mathfrak{L}[-\infty,\infty]$ implies $|f|\in\mathfrak{L}[-\infty,\infty]$ makes the Lebesgue integral over $(-\infty,\infty)$ very different from the improper Riemann integral over $(-\infty,\infty)$.

It may be shown that $\mathfrak{L}[-\infty,\infty]$ has all the important properties of $\mathfrak{L}[a,b]$. Theorems such as 11.8B may be demonstrated without undue difficulty for $\mathfrak{L}[-\infty,\infty]$ by use of the theorem for $\mathfrak{L}[a,b]$. Note, however, that a function f may not be in $\mathfrak{L}[-\infty,\infty]$ even though f is bounded on $(-\infty,\infty)$ and f is measurable on every bounded interval. For example, $1\notin\mathfrak{L}[-\infty,\infty]$.

11.10C. An integral of the form $\int_0^\infty f$ may be defined as in 11.10A. That is,

$$\int_0^\infty f = \lim_{N\to\infty} \int_0^N f$$

for nonnegative-valued f, and so forth.

11.10D. We next outline the theory of measure and integration in the plane R^2. The procedure is very much analogous to that for the line. Rectangles in R^2 play the role of intervals in R^1.

A rectangle K in R^2 is, by definition, the Cartesian product of two bounded intervals I_1 and I_2 in R^1. The rectangle K may contain all, some, or none of its edges, depending on whether I_1 and I_2 are open, half-open, or closed. For example, if $K=(0,1)\times[1,3)$, then $K=\{\langle x,y\rangle | 0<x<1, 1\leqslant y<3\}$. Here K contains its base but none of its other three edges. For any rectangle K we define $|K|$ to be the area of K. This corresponds to the length of an interval in R^1.

We henceforth assume that all sets under consideration are subsets of a fixed closed rectangle $T=[a,b]\times[c,d]$. A subset G of T is called an *elementary* set if G can be written $G=\cup_n K_n$ where the K_n are a finite or countable number of rectangles and are pairwise disjoint. (It may be shown that every open subset of T is an elementary set.) For such a G we define $|G|$ as $|G|=\Sigma_n |K_n|$.

For any subset E of T we define the outer measure $\overline{m}E$ as

$$\overline{m}E = \text{g.l.b.}|G|$$

where the g.l.b. is taken over all *elementary* sets G such that $G\supset E$. (Compare with 11.2.) We define the inner measure $\underline{m}E$ as

$$\underline{m}E=(b-a)(d-c)-\overline{m}E'$$

where $E'=T-E$. [Note that $(b-a)(d-c)=|T|$.] This corresponds to theorem 11.2D.

Finally, the set E is said to be measurable if $\underline{m}E=\overline{m}E$. If E is measurable, we define the measure mE as

$$mE=\underline{m}E=\overline{m}E.$$

It may then be shown that the measure m on T has all the properties that we established for the measure on a closed bounded interval.

11.10E. Our definition of measurable function is akin to that in 0.3A. If f is a real-valued function on T, then f is said to be measurable if for each $s\in R$, the set $\{\langle x,y\rangle \in T | f(x,y)<s\}$ is a measurable subset of T. Properties of measurable functions may then be developed as in Section 11.4.

A partition P of T is defined in exactly the same way as a partition of $[a,b]$. That is, $P=\{E_1,\ldots,E_n\}$ where the E_i are measurable subsets of T and any two distinct E_i intersect in (at most) a set of measure zero. We may then define upper sums, lower sums, upper integrals, lower integrals, and finally the Lebesgue integral for bounded functions. Moreover, the proof in 11.5I that bounded measurable functions are Lebesgue integrable carries over to bounded measurable functions on T. The extension of the definition to unbounded functions proceeds as in Section 11.6.

If f is Lebesgue integrable on T, we write $f\in\mathcal{L}(T)$. If $f\in\mathcal{L}(T)$, the integral of f over T is denoted by

$$\iint_T f(x,y)\,dx\,dy.$$

Such an integral is often called a double integral.

If E is a measurable subset of T, we may define

$$\iint_E f(x,y)\,dx\,dy \quad \text{as} \quad \iint_T f(x,y)\chi_E(x,y)\,dx\,dy,$$

which is analogous to 11.7F. Moreover, we can extend the double integral to functions on all of R^2 in a manner analogous to that in 11.10A.

11.10F. We now take up the connection between measurable functions on

$$T=[a,b]\times[c,d]$$

and measurable functions on an interval in R^1. The main result is as follows.

THEOREM. If f is a measurable function on $T=[a,b]\times[c,d]$, then

1. for almost all $x\in[a,b]$ the function $f(x,y)$ is a measurable function of y, and
2. for almost all $y\in[c,d]$ the function $f(x,y)$ is a measurable function of x.

That is, if $x\in[a,b]$, the function g_x defined by

$$g_x(y)=f(x,y) \qquad (c\leqslant y\leqslant d)$$

is a real-valued function on $[c,d]$. According to (1) of the theorem, if f is measurable on T, then, for almost all x, g_x is a measurable function on $[c,d]$. Conclusion (2) has similar meaning.

In practice, an integral

$$\iint_T f(x,y)\,dx\,dy$$

is computed by integrating first with respect to x and then with respect to y (or vice versa). The fact that a double integral may be evaluated by this method of iterated integration does not follow immediately from the definition of

$$\iint_T f(x,y)\,dx\,dy,$$

but rather is a famous and difficult theorem called Fubini's theorem.

FUBINI'S THEOREM. Let $T=[a,b]\times[c,d]$ and suppose $f\in\mathfrak{L}(T)$. Then

$$\text{for almost all} \quad x\in[a,b], \quad f(x,y) \quad \text{is in} \quad \mathfrak{L}[c,d] \qquad (\text{as a function of } y). \tag{1}$$

Hence,

$$\int_c^d f(x,y)\,dy$$

is defined for almost all x. Moreover,

$$\int_c^d f(x,y)\,dy \quad \text{is in} \quad \mathfrak{L}[a,b] \qquad (\text{as a function of } x), \tag{2}$$

and

$$\iint_T f(x,y)\,dx\,dy=\int_a^b\left[\int_c^d f(x,y)\,dy\right]dx. \tag{3}$$

Similarly,

$$\iint_T f(x,y)\,dx\,dy = \int_c^d \left[\int_a^b f(x,y)\,dx\right] dy. \tag{4}$$

That is, if f is Lebesgue integrable as a function of two variables, then $\iint_T f$ can be evaluated by performing an iterated integration. It follows that if $f \in \mathcal{L}(T)$, then the right sides of (3) and (4) are equal.

Very often an iterated integral is given and it is necessary to change the order of integration. This procedure can be justified by the Tonelli-Hobson theorem.

TONELLI-HOBSON THEOREM. Let f be a measurable function on T. Then, if either of the iterated integrals

$$\int_a^b \left[\int_c^d |f(x,y)|\,dy\right] dx$$

or

$$\int_c^d \left[\int_a^b |f(x,y)|\,dx\right] dy$$

exists, then $f \in \mathcal{L}(T)$ and hence, (by Fubini's theorem)

$$\int_a^b \left[\int_c^d f(x,y)\,dy\right] dx = \int_c^d \left[\int_a^b f(x,y)\,dx\right] dy.*$$

Thus the order of integration in iterated integrals may be interchanged if either of the iterated integrals "converges absolutely." Both the Fubini theorem and the Tonelli-Hobson theorem hold for integrals over all of R^2. Indeed, all the above theory of integration on T may be extended on all of R^2, or to integrals $\iint_E f$ where E is a measurable subset of R^2.

11.10G. Let us give an illustration of the results of this section. If $f \in \mathcal{L}[0,\infty)$, then, since $e^{-a} \leqslant 1$ if $a \geqslant 0$, the function $e^{-xt}f(t)$ is also in $\mathcal{L}[0,\infty)$ for every $x \geqslant 0$. The function $\hat{f}$ defined by

$$\hat{f}(x) = \int_0^\infty e^{-xt} f(t)\,dt \qquad (0 \leqslant x < \infty)$$

is called the Laplace transform of f.

If f and g are in $\mathcal{L}[0,\infty)$, then the integral

$$\int_0^x f(x-t)\,g(t)\,dt \tag{1}$$

exists for almost all $x \in [0,\infty)$ and is an integrable function of x on $[0,\infty)$. The proof of

* Saying that $\int_a^b [\int_c^d |f(x,y)|\,dy]\,dx$ exists means

1. For almost all x in $[a,b]$, the function $|f(x,y)|$ is in $\mathcal{L}[c,d]$, and
2. The function $\int_c^d |f(x,y)|\,dy$ [which is defined for almost all x by 1] is in $\mathcal{L}[a,b]$.

Thus, under these conditions, $\int_a^b [\int_c^d |f(x,y)|\,dy]\,dx$ is a real number.

this is as follows: Define $f(t)=0=g(t)$ for $-\infty<t<0$. For each real t we have, by a change of variable $u=x-t$,

$$\int_{-\infty}^{\infty}|f(x-t)|dx=\int_{-\infty}^{\infty}|f(u)|du,$$

and hence,

$$\int_{-\infty}^{\infty}\left[\int_{-\infty}^{\infty}|f(x-t)g(t)|dx\right]dt=\int_{-\infty}^{\infty}|g(t)|dt\int_{-\infty}^{\infty}|f(u)|du$$

$$=\int_{0}^{\infty}|g(t)|dt\cdot\int_{0}^{\infty}|f(u)|du.$$

The last quantity is finite, since $f,g\in\mathcal{L}[0,\infty]$. Hence, by the Tonelli-Hobson theorem, the function $f(x-t)g(t)$ is in $\mathcal{L}(R^2)$. It follows from the Fubini theorem that

$$\int_{-\infty}^{\infty}f(x-t)g(t)dt \tag{2}$$

exists for almost all x and is integrable. However, $g(t)=0$ for $t<0$ and $f(x-t)=0$ for $t>x$. The integral (2) is therefore equal to the integral (1) and our assertion is thus proved. If $f,g\in\mathcal{L}[0,\infty)$ and if

$$h(x)=\int_0^x f(x-t)g(t)dt,$$

we say that h is the convolution of f and g and write $h=f_*g$. We have shown that $h(x)$ exists for almost all x and that $h\in\mathcal{L}[0,\infty)$.

We next prove a famous result involving Leplace transforms and convolutions.

THEOREM. Let f and g be functions in $\mathcal{L}[0,\infty)$ with Laplace transforms $\hat{f}$ and $\hat{g}$. If $h=f*g$, then $\hat{h}=\hat{f}\hat{g}$. That is, the Laplace transform of the convolution of f and g is the product of the Laplace transforms of f and g.

PROOF: For $y\geqslant 0$ we have

$$\hat{h}(y)=\int_0^{\infty}e^{-yx}h(x)dx=\int_0^{\infty}e^{-yx}\left[\int_0^x f(x-t)g(t)dt\right]dx.$$

We are integrating over a triangle in R^2—namely, $E=\{\langle x,t\rangle|0\leqslant t\leqslant x, 0\leqslant x<\infty\}$. Reversing the order of integration we have $E=\{\langle x,t\rangle|t\leqslant x<\infty, 0\leqslant t<\infty\}$. Thus

$$\hat{h}(y)=\int_0^{\infty}g(t)\left[\int_t^{\infty}e^{-yx}f(x-t)dx\right]dt$$

$$=\int_0^{\infty}g(t)\left[\int_0^{\infty}e^{-y(u+t)}f(u)du\right]dt$$

$$=\int_0^{\infty}e^{-yt}g(t)dt\cdot\int_0^{\infty}e^{-yu}f(u)du=\hat{g}(y)\hat{f}(y).$$

We have shown that $\hat{h}(y)=\hat{f}(y)\hat{g}(y)$ for all $y\geqslant 0$. Hence, $\hat{h}=\hat{f}\hat{g}$, which is what we wished to prove. The change in order of integration may be justified by the Tonelli-Hobson theorem (verify).

11.10H. We close this chapter by remarking that the theory of measure and integration can be carried out on a much more general class of spaces than R^1, R^2, and so forth.

For this more general treatment we recommend as a reference the book *Real Analysis* by H. L. Royden (Macmillan, New York, 1968).

For the differentiation theory that goes with Lebesgue integration, and which we summarily dispatched with the statement of 11.8E, we refer the reader to either *Functional Analysis* by F. Riesz and B. von Sz. Nagy (New York: Ungar, 1955) or *Theory of Functions* (2nd ed.) by E. C. Titchmarsh (Oxford, New York, 1939).

We might add that theorems on change of variable, integration by parts, and so forth, for the Lebesgue integral, which we have not touched on, can be found in these references.

Exercises 11.10

1. Which of the following functions are in $\mathfrak{L}[0,\infty)$?
   (a) $f(x)=\dfrac{\sin x}{x} \qquad (0<x<\infty)$,
   (b) $f(x)=\dfrac{1}{1+x^2} \qquad (0\leqslant x<\infty)$,
   (c) the characteristic function of
$$\bigcup_{n=1}^{\infty}\left[n,n+\frac{1}{n^2}\right],$$
   (d) the characteristic function of the rationals in $[0,\infty)$,
   (e) the characteristic function of the irrationals in $[0,\infty)$.
2. If f is a nonnegative function on $(0,\infty)$, if $f\in\mathfrak{R}[\epsilon,N]$ for every $\epsilon>0, N>0$, and if the improper Riemann integral $\int_0^\infty f$ exists, prove that $f\in\mathfrak{L}[0,\infty]$.
3. Show that if $x>0$, then $e^{-t}t^{x-1}$ is Lebesgue integrable on $[0,\infty)$. Let
$$\Gamma(x)=\int_0^\infty e^{-t}t^{x-1}\,dt \qquad (0<x<\infty).$$
   This function is known as the Gamma function.
   (a) Using integration by parts over the integral $[\epsilon,N]$, and letting $\epsilon\to 0+, N\to\infty$, show that
$$\Gamma(x+1)=x\Gamma(x) \qquad (0<x<\infty).$$
   (b) Show that $\Gamma(1)=1$,
$$\Gamma(n+1)=n! \qquad (n=1,2,\ldots).$$
4. If $T=[a,b]\times[c,d]$ and if f is continuous on T, prove that
$$\iint_T f(x,y)\,dx\,dy=\int_a^b\left[\int_c^d f(x,y)\,dy\right]dx=\int_c^d\left[\int_a^b f(x,y)\,dx\right]dy.$$
5. Show that
$$\int_0^1\left[\int_0^x f(x,y)\,dy\right]dx=\int_0^1\left[\int_y^1 f(x,y)\,dx\right]dy$$
   provided f is Lebesgue integrable on $[0,1]\times[0,1]$.
6. Let $T=[0,1]\times[0,1]$ and let
$$f(x,y)=\frac{x^2-y^2}{(x^2+y^2)^2} \qquad (\langle x,y\rangle\in T;\langle x,y\rangle\neq\langle 0,0\rangle),$$
$$f(0,0)=0.$$

(a) Show that

$$\int_0^1 f(x,y)\,dy = \frac{y}{x^2+y^2}\Bigg|_0^1 = \frac{1}{1+x^2} \qquad (0 < x \leqslant 1),$$

and hence, that

$$\int_0^1 \left[\int_0^1 f(x,y)\,dy \right] dx = \frac{\pi}{4}.$$

(b) Similarly, show that

$$\int_0^1 \left[\int_0^1 f(x,y)\,dx \right] dy = -\frac{\pi}{4}.$$

(c) Conclude from the Fubini theorem that f is not integrable over T.

(d) Conclude from the Tonelli-Hobson theorem that

$$\int_0^1 \left[\int_0^1 |f(x,y)|\,dy \right] dx \qquad \text{does not exist.}$$

7. Let $T = [-1,1] \times [-1,1]$ and define f as

$$f(x,y) = \frac{xy}{(x^2+y^2)^2} \qquad (\langle x,y\rangle \in T, \langle x,y\rangle \neq \langle 0,0\rangle),$$

$$f(0,0) = 0.$$

(a) Show that

$$\int_{-1}^1 f(x,y)\,dy = 0 \qquad (-1 \leqslant x \leqslant 1)$$

and hence, that

$$\int_{-1}^1 \left[\int_{-1}^1 f(x,y)\,dy \right] dx = 0.$$

(b) If f were integrable over T, then f would be integrable over $T^* = [0,1] \times [0,1]$. In this case

$$\int_0^1 \left[\int_0^1 f(x,y)\,dy \right] dx \qquad (*)$$

would exist (why?).

(c) Show that

$$\int_0^1 f(x,y)\,dy = \frac{1}{2x} - \frac{x}{2(x^2+1)} \qquad (0 < x \leqslant 1),$$

and hence, that the iterated integral (*) does not exist.

(d) Conclude that f is not integrable over T even though

$$\int_{-1}^1 \left[\int_{-1}^1 f(x,y)\,dy \right] dx$$

exists and is equal to 0.

12

FOURIER SERIES

12.1 DEFINITION OF FOURIER SERIES

We will be discussing the expansion of real-valued functions in terms of the functions $1, \cos x, \sin x, \cos 2x, \sin 2x, \ldots$. We must first evaluate some important definite integrals of trigonometric functions.

12.1A. THEOREM

$$\int_{-\pi}^{\pi} \cos kx \cos nx \, dx = 0 \qquad (k, n = 0, 1, 2, \ldots;\ k \neq n) \tag{1}$$

$$\int_{-\pi}^{\pi} \cos^2 nx \, dx = \begin{cases} \pi & (n = 1, 2, \ldots) \\ 2\pi & (n = 0) \end{cases} \tag{2}$$

$$\int_{-\pi}^{\pi} \sin kx \sin nx \, dx = 0 \qquad (k, n = 1, 2, \ldots;\ k \neq n) \tag{3}$$

$$\int_{-\pi}^{\pi} \sin^2 nx \, dx = \pi \qquad (n = 1, 2, \ldots) \tag{4}$$

$$\int_{-\pi}^{\pi} \cos kx \sin nx \, dx = 0 \qquad (k, n = 0, 1, 2, \ldots). \tag{5}$$

PROOF: All these results may be easily derived from appropriate trigonometric identities. For example,

$$\cos(kx + nx) = \cos kx \cos nx - \sin kx \sin nx,$$
$$\cos(kx - nx) = \cos kx \cos nx + \sin kx \sin nx.$$

Adding, we have

$$\cos kx \cos nx = \tfrac{1}{2}\left[\cos(k+n)x + \cos(k-n)x\right].$$

Suppose $k \neq n$. Integrating from $-\pi$ to π we obtain

$$\int_{-\pi}^{\pi} \cos kx \cos nx\,dx = \frac{1}{2}\int_{-\pi}^{\pi} \cos(k+n)x\,dx + \frac{1}{2}\int_{-\pi}^{\pi} \cos(k-n)x\,dx$$

$$= \frac{1}{2}\cdot\frac{\sin(k+n)x}{k+n}\bigg|_{-\pi}^{\pi} + \frac{1}{2}\cdot\frac{\sin(k-n)x}{k-n}\bigg|_{-\pi}^{\pi} = 0.$$

This proves (1). The proof of (2)–(5) is left to the reader.

12.1B. Suppose we have a real-valued function f on $[-\pi,\pi]$ that can be expressed

$$f(x) = \frac{a_0}{2} + \sum_{k=1}^{\infty} (a_k \cos kx + b_k \sin kx) \qquad (-\pi \leqslant x \leqslant \pi). \tag{1}$$

Let us see *formally* what we should expect the coefficients $a_0, a_1, b_1, a_2, b_2, \ldots$ to be.

If we integrate in (1), we obtain

$$\int_{-\pi}^{\pi} f(x) = \pi a_0 + \sum_{k=1}^{\infty}\left(a_k\int_{-\pi}^{\pi} \cos kx\,dx + b_k\int_{-\pi}^{\pi} \sin kx\,dx\right) = \pi a_0.$$

Hence, assuming the validity of our term-by-term integration, we have

$$a_0 = \frac{1}{\pi}\int_{-\pi}^{\pi} f(x)\,dx. \tag{2}$$

Now, for a fixed $n \in I$, let us multiply (1) by $\cos nx$ and integrate. We have

$$\int_{-\pi}^{\pi} f(x)\cos nx\,dx = \frac{a_0}{2}\int_{-\pi}^{\pi} \cos nx\,dx$$

$$+ \sum_{k=1}^{\infty}\left(a_k\int_{-\pi}^{\pi} \cos kx \cos nx\,dx + b_k\int_{-\pi}^{\pi} \sin kx \cos nx\,dx\right).$$

Using 12.1A, we see that only one of the integrals on the right is not zero—namely, the integral

$$\int_{-\pi}^{\pi} \cos kx \cos nx\,dx$$

for $k = n$. Hence,

$$\int_{-\pi}^{\pi} f(x)\cos nx\,dx = a_n\int_{-\pi}^{\pi} \cos^2 nx = \pi a_n,$$

and so, for $n = 1, 2, \ldots,$

$$a_n = \frac{1}{\pi}\int_{-\pi}^{\pi} f(x)\cos nx\,dx. \tag{3}$$

Note that if we put $n=0$ in (3), we obtain (2). That is why we use $a_0/2$ in (1) instead of a_0. It may be similarly shown that

$$b_n = \frac{1}{\pi}\int_{-\pi}^{\pi} f(x)\sin nx\,dx. \tag{4}$$

Thus if a function f is representable in the form (1), we would expect the coefficients to be found from (2), (3), (4). This leads us to the following definition.

12.1C. DEFINITION. If $f \in \mathcal{L}[-\pi,\pi]$, then the Fourier series for f is the series

$$\frac{a_0}{2} + \sum_{k=1}^{\infty} (a_k \cos kx + b_k \sin kx) \qquad (-\pi \leqslant x \leqslant \pi),$$

where

$$a_k = \frac{1}{\pi}\int_{-\pi}^{\pi} f(x)\cos kx\,dx \qquad (k=0,1,2,\ldots),$$

$$b_k = \frac{1}{\pi}\int_{-\pi}^{\pi} f(x)\sin kx\,dx \qquad (k=1,2,\ldots).$$

The a_k and b_k are called the Fourier coefficients for f. We write

$$f \sim \frac{a_0}{2} + \sum_{k=1}^{\infty} (a_k \cos kx + b_k \sin kx).$$

Note that we use $\sim$ and not $=$. We have not as yet shown that the Fourier series for a function $f \in \mathcal{L}[-\pi,\pi]$ will converge, much less converge to $f(x)$ for some or all values of x.

For example, let

$$f(x)=0 \qquad (-\pi \leqslant x < 0),$$
$$f(x)=1 \qquad (0 \leqslant x \leqslant \pi).$$

Then

$$a_k = \frac{1}{\pi}\int_{-\pi}^{\pi} f(x)\cos kx\,dx = \frac{1}{\pi}\int_{0}^{\pi} \cos kx\,dx.$$

Hence, $a_0 = 1$, $a_k = 0$ $(k=1,2,\ldots)$. Also

$$b_k = \frac{1}{\pi}\int_{-\pi}^{\pi} f(x)\sin kx\,dx = \frac{1}{\pi}\int_{0}^{\pi} \sin kx\,dx = \frac{1-\cos k\pi}{k\pi}.$$

Thus $b_k = 2/k\pi$ $(k=1,3,5,\ldots)$, $b_k = 0$ $(k=2,4,6,\ldots)$. We thus have

$$f \sim \frac{1}{2} + \frac{2}{\pi}\left[\frac{\sin x}{1} + \frac{\sin 3x}{3} + \frac{\sin 5x}{5} + \cdots\right].$$

Note that at $x=0$ the sum of the Fourier series for f is $\frac{1}{2}$ and hence, is not equal to $f(0)$.

Exercises 12.1

1. A function f on $[-\pi,\pi]$ is said to be an even function if

$$f(-x) = f(x) \qquad (-\pi \leqslant x \leqslant \pi).$$

For example, the cosine function is an even function.

A function f on $[-\pi,\pi]$ is said to be an odd function if

$$f(-x) = -f(x) \qquad (-\pi \leqslant x \leqslant \pi).$$

For example, the sine function is an odd function.

(a) Show that if f is an even function, then

$$\int_{-\pi}^{\pi} f(x)\,dx = 2\int_{0}^{\pi} f(x)\,dx.$$

(b) Show that if f is an odd function, then

$$\int_{-\pi}^{\pi} f(x)\,dx = 0.$$

(c) Show that if f is an even function and g is an odd function, then fg is an odd function.
(d) Show that fg is even if either f and g are both even or if f and g are both odd.
(e) Suppose $f \in \mathcal{L}[-\pi,\pi]$ and

$$f(x) \sim \frac{a_0}{2} + \sum_{k=1}^{\infty} (a_k \cos kx + b_k \sin kx).$$

Show that if f is even, then $b_1 = b_2 = \cdots = 0$, while if f is odd, then

$$a_0 = a_1 = a_2 = \cdots = 0.$$

2. If f is any real-valued function on $[-\pi,\pi]$ and

$$g(x) = f(x) + f(-x) \qquad (-\pi \leqslant x \leqslant \pi),$$
$$h(x) = f(x) - f(-x) \qquad (-\pi \leqslant x \leqslant \pi),$$

show that g is an even function and h is an odd function. Deduce that any real-valued function on $[-\pi,\pi]$ may be written as the sum of an even function and and odd function.

3. Find the Fourier series for each of the following functions f.

(a) $f(x) = -1 \quad (-\pi \leqslant x < 0),$
$f(x) = 1 \quad (0 \leqslant x \leqslant \pi).$
(b) $f(x) = x \quad (-\pi \leqslant x \leqslant \pi).$
(c) $f(x) = |x| \quad (-\pi \leqslant x \leqslant \pi).$
(d) $f(x) = e^x \quad (-\pi \leqslant x \leqslant \pi).$
(e) $f(x) = \sin x + \cos 2x \quad (-\pi \leqslant x \leqslant \pi).$

Use (e) of exercise 1 when possible.

4. (a) For each of the functions in (a) and (b) of exercise 3 check whether the Fourier series at $x = 0$ converges to $f(0)$.
(b) Given that the Fourier series of

$$f(x) = |x| \qquad (-\pi \leqslant x \leqslant \pi)$$

at the point $x = 0$ converges to $f(0)$, prove that

$$\frac{1}{1^2} + \frac{1}{3^2} + \frac{1}{5^2} + \cdots = \frac{\pi^2}{8}.$$

5. If the series

$$\frac{a_0}{2} + \sum_{k=1}^{\infty} (a_k \cos kx + b_k \sin kx) \tag{1}$$

converges uniformly to f on $[-\pi,\pi]$, prove that (1) is the Fourier series for f. [*Hint*: For $n \in I$ show that

$$\frac{a_0 \cos nx}{2} + \sum_{k=1}^{\infty} (a_k \cos kx \cos nx + b_k \sin kx \cos nx)$$

converges uniformly to $f(x)\cos nx$ on $[-\pi,\pi]$. Then integrate to show that a_n is the nth Fourier cosine coefficient for f. Do similarly for b_n.]

12.2 FORMULATION OF CONVERGENCE PROBLEMS

12.2A. To see if the Fourier series of a function $f \in \mathcal{L}[-\pi,\pi]$ actually converges to the value of f at some fixed $x \in [-\pi,\pi]$, we must investigate whether

$$\lim_{n \to \infty} s_n(x) = f(x),$$

where the s_n are the partial sums of the Fourier series:

$$s_n(t) = \frac{a_0}{2} + \sum_{k=1}^{n} (a_k \cos kt + b_k \sin kt) \qquad (-\pi \leqslant t \leqslant \pi). \tag{1}$$

To express $s_n(x)$ in a more manageable form we use the definition of a_k, b_k to obtain

$$\begin{aligned} s_n(x) &= \frac{a_0}{2} + \sum_{k=1}^{n} (a_k \cos kx + b_k \sin kx) \\ &= \frac{1}{2\pi}\int_{-\pi}^{\pi} f(t)\,dt + \frac{1}{\pi}\sum_{k=1}^{n}\left(\cos kx \int_{-\pi}^{\pi} f(t)\cos kt\,dt + \sin kx \int_{-\pi}^{\pi} f(t)\sin kt\,dt\right) \\ &= \frac{1}{\pi}\int_{-\pi}^{\pi} f(t)\left[\frac{1}{2} + \sum_{k=1}^{n} (\cos kx \cos kt + \sin kx \sin kt)\right] dt \\ &= \frac{1}{\pi}\int_{-\pi}^{\pi} f(t)\left[\frac{1}{2} + \sum_{k=1}^{n} \cos k(x-t)\right] dt. \end{aligned}$$

Using the identity (39) of Section 8.4 we then have

$$s_n(x) = \frac{1}{\pi}\int_{-\pi}^{\pi} f(t) D_n(x-t)\,dt$$

where*

$$D_n(t) = \frac{1}{2} + \sum_{k=1}^{n} \cos kt = \frac{\sin(n+\frac{1}{2})t}{2\sin(t/2)} \qquad (-\infty < t < \infty). \tag{2}$$

(The function D_n is called the Dirichlet kernel.)
If we set $u = x - t$, we have

$$s_n(x) = \frac{1}{\pi}\int_{x-\pi}^{x+\pi} f(x-u) D_n(u)\,du. \tag{3}$$

It is now convenient to extend f to all of $(-\infty, \infty)$ by requiring that

$$f(u + 2\pi) = f(u). \tag{4}$$

This defines $f(u)$ uniquely for all values of u except [in case $f(-\pi) \neq f(\pi)$] for $u = \pm 3\pi, \pm 5\pi, \ldots$. For these exceptional values of u we can define $f(u)$ in any manner without affecting anything concerning integrals of f. Functions satisfying (4) are called periodic functions (of period 2π). Since D_n is also a periodic function of period 2π, it is easy to show from (3) that

$$s_n(x) = \frac{1}{\pi}\int_{-\pi}^{\pi} f(x-u) D_n(u)\,du.$$

(See exercise 1 of this section.) Thus

$$\begin{aligned} s_n(x) &= \frac{1}{\pi}\int_{-\pi}^{0} f(x-u) D_n(u)\,du + \frac{1}{\pi}\int_{0}^{\pi} f(x-u) D_n(u)\,du \\ &= \frac{1}{\pi}\int_{0}^{\pi} f(x+t) D_n(-t)\,dt + \frac{1}{\pi}\int_{0}^{\pi} f(x-t) D_n(t)\,dt. \end{aligned}$$

Since $D_n(-t) = D_n(t)$ for all t, we have

$$s_n(x) = \frac{1}{\pi}\int_{0}^{\pi} [f(x+t) + f(x-t)] D_n(t)\,dt. \tag{5}$$

* For $t = 0, \pm 2\pi, \pm 4\pi, \ldots$ we interpret $D_n(t)$ as $n + \frac{1}{2}$ so that D_n will be continuous on $(-\infty, \infty)$.

In the special case $f=1$ we have $a_0=2$, $a_k=b_k=0$ $(k \geqslant 1)$. Hence, $s_n(x)=1$ for all n. From (5) we then have

$$1=\frac{2}{\pi}\int_0^{\pi} D_n(t)\,dt. \tag{6}$$

Going back to an arbitrary $f \in \mathcal{L}[-\pi,\pi]$, we multiply (6) by $f(x)$ and subtract from (5) to obtain

$$s_n(x)-f(x)=\frac{2}{\pi}\int_0^{\pi}\left[\frac{f(x+t)+f(x-t)}{2}-f(x)\right]D_n(t)\,dt.$$

We have thus shown that

$$\frac{a_0}{2}+\sum_{k=1}^{\infty} a_k \cos kx + b_k \sin kx = f(x)$$

[that is, the Fourier series for f at x will converge to $f(x)$] if and only if

$$\lim_{n\to\infty}\left[s_n(x)-f(x)\right]=\lim_{n\to\infty}\frac{2}{\pi}\int_0^{\pi}\left[\frac{f(x+t)+f(x-t)}{2}-f(x)\right]D_n(t)\,dt=0,$$

where D_n is as in (2).

12.2B. If $f \in \mathcal{L}^2[-\pi,\pi]$, then $f \in \mathcal{L}[-\pi,\pi]$ (why?), so that f has a Fourier series. Since each function $\cos kt$, $\sin kt$ is in $\mathcal{L}^2[-\pi,\pi]$, it follows from 11.9C that $s_n \in \mathcal{L}^2[-\pi,\pi]$, where s_n is the nth partial sum of the Fourier series for f defined in (1) of 12.2A. In addition to asking whether $\{s_n(x)\}_{n=1}^{\infty}$ converges to $f(x)$ at individual points x, it makes sense (and is very fruitful) to inquire if the sequence of $\mathcal{L}^2[-\pi,\pi]$ functions $\{s_n\}_{n=1}^{\infty}$ converges to f in the metric of $\mathcal{L}^2[-\pi,\pi]$. That is, does

$$\lim_{n\to\infty}\|s_n-f\|_2=0?$$

We will see in Section 12.4 that the answer is "yes" for all $f \in \mathcal{L}^2[-\pi,\pi]$.

12.2C. As a final problem we ask: For what x is it true that the Fourier series of $f \in \mathcal{L}[-\pi,\pi]$ is $(C,1)$ summable to $f(x)$? That is, for what $x \in [-\pi,\pi]$ is it true that

$$\lim_{n\to\infty} s_n(x)=f(x) \qquad (C,1),$$

or, equivalently,

$$\lim_{n\to\infty}\sigma_n(x)=f(x)$$

where

$$\sigma_n(x)=\frac{s_0(x)+s_1(x)+\cdots+s_{n-1}(x)}{n}=\frac{1}{n}\sum_{k=0}^{n-1}s_k(x)?$$

Using (5) of 12.2A we have

$$\sigma_n(x)=\frac{1}{n}\sum_{k=0}^{n-1}\cdot\frac{1}{\pi}\int_0^{\pi}\left[f(x+t)+f(x-t)\right]D_k(t)\,dt,$$

and so

$$\sigma_n(x)=\frac{1}{\pi}\int_0^{\pi}\left[f(x+t)+f(x-t)\right]K_n(t)\,dt \tag{1}$$

where

$$K_n(t) = \frac{1}{n}\sum_{k=0}^{n-1} D_k(t) = \frac{1}{2n\sin(t/2)}\sum_{k=0}^{n-1}\sin(k+\tfrac{1}{2})t \qquad (-\infty < t < \infty). \tag{2}$$

Now

$$\sum_{k=0}^{n-1}\sin(k+\tfrac{1}{2})t = \sum_{j=1}^{n}\sin(j-\tfrac{1}{2})t = \sum_{j=1}^{n}\sin(2j-1)\frac{t}{2},$$

and so, by exercise 8 of Section 8.4,

$$\sum_{k=0}^{n-1}\sin(k+\tfrac{1}{2})t = \frac{\sin^2(nt/2)}{\sin(t/2)}.$$

Thus from (2),

$$K_n(t) = \frac{\sin^2(nt/2)}{2n\sin^2(t/2)} \qquad (-\infty < t < \infty). \tag{3}$$

[In particular, note that $K_n(t) \geqslant 0$ for all t.] In the special case $f=1$, then $s_0(x) = s_1(x) = \cdots = s_{n-1}(x) = 1$. Hence, from (1) we have

$$1 = \frac{2}{\pi}\int_0^{\pi} K_n(t)\,dt. \tag{4}$$

Going back to an arbitrary $f \in \mathcal{L}[-\pi,\pi]$, we multiply (4) by $f(x)$ and subtract from (1). We then have

$$\sigma_n(x) - f(x) = \frac{2}{\pi}\int_0^{\pi}\left[\frac{f(x+t)+f(x-t)}{2} - f(x)\right]K_n(t)\,dt.$$

We have thus shown that

$$\frac{a_0}{2} + \sum_{k=1}^{\infty} a_k\cos kx + b_k\sin kx = f(x) \qquad (C,1)$$

if and only if

$$\lim_{n\to\infty}\left[\sigma_n(x) - f(x)\right] = \lim_{n\to\infty}\frac{2}{\pi}\int_0^{\pi}\left[\frac{f(x+t)+f(x-t)}{2} - f(x)\right]K_n(t)\,dt = 0,$$

where K_n is as in (3).

Exercises 12.2

1. If $f \in \mathcal{L}[-\pi,\pi]$ and if

$$f(u+2\pi) = f(u) \qquad (-\infty < u < \infty),$$

show that, for any real x,

$$\int_{-\pi+x}^{\pi+x} f(u)\,du = \int_{-\pi}^{\pi} f(u)\,du.$$

[*Hint*: Write

$$\int_{-\pi}^{\pi} f(u)\,du = \int_{-\pi}^{-\pi+x} f(u)\,du + \int_{-\pi+x}^{\pi} f(u)\,du = I_1 + I_2.$$

Then show that

$$I_1=\int_{\pi}^{\pi+x} f(t-2\pi)\,dt=\int_{\pi}^{\pi+x} f(t)\,dt=\int_{\pi}^{\pi+x} f(u)\,du$$

and add to I_2.] A picture may help.

2. If $f\in\mathcal{L}[-\pi,\pi]$ and if

$$f\sim\frac{a_0}{2}+\sum_{k=1}^{\infty}(a_k\cos kt+b_k\sin kt)\qquad(-\pi\leqslant t\leqslant\pi),$$

show that

$$\sigma_n(t)=\frac{a_0}{2}+\sum_{k=1}^{n-1}\left(1-\frac{k}{n}\right)(a_k\cos kt+b_k\sin kt)$$

for $n=1,2,\ldots;\ -\pi\leqslant t\leqslant\pi$.

3. True or false?

(a) $1=\dfrac{2}{\pi}\displaystyle\int_0^{\pi}|K_n(t)|\,dt\qquad(n\in I)$.

(b) $1=\dfrac{2}{\pi}\displaystyle\int_0^{\pi}|D_n(t)|\,dt\qquad(n\in I)$.

4. If f is continuous on $[-\pi,\pi]$ and

$$\max_{-\pi\leqslant t\leqslant\pi}|f(t)|=M,$$

use (1) of 12.2C to prove that

$$|\sigma_n(t)|\leqslant M\qquad(-\pi\leqslant t\leqslant\pi).$$

Does your method of proof enable you to prove that

$$|s_n(t)|\leqslant M\qquad(-\pi\leqslant t\leqslant\pi)?$$

12.3 $(C,1)$ SUMMABILITY OF FOURIER SERIES

We next show that continuity at x is sufficient to ensure that the Fourier series of $f\in\mathcal{L}[-\pi,\pi]$ at x be $(C,1)$ summable to $f(x)$. The following theorem is due to Fejer (and K_n is often called the Fejer kernel).

12.3A. THEOREM. If $f\in\mathcal{L}[-\pi,\pi]$ and if f is continuous at $x\in[-\pi,\pi]$,* then

$$\frac{a_0}{2}+\sum_{k=1}^{\infty}(a_k\cos kx+b_k\sin kx)=f(x)\qquad(C,1).\tag{1}$$

PROOF: Fix $\epsilon>0$. According to 12.2C, to prove the theorem it is sufficient to find $N\in I$ such that

$$|\sigma_n-f(x)|=\left|\frac{2}{\pi}\int_0^{\pi}\left[\frac{f(x+t)+f(x-t)-2f(x)}{2}\right]K_n(t)\,dt\right|<\epsilon\qquad(n\geqslant N).\tag{2}$$

Since f is continuous at x we can find δ with $0<\delta<\pi$ such that

$$|f(y)-f(x)|<\frac{\epsilon}{2}\qquad(|y-x|<\delta).$$

* If $x=\pi$ or $x=-\pi$, by continuity of f at x we mean the continuity of f as extended via the equation $f(u)=f(u+2\pi)$.

Thus if $0 \leqslant t < \delta$, then

$$\left| \frac{f(x+t)+f(x-t)-2f(x)}{2} \right| \leqslant \frac{1}{2}\left[|f(x+t)-f(x)| + |f(x-t)-f(x)| \right]$$

$$< \frac{1}{2}\left(\frac{\epsilon}{2} + \frac{\epsilon}{2} \right) = \frac{\epsilon}{2}.$$

Consequently,

$$\left| \frac{2}{\pi} \int_0^{\delta} \left[\frac{f(x+t)+f(x-t)-2f(x)}{2} \right] K_n(t)\,dt \right| \leqslant \frac{\epsilon}{2} \cdot \frac{2}{\pi} \int_0^{\delta} K_n(t)\,dt,$$

and so, by (4) of 12.2C,

$$\left| \frac{2}{\pi} \int_0^{\delta} \left[\frac{f(x+t)+f(x-t)-2f(x)}{2} \right] K_n(t)\,dt \right| < \frac{\epsilon}{2} \qquad (n \in I). \tag{3}$$

If $t \geqslant \delta$, then $K_n(t) \leqslant \dfrac{1}{2n\sin^2(\delta/2)}$. Thus

$$\left| \frac{2}{\pi} \int_{\delta}^{\pi} \left[\frac{f(x+t)+f(x-t)-2f(x)}{2} \right] K_n(t)\,dt \right|$$

$$\leqslant \frac{2}{4n\pi \sin^2(\delta/2)} \int_{\delta}^{\pi} \left[|f(x+t)| + |f(x-t)| + 2|f(x)| \right] dt, \tag{4}$$

and so, for some $N \in I$,

$$\left| \frac{2}{\pi} \int_{\delta}^{\pi} \left[\frac{f(x+t)+f(x-t)-2f(x)}{2} \right] K_n(t) \right| < \frac{\epsilon}{2} \qquad (n \geqslant N). \tag{5}$$

Inequality (2) then follows from (3) and (5), and the proof is complete.

12.3B. A stronger result may be proved. Indeed, with more careful estimates it may be shown that if $f \in \mathcal{L}[-\pi,\pi]$, then (1) must hold for almost every x in $[-\pi,\pi]$ (even if f is not continuous at any x). The set E of points x for which (1) holds contains all points at which $f(x)$ is the derivative of

$$F(x) = \int_0^x f(t)\,dt$$

and hence (11.8E), E contains almost every point in $[-\pi,\pi]$. We omit the proof.

If we assume that f is continuous on $[-\pi,\pi]$, and that $f(-\pi)=f(\pi)$, we can establish the uniform convergence of $\{\sigma_n\}_{n=1}^{\infty}$.

12.3C. THEOREM. If $f \in C[-\pi,\pi]$ and if $f(-\pi)=f(\pi)$, then $\{\sigma_n\}_{n=1}^{\infty}$ converges uniformly to f on $[-\pi,\pi]$, where

$$\sigma_n = \frac{s_0 + s_1 + \cdots + s_{n-1}}{n}$$

and

$$s_n(t) = \frac{a_0}{2} + \sum_{k=1}^{n} (a_k \cos kt + b_k \sin kt).$$

PROOF: Since f is continuous on the closed bounded interval $[-\pi,\pi]$, we know from 6.8C that f is uniformly continuous there. Since $f(-\pi)=f(\pi)$, it follows that f [as extended to $(-\infty,\infty)$ by $f(u+2\pi)=f(u)$] is uniformly continuous on $(-\infty,\infty)$. Consequently, the δ in the proof of 12.3A may be chosen independent of x. Moreover, from (4) of 12.3A, we have

$$\left|\frac{2}{\pi}\int_{\delta}^{\pi}\left[\frac{f(x+t)+f(x-t)-2f(x)}{2}K_n(t)\,dt\right|\leqslant\frac{2\|f\|(\pi-\delta)}{n\pi\sin^2(\delta/2)}\right.$$

where $\|f\|=\max_{-\pi\leqslant x\leqslant\pi}|f(x)|$. Consequently, the N in 12.3A such that (5) holds may be chosen to depend only on δ and not on x. It will follow that

$$|\sigma_n(x)-f(x)|<\epsilon \qquad (n\geqslant N;\ -\pi\leqslant x<\pi),$$

which proves the theorem.

Exercises 12.3

1. If $f\in\mathcal{L}[-\pi,\pi]$ and if f is continuous at $x\in[-\pi,\pi]$, prove that

$$f(x)=\lim_{r\to 1-}\left[\frac{a_0}{2}+\sum_{k=1}^{\infty}(a_k\cos kx+b_k\sin kx)r^k\right].$$

12.4. THE $\mathcal{L}^2$ THEORY OF FOURIER SERIES

12.4A. If $n\in I$, then a trigonometric polynomial of degree n is a function T_n of the form

$$T_n(t)=A_0+\sum_{k=1}^{n}(A_k\cos kt+B_k\sin kt) \qquad (-\pi\leqslant t\leqslant\pi), \tag{1}$$

where $A_0,\ldots,A_n$ and $B_1,\ldots,B_n$ are real numbers. Every trigonometric polynomial is in $\mathcal{L}^2[-\pi,\pi]$. We will now prove that if $f\in\mathcal{L}[-\pi,\pi]$, then the trigonometric polynomial closest to f (in the metric for $\mathcal{L}^2[-\pi,\pi]$) is s_n—the nth partial sum of the Fourier series for f.

THEOREM. Let $f\in\mathcal{L}^2[-\pi,\pi]$ and let T_n be any trigonometric polynomical of degree n. Then

$$\|f-T_n\|_2\geqslant\|f-s_n\|_2 \tag{2}$$

where

$$s_n(t)=\frac{a_0}{2}+\sum_{k=1}^{n}(a_k\cos kt+b_k\sin kt) \qquad (-\pi\leqslant t\leqslant\pi)$$

and the a_k,b_k are the Fourier coefficients of f.

PROOF: For any T_n let

$$J=\frac{1}{\pi}\int_{-\pi}^{\pi}[f(t)-T_n(t)]^2\,dt.$$

Then $\sqrt{\pi J}=\|f-T_n\|_2$, and so we must prove that J is a minimum when $T_n=s_n$. We have

$$J=\frac{1}{\pi}\int_{-\pi}^{\pi} f^2-\frac{2}{\pi}\int_{-\pi}^{\pi} fT_n+\frac{1}{\pi}\int_{-\pi}^{\pi} T_n^2. \tag{3}$$

If T_n is as in (1), then

$$\frac{1}{\pi}\int_{-\pi}^{\pi} fT_n=A_0\frac{1}{\pi}\int_{-\pi}^{\pi} f+\sum_{k=1}^{n}\left(A_k\cdot\frac{1}{\pi}\int_{-\pi}^{\pi} f(t)\cos kt\,dt+B_k\cdot\frac{1}{\pi}\int_{-\pi}^{\pi} f(t)\sin kt\,dt\right)$$

and so

$$\frac{1}{\pi}\int_{-\pi}^{\pi} fT_n=A_0a_0+\sum_{k=1}^{n}(A_ka_k+B_kb_k). \tag{4}$$

Also,

$$\frac{1}{\pi}\int_{-\pi}^{\pi} T_n^2=\frac{1}{\pi}\int_{-\pi}^{\pi} T_n(t)T_n(t)\,dt=\frac{1}{\pi}\int_{-\pi}^{\pi}\left[A_0+\sum_{k=1}^{n}(A_k\cos kt+B_k\sin kt)\right]$$
$$\cdot\left[A_0+\sum_{j=1}^{n}(A_j\cos jt+B_j\sin jt)\right]dt.$$

We can compute the integral on the right by use of 12.2A. This yields

$$\frac{1}{\pi}\int_{-\pi}^{\pi} T_n^2=2A_0^2+\sum_{k=1}^{n}(A_k^2+B_k^2). \tag{5}$$

Substituting (4) and (5) into (3) we obtain

$$J=\frac{1}{\pi}\int_{-\pi}^{\pi} f^2-2A_0a_0-2\sum_{k=1}^{n}(A_ka_k+B_kb_k)+2A_0^2+\sum_{k=1}^{n}(A_k^2+B_k^2).$$

Adding and subtracting

$$\frac{a_0^2}{2}+\sum_{k=1}^{n}(a_k^2+b_k^2)$$

and doing some algebra, we have

$$\begin{aligned}J&=\frac{1}{\pi}\int_{-\pi}^{\pi}(f-T_n)^2\\&=\frac{1}{\pi}\int_{-\pi}^{\pi} f^2-\left[\frac{a_0^2}{2}+\sum_{k=1}^{n}(a_k^2+b_k^2)\right]\\&\quad+\left\{2\left(A_0-\frac{a_0}{2}\right)^2+\sum_{k=1}^{n}\left[(A_k-a_k)^2+(B_k-b_k)^2\right]\right\}.\end{aligned} \tag{6}$$

The quantity in braces cannot be negative, and so J will be a minimum if we let $A_0=a_0/2, A_k=a_k$ $(k=1,\ldots,n), B_k=b_k$ $(k=1,\ldots,n)$. That is, J will be a minimum if $T_n=s_n$, which is what we wished to show.

The following corollary is very important.

12.4B. COROLLARY (BESSEL'S INEQUALITY). If $f\in\mathcal{L}^2[-\pi,\pi]$ has Fourier coefficients

a_k, b_k, then

$$\frac{a_0^2}{2} + \sum_{k=1}^{\infty} (a_k^2 + b_k^2) \leq \frac{1}{\pi} \int_{-\pi}^{\pi} f^2. \tag{1}$$

In particular, the series $\sum_{k=1}^{\infty} (a_k^2 + b_k^2)$ converges.

PROOF: If we set $A_0 = a_0/2$, $A_k = a_k$, $B_k = b_k$ in (6) of 12.4A, we obtain

$$J = \frac{1}{\pi} \int_{-\pi}^{\pi} (f - s_n)^2 = \frac{1}{\pi} \int_{-\pi}^{\pi} f^2 - \left[\frac{a_0^2}{2} + \sum_{k=1}^{n} (a_k^2 + b_k^2) \right]. \tag{2}$$

The integral on the left is nonnegative, and hence, the right-hand side is nonnegative. It follows that

$$\frac{a_0^2}{2} + \sum_{k=1}^{n} (a_k^2 + b_k^2) \leq \frac{1}{\pi} \int_{-\pi}^{\pi} f^2. \tag{3}$$

Since the right side of (3) is independent of n, we may let $n \to \infty$ and obtain (1).

We next show that if $f \in \mathcal{L}^2[-\pi, \pi]$, then the Fourier series of f converges to f in the metric for $\mathcal{L}^2[-\pi, \pi]$. That is,

12.4C. THEOREM. If $f \in \mathcal{L}^2[-\pi, \pi]$, then

$$\lim_{n \to \infty} \|s_n - f\|_2 = 0, \tag{1}$$

where

$$s_n(t) = \frac{a_0}{2} + \sum_{k=1}^{n} (a_k \cos kt + b_k \sin kt) \qquad (-\pi \leq t \leq \pi).$$

PROOF: By 11.9K, given $\epsilon > 0$ there exists a continuous function f^* such that $f^*(-\pi) = f^*(\pi)$, and such that

$$\|f - f^*\|_2 < \frac{\epsilon}{2}. \tag{2}$$

By 12.3C we know that $\{\sigma_n^*\}_{n=1}^{\infty}$ converges uniformly on $[-\pi, \pi]$ to f^*, where

$$\sigma_n^* = \frac{s_0^* + \cdots + s_{n-1}^*}{n}$$

and the s_k^* are the partial sums of the Fourier series for f^*.This uniform convergence implies that

$$\lim_{n \to \infty} \|\sigma_{n+1}^* - f^*\|_2 = 0,$$

and so, for some $N \in I$,

$$\|\sigma_{n+1}^* - f^*\|_2 < \frac{\epsilon}{2} \qquad (n \geq N). \tag{3}$$

From (2) and (3) we have

$$\|f - \sigma_{n+1}^*\|_2 < \epsilon \qquad (n \geq N). \tag{4}$$

But

$$\sigma_{n+1}^* = \frac{s_0^* + \cdots + s_n^*}{n+1}$$

is a trigonometric polynomial of degree n. Hence, by 12.4A,

$$\|f-s_n\|_2 \leqslant \|f-\sigma_{n+1}^*\|_2.$$

From (4) we thus have

$$\|f-s_n\|_2 < \epsilon \qquad (n \geqslant N),$$

which implies (1), and the proof is complete.

We emphasize that theorem 12.4C deals with convergence in the metric of $\mathcal{L}^2[-\pi,\pi]$. From (1) it does not necessarily follow that $\lim_{n\to\infty} s_n(x) = f(x)$ for any particular x.

Here are two important consequences of 12.4C.

12.4D. COROLLARY. If the Fourier coefficients a_k $(k=0,1,2,\ldots)$ and b_k $(k=1,2,\ldots)$ of $f\in\mathcal{L}^2[-\pi,\pi]$ are all 0, then $f=0$.*

PROOF: Since all the a_k and b_k are equal to 0, it follows that $s_n=0$ for every $n\in I$. The corollary then follows immediately from 12.4C.

12.4E. COROLLARY. If f and g are two functions in $\mathcal{L}^2[-\pi,\pi]$ that have the same Fourier coefficients, then $f=g$.†

PROOF: Under these hypotheses, $f-g$ is in $\mathcal{L}^2[-\pi,\pi]$ and the Fourier coefficients of $f-g$ are all 0. Apply 12.4D.

In 12.4B we showed that if a_k $(k=0,1,\ldots)$ and b_k $(k=1,2,\ldots)$ are the Fourier coefficients of $f\in\mathcal{L}^2[-\pi,\pi]$, then

$$\frac{a_0^2}{2} + \sum_{n=1}^{\infty}(a_k^2+b_k^2) < \infty. \qquad (*)$$

We now prove the converse—namely, that if $\{a_k\}_{k=1}^{\infty}$ and $\{b_k\}_{k=1}^{\infty}$ are sequences of numbers satisfying $(*)$, then there exists a function $f\in\mathcal{L}^2[-\pi,\pi]$ whose Fourier coefficients are the a_k and b_k. In somewhat different terminology, 12.4B states that if $f\in\mathcal{L}^2[-\pi,\pi]$, then $\{a_k\}_{k=0}^{\infty}$ and $\{b_k\}_{k=1}^{\infty}$ are in ℓ^2. What we now prove is that if $\langle a_k\}_{k=0}^{\infty}$ and $\{b_k\}_{k=1}^{\infty}$ are any sequences in ℓ^2, then they are the Fourier coefficients of some $f\in\mathcal{L}^2[-\pi,\pi]$. Note that the proof makes use of the completeness of $\mathcal{L}^2[-\pi,\pi]$ (theorem 11.9G). There is no corresponding theorem for Riemann-square-integrable functions.

12.4F. THEOREM.‡ If $\{a_k\}_{k=0}^{\infty}$ and $\{b_k\}_{k=1}^{\infty}$ are any sequences of real numbers such that

$$\frac{a_0^2}{2} + \sum_{n=1}^{\infty}(a_k^2+b_k^2) < \infty, \qquad (1)$$

then there exists $f\in\mathcal{L}^2[-\pi,\pi]$ whose Fourier coefficients are precisely the a_k and b_k.

PROOF: For each $n\in I$ define the trigonometric polynomial s_n as

$$s_n(t) = \frac{a_0}{2} + \sum_{k=1}^{n}(a_k\cos kt + b_k\sin kt) \qquad (-\pi\leqslant t\leqslant\pi). \qquad (2)$$

* That is, f is the zero elements of $\mathcal{L}^2[-\pi,\pi]$ or, equivalently, $f(x)=0$ for almost every x in $[-\pi,\pi]$.
† That is, $f(x)=g(x)$ for almost all x.
‡ This theorem is also known as the Riesz-Fischer theorem.

If $m<n$, then

$$\frac{1}{\pi}\int_{-\pi}^{\pi}(s_n-s_m)^2=\frac{1}{\pi}\int_{-\pi}^{\pi}\left[\sum_{k=m+1}^{n}(a_k\cos kt+b_k\sin kt)\right]^2 dt,$$

and application of 12.1A yields

$$\frac{1}{\pi}\int_{-\pi}^{\pi}(s_n-s_m)^2=\sum_{k=m+1}^{n}(a_k^2+b_k^2).$$

The left side is equal to $(1/\pi)\,\|s_n-s_m\|_2^2$. In view of (1) we see that $\{s_n\}_{n=1}^{\infty}$ is a Cauchy sequence in $\mathfrak{L}^2[-\pi,\pi]$. Hence, by the Riesz-Fischer theorem 11.9G, there exists a function $f\in\mathfrak{L}^2[1-\pi,\pi]$ such that

$$\lim_{n\to\infty}\|s_n-f\|_2=0.$$

For any $j=0,1,2,\ldots$ it follows that

$$\lim_{n\to\infty}\frac{1}{\pi}\int_{-\pi}^{\pi}s_n(t)\cos jt\,dt=\frac{1}{\pi}\int_{-\pi}^{\pi}f(t)\cos jt\,dt \tag{3}$$

(see exercise 2 of Section 11.9). But if $n\geqslant j$, it follows from (2) and 12.1A that

$$\frac{1}{\pi}\int_{-\pi}^{\pi}s_n(t)\cos jt\,dt=a_j.$$

Hence, from (3) we have

$$a_j=\frac{1}{\pi}\int_{-\pi}^{\pi}f(t)\cos jt\,dt \qquad (j=0,1,2,\ldots).$$

This shows that the a_j are the Fourier cosine coefficients of f. It may be shown similarly that the b_j are the Fourier sine coefficients of f, and this will complete the proof.

Exercises 12.4

1. If $f\in\mathfrak{L}^2[-\pi,\pi]$ and

$$f\sim\frac{a_0}{2}+\sum_{k=1}^{\infty}(a_k\cos kt+b_k\sin kt),$$

prove that

$$\frac{1}{\pi}\int_{-\pi}^{\pi}f^2=\frac{a_0^2}{2}+\sum_{k=1}^{\infty}(a_k^2+b_k^2).$$

(This is called Parseval's equality.) (*Hint*: Apply 12.4C to (2) of 12.4B.)

2. Apply the result of exercise 1 to

$$f(t)=t \qquad (-\pi\leqslant t\leqslant\pi).$$

Deduce that

$$\frac{1}{1^2}+\frac{1}{2^2}+\frac{1}{3^2}+\cdots+\frac{1}{n^2}+\cdots=\frac{\pi^2}{6}.$$

3. Apply Parseval's equality to the function after 12.1C to show that

$$\frac{1}{1^2}+\frac{1}{3^2}+\frac{1}{5^2}+\cdots=\frac{\pi^2}{8}.$$

4. Is the series

$$\sum_{n=1}^{\infty} \frac{\cos nx}{\sqrt{n}} \qquad (-\pi \leqslant x \leqslant \pi)$$

the Fourier series for some function in $\mathcal{L}^2[-\pi,\pi]$?

5. Is the series

$$\sum_{n=1} \frac{\cos nx}{n^{3/2}} \qquad (-\pi \leqslant x \leqslant \pi)$$

the Fourier series for some function in $\mathcal{L}^2[-\pi,\pi]$?

6. (a) If $f \in \mathcal{L}^2[-\pi,\pi]$ and

$$f \sim \frac{a_0}{2} + \sum_{k=1}^{\infty} (a_k \cos kt + b_k \sin kt),$$

prove that

$$\lim_{k\to\infty} a_k = 0 = \lim_{k\to\infty} b_k. \qquad (*)$$

(b) Prove that (*) holds even if we assume only that $f \in \mathcal{L}[-\pi,\pi]$.

12.5 CONVERGENGE OF FOURIER SERIES

We have not yet established anything about the convergence at a point x of the Fourier series of a function $f \in \mathcal{L}[-\pi,\pi]$. The conditions sufficient for convergence that we will give involve the existence of the left- and right-hand limits $\lim_{t\to x+} f(t)$ and $\lim_{t\to x-} f(t)$ *and* the existence of generalized right-and left-hand derivatives of f at x, which we now proceeed to define.

12.5A. For any real-valued function f on R^1 the definition of $\lim_{t\to x+} f(t)$ and $\lim_{t\to x-} f(t)$ was given in 4.1F. We recall from 6.9E the notations $f(x+)$ and $f(x-)$ defined as

$$f(x+) = \lim_{t\to x+} f(t); \quad f(x-) = \lim_{t\to x-} f(t),$$

provided the one-sided limits in question exist. [Thus f is continuous as x if and only if $f(x+)$ and $f(x-)$ both exist and are equal to $f(x)$.]

12.5B. DEFINITION. If $x \in R^1$ and if f is a real-valued function such that $f(x+)$ exists, we define $f_r'(x)$, the generalized right-hand derivative of f at x, as

$$f_r'(x) = \lim_{t\to x+} \frac{f(t) - f(x+)}{t-x}$$

provided the limit exists. Similarly, $f_l'(x)$, the generalized left-hand derivative of f at x, is defined as

$$f_l'(x) = \lim_{t\to x-} \frac{f(t) - f(x-)}{t-x}.$$

For example, suppose

$$f(t) = 1 + t \qquad (t > 1),$$
$$f(1) = 17,$$
$$f(t) = 3t^2 \qquad (t < 1).$$

Then $f(1+)=\lim_{t\to 1+} f(t)=\lim_{t\to 1+}(1+t)=2$. Moreover,

$$f_r'(1)=\lim_{t\to 1+}\frac{(1+t)-2}{t-1}=1.$$

Similarly, $f(1-)=3$ and $f_l'(1)=6$ (verify). In this example, both $f_r'(x)$ and $f_l'(x)$ exist at $x=1$. Note that f is not continuous on the right or on the left at $x=1$. Hence, f does not have an ordinary derivative on the left or on the right at $x=1$.

Before we can prove our theorem on the convergence of Fourier series, we need a well-known and important result. It states that the Fourier coefficients a_k, b_k of a Lebesgue integrable function must approach zero as $k\to\infty$.

12.5C. THEOREM (RIEMANN-LEBESGUE). If $f\in\mathcal{L}[-\pi,\pi]$ and if $\{a_k\}_{k=0}^{\infty}$ and $\{b_k\}_{k=1}^{\infty}$ are the Fourier coefficients of f, then

$$\lim_{k\to\infty} a_k=\lim_{k\to\infty}\frac{1}{\pi}\int_{-\pi}^{\pi} f(t)\cos kt\,dt=0 \tag{1}$$

and

$$\lim_{k\to\infty} b_k=\lim_{k\to\infty}\frac{1}{\pi}\int_{-\pi}^{\pi} f(t)\sin kt\,dt=0. \tag{2}$$

PROOF: Fix $\epsilon>0$. From the definition of $\int_{-\pi}^{\pi} f$ it is easy to show that there is a bounded measurable function g on $[-\pi,\pi]$ such that

$$\int_{-\pi}^{\pi}|f(t)-g(t)|\,dt<\frac{\epsilon}{2\pi}. \tag{3}$$

(See exercise 9, Section 11.7.) Now $g\in\mathcal{L}^2[-\pi,\pi]$ since g is bounded and measurable. Hence, by 12.4B,

$$\sum_{k=1}^{\infty}(A_k^2+B_k^2)<\infty, \tag{4}$$

where

$$A_k=\frac{1}{\pi}\int_{-\pi}^{\pi} g(t)\cos kt\,dt,\quad B_k=\frac{1}{\pi}\int_{-\pi}^{\pi} g(t)\sin kt\,dt.$$

From (4) it follows that $\sum_{k=1}^{\infty}A_k^2<\infty$ and hence, that $\lim_{k\to\infty}A_k=0$. Thus there exists $N\in I$ such that

$$|A_k|<\frac{\epsilon}{2}\qquad (k\geqslant N). \tag{5}$$

But, for any k,

$$|a_k-A_k|=\left|\frac{1}{\pi}\int_{-\pi}^{\pi}[f(t)-g(t)]\cos kt\,dt\right|\leqslant\frac{1}{\pi}\int_{-\pi}^{\pi}|f(t)-g(t)|\,dt,$$

and hence, by (2),

$$|a_k-A_k|<\frac{\epsilon}{2}\qquad (k\in I). \tag{6}$$

Now

$$a_k=(a_k-A_k)+A_k,$$
$$|a_k|\leqslant|a_k-A_k|+|A_k|,$$

and hence, using (5) and (6) we obtain

$$|a_k|<\epsilon\qquad (k\geqslant N).$$

This proves (1). Equation (2) may be established in the same manner to complete the proof.

As an easy consequence we have the following corollary.

12.5D. COROLLARY. If $\varphi \in \mathcal{L}[0,\pi]$, then

$$\lim_{n\to\infty} \int_0^{\pi} \varphi(t)\sin(k+\tfrac{1}{2})t\,dt = 0.$$

PROOF: Define $\varphi(t)=0$ for $-\pi \leqslant t < 0$. Then $\varphi \in \mathcal{L}[-\pi,\pi]$. Since

$$\sin(k+\tfrac{1}{2})t = \sin kt \cos\frac{t}{2} + \cos kt \sin\frac{t}{2}$$

we have

$$\int_0^{\pi} \varphi(t)\sin(k+\tfrac{1}{2})t\,dt = \int_{-\pi}^{\pi}\left[\varphi(t)\cos\frac{t}{2}\right]\sin kt\,dt + \int_{-\pi}^{\pi}\left[\varphi(t)\sin\frac{t}{2}\right]\cos kt\,dt.$$

But $\varphi(t)\cos(t/2)$ and $\varphi(t)\sin(t/2)$ are functions in $\mathcal{L}[-\pi,\pi]$. Hence, by 12.5C, both integrals on the right approach 0 as $k\to\infty$. The corollary follows.

Here is our theorem on the convergence of Fourier series.

12.5E. THEOREM. Let $f\in\mathcal{L}[-\pi,\pi]$, and let x be any point in $[-\pi,\pi]$. If $f(x+)$ and $f(x-)$ exist, if*

$$f(x) = \frac{f(x+)+f(x-)}{2}, \tag{1}$$

and if $f_r'(x)$ and $f_l'(x)$ exist, then the Fourier series for f at x will converge to $f(x)$.

PROOF: According to what we showed in 12.2A, to prove that the Fourier series for f will converge to $f(x)$ we must show that

$$\lim_{n\to\infty} \frac{2}{\pi}\int_0^{\pi}\left[\frac{f(x+t)+f(x-t)}{2} - f(x)\right]D_n(t)\,dt = 0$$

where

$$D_n(t) = \frac{\sin(n+\frac{1}{2})t}{2\sin(t/2)} \qquad (-\infty < t < \infty).$$

In view of (1), we must show

$$\lim_{n\to\infty}\frac{1}{\pi}\int_0^{\pi}\{[f(x+t)-f(x+)]+[f(x-t)-f(x-)]\}D_n(t)\,dt = 0,$$

or

$$\lim_{n\to\infty}\frac{1}{\pi}\int_0^{\pi}\varphi(t)\sin(n+\tfrac{1}{2})t\,dt = 0 \tag{2}$$

where

$$\varphi(t) = \{[f(x+t)-f(x+)]+[f(x-t)-f(x-)]\}\cdot\frac{1}{2\sin(t/2)} \qquad (0<t\leqslant\pi). \tag{3}$$

* If $x=\pi$, then $f(x+)=f(\pi+)$ is computed using the values $f(t)$ for $t>\pi$ obtained by our extension of f via the equation $f(u+2\pi)=f(u)$. A similar remark applies to $x=-\pi$.
Note that (1) will hold if f is continuous at x.

Writing $\varphi(t)$ in the form

$$\varphi(t)=\left[\frac{f(x+t)-f(x+)}{t}+\frac{f(x-t)-f(x-)}{t}\right]\cdot\frac{t}{2\sin(t/2)}\qquad(0<t\leqslant\pi),$$

we see that $\lim_{t\to 0+}\varphi(t)=f_r'(x)-f_l'(x)$, and so φ is "well behaved" near $t=0$. More precisely, φ is bounded on $(0,\delta]$ for some $\delta>0$. But from (3) it is clear that φ is Lebesgue integrable on $[\delta,\pi]$ since $f\in\mathfrak{L}[-\pi,\pi]$, and $1/[2\sin(t/2)]$ is bounded on $[\delta,\pi]$. Hence, φ must be Lebesgue integrable on $[0,\pi]$. Corollary 12.5D thus implies that (2) holds, which is what we wished to show.

Actually we have proved a little more than was stated in 12.5E. For 12.5E states that if $f(x+),f(x-),f_r'(x),f_l'(x)$ exist, then

$$\frac{a_0}{2}+\sum_{k=1}^{\infty}(a_k\cos kx+b_k\sin kx=\frac{f(x+)+f(x-)}{2}\qquad(*)$$

provided that $f(x)=[f(x+)+f(x-)]/2$. However, the value of f at the single point x cannot affect the values of a_k and b_k, and hence, cannot affect the left side of $(*)$. Hence, $(*)$ must hold even if $f(x)\neq[f(x+)+f(x-)]/2$. That is,

12.5F. COROLLARY. Let $f\in\mathfrak{L}[-\pi,\pi]$ and let x be any point in $[-\pi,\pi]$. If $f(x+),f(x-),f_r'(x),f_l'(x)$ exist, then the Fourier series for f at x converges to the value $[f(x+)+f(x-)]/2$. [That is, $(*)$ holds.]

For example, let

$$f(x)=0\qquad(-\pi\leqslant x<0),$$
$$f(x)=1\qquad(0\leqslant x\leqslant\pi),$$

and suppose $f(u+2\pi)=f(u)$. Then $f_r'(\pi)$ and $f_l'(\pi)$ exist and $f(\pi+)=0,f(\pi-)=1$. According to 12.5F, The Fourier series for f at $x=\pi$ will converge to $(0+1)/2=\frac{1}{2}$. Indeed, at the end of Section 12.1 we saw that the Fourier series for f is

$$\frac{1}{2}+\frac{2}{\pi}\left[\frac{\sin x}{1}+\frac{\sin 3x}{3}+\frac{\sin 5x}{5}+\cdots\right],$$

which *does* converge to $\frac{1}{2}$ when $x=\pi$.

Note also that if $x=\pi/2$, then the series becomes

$$\frac{1}{2}+\frac{2}{\pi}\left[1-\frac{1}{3}+\frac{1}{5}-\cdots\right].$$

By 12.5F, this series must converge to $[f(\pi/2+)+f(\pi/2-)]/2=1$. That is,

$$1=\frac{1}{2}+\frac{2}{\pi}\left[1-\frac{1}{3}+\frac{1}{5}-\cdots\right],$$

or

$$\frac{\pi}{4}=1-\frac{1}{3}+\frac{1}{5}-\frac{1}{7}+\cdots.$$

12.5G. On the subject of convergence and summability of Fourier series, there are theorems much stronger than the ones we have presented. These theorems require results on the Lebesgue integral that we have not developed. We refer the reader to G. H. Hardy and W. W. Rogosinski *Fourier Series*, Cambridge, 1944.

12.5H. We emphasize that continuity at a point s is *not* sufficient for the Fourier series of a function $f \in \mathfrak{L}[-\pi,\pi]$ to converge to $f(x)$. Indeed, there exists a function f which is continuous on $[-\pi,\pi]$ but whose Fourier series *diverges* at each point in a *dense* subset of $[-\pi,\pi]$! See Section 12.7.

Exercises 12.5

1. Suppose $f,g \in \mathfrak{L}[-\pi,\pi]$ and suppose there exists $\delta > 0$ such that

$$f(t) = g(t) \qquad (x-\delta < t < x+\delta)$$

for some $x \in [-\pi,\pi]$.

Show that the Fourier series for f and g either both converge at x or both diverge at x.

2. Show that the Fourier series for

$$f(t) = \frac{t^2}{4} \qquad (-\pi \leqslant t \leqslant \pi)$$

is

$$\frac{\pi^2}{12} - \sum_{n=1}^{\infty} \frac{(-1)^{n+1}\cos nt}{n^2}.$$

(a) Use 12.5E to show that the Fourier series at $t=0$ converges to $f(0)$. Deduce that

$$\frac{\pi^2}{12} = \frac{1}{1^2} - \frac{1}{2^2} + \frac{1}{3^2} - \frac{1}{4^2} + \cdots.$$

(b) Use Parseval's equality to show that

$$\sum_{n=1}^{\infty} \frac{1}{n^4} = \frac{\pi^4}{90}.$$

3. (a) Use the Fourier series for $f(t) = |t| (-\pi \leqslant t \leqslant \pi)$ to find the sum of

$$\frac{1}{1^2} + \frac{1}{3^2} + \frac{1}{5^2} + \frac{1}{7^2} + \cdots.$$

(b) Calculate the sum of

$$\frac{1}{1^4} + \frac{1}{3^4} + \frac{1}{5^4} + \cdots.$$

12.6 ORTHONORMAL EXPANSIONS IN $\mathfrak{L}^2[a,b]$

In this section we put some of the results on the $\mathfrak{L}^2[-\pi,\pi]$ theory of Fourier series in a more general setting. We first introduce the notion of the inner product of two functions in $\mathfrak{L}^2[a,b]$. This inner product has many of the properties of the inner (or dot) product of two vectors.

12.6A. DEFINITION. If $f,g \in \mathfrak{L}^2[a,b]$, then we define (f,g), called the inner product of f and g, as

$$(f,g) = \int_a^b f(t)\,g(t)\,dt.$$

Thus $\|f\|_2=\sqrt{(f,f)}$ for any $f\in\mathcal{L}[a,b]$. (This corresponds to the fact that the square root of the inner product of a vector with itself is the length of the vector.) It is easy to verify the following result.

12.6B. THEOREM. The inner product has the following properties:

$$(f,g)=(g,f) \qquad (f,g\in\mathcal{L}^2[a,b]), \tag{1}$$

$$(\lambda f,g)=\lambda(f,g) \qquad (\lambda\in R; f,g\in\mathcal{L}^2[a,b]), \tag{2}$$

$$(f+g,h)=(f,h)+(g,h) \qquad (f,g,h\in\mathcal{L}^2[a,b]), \tag{3}$$

$$(f,f)\geqslant 0 \qquad (f\in\mathcal{L}^2[a,b]), \tag{4}$$

$$(f,f)=0 \quad \text{if and only if} \quad f=0.^* \tag{5}$$

We next define orthogonality in the same manner as for vectors.

12.6C. DEFINITION. If $f,g\in\mathcal{L}^2[a,b]$, then we say that f and g are orthogonal if $(f,g)=0$.

Thus from 12.1A we see that the functions $\cos kx$ and $\cos nx$ are orthogonal on $[-\pi,\pi]$ if $k\neq n$. For

$$(\cos kx,\cos nx)=\int_{-\pi}^{\pi}\cos kx\cos nx\,dx=0.$$

The object in $\mathcal{L}^2[a,b]$ that corresponds to a set of unit vectors that are mutually orthogonal is called an orthonormal family.

12.6D. DEFINITION. The countable family $\Phi=\{\varphi_1,\varphi_2,\ldots\}$ of functions in $\mathcal{L}^2[a,b]$ is called an orthonormal family in $\mathcal{L}^2[a,b]$ if

$$(\varphi_k,\varphi_n)=\int_a^b\varphi_k\varphi_n=0 \quad (k,n=1,2,\ldots;k\neq n),$$

$$(\varphi_n,\varphi_n)=\int_a^b\varphi_n^2=1 \quad (n=1,2,\ldots).$$

Thus if $\|\varphi_n\|_2=1$ for all n, and if every two distinct members of Φ are orthogonal, then Φ is an orthonormal family.

For example, if T denotes the family

$$T=\left\{\frac{1}{\sqrt{2\pi}},\frac{\cos x}{\sqrt{\pi}},\frac{\sin x}{\sqrt{\pi}},\frac{\cos 2x}{\sqrt{\pi}},\frac{\sin 2x}{\sqrt{\pi}},\frac{\cos 3x}{\sqrt{\pi}},\ldots\right\},$$

then theorem 12.1A shows that T is an orthonormal family in $\mathcal{L}^2[-\pi,\pi]$.

As another example, consider the family $L=\{L_0,L_1,L_2,\ldots\}$ of Legendre functions. Here

$$L_n(x)=\frac{2n+1}{2}\cdot\frac{1}{2^n n!}\frac{d^n}{dx^n}(x^2-1)^n \qquad (-1\leqslant x\leqslant 1; n=0,1,2,\ldots).$$

It may be shown that L is an orthonormal family in $\mathcal{L}^2[-1,1]$.

Another well-known example is the family $R=\{R_0,R_1,\ldots\}$ of Rademacher functions. Here

$$R_n(x)=\operatorname{sgn}(\sin 2^n\pi x) \qquad (0\leqslant x\leqslant 1; n=0,1,2,\ldots),$$

* Remember that $f=0$ means $f(x)=0$ for almost all x in $[a,b]$.

where the sgn function is defined as

$$\operatorname{sgn} a = 1 \qquad (a>0),$$
$$\operatorname{sgn} a = -1 \qquad (a<0),$$
$$\operatorname{sgn} 0 = 0.$$

For example, $R_2(x)$ takes the value $+1$ on the intervals $(0,\frac{1}{4})$ and $(\frac{2}{4},\frac{3}{4})$ and takes the value -1 on the intervals $(\frac{1}{4},\frac{2}{4})$ and $(\frac{3}{4},1)$, while $R_2(0)=R_2(\frac{1}{4})=R_2(\frac{2}{4})=R_2(\frac{3}{4})=R_2(1)=0$. It may be shown that R is an orthonormal family in $\mathcal{L}^2[0,1]$.

12.6E. Let $\Phi=\{\varphi_1,\varphi_2,\dots\}$ be an orthonormal family in $\mathcal{L}^2[a,b]$. Suppose that a function $f\in\mathcal{L}^2[a,b]$ may be represented as

$$f=\sum_{k=1}^{\infty} c_k\varphi_k = c_1\varphi_1 + c_2\varphi_2 + \cdots$$

where $c_1,c_2,\dots$ are real numbers. The series must be intrepreted as $\lim_{n\to\infty}\sum_{k=1}^{n} c_k\varphi_k$ where the limit is taken in the metric for $\mathcal{L}^2[a,b]$.

Let us see formally how to compute the coefficients $c_1,c_2,\dots,$ in terms of f. Taking the inner product with φ_n we have

$$(f,\varphi_n)=\sum_{k=1}^{\infty} c_k(\varphi_k,\varphi_n).$$

But since Φ is an orthonormal family, we have $(\varphi_k,\varphi_n)=0$ for $k\neq n$, and $(\varphi_k,\varphi_n)=1$ for $k=n$. Hence,

$$(f,\varphi_n)=c_n.$$

This leads us to the following definition.

DEFINITION. Let $\Phi=\{\varphi_1,\varphi_2,\dots\}$be an orthonormal family in $\mathcal{L}^2[a,b]$. If $f\in\mathcal{L}^2[a,b]$ and if

$$c_k=(f,\varphi_k) \qquad (k\in I),$$

we call the c_k the generalized Fourier coefficients of f. The series $\sum_{k=1}^{\infty} c_k\varphi_k$ is called the generalized Fourier series for f, and we write

$$f\sim\sum_{k=1}^{\infty} c_k\varphi_k.$$

For example, if $[a,b]=[-\pi,\pi]$ and we take

$$\Phi=T=\left\{\frac{1}{\sqrt{2\pi}},\frac{\cos x}{\sqrt{\pi}},\frac{\sin x}{\sqrt{\pi}},\dots\right\},$$

then the generalized Fourier series for $f\in\mathcal{L}^2[-\pi,\pi]$ is the series

$$\left(f,\frac{1}{\sqrt{2\pi}}\right)\cdot\frac{1}{\sqrt{2\pi}}+\left(f,\frac{\cos x}{\sqrt{\pi}}\right)\frac{\cos x}{\sqrt{\pi}}+\left(f,\frac{\sin x}{\sqrt{\pi}}\right)\frac{\sin x}{\sqrt{\pi}}+\cdots$$

or

$$\left(\int_{-\pi}^{\pi} f(x)\cdot\frac{1}{\sqrt{2\pi}}\,dx\right)\frac{1}{\sqrt{2\pi}}+\left(\int_{-\pi}^{\pi} f(x)\frac{\cos x}{\sqrt{\pi}}\,dx\right)\frac{\cos x}{\sqrt{\pi}}$$
$$+\left(\int_{-\pi}^{\pi} f(x)\frac{\sin x}{\sqrt{\pi}}\,dx\right)\frac{\sin x}{\sqrt{\pi}}+\cdots$$

or

$$\frac{a_0}{2} + a_1 \cos x + b_1 \sin x + \cdots.$$

That is, in this special case where $\Phi = T$, the generalized Fourier series for $f \in \mathcal{L}^2[-\pi,\pi]$ is precisely the (ordinary) Fourier series defined in 12.1C. Thus the preceding definition is a bona-fide generalization iof 12.1C. Note, however, that with respect to T the generalized Fourier coefficients of $f \in \mathcal{L}^2[-\pi,\pi]$ are

$$a_0\sqrt{\pi/2}\ , a_1\sqrt{\pi}\ , b_1\sqrt{\pi}\ , a_2\sqrt{\pi}\ , b_2\sqrt{\pi}\ , \ldots.$$

12.6F. We will now generalize 12.4B. If $\Phi = \{\varphi_1, \varphi_2, \ldots\}$ is an orthonormal family in $\mathcal{L}^2[a,b]$, then a function T_n of the form

$$T_n = d_1\varphi_1 + d_2\varphi_2 + \cdots + d_n\varphi_n \tag{1}$$

(where $d_1, \ldots, d_n$ are real numbers) will be called a Φ polynomial of degree n. We first show that if $f \in \mathcal{L}^2[a,b]$, then the Φ polynomial that is closest to f (in the metric of $\mathcal{L}^2[a,b]$), is s_n—the nth partial sum of the generalized Fourier series f. This generalizes 12.4A.

THEOREM. Let $f \in \mathcal{L}[a,b]$ and let T_n be any Φ polynomial of degree n. Then

$$\|f - T_n\|_2 \geqslant \|f - s_n\|_2 \tag{2}$$

where $s_n = c_1\varphi_1 + \cdots + c_n\varphi_n$ and the c_k are the Fourier coefficients of f,

$$c_k = (f, \varphi_k) \qquad (k = 1, \ldots, n).$$

PROOF: For any T_n let $J = \|f - T_n\|_2^2$. We must prove that J is a minimum when $T_n = s_n$. We have, using 12.6B,

$$J = (f - T_n, f - T_n) = (f,f) - 2(f, T_n) + (T_n, T_n). \tag{3}$$

If T_n is as in (1), then

$$(f, T_n) = (f, d_1\varphi_1 + \cdots + d_n\varphi_n) = d_1(f, \varphi_1) + \cdots + d_n(f, \varphi_n),$$

and so

$$(f, T_n) = \sum_{k=1}^{n} d_k c_k. \tag{4}$$

Also

$$\begin{aligned}(T_n, T_n) &= (d_1\varphi_1 + \cdots + d_n\varphi_n, d_1\varphi_1 + \cdots + d_n\varphi_n)\\ &= d_1(\varphi_1, d_1\varphi_1 + \cdots + d_n\varphi_n) + \cdots + d_n(\varphi_n, d_1\varphi_1 + \cdots + d_n\varphi_n).\end{aligned}$$

Since $(\varphi_j, \varphi_k) = 0$ for $j \neq k$, and $(\varphi_k, \varphi_k) = 1$, we obtain

$$(T_n, T_n) = \sum_{k=1}^{n} d_k^2. \tag{5}$$

Substituting (4) and (5) into (3) we have

$$J = (f,f) - 2\sum_{k=1}^{n} d_k c_k + \sum_{k=1}^{n} d_k^2.$$

Adding and subtracting $\Sigma_{k=1}^{n} c_k^2$ we obtain

$$J = (f,f) - \sum_{k=1}^{n} c_k^2 + \sum_{k=1}^{n} (d_k - c_k)^2. \tag{6}$$

From (6) it is clear that J will be a minimum if $d_k = c_k (k=1,\ldots,n)$. That is, J will be a minimum when $T_n = s_n$, which is what we wished to show.

As an immediate corollary we have the following inequality.

12.6G. BESSEL'S INEQUALITY FOR GENERALIZED FOURIER SERIES. Let $\Phi = \{\varphi_1, \varphi_2, \ldots\}$ be an orthonormal family in $\mathfrak{L}^2[a,b]$. If $f \in \mathfrak{L}^2[a,b]$ and if

$$c_k = (f, \varphi_k) \qquad (k \in I),$$

then

$$\sum_{k=1}^{\infty} c_k^2 \leqslant \|f\|_2^2.$$

PROOF: If we set $d_k = c_k$ $(k=1,\ldots,n)$ in (6) of 12.6F, we obtain

$$J = \|f - s_n\|_2^2 = (f,f) - \sum_{k=1}^{n} c_k^2. \tag{$*$}$$

Since $\|f - s_n\|_2^2 \geqslant 0$, this implies

$$\sum_{k=1}^{n} c_k^2 \leqslant (f,f) = \|f\|_2^2.$$

The conclusion follows on letting $n \to \infty$.

If $[a,b] = [-\pi, \pi]$ and

$$\Phi = T = \left\{ \frac{1}{\sqrt{2\pi}}, \frac{\cos x}{\sqrt{\pi}}, \frac{\sin x}{\sqrt{\pi}}, \ldots \right\},$$

then we have shown in 12.6E that the generalized Fourier coefficients of $f \in \mathfrak{L}^2[-\pi, \pi]$ become

$$a_0 \sqrt{\frac{\pi}{2}}, a_1 \sqrt{\pi}, b_1 \sqrt{\pi}, a_2 \sqrt{\pi}, b_2 \sqrt{\pi}, \ldots.$$

By 12.6G, it follows that the sum of the squares of these coefficients is less than or equal to $\|f\|_2^2$. That is,

$$\frac{\pi}{2} a_0^2 + \pi \sum_{k=1}^{\infty} (a_k^2 + b_k^2) \leqslant \int_{-\pi}^{\pi} f^2.$$

But this is precisely 12.4B!

Theorem 12.4F can be generalized as follows.

12.6H. THEOREM. Let $\Phi = \{\varphi_1, \varphi_2, \ldots\}$ be an orthonormal family in $\mathfrak{L}^2[a,b]$. If $\{c_k\}_{k=1}^{\infty}$ is any sequence of numbers such that

$$\sum_{k=1}^{\infty} c_k^2 < \infty, \tag{1}$$

then there exists $f \in \mathfrak{L}^2[a,b]$ such that the generalized Fourier coefficients of f are precisely the c_k—that is,

$$c_k = (f, \varphi_k) \qquad (k \in I).$$

Moreover, for this f, $\lim_{n\to\infty}\|s_n - f\|_2 = 0$, where

$$s_n = \sum_{k=1}^{n} c_k \varphi_k.$$

PROOF: If $m < n$, then the fact that Φ is orthonormal shows that

$$\|s_n - s_m\|_2^2 = \left\| \sum_{k=m+1}^{n} c_k \varphi_k \right\|_2^2 = \sum_{k=m+1}^{n} c_k^2.$$

This and (1) show that $\{s_n\}_{n=1}^{\infty}$ is a Cauchy sequence in $\mathcal{L}^2[-\pi,\pi]$. Hence, by 11.9G, there exists $f \in \mathcal{L}^2[a,b]$ such that

$$\lim_{n\to\infty} \|s_n - f\|_2 = 0.$$

By exercise 2 of Section 11.9 it follows that

$$\lim_{n\to\infty} (s_n, \varphi_k) = (f, \varphi_k) \qquad (k \in I). \tag{2}$$

But, if $n > k$, then $(s_n, \varphi_k) = (c_1\varphi_1 + \cdots + c_k\varphi_k + \cdots + c_n\varphi_n, \varphi_k) = c_k$. Hence, from (2), we have

$$c_k = (f, \varphi_k) \qquad (k \in I),$$

and the theorem is proved.

We have not generalized theorem 12.4C to arbitrary orthonormal families for the simple reason that it is not true for arbitrary orthonormal families. If Φ is an orthonormal family in $\mathcal{L}^2[a,b]$, then Φ must have a special property to ensure that the generalized Fourier series of *every* $f \in \mathcal{L}^2[a,b]$ will converge to f in the metric of $\mathcal{L}^2[a,b]$. This property, which we now define, states that Φ must not be a subfamily of a larger orthonormal family.

12.6I. DEFINITION. If $\Phi = \{\varphi_1, \varphi_2, \ldots\}$ is an orthonormal family in $\mathcal{L}^2[a,b]$, then Φ is said to be complete* if the only function $h \in \mathcal{L}^2[a,b]$ that is orthogonal to all the φ_k $(k = 1, 2, \ldots)$ is $h = 0$.

That is, Φ is complete if

$$(h, \varphi_k) = 0 \qquad (k \in I)$$

implies $h = 0$. In other words, if Φ is a complete orthonormal family in $\mathcal{L}^2[a,b]$, then the only function in $\mathcal{L}^2[a,b]$ whose generalized Fourier coefficients are all zero is the function that is equal to zero (almost everywhere).

From 12.4D it follows that the family

$$T = \left\{ \frac{1}{\sqrt{2\pi}}, \frac{\cos x}{\sqrt{\pi}}, \frac{\sin x}{\sqrt{\pi}}, \ldots \right\}$$

is complete. It may be shown that the family L of Legendre functions is also complete (in $\mathcal{L}^2[-1,1]$) but that the family R of Rademacher functions is *not* complete (in $\mathcal{L}^2[0,1]$).

For *complete* orthonormal families we may generalize 12.4C. This generalization is a corollary to the following result.

12.6J. THEOREM. Let $\Phi = \{\varphi_1, \varphi_2, \ldots\}$ be an orthonormal family in $\mathcal{L}^2[a,b]$. Then if Φ

* This use of the word "complete" as applied to orthonormal families has nothing to do with "complete" as applied to metric spaces in 6.4A.

has any one of the following properties, it has them all.

(a) Φ is complete.
(b) The set of all Φ polynomials (of all degrees) is dense in $\mathfrak{L}^2[a,b]$.
(c) For any $f \in \mathfrak{L}^2[a,b]$ the generalized Fourier series for f converges to f in the metric of $\mathfrak{L}^2[a,b]$. [That is,

$$\lim_{n\to\infty} \|s_n - f\|_2 = 0$$

where $s_n = \sum_{k=1}^{n} c_k \varphi_k$ and $c_k = (f, \varphi_k)$.]
(d) If $f \in \mathfrak{L}^2[a,b]$, then

$$\|f\|_2^2 = \sum_{k=1}^{\infty} c_k^2$$

where $c_k = (f, \varphi_k)$. (This is called Parseval's equality.)

PROOF: I. (a) implies (b).
Assume (a) is true. For any $f \in \mathfrak{L}^2[a,b]$ let $c_k = (f, \varphi_k)$ $(k \in I)$. Then

$$\sum_{k=1}^{\infty} c_k^2 < \infty,$$

by 12.6G. Theorem 12.6H then shows that there exists $g \in \mathfrak{L}^2[a,b]$ such that

$$\lim_{n\to\infty} \|s_n - g\|_2 = 0 \tag{1}$$

where $s_n = \sum_{k=1}^{n} c_k \varphi_k$, and such that

$$c_k = (g, \varphi_k) \qquad (k \in I).$$

Thus $c_k = (f, \varphi_k) = (g, \varphi_k)$ and so

$$(f - g, \varphi_k) = 0 \qquad (k \in I).$$

Since, by assumption, Φ is complete, it follows that $f - g = 0$. Hence, from (1), $\lim_{n\to\infty} \|s_n - f\|_2 = 0$. This shows that f is a limit point of the set of all Φ polynomials, which implies (b).

II. (b) implies (c).

If (b) is true, then for any $f \in \mathfrak{L}^2[a,b]$ and any $\epsilon > 0$ there exists a Φ polynomial T_N of degree N such that $\|T_N - \varphi\|_2 < \epsilon$. From 12.6F it follows that

$$\|s_N - f\|_2 < \epsilon.$$

If $n \geqslant N$, the $s_N = s_N + 0 \cdot \varphi_{N+1} + \cdots + 0 \cdot \varphi_n$ is also a Φ polynomial of degree n. Hence, by 12.6F,

$$\|s_n - f\|_2 \leqslant \|s_N - f\|_2 \qquad (n \geqslant N).$$

We thus have $\|s_n - f\| < \epsilon$ $(n \geqslant N)$, which proves that (c) holds.

III. (c) implies (d).

Suppose (c) is true. If $f \in \mathfrak{L}^2[a,b]$, then (c) implies

$$\lim_{n\to\infty} \|s_n - f\|_2^2 = 0. \tag{2}$$

But, by (*) of 12.6G, we have

$$\|f - s_n\|_2^2 = \|f\|_2^2 - \sum_{k=1}^{n} c_k^2.$$

Letting $n \to \infty$ and using (2) we prove (d).

IV. (d) implies (a).

Suppose (d) is true. If h is any function in $\mathcal{L}^2[a,b]$ such that

$$(h,\varphi_k)=0 \qquad (k\in I),$$

then, by (d), we have $\|h\|_2^2=\sum_{k=1}^{\infty}(h,\varphi_k)^2=0$. Hence, $h=0$. This proves that (a) is true.
The proof of the theorem is now complete.

Exercises 12.6

1. Calculate the Legendre functions L_1, L_2, L_3 and show that they are orthogonal to one another on $[-1,1]$ and that each has norm equal to 1.
2. Do the same for the Rademacher functions R_1, R_2, R_3.
3. Let $\Phi=\{\varphi_1,\varphi_2,\dots\}$ be a complete orthonormal family in $\mathcal{L}^2[a,b]$. Define the function A from $\mathcal{L}^2[a,b]$ into ℓ^2 as

$$A(f)=\{c_k\}_{k=1}^{\infty} \qquad (f\in\mathcal{L}^2[a,b]),$$

where $c_k=(f,\varphi_k)$ $(k\in I)$.
   (a) Show that A is 1–1.
   (b) Show that A maps $\mathcal{L}^2[a,b]$ *onto* ℓ^2.
   (c) If $f\in\mathcal{L}^2[a,b]$, show that $\|f\|_2=\|A(f)\|_2$ (where the second norm refers, of course, to ℓ^2.)

12.7 NOTES AND ADDITIONAL EXERCISES ON CHAPTERS 11 AND 12

I. A continuous function whose Fourier series diverges at a point

12.7A We will need to use notation a bit more complicated than in previous sections. Suppose $f\in\mathcal{L}[-\pi,\pi]$ and

$$f\sim\frac{a_0}{2}+\sum_{k=1}^{\infty}(a_k\cos kt+b_k\sin kt). \qquad (*)$$

We let

$$s_n[f;x]=\frac{a_0}{2}+\sum_{k=1}^{n}(a_k\cos kx+b_k\sin kx)$$

so that $s_n[f;x]$ is precisely what we have been previously denoting as $s_n(x)$. Similarly, let

$$\sigma_n[f;x]=\frac{1}{n}\sum_{k=0}^{n-1}s_k[f;x].$$

Thus

$$s_n[f;x]=\frac{1}{\pi}\int_{-\pi}^{\pi}f(u)D_n(x-u)\,du,$$

$$\sigma_n[f;x]=\frac{1}{\pi}\int_{-\pi}^{\pi}f(u)K_n(x-u)\,du$$

where D_n and K_n are defined in Section 12.2.

EXERCISE: If f is as in (*), show that

$$\sigma_n[f;x]=\frac{a_0}{2}+\sum_{k=1}^{n}\left(1-\frac{k}{n}\right)\quad(a_k\cos kx+b_k\sin kx).$$

12.7B Here is the result on nonconvergence:

THEOREM. The exists $f\in C[-\pi,\pi]$ whose Fourier series diverges at 0. This will follow from a sequence of lemmas.

EXERCISE: Prove in detail the following lemmas.

LEMMA 1. The sequence $\{\int_0^\pi|D_n(t)|\,dt\}_{n=1}^{\infty}$ is unbounded. Specifically,

$$\int_0^\pi|D_n(t)|\,dt>\frac{2}{\pi}\log n\qquad(n\in I).$$

SKETCH OF PROOF:

$$\int_0^\pi|D_n(t)|\,dt\geqslant\int_0^\pi\frac{|\sin(n+\frac{1}{2})t|}{t}\,dt$$

$$=\int_0^{\pi(n+\frac{1}{2})}\frac{|\sin u|}{u}\,du>\sum_{k=1}^{n}\frac{1}{k\pi}\int_{(k-1)\pi}^{k\pi}|\sin u|\,du>\frac{2}{\pi}\log n.$$

This behavior of the D_n, in contrast with that of the K_n, is what sometimes causes convergence to fail, whereas $(C,1)$ summability for continuous functions always succeeds.

LEMMA 2. For each $n\in I$ there exists $g_n\in C[-\pi,\pi]$ such that

$$\|g_n\|=\max_{x\in[-\pi,\pi]}|g_n(x)|=1$$

and

$$|s_n[g_n;0]|>\frac{1}{\pi^2}\log n.$$

SKETCH OF PROOF: Fix n. Let

$$h(x)=1\quad\text{if}\quad D_n(x)\geqslant 0,$$

$$h(x)=-1\quad\text{if}\quad D_n(x)<0.$$

Then h is a "step function" and $\int_{-\pi}^{\pi}h(t)D_n(t)\,dt=\int_{-\pi}^{\pi}|D_n(t)|\,dt$. Let $g_n=h$ except in small intervals around the discontinuities of h. In these intervals let g_n be linear so as to make g_n continuous. If the total measure of these intervals is ϵ, then

$$\left|\int_{-\pi}^{\pi}g_n(t)D_n(t)\,dt-\int_{-\pi}^{\pi}h_n(t)D_n(t)\,dt\right|<\epsilon(n+\tfrac{1}{2})$$

so that, if ϵ is sufficiently small,

$$\int_{-\pi}^{\pi}g_n(t)D_n(t)>\frac{1}{\pi}\log n.$$

Finally, note that

$$s_n[g_n;0]=\frac{1}{\pi}\int_{\pi}^{\pi} g_n(t)\,D_n(t)\,dt.$$

LEMMA 3. For each $n\in I$ there exists a trigonometric polynomial ϕ_n of degree n^2 such that

$$\|\phi_n\| = \max_{x\in[-\pi,\pi]} |\phi_n(x)| \leqslant 1$$

and

$$s_n[\phi_n;0] > \frac{1}{\pi^2}\log n - 2.$$

SKETCH OF PROOF: Fix n. Let

$$\phi_n(t)=\sigma_{n^2}[g_n;t]$$

where g_n is as in lemma 2. If

$$g_n \sim \frac{A_0}{2} + \sum_{k=1}^{\infty}(A_k\cos kt + B_k\sin kt),$$

then

$$\phi_n(t)=\frac{A_0}{2}+\sum_{k=1}^{n^2}\left(1-\frac{k}{n^2}\right)(A_k\cos kt + B_k\sin kt)$$

so that

$$s_n[\phi_n;t]=\frac{A_0}{2}+\sum_{k=1}^{n}\left(1-\frac{k}{n^2}\right)(A_k\cos kt + B_k\sin kt).$$

Hence,

$$|s_n[\phi_n;0]-s_n[g_n;0]|=\left|\frac{1}{n^2}\sum_{k=1}^{n} kA_k\right|\leqslant 2$$

from which the lemma follows.

Now let $\lambda_n = 2^{3^n}$ and define

$$f(t)=\sum_{n=1}^{\infty}\frac{1}{n^2}\cdot\phi_{\lambda_n}(\lambda_n t) \qquad (-\pi\leqslant t\leqslant\pi).$$

Then $f\in C[-\pi,\pi]$ (why?) and so the theorem will follow once we have proved

LEMMA 4. We have

$$s_{\lambda_n^2}[f;0]\to\infty \quad \text{as} \quad n\to\infty.$$

SKETCH OF PROOF: Fix n. We will consider

$$s_{\lambda_n^2}[\phi_{\lambda_j}(\lambda_j t);t]$$

first for $j \geqslant n+1$, then for $j \leqslant n-1$, and finally for $j=n$. Now if

$$\phi_{\lambda_j}(t) \sim \frac{a_0}{2} + \sum_{m=1}^{\lambda_j^2} (a_m \cos mt + b_m \sin mt)$$

(where the a_k, b_k depend, of course, on j), then

$$\phi_{\lambda_j}(\lambda_j t) \sim \frac{a_0}{2} + \sum_{m=1}^{\lambda_j^2} (a_m \cos m\lambda_j t + b_m \sin m\lambda_j t).$$

If $j \geqslant n+1$, then the Fourier series for $\phi_{\lambda_j}(\lambda_j t)$ consists of the constant term and terms involving $\cos kt$, $\sin kt$ for some $k \geqslant \lambda_j > \lambda_n^2$. Hence,

$$s_{\lambda_n^2}\left[\phi_{\lambda_j}(\lambda_j t); 0\right] = \text{the constant term},$$

so that

$$\left| s_{\lambda_n^2}\left[\sum_{j=n+1}^{\infty} \frac{1}{j^2} \phi_{\lambda_j}(\lambda_j t); 0 \right] \right| \leqslant \sum_{j=n+1}^{\infty} \frac{1}{j^2} \tag{1}$$

Next, if $j \leqslant n-1$, then $\phi_{\lambda_j}(\lambda_j t)$ is a trigonometric polynomial of degree λ_j^3. Since $\lambda_j^3 \leqslant (2^{3^{n-1}})^3 = 2^{3^n} < \lambda_n^2$ we have

$$s_{\lambda_n^2}\left[\phi_{\lambda_j}(\lambda_j t); t\right] = \phi_{\lambda_j}(\lambda_j t).$$

Hence,

$$\left| s_{\lambda_n^2}\left[\sum_{j=1}^{n-1} \frac{1}{j^2} \phi_{\lambda_j}(\lambda_j t); 0 \right] \right| \leqslant \sum_{j=1}^{n} \frac{1}{j^2}. \tag{2}$$

Finally we consider $j=n$. If

$$\phi_{\lambda_n}(t) = A_0 + \sum_{m=1}^{\lambda_n^2} (A_m \cos mt + B_m \sin mt),$$

then

$$\phi_{\lambda_n}(\lambda_n t) = A_0 + \sum_{m=1}^{\lambda_n^2} (A_m \cos m\lambda_n t + B_m \sin \lambda_n t).$$

Hence,

$$s_{\lambda_n^2}\left[\phi_{\lambda_n}(\lambda_n t); 0\right] = A_0 + \sum_{m=1}^{\lambda_n} A_m = s_{\lambda_n}\left[\phi_{\lambda_n}; 0\right]. \tag{3}$$

The lemma follows from (1), (2), (3), and lemma 3.

EXERCISE: For each $j \in I$ let X_j be the set of values of $k \geqslant 1$ such either $\cos kt$ or $\sin kt$ has a nonzero coefficient in the Fourier series for $\phi_{\lambda_j}(\lambda_j t)$. Show that the X_j are disjoint. This is the key to the above proof.

EXERCISE: Show that there is a continuous function whose Fourier series vanishes at every rational multiple of 2π.

Start by letting f be as before but with $\lambda_n = n! \cdot 2^{3^n}$. Show that for $m \in I$, f differs by a trigonometric polynomial from a function that is periodic with period $(2\pi)/(\lambda_m)$. Hence, the Fourier series for f diverges at $2\pi(j)/(\lambda_m)$ for every integer j.

II. Egoroff's theorem and Lusin's theorem

12.7C J. E. Littlewood formulated three principles that underline the theory of Lebesgue integration—namely,

1. Measurable sets are "almost" a union of a finite number of intervals.
2. Pointwise convergence of a sequence of measurable functions is "almost" uniform convergence.
3. Measureable functions are "almost" continuous.

Statement 1 is simply a way of looking at

$$mE = \text{g.l.b.}\, mG$$

for open sets G containing the measurable set E, together with 5.4 F. Statements 2 and 3 are ways of looking at Egoroff's theorem and Lusin's theorem, which we now present.

We will abbreviate "almost everywhere" as a.e.

THEOREM. (Egoroff) Let $\{f_n\}_{n=1}^{\infty}$ be a sequence of measurable functions on $[a,b]$ such that

$$\lim_{n\to\infty} f_n(x) = f(x) \quad \text{a.e.} \quad (a \leqslant x \leqslant b).$$

Then given $\epsilon > 0$ there exists a closed set $F \subset [a,b]$ such that $mF' < \epsilon$ and such that $\{f_n\}_{n=1}^{\infty}$ converges uniformly to f on F. (Here, as usual, $F' = [a,b] - F$.)

SKETCH OF PROOF: For each $n \in I$ let $g_n = |f_n - f|$. Given $n,p \in I$ let

$$E_{n,p} = \left\{ x \in [a,b] \,\middle|\, g_k(x) < \frac{1}{p} \ (k \geqslant n) \right\}.$$

Then $E_{n,p}$ is measurable,

$$E_{1,p} \subset E_{2,p} \subset \ldots,$$

and

$$m\left[\bigcup_{n=1}^{\infty} E_{n,p}\right] = b - a.$$

Hence, for each p there exists $n(p)$ such that

$$mE'_{n(p),p} < \frac{\epsilon}{2^p}.$$

Let

$$E = \bigcap_{p=1}^{\infty} E_{n(p),p}.$$

Then $mE' < \epsilon$ and $\{f_n\}_{n=1}^{\infty}$ converges uniformly on E to f.

EXERCISE: Fill in the details and finish the proof.

12.7D If the range of the real-valued function f on $[a,b]$ is finite, we call f a simple function. Thus, if f is simple and the range of f is the set $\{c_1, c_2, \ldots, c_n\}$, then

$$f = \sum_{k=1}^{n} c_k \chi_k$$

where χ_k is the characteristic function of the set

$$E_k = \{x \in [a,b] \mid f(x) = c_k\}.$$

There is an approach to the study of the Lebesgue integral that makes great use of simple functions. In this approach the following theorem is crucial. (We will use it in proving Lusin's theorem.)

THEOREM. Let f be a nonnegative-valued measurable function on $[a,b]$. Then there exists a sequence $\{s_n\}_{n=1}^{\infty}$ of measurable simple functions on $[a,b]$ such that

$$0 \leqslant s_1(x) \leqslant s_2(x) \leqslant \cdots \leqslant s_n(x) \leqslant \cdots \qquad (a \leqslant x \leqslant b)$$

and

$$\lim_{n\to\infty} s_n(x) = f(x) \qquad (a \leqslant x \leqslant b).$$

SKETCH OF PROOF: For each $N \in I$ let

$$E_{N,k} = \left\{ x \in [a,b] \,\middle|\, \frac{k-1}{2^N} \leqslant f(x) < \frac{k}{2^N} \right\} \qquad (k = 1, 2, \ldots, N2^N),$$

and let

$$E_N = \{x \in [a,b] \mid f(x) \geqslant N\}.$$

Define s_N as

$$s_N(x) = \frac{k-1}{2^N} \qquad (x \in E_{N,k}; k = 1, 2, \ldots, N2^N).$$

$$s_N(x) = N \qquad (x \in E_N).$$

Then if $f(x) < N$, we have $|s_N(x) - f(x)| < (1)/(2^N)$.

EXERCISE: Fill in the details and finish the proof.

12.7E Now for Lusin's theorem. Let f be a real-valued function on $[a,b]$. We say that f has property C if, given $\epsilon > 0$, there exists a closed set $F \subset [a,b]$ such that $mF' < \epsilon$ and such that the restriction of f to F is continuous.

EXERCISE: Show that every measurable simple function has property C.

THEOREM. (Lusin) Every measurable function on $[a,b]$ has property C.

SKETCH OF PROOF: If f is measurable, then there exists a sequence $\{s_n\}_{n=1}^{\infty}$ of measurable simple functions that converges to f on $[a,b]$. Each s_n has property C. Fix $\epsilon > 0$. For each n there exists a closed set $F_n \subset [a,b]$ such that $mF_n' < (\epsilon)/(2^{n+1})$ and such that the restriction of s_n to F_n is continuous.

There also exists a closed set $F_0 \subset [a,b]$ such that $mF_0' < \epsilon/2$ and such that $\{s_n\}_{n=1}^{\infty}$ converges uniformly to f on F_0. Let $F = F_0 \cap F_1 \cap F_2 \cap \cdots$.

EXERCISE: Fill in the details and finish the proof.

12.7F The converse of Lusin's theorem is also true.

THEOREM. Let f be a real-valued function on $[a,b]$. If f has property C, then f is measurable.

SKETCH OF PROOF: Suppose f has property C. For each $n \in I$ choose a closed set F_n such that $mF_n' < 1/n$ and such that the restriction of f to F_n is continuous. Then there exists a set $A \subset [a,b]$ of measure zero such that

$$[a,b] = \bigcup_{n=1}^{\infty} F_n \cup A.$$

Given $s \in R$ the set

$$E = \{x \in [a,b] \,|\, f(x) \geqslant s\}$$

is the union of the sets

$$E_n = \{x \in F_n \,|\, f(x) \geqslant s\}$$

together with a set of measure zero.

EXERCISE: Fill in the details and finish the proof. When do you use the fact that each F_n is closed?

MISCELLANEOUS EXERCISES

1. Show that there exists a function f on $(0,1]$ that is continuous and for which the improper integral $\int_{0+}^{1} f$ exists but such that $f \not\in \mathcal{L}[0,1]$.
2. Prove that there exists a closed set $F \subset [0,1]$ such that F is nowhere dense and such that $0 < mF < 1$.
3. Find a bounded measurable function f on $[0,1]$ such that
$$\int_0^1 |f(x) - g(x)|\,dx > 0$$
for every $g \in \mathcal{R}[0,1]$.
4. Let $\{f_n\}_{n=1}^{\infty}$ be a Cauchy sequence in $\mathcal{L}^2[a,b]$. Prove that $\{f_n\}_{n=1}^{\infty}$ has a subsequence that converges almost everywhere on $[a,b]$.
5. Let $\{f_n\}_{n=1}^{\infty}$ be a sequence in $\mathcal{L}^2[a,b]$. Suppose that
$$\lim_{n\to\infty} f_n(x) = f(x) \qquad \text{a.e.} \quad (a \leqslant x \leqslant b)$$
for some (measurable) function f on $[a,b]$, and that
$$\lim_{n\to\infty} \|f_n - g\|_2 = 0$$
for some $g \in \mathcal{L}^2[a,b]$. Prove that
$$f(x) = g(x) \qquad \text{a.e.} \quad (a \leqslant x \leqslant b).$$

6. Let $a = a_0 < a_1 < \cdots < a_n = b$. If $f: [a,b] \to R$ and f is constant on each open interval (a_{j-1}, a_j) for $j = 1, \cdots, n$, we call f a step function. (We make no restriction on the values $f(a_j)$.)

   Prove that the set of all step functions is dense in $\mathcal{L}^2[a,b]$.

7. If $f \in \mathcal{L}^2[a,b]$, prove that there exists a sequence of step functions $\{\Phi_n\}_{n=1}^{\infty}$ such that

$$\lim_{n \to \infty} \Phi_n(x) = f(x) \qquad \text{a.e.} \quad (a \leqslant x \leqslant b).$$

8. Suppose f, F are as in Lusin's theorem. Prove that there exists $g \in C[a,b]$ such that the set

$$\{x \in [a,b] \,|\, f(x) \neq g(x)\}$$

   has measure $< \epsilon$, and such that

$$\max_{a \leqslant x \leqslant b} |g(x)| = \max_{x \in F} |f(x)|.$$

9. Recall the properties of $\limsup_{n \to \infty} E_n$ from Section 2.12.

   Let $\{E_n\}$ be a sequence of measurable subsets of $[a,b]$ such that

$$\sum_{n=1}^{\infty} mE_n < \infty.$$

   Prove that the set $\limsup_{n \to \infty} E_n$ has measure zero.

   Deduce that almost all points of $[a,b]$ belong to only a finite number of the E_n.

10. Let f be a measurable function on $[a,b]$. Prove that there exists a sequence $\{g_n\}_{n=1}^{\infty}$ of continuous functions on $[a,b]$ such that

$$\lim_{n \to \infty} g_n(x) = f(x) \qquad \text{a.e.} \quad (a \leqslant x \leqslant b).$$

11. Let f be any measurable function on $[a,b]$. Prove that there exists a sequence $\{\Phi_n\}_{n=1}^{\infty}$ of step functions on $[a,b]$ such that

$$\lim_{n \to \infty} \Phi_n(x) = f(x) \qquad \text{a.e.} \quad (a \leqslant x \leqslant b).$$

   (This improves on exercise 7.)

12. Let f and $\{f_n\}_{n=1}^{\infty}$ be functions in $\mathcal{L}^2[a,b]$. Assume that

$$\lim_{n \to \infty} f_n(x) = f(x) \qquad \text{a.e.} \quad (a \leqslant x \leqslant b)$$

   and that

$$\lim_{n \to \infty} \|f_n\|_2 = \|f\|_2.$$

   Prove that

$$\lim_{n \to \infty} \|f_n - f\|_2 = 0.$$

APPENDIX

THE ALGEBRAIC AND ORDER AXIOMS FOR THE REAL NUMBERS

I ALGEBRA

The real number system is an ordered triple $\langle R, +, \cdot \rangle$ where R is a set and $+$ and $\cdot$ are functions from $R \times R$ into R.

For $a, b \in R$ we will write $+\langle a, b \rangle$ (i.e., the image of the pair $\langle a, b \rangle$ under $+$) in the customary form $a+b$, or, in some cases, $(a)+b, a+(b)$, or $(a)+(b)$. Similarly, the image of $\langle a, b \rangle$ under $\cdot$ will be written as $ab, a \cdot b$, $a(b), (a)b$, or $(a)(b)$, depending on the context.

We assume that the following axioms are satisfied.

For $+$:

A1. (THE COMMUTATIVE LAW FOR ADDITION.)

$$a+b=b+a \qquad (a, b \in R).$$

A2. (THE ASSOCIATIVE LAW FOR ADDITION.)

$$a+(b+c)=(a+b)+c \qquad (a, b, c \in R).$$

A3. (THE EXISTENCE OF AN IDENTITY FOR ADDITION.) There exists an element 0 in R such that

$$a+0=a \qquad (a \in R).$$

A4. (THE EXISTENCE OF INVERSE ELEMENTS FOR ADDITION (NEGATIVES).) If $a \in R$, then there exists $b \in R$ such that $a+b=0$, where 0 is as in A3.

For $\cdot$:

M1. (THE COMMUTATIVE LAW FOR MULTIPLICATION.)

$$ab=ba \qquad (a, b \in R).$$

M2. (THE ASSOCIATIVE LAW FOR MULTIPLICATION.)

$$a(bc)=(ab)c \qquad (a,b,c\in R).$$

M3. (THE EXISTENCE OF AN IDENTITY FOR MULTIPLICATION.) There exists an element 1 in R such that

$$1\cdot a=a \qquad (a\in R).$$

M4. (THE EXISTENCE OF INVERSE ELEMENTS FOR MULTIPLICATION (RECIPROCALS).) If $a\in R$ and $a\neq 0$, there exists $b\in R$ such that $ab=1$, where 1 is as in M3.

The final axiom is the distributive law.

D. If $a,b,c\in R$, then

$$a(b+c)=ab+ac.$$

All of the familiar laws of elementary algebra can be deduced from these nine axioms. We shall prove some samples. We first give five theorems concerning multiplication.

THEOREM 1. If $a\neq 0$ and $ab=ac$, then $b=c$.

PROOF: By M4 there exists $x\in R$ such that $ax=1$. Then, by M1, we have $xa=1$. From $ab=ac$ we then have $x(ab)=x(ac)$, since $\cdot$ is a function. Therefore, by M2, $(xa)b=(xa)c$. Since $xa=1$, this yields $1\cdot b=1\cdot c$. Hence, $b=c$ by M3 and the proof is complete.

THEOREM 2. The element 1 in M3 is unique.

PROOF: Suppose there were an element $1'$ satisfying $1'\cdot a=a$ for all a. Then $1'\cdot 1=1=1\cdot 1$. Hence, by M1, $1\cdot 1'=1\cdot 1$. By theorem 1 this implies $1'=1$ so that 1 is unique as asserted.

THEOREM 3. If $a,b\in R$ and $b\neq 0$, there exists a unique $x\in R$ such that $ax=b$.

PROOF: By M4 there exists $y\in R$ such that $ay=1$. Let $x=yb$. Using M2 and M3 we then have $ax=a(yb)=(ay)b=1\cdot b=b$.

This shows that an x exists such that $ax=b$. If, also, $ax'=b$, then $ax=ax'$. Hence, $x=x'$ by theorem 1, which shows that x *is* unique.

As is customary, we denote the x for which $ax=b$ by b/a. We denote the x for which $ax=1$ by a^{-1}. That is, $a^{-1}=1/a$.

THEOREM 4. If $a,b\in R$ and $a\neq 0$, then $b/a=b\cdot a^{-1}$.

PROOF: Let $x=b/a$ and let $y=b\cdot a^{-1}$. By definition of b/a we have $ax=b$. Also we have $ay=ya=(b\cdot a^{-1})a=b(a^{-1}\cdot a)=b(a\cdot a^{-1})$. But $a\cdot a^{-1}=1$, by definition of a^{-1}. Hence, $ay=b\cdot 1=b$. So $ax=ay=b$. By theorem 1 we have $x=y$, which is what we wished to prove.

THEOREM 5. If $a\neq 0$, then $(a^{-1})^{-1}=a$.

PROOF: Let $b=(a^{-1})^{-1}$. We have $a \cdot a^{-1}=1$, by definition of a^{-1}. Also $a^{-1} \cdot b=1$ by definition of $(a^{-1})^{-1}$. Therefore $a^{-1} \cdot a=1=a^{-1} \cdot b$. Theorem 1 then implies that $a=b$.

Theorems 6–10 concern addition and may be proved in exactly the same manner as theorems 1–5. Simply substitute $+$ for $\cdot$ and use the A axioms instead of the corresponding M axioms.

THEOREM 6. If $a+b=a+c$, then $b=c$.

THEOREM 7. The element 0 in A3 is unique.

THEOREM 8. If $a,b \in R$, there exists a unique $x \in R$ such that $a+x=b$.

As is customary we denote the x for which $a+x=b$ by $b-a$. We denote the x for which $a+x=0$ by $-a$. That is, $-a=0-a$.

THEOREM 9. If $a,b \in R$, then $b-a=b+(-a)$.

THEOREM 10. If $a \in R$, then $-(-a)=a$.

Next we prove results involving (at least implicitly) both addition and multiplication. The reader should provide support for the assertions in the proofs.

THEOREM 11. If $a,b,c \in R$, then $a(b-c)=ab-ac$.

PROOF: Let $x=a(b-c), y=ab-ac$. Then $ac+x=ac+a(b-c)=a[c+(b-c)]=ab$. Also $ac+y=ac+(ab-ac)=ab$. So $ac+x=ac+y$. Hence, $x=y$.

THEOREM 12. If $a \in R$, then $0 \cdot a=a \cdot 0=0$.

PROOF: We have $0+0=0$. Hence, $a \cdot (0+0)=a \cdot 0$. Therefore, $(a \cdot 0)+(a \cdot 0)=(a \cdot 0)+0$. Hence, $a \cdot 0=0$.

THEOREM 13. If $ab=0$, then either $a=0$ or $b=0$.

PROOF: Suppose $a \neq 0$. We must show that $b=0$. We have $b=b \cdot 1=b(a \cdot a^{-1})=(ba)a^{-1}$ $=(ab)a^{-1}=0 \cdot a^{-1}=0$.

THEOREM 14. If $a,b \in R$, then $(-a)b=-(ab)$.

PROOF: Let $x=(-a)b, y=-(ab)$. Then $ab+x=ab+(-a)b=ba+b(-a)=b[a+(-a)]=b \cdot 0=0$. Also, $ab+y=ab+[-(ab)]=0$. So $ab+x=0=ab+y$. Hence, $x=y$.

THEOREM 15. If $a,b \in R$, then $(-a)(-b)=ab$.

PROOF: Let $x=(-a)(-b)$. Then $x=-[a(-b)]=-[(-b)a]=-[-(ba)]=ba$. So $x=ab$.

THEOREM 16. If $a,b,c,d \in R$, and if $b \neq 0, d \neq 0$, then

$$\frac{a}{b}+\frac{c}{d}=\frac{ad+bc}{bd}.$$

PROOF: Let $x=a/b+c/d$, $y=(ad+bc)/bd$. Then

$$(bd)x=(bd)\left(\frac{a}{b}+\frac{c}{d}\right)=(bd)\left(\frac{a}{b}\right)+(bd)\left(\frac{c}{d}\right)=(db)\left(\frac{a}{b}\right)+(bd)\left(\frac{c}{d}\right)$$

$$=d\left[b\cdot\left(\frac{a}{b}\right)\right]+b\left[d\cdot\left(\frac{c}{d}\right)\right]=da+bc=ad+bc.$$

Also, $(bd)y=(bd)\cdot((ad+bc)/bd)=ad+bc$. Hence, $(bd)x=(bd)y$, so $x=y$.

These results should be enough to convince the reader of the truth of our assertion that all formulae from elementary algebra can be derived from the nine axioms. For an exercise, try to prove the following:

$$1^{-1}=1,$$

$$(ab)^{-1}=a^{-1}b^{-1},$$

$$\left(\frac{a}{b}\right)\cdot\left(\frac{c}{d}\right)=\frac{ac}{bd} \qquad \text{if} \quad b\neq 0, d\neq 0,$$

$$-(a+b)=-a-b.$$

II ORDER

We now turn to the question of order. We impose the additional assumption on R:

ORDER AXIOM: There exists a subset R^+ of R such that

O1. If $x,y\in R^+$, then $x+y, xy\in R^+$.
O2. If $x\in R$ and $x\neq 0$, then either $x\in R^+$ or $-x\in R^+$.
O3. $0\notin R^+$.

We call the numbers of R^+ *positive*.

With regard to O2, if $x\neq 0$, then it is not possible for both x and $-x$ to belong to R^+. For if $x\in R^+$ and $-x\in R^+$, then, by O1, $0=x+(-x)$ would belong to R^+, contradicting O3.

We now define the traditional inequality signs.

DEFINITION. Let $x,y\in R$. Then

$$x<y \quad \text{or} \quad y>x$$

means that $y-x\in R^+$.

Also,

$$x\leqslant y \quad \text{of} \quad y\geqslant x$$

means that either $x=y$ or $x<y$.

Here are a few of the familiar inequality laws.

THEOREM 1. If $a,b\in R$, then precisely one of the following statements holds:

$$a=b,$$

$$a<b,$$

$$b<a.$$

PROOF: First suppose $a=b$. Then $b-a=0$ so that $b-a\notin R^+$. Hence, $a<b$ does not hold. Similarly, $a-b=0$ so that $a-b\notin R^+$. Hence, $b<a$ does not hold.

Next suppose $a\neq b$. Then $b-a\neq 0$. By O2, either $b-a\in R^+$ or $a-b=-(b-a)\in R^+$, but not both. Hence, either $a<b$ or $b<a$, but not both. This completes the proof.

THEOREM 2. If $a,b,c\in R$ and $a<b$, then $a+c<b+c$.

PROOF: Since $a<b$, we have $b-a\in R^+$. Hence, $(b+c)-(a+c)=b-a\in R^+$, so that $a+c<b+c$.

THEOREM 3. If $a<b$ and $b<c$, then $a<c$.

PROOF: By assumption both $b-a$ and $c-b$ belong to R^+. By O1, so does the sum $(b-a)+(c-b)=c-a$. Hence, $a<c$.

THEOREM 4. If $a<b$ and $c>0$, then $ac<bc$.

PROOF: By assumption $b-a$ and $c=c-0$ are in R^+. By O1 so is their product $c(b-a)=bc-ac$. Hence, $ac<bc$.

THEOREM 5. If $a\in R$ and $a\neq 0$, then $a^2>0$.

PROOF: Since $a\neq 0$, by theorem 1, either $0<a$ or $a<0$ but not both. If $0<a$, then, by theorem 4, $0\cdot a<a\cdot a$ so that $0<a^2$. On the other hand, if $a<0$, then $0-a=-a$ belongs to R^+. Hence, by O1, so does $(-a)(-a)=a^2$. Thus $a^2=a^2-0\in R^+$ and so $0<a^2$.

THEOREM 6. If $a<b$ and $c<0$, then $ac>bc$.

PROOF: Since $c<0$ we have $-c=0-c\in R^+$. Hence, $-c>0$. By theorem 4 it follows that $a(-c)<b(-c)$. Hence, $-ac<-bc$ so that $-bc-(-ac)\in R^+$. That is, $ac-bc\in R^+$ so that $bc<ac$.

These should suffice to show that all inequality formulae can be derived from the order axiom (together with the algebra axioms). For an exercise the reader should try to prove the following:

$1>0$.
If $ab>0$, then either $a>0,b>0$ or $a<0,b<0$.
If $x\in R$, then $x^2+1>0$. (Hence, there is no $x\in R$ such that $x^2=-1$.)
If $0<a<b$, then $0<b^{-1}<a^{-1}$.

III THE INTEGERS AND THE RATIONAL NUMBERS

If $A\subset R^+$, we say that A is a set of induction if

$$1\in A$$

and if

$$x\in A \quad \text{implies} \quad x+1\in A.$$

For example, R^+ is a set of induction.

THEOREM. Let I be the intersection of all sets of induction. Then I is a set of induction.

PROOF: Since the number 1 is in every set of induction, it follows that 1 is in the intersection of all sets of induction. Thus

$$1 \in I.$$

Now suppose $x \in I$. We will show that $x+1 \in I$. If A is any set of induction, then, by definition of I, we have $x \in A$ so that $x+1 \in A$. Thus if $x \in I$, then $x+1$ is in every set of induction and so $x+1$ is in the intersection of all sets of induction. That is,

$$x \in I \quad \text{implies} \quad x+1 \in I.$$

This completes the proof.

DEFINITION. The elements of I are called positive integers. The number $n \in R$ is called a negative integer if $-n \in I$. Finally, we say that n is an integer if either $n=0$ or n is a positive or negative integer.

DEFINITION. If $x \in R$, we say that x is rational if there exist integers a, b with $b \neq 0$ such that $x = a/b$. If $x \in R$ but x is not rational, we say that x is irrational.

If a is an integer, then $a = a/1$ and so a is rational. That is, every integer is a rational number.

IV COMPLETENESS

It is not difficult to show that the set of rational numbers obeys the same algebra and order axioms as the set of reals does. Another axiom is therefore required to distinguish the reals from the rationals. Such an axiom (called a completeness axiom) can take many forms. We present it as the Least Upper Bound axiom in Section 1.7. The reader who wishes to proceed in strictly logical order should turn to that section, after finishing this Appendix.

V ABSOLUTE VALUES

If $x \in R$ and $x > 0$, we define $|x|$ to be x. If $x < 0$, we define $|x|$ to be $-x$. Finally, we define $|0|$ to be 0. Thus $|x| \geqslant 0$ for all $x \in R$ and $-|x| \leqslant x \leqslant |x|$.

THEOREM. If $x, y \in R$, then

$$|x+y| \leqslant |x| + |y| \tag{1}$$

and

$$|xy| = |x| \cdot |y|. \tag{2}$$

PROOF: Since $x \leqslant |x|, y \leqslant |y|$ we have

$$x+y \leqslant |x| + |y|. \tag{3}$$

Since $-|x| \leqslant x$, $-|y| \leqslant y$ we have

$$-|x| - |y| \leqslant x+y$$

or

$$-(x+y) \leqslant |x|+|y|. \tag{4}$$

If $x+y \geqslant 0$, then $|x+y|=x+y$, so (1) follows from (3). Otherwise, $x+y<0$ so that $|x+y|=-(x+y)$ and (1) follows from (4). We leave the proof of (2) as an exercise.

For $a,b,c \in R$, let $x=a-c$ and let $y=c-b$. Then $x+y=a-b$ and we deduce from the preceding theorem the inequality,

$$|a-b| \leqslant |a-c|+|c-b|.$$

Since the geometric interpretation of $|x-y|$ is the distance from x to y, this last inequality says that the distance from a to b is never more than the distance from a to c plus the distance from c to b.

Special Symbols

Symbol	Description
$\|x+y\| \leqslant \|x\| + \|y\|$	absolute value bars; less than or equal to
$3/2$ and $-9/276$; $\frac{1}{2}$	shilling fractions; case fractions
$x<y$ then $-x>-y$	less than; greater than
$x^2=2$; $a^{1/2}$; $a^x \cdot a^y$	superscripts
$\sqrt{a}$	radical sign
$(-\infty, \infty)$	infinity
$[a,b]$	brackets
$\langle a,b \rangle$	angle brackets
$\{\langle x,y \rangle \mid x \geqslant 0, y \geqslant 0\}$	set braces; greater than or equal to
$a \in \{a,b,c\}$ but $d \notin \{a,b,c\}$	is an element of; also with cancel
$A \cup B$; $A \cap B$	union; intersection
$\varnothing$	empty set
$A \subset B$; $B \supset A$; $B \not\subset C$	included in; contained in; also with cancel

$\rightarrow$; $\Rightarrow$	single arrow; double-shaft arrow
$g \circ f$	open dot (composition)
π, χ_A, ϵ, θ, Δ; σ_n; and so on	cap and l.c. Greek alphabet; also as superscripts and subscripts
$\frac{\pi}{2}$; $\frac{f(x)}{g(x)}$	built-up fraction
$\cup_{n=1}^{\infty} A_n$	boldface union symbol with side heads; also intersection
a_{n_2}; a_4^2; $I_{k_r}^{n_r}$	combinations of superscripts and subscripts
$b_n \neq a_n^n$	not equal to
$\lim\limits_{n \to \infty} s_n = L$	limit (in display)
$\lim_{n \to \infty} s_n = L$	limit (in text line)
$\mathcal{S} = \mathcal{C} \cup \mathcal{D}$; $H = \cup_{G \in \mathcal{F}} G$	cap and l.c. sript letters; also as subscripts
$\binom{n+k-2}{n-1}$	2-or-more-line matrix
$\sum\limits_{j=1}^{n}$	summation (in display)
$\Sigma_{n=1}^{\infty} a_n$	summation (in text line)
$\Sigma_{n=1}^{\infty} a_n \ll \Sigma_{n=1}^{\infty} b_n$	nested angle symbol
$\int_c^d b\,ds = sb\Big\vert_c^d - \int_c^d s\,db$	integral with side heads; vertical bar with side heads
$\Vert s \Vert_2 \geqslant 0$	norm (double bar)
$e^{i \propto \theta} = \cos \propto \theta + \sin \propto \theta$	variation
$\overline{E}$	letter with overline
$\frac{dy}{dx}\Big\vert_{x=c} = f'(c)$; $dy/dx\vert_{x=c}$	differential in display; in text line

$x \sim x$	approximation
$\int_E f$; Then $\int_E f$ is; $\int_{E_1 \cup E_2} f$	integral with subscript; in text line; with intersection subscript
$\hat{h}(y)$	letter with accent

INDEX